Cognitive Science

Series Editors:
Marc M. Sebrechts
Gerhard Fischer
Peter M. Fischer

Michael Smithson
Department of Behavioral Sciences
James Cook University of North Queensland
Townsville Q 4811
Australia

Library of Congress Cataloging-in-Publication Data
Ignorance and uncertainty : emerging paradigms
    / Michael Smithson.
        p. cm.-- (cognitive science)
    Bibliography: p.
    Includes indexes.
        1. Ignorance (Theory of knowledge). 2. Uncertainty.
3. Probabilities. I. Title. II. Series: Cognitive science (New York, N.Y.)
BD221.S63 1988   88-36716
121--dc19

Printed on acid-free paper

© 1989 by Springer-Verlag New York Inc.
All rights reserved. This work may not be translated or copied in whole or in part without the written permission of the publisher (Springer-Verlag, 175 Fifth Avenue, New York, NY 10010, USA), except for brief excerpts in connection with reviews or scholarly analysis. Use in connection with any form of information storage and retrieval, electronic adaptation, computer software, or by similar or dissimilar methodology now known or hereafter developed is forbidden.
The use of general descriptive names, trade names, trademarks, etc. in this publication, even if the former are not especially identified, is not to be taken as a sign that such names, as understood by the Trade Marks and Merchandise Marks Act, may accordingly be used freely by anyone.

Camera-ready copy prepared by the author using Microsoft® Word.
Printed and bound by R.R. Donnelley and Sons, Harrisonburg, Virginia.
Printed in the United States of America.

9 8 7 6 5 4 3 2 1

ISBN 0-387-96945-4 Springer-Verlag New York Berlin Heidelberg
ISBN 3-540-96945-4 Springer-Verlag Berlin Heidelberg New York

To my parents, Nancy and Robert Solibakke, who not only taught me most of what I know, but also a great deal of what I know I don't know.

Preface

Like many other people, I have been ignorant for as long as I can remember. I became consciously interested in ignorance itself as an undergraduate mathematics student in my first course on probability, but that interest was limited to 'uncertainty', which I took to be a feature of the outside world. A seminar on Bayesian statistical inference and decision theory persuaded me that uncertainty in the mind also could be formalized and analyzed. My vision broadened further when my studies introduced me to Godel's work; suddenly uncertainty was inside the citadel of mathematics itself, and it appeared irreducible.

Studying for a Ph.D. in sociology exposed me to new varieties of 'uncertainty', the indeterminacies of human thought and behavior. I began to doubt whether probability theory could adequately represent these indeterminacies, but I was ignorant of any alternative frameworks. I encountered the sociology of knowledge and its fundamental insight that all of what passes for knowledge is socially constructed and negotiated. It occurred to me to wonder what a 'sociology of ignorance' would require and what insights could be gained from it. Would it make sense to say that what passes for ignorance also is socially constructed and negotiated?

An embarrassingly long time passed before I realized that my involvement with nonprobabilistic frameworks (e.g., fuzzy set theory) and attempts to construct a sociology of ignorance derived from the same source, that they were connected, and that both activities amounted to my participation in a widespread but unobvious trend in Western intellectual culture. This book is an attempt to elucidate that connection and to describe the trend.

Broadly speaking, the trend involves recent changes in how we think about and respond to ignorance. We are currently witnessing the greatest creative work on ignorance since the decade of 1660 (when probability theory emerged). Several new frameworks for representing nonprobabilistic uncertainty have been invented, and studies in the psychology of judgment and decision under uncertainty and ignorance have multiplied. The representation and management of ignorance has become a more central concern in older fields such as cognitive psychology, economics, management science, and the sociology of organizations. It is virtually a hot topic in newer fields such as expert systems, artificial intelligence, and risk analysis.

Moreover, the very style of our responses to ignorance appears to be changing. Not long ago, the dominant methods of coping with ignorance were to try eliminating it or absorbing it. The emerging frameworks now seem to have jettisoned the assumption that ignorance is ultimately reducible, and the new style is 'managerial' in the sense of attempting to understand, tolerate, and even utilize certain kinds of ignorance.

These sudden changes, and the prospect of some sort of revolution in Western orientations towards the unknown, pose tantalizing questions. What, exactly, is the nature of the new ignorance frameworks? How revolutionary are they, and do they amount to a paradigmatic shift? Why is this all happening now? What does it mean and where will it lead? Numerous articles and books on uncertainty and other kinds of ignorance have appeared in recent years. The literature in any of the several fields dealt with in this book is voluminous and expanding very rapidly. Yet, few of these publications have addressed these questions; most writing in this area consists of attempts to advance or influence a particular area, rather than trying to obtain an overview.

This book is an attempt at such an overview. Of course, it does not fully answer all the questions given above, although it squarely addresses each of them. There can be no complete answers now in any case, since much of the current work on ignorance is still in mid-construction. Nevertheless, it seems timely to suggest some tentative explanations for the recent ferment over ignorance.

The perspective which I utilize is based on a mixture of ideas and concepts from the social sciences and modern philosophies of science. It starts from the postulate that there is no intelligible way to discuss ignorance without placing it in a social frame of reference. But it is also informed by my participation inside the frame itself. I am not a detached observer of the debates over probability, fuzzy sets, certainty factors, and so on. I have used several of these frameworks in my own research, tried to extend them, and even had my say in a few of the debates. I hope these influences have enabled me to attain a balanced and genuinely interdisciplinary viewpoint.

I also hope that researchers from several disciplines will read this book. It is my conviction that we need a much more sophisticated dialog between the various fields in which creative work on ignorance is being pursued. Perhaps the greatest barrier is between the computer scientists-engineers and those in the behavioral and social sciences. Yet, that is precisely the nexus where some of the best work is being done. If this book brings the possibility of a creative dialog between these researchers a little closer to realization, it will have done its job well.

In the interests of readers from multiple disciplines, I have tried to strike an effective compromise between technical rigor and exhaustiveness, on the one hand, and comprehensibility and brevity, on the other. Several of the fields which feature in this book are producing literature so rapidly that I have been sorely tempted to delay completing the manuscript until "X's paper" or "Y's book" comes out. I can only utter the same rationalization that countless authors have used: A complete and rigorous treatment of this subject matter is possible only once it has ceased to expand, by which time it also may have ceased to be interesting.

The completion of this book finds me in considerable intellectual and emotional debt to a number of people. Robert Alun Jones introduced me to sociology generally, and to the sociology of knowledge in particular. I am grateful to Benton Johnson and Jay Jackson, fine teachers both, for stimulating discussions and guidance when, in the midst of my Ph.D. studies, I started pondering the 'sociology of ignorance'. In recent years, I have benefited from discussions and collaborative work with several colleagues and friends, especially Margaret Foddy, Bernard Guerin, Nick Pidgeon, and Joseph Reser. They have kindly read and commented on various drafts of chapters from this book. My wife, Susan, has offered not only support and encouragement, but also valuable discussion, criticisms, and suggestions. Through their efforts, I have become considerably more ignorant, but somewhat less meta-ignorant.

Table of Contents

Preface

Chapter 1: A Vocabulary of Ignorance — 1
  1.1 Ignoring Ignorance — 1
  1.2 A Framework for Ignorance — 5
  1.3 The Rest of this Book — 10

Part I: Normative Paradigms

Chapter 2: Full Belief and the Pursuit of Certainty — 14
  2.1 Ignorance: The Views from Dogmatism and Skepticism — 14
  2.2 Examples of Traditional Normative Pragmatism — 20
  2.3 The Ultimate Mental Health Project — 28

Chapter 3: Probability and the Cultivation of Uncertainty — 41
  3.1 Probabilists as Strategists — 41
  3.2 Definitions of Cultivatable Uncertainty — 43
  3.3 Classical Theory and the Unifying Impulse — 47
  3.4 Relative Frequency Theory: Banishing the Monsters — 52
  3.5 Subjective Probability and the New Imperialists — 58
  3.6 Neoclassicism and the Logical Probabilists — 69
  3.7 Nonquantitative Probability — 70
  3.8 Combining Probability Judgments — 77
  3.9 Probability and Risk Assessment — 83

Chapter 4: Beyond Probability: New Normative Paradigms — 92
  4.1 The Modern Managerial Approach to Ignorance — 92
  4.2 Normative and Pragmatic Objections to Probability — 94
  4.3 The New Uncertainties — 99
    4.3.1 The Certainty Factors Debate — 99
    4.3.2 Fuzzy Sets, Vagueness, and Ambiguity — 108
    4.3.3 Possibility and Belief — 118
  4.4 The Leap to Second-Order Relations — 125
    4.4.1 Probability of Probabilities, Saith the Preacher — 125

    4.4.2 Rough Sets, Shaferian Belief,
          and Possibility Theory   128
    4.4.3 Measures of Uncertainty   131
  4.5 Imposing Order on the New Chaos   135
    4.5.1 The Probabilists' Rejoinders   135
    4.5.2 Reductionists: Logic, Topoi, and
          Probability Again   138
    4.5.3 Pluralists and Synthesists   141
  4.6 A Normative Paradigm Shift?   145

Part II: Descriptive and Explanatory Paradigms

Chapter 5: Psychological Accounts: Biases, Heuristics,
          and Control   152
  5.1 The Normative View from Psychology   152
    5.1.1 Three Traditions   152
    5.1.2 Coping, Defending, and Ignorance   154
    5.1.3 The Control and Predictability Thesis   158
  5.2 The Bayesian Inquisition   161
    5.2.1 A Paradigm for Studying Statistical Intuitions 161
    5.2.2 From Words to Numbers and Vice-Versa   164
    5.2.3 Selective Attention to Evidence   168
    5.2.4 Aggregation and the Conjunction Effect   172
    5.2.5 Randomicity and the Illusion of Control   174
    5.2.6 Reframing Biases and the
          Independence Axiom   176
    5.2.7 Normative Violations: Is the
          Evidence Clearcut?   181
  5.3 Explaining Human Heuristics   187
    5.3.1 Availability, Relevance, and Specificity   187
    5.3.2 Representativeness, Anchoring,
          and Adjustment   188
    5.3.3 Nonstandard Decision Theories   190
    5.3.4 Criticisms of the Explanations   193
  5.4 New Directions and New Uncertainties   197
    5.4.1 Fuzzy Set Theory as a Behavioral Science?   198
    5.4.2 Vagueness and Ambiguity   201
    5.4.3 The New Reductionists   204
  5.5 The Rationality Debate   208
    5.5.1 The Bayesians and Rival Rationalists   208
    5.5.2 The Normative and the Descriptive   213

Chapter 6: The Social Construction of Ignorance   216
  6.1 The View from Social Science   216
  6.2 Ignorance and the Sociology of Knowledge   219
  6.3 Ignorance in the Micro-Order   226
    6.3.1 Language, Politeness, and Ignorance   227
    6.3.2 Norms Inhibiting Communication   229
    6.3.3 Privacy, Secrecy, and the Valuation of Information   231
    6.3.4 Problems and Limitations   237
  6.4 Ignorance in Organizational Life   239
    6.4.1 Definitions of Uncertainty and Related Concepts   241
    6.4.2 Uncertainty, Organizational Structure, and Process   246
    6.4.3 Creating and Using Ignorance in Organizations   250
  6.5 Coda: Ignorance and the Sociology of Science   259

Chapter 7: A Dialog with Ignorance   264
  7.1 The New Preoccupation with Ignorance   264
    7.1.1 Techno-Rational Explanations   264
    7.1.2 The Stage Model and Motivational Accounts   266
    7.1.3 The Deviancy Analogy   270
  7.2 An Example of a Normative-Descriptive Dialog: The Study of Second-Order Uncertainty   275
    7.2.1 The Normative versus the Descriptive   275
    7.2.2 A Study of Preference Patterns   278
    7.2.3 Framing Effects in Nonprobabilistic Uncertainty   284
    7.2.4 Expanding the Dialog   288
  7.3 What Goals Are Served by Which Normative Frameworks?   290
  7.4 The Social Nature of Rationality   297

Bibliography   308

Name Index   368

Subject Index   386

Chapter 1:
A Vocabulary of Ignorance

"...as a corollary to writing about what we know, maybe we should add getting familiar with our ignorance, and the possibilities therein for ruining a good story." Thomas Pynchon.

1.1 Ignoring Ignorance

Until recently, ignorance and uncertainty were neglected topics in the human sciences and even in philosophy. Even now, they do not share center stage with knowledge in those disciplines, but remain sideshow oddities for the most part. Instead, Western intellectual culture has been preoccupied with the pursuit of absolutely certain knowledge or, barring that, the nearest possible approximation to it. This preoccupation is worth investigating, since it appears to be responsible not only for the neglect of ignorance, but also the absence of a conceptual framework for seriously studying it. Accordingly, in this section I will briefly outline the orientations in Western science and philosophy which have motivated the neglect of ignorance. I will also take the position that these orientations have begun to change recently, and that one of the spinoffs has been an unprecedented interest in ignorance, uncertainty, and related topics.

Ignorance usually is treated as either the absence or the distortion of 'true' knowledge, and uncertainty as some form of incompleteness in information or knowledge. To some extent these commonsense conceptions are reasonable, but they may have deflected attention away from ignorance by defining it indirectly as *non*knowledge. From at least the time of Plato, Western philosophers and scientists have worked as if infallible, demonstrable knowledge were an attainable goal. Modern epistemology notwithstanding, vestiges of the Thomistic distinction between "scientia" and "opinio" remain with us, and the most common approach to ignorance has been either elimination or absorption by exercising some version of the 'scientific method'.

The dominant traditional methods for acquiring justified beliefs or making 'rational' decisions are predicated on the twin goals of complete information and absolute epistemological standards for transforming that information into knowledge and thence into appropriate action. Good information and true knowledge are nearly universally treated as desirable, even scarce, commodities. While not everyone goes as far as Samuel Johnson's declaration that all knowledge is of itself valuable, even the most hard-headed pragmatist prefers knowledge to ignorance if no cost is entailed in obtaining either. Prior to any commitment to decisions or actions, almost all Western intellectual traditions direct us to maximize information and/or knowledge, and so reduce uncertainty or ignorance.

It is therefore no accident that theories of knowledge far outnumber those of ignorance. This is so for both normative and explanatory perspectives. The 1967 Encyclopedia of Philosophy, for instance, has no major headings for "ignorance" or "uncertainty" although it does for "vagueness", "skepticism", "error", and "doubt". The tone in these sections is prescriptive and corrective. The dominating assumption appears to be that ultimately vagueness and error are corrigible, and that radical skepticism or doubt are impossible or destructive attitudes. The only framework for dealing with incorrigibly or irreducibly incomplete knowledge that gets an airing in this authoritative sourcebook is probability theory, which has been the landmark normative paradigm of its kind for three centuries.

Unger (1975) has observed that ordinary language is more direct and clear in matters of fact and knowledge than it is in ignorance. While we may say "Joe knows that Clara is pregnant" we cannot express ignorance in a similarly direct way. To phrase ignorance in the active voice we must use negation ("Joe does not know that Clara is pregnant"). The sole active voice construction for ignorance does not merely indicate a state of nonknowledge. The statements "Joe ignores Clara's pregnancy" or "Joe ignores the fact that Clara is pregnant" fails to convey the simple opposite of knowledge about Clara's pregnancy. A passive voice construction for ignorance is possible, but even then it is not as direct as its knowledge-oriented counterpart. Compare "Joe is informed that Clara is pregnant" with "Joe is ignorant *of the fact* that Clara is pregnant". To omit the italicized words in the second sentence makes it a deviant construction in English.

The human sciences are no less biased in favor of discussing knowledge rather than ignorance. While knowledge itself often has formed subfields in psychology, sociology, and anthropology, ignorance traditionally has not. Thus, for instance, there is a sociology of knowledge but no sociology of ignorance (proposals have been made for a "sociology of nonknowledge" (Weinstein and Weinstein 1978) and a "social theory of ignorance" (Smithson 1985), but these remain outside mainstream theory). Most explanatory accounts from the human sciences treat the latter marginally or implicitly if at all, and often negatively (usually in reference to their supposedly destructive consequences for human social or mental functioning). In psychology the traditional interests have been directed towards performative models of human cognition and intelligence, with ignorance treated largely as a matter of corrigible bias or error on the part of lay thinkers.

The last 40 years, however, and especially the last two decades, have seen a flurry of new perspectives on uncertainty and ignorance whose magnitude arguably eclipses anything since the decade of 1660 which saw the emergence of modern probability theory. This activity has arisen in several fields at once, but most notably in those concerned with the interactions between people and modern complex technologies (e.g., economics, large-scale process and energy technologies, systems engineering, management science, computer science, and artificial intelligence). Many explanations could be adduced for this sudden upsurge in interest, and it may be too early to decide in favor of one above all others. By way of introducing one of the central themes of this book, however, I should like to briefly allude to the more popular explanations here.

One obvious explanation appeals to such matters as the increases in complexity and sheer scale of the technologies and organizations developed since World War II and the widely acknowledged failures of deterministic, mechanical models to cope with the uncertainties arising in such systems. Whole new areas of professional expertise such as risk assessment and decision theory have appeared in response to demands for ways of coping with these new uncertainties. More recently technological aids have been combined with human expertise in the development of expert systems, whose commercial appeal is based partly on their alleged capacity to handle both high complexity and uncertainty.

The results of these developments have been as bewildering as they are intriguing. The past 20 years alone have seen several nonprobabilistic formalisms proposed for dealing with uncertainty, surprise, doubt, fuzziness, vagueness, and possibility. Probability itself has undergone major shifts in emphasis and the Bayesian school has ascended from relative disreputability to a point whereby it effectively challenges the frequentist school for dominance. New theories and accounts of uncertainty have been forthcoming from cognitive psychology, decision theory, management science, the sociology of organizations, risk assessment, studies of disasters and accidents, and economics. Managers and government leaders have altered their methods for dealing with uncertainty, moving away from the traditional attempts to eliminate or ignore it. The signs point clearly toward the emergence of new normative and explanatory paradigms of uncertainty and ignorance in response to the increasing complexity and uncertainty of the artificial environment.

While this explanation is intuitively appealing, it fails to account for a substantial number of developments that emerged independently of immediate commercial or other practical utility. Furthermore, several current innovations were clearly foreshadowed by work earlier in this century, particularly in mathematics, physics, and philosophy. Even a cursory investigation into any of these fields reveals that they underwent crises of certainty spanning the latter part of the 19th and the first half of the 20th centuries, from which they have only recently begun to recover. These crises were marked not only by increasing specialization and fragmentation, but also a loss of consensus concerning fundamental criteria of truth. The end results were the relativization of truth and the recognition of sources of incorrigible uncertainty or ignorance.

Those earlier developments provide several clues for what may be happening now, not only by pointing to a connection between the short term pragmatic concerns with technical complexity and wider intellectual phenomena such as specialization and relativization, but also in the strategies utilized by those in the thrall of these crises. In the next chapter I will compare the aptly termed "loss of certainty" (Kline 1980) in pure mathematics with the recent activities in theories of uncertainty. These comparisons will, I believe, point towards an understanding of crucial aspects of the Western intellectual response to a uniquely Western creation: The ignorance explosion.

Before we can make these comparisons, however, we must have a set of reasonable terms and definitions by which uncertainty and ignorance may be discussed intelligibly. There are, after all, a number of pitfalls in commonsense and even scholarly frameworks for thinking about ignorance, and so the next section introduces some necessary conceptual machinery.

1.2 A Framework for Ignorance

Both the prevailing normative and explanatory frameworks for knowledge tend to treat ignorance as either the absence or distortion of 'truth'. This position immediately poses a problem because it requires the assumption that in order to talk of ignorance we must have established criteria for absolute knowledge and epistemology. A related problem is that from an absolutist standpoint, it seems as though we must fully enumerate the field of 'nonknowledge' in order to exhaustively describe ignorance. Clearly, such a task leads to the usual self-referential paradoxes. It might seem that the commonsense distinction between 'objective' (real) and 'subjective' (perceived) ignorance could get us by. After all, this is the duality that occurs both in probability theory and several other modern formalisms for dealing with uncertainty. But this distinction merely corrals the problem into a smaller enclosure. We are still faced with the problem of defining what objective ignorance is.

Even in the sociology of knowledge, where knowledge is viewed as a social construct, the tendency to fall into naive absolutism surfaces in a number of major theoretical perspectives. In an early (and rare) article on ignorance from a sociological point of view (Moore and Tumin 1949), the definition of ignorance as the absence or distortion of truth is advocated explicitly. Despite their unfounded dogmatism, at least the sociologists have discussed this problem and it is worth taking a lesson from them on how to evade it.

The most blatant version of sociological absolutism occurs in both the functionalist and Marxist concepts of ideology, which boils down to erroneous thought with some version of 'science' providing the template for correct thought (for an unsuccessful attempt by a modern Marxist theorist to extricate himself from this trap, see Althusser's 1976 *Essays in Self Criticism*). Somewhat more consistent sociologists have fallen into the 'fallacy of dual residentialism' (Dixon 1980), which arises from assuming that correct thought is possible only for certain social groups (usually intellectuals). By contrast, the more recent 'strong programme' in the sociology of knowledge does not assume that what passes for knowledge in a given group of people at a particular time is valid or invalid. Instead, the object of study is the social relations that give rise to the creation and maintenance of knowledge systems (for an early version of this position, see Berger and Luckmann 1967; for an articulation of the strong program itself, see Bloor 1976).

Only by bracketing questions of epistemology has the sociology of knowledge been able to make substantial progress, as is most evident in its subdiscipline, the sociology of science (cf. Mulkay 1979, Knorr-Cetina 1981, Whitley 1984). Once we decide that the study of what passes for scientific knowledge does not require that we pass judgments on the validity of such knowledge, then we are able to find out how scientists themselves come to agree on what is scientific truth. Precisely the same strategy may be applied to the study of ignorance. Ignorance is a social creation, like knowledge. Indeed, we cannot even talk about particular instances of ignorance without referring to the standpoint of some group or individual. Ignorance, like knowledge, is socially constructed and negotiated.

A working definition of ignorance, then, is this: "A is *ignorant* from B's viewpoint if A fails to agree with or show awareness of ideas which B defines as actually or potentially valid." This definition avoids the absolutist problem by placing the onus on B to define what she or he means by ignorance. It also permits self-attributed ignorance, since A and B may be the same person. Most importantly, it incorporates anything B thinks A could or should know (but doesn't) and anything which B thinks A must not know (and doesn't). B may be a perpetrator as well as an attributor of ignorance.

A single definition is insufficient, however, to send us on our way. A second major pitfall in conventional approaches to ignorance is to view it as unitary. Ignorance is multiple, and has distinct levels. Some such distinctions have been known for some time. For instance in his dialog with Meno, Socrates pointed out the difference between what he called "ignorance" and "error". The person in error believes he knows that he doesn't know, while the ignoramus is conscious of his lack of knowledge. I prefer to term the first *conscious ignorance* and the second *meta-ignorance*, since Socrates' terms do not correspond to modern usage (cf. Smithson 1985). It is noteworthy that even meta-ignorance requires some person's viewpoint to exist, and so also is a social construct.

A second distinction that often is made in ordinary discourse is between *informational* and *epistemological* ignorance. This distinction pertains to whether the ignoramus is in error about factual matters or, even having the facts at hand, does not process them appropriately. This is a useful distinction if for no other reason than that it enables us to deal with 'erroneous' information separately from information processing.

The most important distinctions, however, are those which refer to different kinds of ignorance, rather than different levels or loci. Another commonsense distinction which arises in common language is between 'ignoring' and 'being ignorant'. The first is in active voice, and connotes overlooking or even deliberate inattention to something. Being ignorant of something, on the other hand, is in the passive voice and connotes distorted or incomplete knowledge. This duality is fundamental and therefore is an appropriate first branching point in a typology of ignorance. The act of ignoring is a declaration of *irrelevance*, which is the term I will use to refer to this kind of ignorance. The state of ignorance, on the other hand, is in one way or another an erroneous cognitive state, and I will refer to it by the term *error*.

Language often reflects social reality, and there is more at stake than the difference between a transitive and intransitive expression. These two kinds of ignorance denote negative and positive strategies for dealing with anomalies and other threats to established cognitive order. The negative way is to ignore them or not perceive them, thereby in effect declaring the anomalies irrelevant and banishing them from reality. The positive way is to revise the framework of reality itself to make a place for the anomalous material, even if only within a declaration of ignorance. The former is a strategy that results in exclusion; the latter ends in inclusion.

Error may arise from either incomplete or distorted views (or both, of course). *Distortion* usually is referred to in terms of 'bias' or 'inaccuracy', on the one hand, and 'confusion' on the other. The former refers to distortion in degree while the latter indicates wrongful substitution in kind. *Incompleteness*, on the other hand, has received considerable attention from philosophers and other scholars, quite possibly because it seems more corrigible than distortion (readers should take care not to confuse this term with 'completeness' in the mathematical sense). As a consequence, it has been subdivided more minutely. For purposes of our discussion, I propose to dichotomize incompleteness in a similar fashion to the bicategorical division of distortion. Incompleteness in kind will be termed *absence*, while incompleteness in degree will be called *uncertainty*. Uncertainty, in turn, includes such concepts as probability, vagueness, and ambiguity. I will delay further dissection of this part of the typology for now, however.

Turning now to irrelevance, the most obvious kind is *untopicality*. For our purposes, it refers to the intuitions people carry with them and negotiate with others about how their cognitive domains fit together. Topical consistency is, for example, one of the unspoken rules guiding ordinary conversation. Folk-wisdom lauds geniuses for their avowed capacity to see connections between matters that appear irrelevant and unrelated to most people. However, a similar attribution is made about the insane. There are two other kinds of irrelevance that pertain to our typology: *undecidability* and *taboo*.

Undecidability is attributed to those matters which people are unable to designate true or false either because they consider the problem insoluble or because the question of validity or verifiability is not pertinent. Certain kinds of fantasy and fiction have this latter quality, as do thoughts that are considered nonsensical or meaningless. It could well be argued that undecidability has two senses, however, one of which belongs outside of the irrelevance domain. The relevant kind of undecidability, however, is to a large extent captured by terms such as 'uncertainty' or even 'incompleteness', and so I will not list it separately under error.

Taboo, on the other hand, is socially enforced irrelevance. Douglas (1973:100-101) was the first to systematically elaborate this point, and her discussion of taboo holds much of interest for anyone who would understand cultural responses to uncertainty. Taboo matters are literally what people must not know or even inquire about. Taboos function as guardians of purity and safety through socially sanctioned rules of (ir)relevance. This concept is particularly rich in its explanatory power for how we deal with anomalous or cognitively threatening material, and Douglas places her concerns with taboos in the center of any explanation concerning how we deal with disorder. As she points out (Douglas 1967: 39), any system for cognitively ordering the environment gives rise to anomalies, and all cultures must therefore confront these anomalies with appropriate strategic defenses. The sophisticated normative paradigms in Western intellectual culture of probability theory, decision theory, and their more recent counterparts and competitors are examples of such strategic defenses. All of them invoke taboos regarding the kinds of uncertainty or ignorance that are beyond the scope of scientific, logical, mathematical, or otherwise 'proper' analysis.

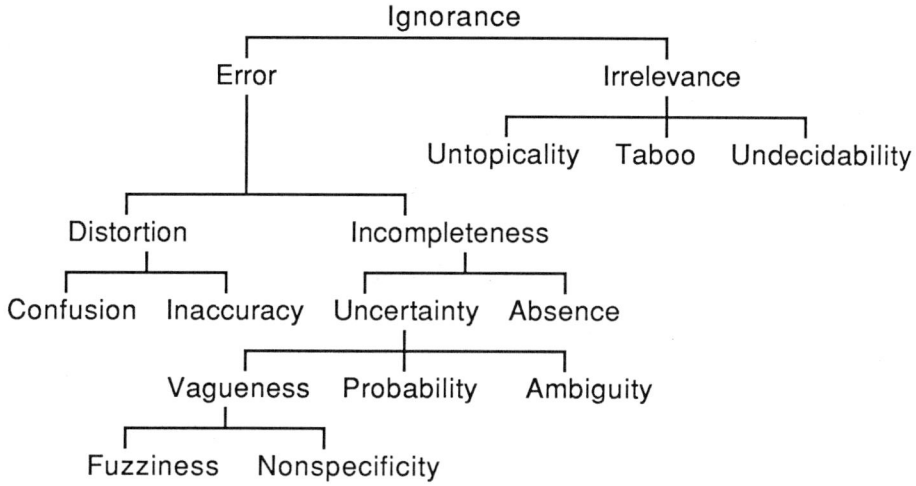

Figure 1.1. Taxonomy of Ignorance

The complete taxonomy is displayed in Figure 1.1. While some readers might wish to use different terms or make other distinctions than those I have provided, I should nevertheless like to argue that this is a viable framework. It is certainly better than the efforts we have reviewed thus far, and where possible adheres to standard usages in the philosophical literature.

Let us return briefly to the topic of uncertainty, which I have not discussed yet and which has been allocated a low position in the Figure 1.1 taxonomy despite its predominance in several fields. Uncertainty, as will become increasingly evident, occupies a special position as one of the most manageable kinds of ignorance. In some fields, the term is employed as a synonym for ignorance. It should be clear by now, however, that uncertainty is not as broad a concept, even though it is the home of probability theory and several other newer normative approaches to ignorance.

I have anticipated some of the arguments in this book by subdividing uncertainty into vagueness, probability, fuzziness, nonspecificity, and ambiguity. As I will argue in Chapter 4, these labels correspond with fairly widespread usages in both philosophical and scientific literature. This is not the only way to subdivide uncertainty. However, most other taxonomies (e.g., Howell and Burnett 1978) are based on well known probabilistic concepts. Kahneman and Tversky (1982), in their article on variants of uncertainty, provide psychological and phenomenological arguments for dichotomizing uncertainty into that which is attributed to the external world and that attributed to our internal state of knowledge. This dichotomy resembles the objective/subjective duality that runs through the history of probability theory. The external variant of uncertainty in turn may be assessed from either a 'distributional' or 'singular' (case-wise) point of view. This division also has probabilistic overtones, resembling the distinctions often drawn between probabilities as relative frequencies in long runs of events and probabilities of unique events. Likewise, their division of internal uncertainty into 'experiential' and 'reasoned' modes echoes another theoretical division between the empirical and a priori schools of classical probability. The authors of course recognise the antecedents of their categories and do not claim great generality for their scheme. The purposes of this book require an approach to uncertainty that is not restricted to probability theory.

1.3 The Rest of this Book

The remainder of this book has a two-part format reflecting the schismatic structure of Western discourse on ignorance. One side of the schism consists of normative paradigms, while the other might be called descriptive or explanatory perspectives. Until quite recently the tendency has been for each side to be unfamiliar with the other, and one of the motivations for writing this book is to provide some bridging concepts and discussions. Applied scientists, statisticians, engineers, and computer scientists could benefit considerably by knowing more about what behavioral and social scientists say about how people perceive and respond to ignorance in its various forms. Likewise, behavioral and social scientists would have even more to contribute if they were more knowledgeable about the controversies over probability theory and the more recent emergence of nonprobabilistic formalisms for representing uncertainty.

Accordingly, I have tried to write each chapter at the level of the ever-elusive 'intelligent nonspecialist'. However, I rather doubt that I have succeeded, and so I wish to offer a map for guiding the reader through the maze of material that lies ahead. Chapters 2, 3, and 4 comprise Part I, which outlines the normative frameworks on ignorance. This material is likely to be more familiar to readers with a grounding in applied mathematics and statistics than those in the behavioral sciences. Part II includes Chapters 5, 6, and 7, and provides overviews of the accounts of ignorance that have come from psychology, sociology, and cognate disciplines. This material will, of course, be more familiar to behavioral scientists and less so to engineers and physical scientists or mathematicians.

Chapter 2 begins with an outline of the philosophical consequences of striking any 'middle ground' between strict dogmatism and radical skepticism, arguing that middle-ground positions necessarily complicate our view of ignorance. Therefore ignorance, for most of us, is multi-faceted and problematic. This theme is carried into a comparison between two professions that have long-standing traditions for coping with uncertainty: Engineering and law. The strategies and standards adopted in these professions are explicable in terms of the primary agendas and practical needs of their practitioners, and this finding suggests that recent ferment about ignorance and uncertainty in several fields may also be linked to social, political, or economic interests. What may be required by way of an explanation is an account of how practitioners and scientists have changed not only their strategies for dealing with ignorance, but also their beliefs in such goals as the eradication of ignorance. Chapter 2 ends with an account of the crisis in pure mathematics at the end of the 19th Century, in which mathematicians were forced not only to abandon the quest for absolute certainty but also to alter their conceptions of mathematical ignorance and their strategies for coping with it.

Chapter 3 applies the framework suggested by the material on the crisis in mathematics to the controversies among competing schools of probability theory. Those readers familiar with the philosophical debates over probability may not find much that is new other than the viewpoint, while the material may strike others as somewhat abstract or technical. I would urge those in the human sciences, especially, to persist; there are riches here that can be mined for their research implications. This chapter is not the only attempt to present the competing schools of probability to nonmathematicians. In fact,

I recommend Oakes' (1986) clear and lucid commentary on statistical inference as a complementary text for behavioral scientist readers.

Some of the most exciting recent developments in uncertainty are prescriptive theories that grew out of critiques of probability, and these are surveyed in Chapter 4. Unfortunately, there seems to be no other attempt to survey these developments in a nontechnical way, so for now readers are stuck with this one. I have endeavored to reach a readable compromise between comprehensiveness and depth. It is not possible for a single chapter to come close to either of those goals. Nonprobabilistic uncertainty formalisms have generated hundreds of publications per year for at least the past decade, and their level of technical sophistication has increased dramatically. Again, I can only urge readers who are unfamiliar with this material to persist, to treat this chapter as an introductory survey, and to follow up references on ideas that seem interesting.

Part II deals with the other side of the schism: Explanatory accounts of how we deal with ignorance. Chapter 5 examines the major trends in research emerging from psychology. The bulk of the psychological research on ignorance is grounded in probabilism, and that perspective therefore flavors most of this chapter. Some space is given, however, to the recent tentative moves by psychologists into nonprobabilistic uncertainty (e.g., fuzzy sets, ambiguity, and vagueness). The research results themselves as well as the debates over rationality and human intuition are instructive for knowledge engineers, expert systems designers, and artificial intelligence researchers. As with Chapter 4, Chapter 5 is a compromise between being thorough and covering topics in depth; the psychological literature in this area is voluminous.

Chapter 6 presents insights from certain branches of sociology, social psychology, and social anthropology. These disciplines have treated ignorance only marginally, but they offer points of view that differ widely from those in psychology. The starting-point for sociologists and social anthropologists is that ignorance and knowledge are social products, and the debates in their classics anticipate many of the modern arguments over rationality and realism in other fields. This chapter selects accounts of ignorance in the social order from the social science literature, primarily from microsociology and anthropology, the sociology of organizations, and the sociology of knowledge and science.

The perspectives described in Chapters 5 and 6 are younger than their normative counterparts, and in some cases do not amount to theories at all, let alone paradigms. Nevertheless, they contain ideas and insights that are crucial for approaching a complete understanding of where we are and where we are headed in our attempts to cope with uncertainty in particular and ignorance generally. Specifically, they may enable us to comprehend the nature of the paradigmatic shift that becomes apparent in the material from Part I. Chapter 7 outlines some possible explanations for that shift and anticipates some future directions in the dialog between normative and explanatory accounts of ignorance. The competing normative frameworks, it is suggested, may serve different social or psychological functions and goals, and the second section of this chapter attempts to outline those functions as well as the tradeoffs involved in selecting one representation of ignorance over another. The chapter includes an example of a dialog between normative and explanatory frameworks in the setting of research on second-order uncertainty, which in turn points toward some fundamental dilemmas in deciding how to represent this kind of uncertainty.

Chapter 2:
Full Belief and The Pursuit of Certainty

"Doubt is the chastity of the mind." R. Zelazny.
"Doubt is not a pleasant mental state but certainly is a ridiculous one." Voltaire.

2.1 Ignorance: The Views from Dogmatism and Skepticism

Without attempting a thoroughgoing survey of classical philosophy, some initial insights may be gained into Western intellectuals' traditional orientations toward ignorance by briefly examining the varieties of ignorance referred to in classical philosophical literature. Every theory of knowledge draws a distinction between knowledge and ignorance, and most between ignorance in the sense of incomplete knowledge and ignorance in the sense of erroneous belief. The earliest forms of philosophy amount to various kinds of dogmatism or skepticism, both of which necessarily refer to and utilize ignorance. I will discuss dogmatism first.

Dogmatic philosophers such as Plato or Locke require absolute certainty about knowledge. In fact, dogmatists equate the two; anything we know we are certain of and vice versa. To a complete dogmatist, everything is known that can or must be known, at least by some ultimate authority. Traditionally this authority is omniscient, although it is possible for the dogmatist to countenance specialists as authorities. Any beliefs or assertions differing from or truncating dogma constitute heresies, and it is noteworthy that the most dogmatic of religions (e.g., many versions of Christianity) attach moral culpability as well as an attribution of error to heresy. To err is not merely human, it is also bad. Dogmatists distinguish doubt from erroneous belief, where doubt usually means to withhold assent both from a proposition and its negation. But doubt also traditionally carries a moral stigma, since the basis for many dogmas rests on faith (Luther, for instance, was adamant that one could not be a Christian and a skeptic simultaneously).

Dogmatists do permit two kinds of ignorance that are not necessarily morally reprehensible. One is conscious ignorance, if accompanied by the appropriate deference to knowledgeable authority. Indeed in many Western and nonWestern knowledge systems, certain kinds of knowledge are not accessible or permitted under various circumstances. The old term 'mystery' at one time referred to both the unknown and the secret. If everything is known, then a mystery can only exist for one who has not been initiated into the inner circle of those who know. Some matters may be taboo, in which case it is knowledge rather than evil which is wicked. So, for the dogmatist, a natural counterpart to conscious ignorance (mystery) is taboo, and in some systems any kind of irrelevant matter (e.g., fantasy or fiction as viewed by Puritans) is considered taboo.

The second kind of morally excusable ignorance is innocence, which usually denotes an unconscious ignorance that nonetheless is free of heresy. This is not to say that all meta-ignorance is excusable to dogmatists; that point is clearly articulated by modern legal codes which deny ignorance of the law as an excuse. Innocence, in some dogmatic systems, is exalted above fire-tested faith and belief, which then leads to a complicated and paradoxical arrangement whereby the innocent must be protected by knowledgeable but tainted guardians.

Ignorance, then, tends to be construed by dogmatists according to the culpability of the ignoramus. Aetheists, heretics, and agnostics all are reproachable, even though the first two believe erroneously while the latter merely doubts or suspends judgment. Innocents and those unitated into the supreme mysteries escape censure by the dogmatist, but only if they keep their places.

Modern dogmatism takes a somewhat mitigated form, although mitigated dogmatism itself is also an ancient style of philosophy. Most modern dogmatists hold only a crucial subset of their beliefs as dogma (e.g., dogmatic rationalists), thereby permitting doubt or critical inquiry into their nonessential beliefs. Not everything is known, and perhaps there is no omniscient being. Indeed, some dogmatists claim the best defense against skeptics is to retain the smallest and most self-evident set of dogmas possible. However, moral sanctions still apply to those in error or doubt regarding the central dogmas, even if those sanctions no longer include being burned at the stake.

Furthermore, while the modern dogmatist may hold that whatever is certain also is known, she or he usually does not require the converse. Mitigated dogmatism admits uncertainty into the realm of knowledge, albeit under careful control. Usually this is done by creating a category for uncertain knowledge or at least knowledge that is not demonstrable, as when Socrates drew his famous distinction between knowledge and opinion. While not denying that one could have correct opinions, philosophers following this line claim that opinion must be accompanied by some degree of probability or doubt.

Now, let us consider how ignorance is viewed by the opposite to radical dogmatism: Pyrrhonist (or complete) skepticism. Complete skepticism maintains that nothing can be certain, nor can anything be known. No one is justified or reasonable in their assertions about reality. Like its dogmatic counterpart, total skepticism is unarguable because any argument requires the disputants to agree on some criterion for determining whose argument is correct, and both dogmatists and skeptics accept no terms other than their own. Those who hope to refute the skeptics (or their modern counterparts, the radical relativists) by pointing out that skepticism itself is unjustified according to its own tenets provoke the rejoinder that skeptics do not see apparent inconsistency as a difficulty since they do not believe in logic anyway.

The complete skeptic therefore has a monolithic view of ignorance which is all-embracing. There is no such thing as error, irrelevance, or innocence. Doubt is universal and overwhelming. All is utter mystery and no one can say what is erroneous to believe. Nor is the complete skeptic attracted to the concept of degrees of uncertainty. As Montaigne pointed out, to acknowledge one proposition to be more likely than another is to incline to one rather than the other, which is what graded measures of uncertainty (e.g., probability) require us to do. Radical skeptics will have none of it.

Restricted or mitigated skepticism, on the other hand, does involve distinctions among different kinds of ignorance. Usually the mitigated skeptic allows at least some comparable assessment of propositions such that one is decidably better than the other. Humean skepticism, for instance, requires only that we find the most probable opinion with a degree of certainty sufficient for action. A modern example is Popper's falsificationism, which holds that scientific hypotheses may not be proved but they can be confirmed in degree or outright disconfirmed. Once this is permitted then most of the classical distinctions among various kinds of ignorance become discussable, specifically doubt, error, and conscious ignorance.

This observation brings us to an interesting point. Ignorance is uncomplicated only for those philosophical positions that come close to embracing total dogmatism or complete skepticism. Moderated skepticism and dogmatism *both* entail more complex descriptions of ignorance. Why should this be the case? The primary reason appears to be the enhancement of strategic control over ignorance. A philosopher who places limits on human knowledge or ignorance must then classify, specify, and describe those limits or else hedge bets on where the limits are. Opinion, doubt, and probability are therefore natural candidates for a buffer zone between that which is certainly true and that which is certainly false.

Regardless of their stance on ignorance and knowledge, traditional Western philosophies universally value knowledge and seldom esteem ignorance. Even radical skeptics do not adopt a moralistic counterpart to the dogmatist's stance on knowledge and ignorance. The closest Western philosophers come to valuing ignorance is either the glorification of specific kinds on moral grounds, as a necessary precursor to gaining knowledge, or as a necessary condition for freedom of action. Thus Rousseau exalts the innocence of the savage but does not contradict the proposition that knowledge itself is beneficial. Awareness of one's own ignorance is widely considered a hallmark of intellectual sobriety, but only as an indication that one is prepared to learn or discover new knowledge. Skepticism itself originated from the Greek term for inquirer and usually plays a gadfly role in philosophical discourse.

The claim that complete knowledge of the universe entails determinism, thereby robbing people of freedom of action, is perhaps the highest valuation of ignorance in Western intellectual traditions. It is one that we shall encounter on several occasions when considering modern accounts of ignorance (especially uncertainty). Freedom, after all, is a positively valued version of uncertainty in conventional terms. Interestingly, chance itself has been argued for on these grounds. The pragmatist James, for example, defined chance as the negation of necessity which in turn creates the opportunities for human free will to exercise itself. Erich Fromm (1947) went even further and declared that the quest for certainty not only restricts freedom but blocks the search for meaning. Only uncertainty impels humanity to develop its full powers.

Now let us return to mitigated dogmatism and skepticism and the treatment of ignorance. Once the position is taken that some matters are certain but others may not be, one question that immediately arises is why people do not know everything in the first place. The Sophists wrestled with what they regarded as a paradox about the existence of erroneous beliefs. If a person believes that which is not, then does that person believe anything at all? If this is the only possible world, then how is ignorance in the sense of belief in nonentities possible? Plato's correspondence theory of truth and error easily answered this question but begged another, namely how erroneous belief arises in the first instance.

Many philosophers (e.g., Descartes), attempting to reconcile human error with the concept of an omniscient and all-loving God, asked how people could possess false beliefs when God not only knows they are false but wishes those same people well. The other side of this coin would have us consider a cosmic deceiver, Descartes' 'evil genius', who deliberately leads human perceptions astray. Descartes' famous and remarkable refutation of these difficulties is that all error in human thought is the product of human volition. This is not to say that Descartes claimed that people deliberately adopt false ideas, but rather that they are capable of assenting to propositions whose validity has not been properly established. This answer to the question of how error arises tosses the ball into the psychologists' court, which we will visit in Chapter 5.

More recent philosophical interests in ignorance have shifted from their traditional moorings in two important respects. One is a renewed interest in what might be generically termed 'nonsense', while the other is an increased moral emphasis on justification of belief (rationality or sanity). Concerns with nonsense were initially raised by the logical positivist theory of verifiability which banished all metaphysical statements as meaningless. Philosophers suddenly had to take stock of whether their own utterances amounted to nonsense. Verifiability theory posed insurmountable problems, and the criteria for nonsense shifted to linguistic analysis, with which philosophers attempted to draw a boundary between the merely false and the outright nonsensical.

The upsurge in rationalist-irrationalist debates during the same period points to a shift in the locus of moral concerns about ignorance. Although rationalisms of various kinds have been proposed since antiquity, it is a relatively modern set of mores which holds that the measure of sanity is not the specific contents of one's belief system, but rather one's way of thinking. Although we still find the older style of heretic hunting in orthodoxy-ridden societies such as the Soviet Union (where anti-Marxist dissidents or the openly religious have been labelled 'schizophrenic' and incarcerated in mental institutions), the modern mental health movements in psychology and psychiatry reflect a corresponding shift in philosophy away from identifying heresy with wrong beliefs and toward diagnosing insanity or irrationality on the basis of wrong thinking. In short, the locus of moral culpability has moved to epistemological ignorance.

Like insanity, irrationality is not considered a philosophical sin, nor do philosophers believe that vice follows from ignorance and virtue follows from knowledge. However, more than one modern philosopher has used sickness and mental illness as a metaphor for improper epistemology or lack of certitude (cf. Wittgenstein 1958: 91, 155; and 1983: 132). The image of Bertrand Russell grasping after certainty as worshippers clutch at faith is supplanted by ignorance as a mental disease and philosophers as therapists who enable themselves and others to "see the world aright." Their clients are the subrational laypeople whose commonsense biases and foibles mislead them into illusions. In this sense, modern attempts to provide secure foundations for knowledge may be viewed as mental health campaigns, and the campaign strategies used are a rich source of information about how Western culture deals with ignorance.

Most thinkers who work with uncertainty or other varieties of ignorance are not philosophers. Often they are scientists or professionals who are trained and motivated to adopt a collection of attitudes and beliefs about ignorance which might well be termed a moderate form of dogmatism. There is a central set of axioms, assumptions, beliefs, or credos which are beyond critical inquiry. These are extended by a combination of empiricist and logical guidelines to form a body of professional knowledge which is revisable but conservative. Like many philosophers, these thinkers and doers distinguish between error and incompleteness, and they have irrelevancy rules as well. Because they are not philosophers, however, they tend to judge the worth of knowledge or ignorance according to extrinsic criteria, and their resemblance to pragmatists is striking. I shall call their orientation toward ignorance 'normative pragmatism'.

The remainder of this chapter, then, consists of a limited survey of normative pragmatists' practices in handling ignorance, followed by an historical sketch of a well-known philosophical 'mental health campaign' in mathematics. These two sections will provide crucial examples, insights, and clues for an understanding of modern trends in the management of ignorance.

2.2 Examples of Traditional Normative Pragmatism: Structural Engineering and Jurisprudence

Most professions in modern society provide services under conditions of less than complete information or knowledge. Medical practitioners, social workers, lawyers, engineers, teachers, and clinical psychologists all risk failure to some extent in their respective practices. Failure, of course, is partly defined by fiat or even social consensus, although it can have an objectively physical component. The mixture of subjective and objective components in definitions of professional malpractice or failure has some bearing on the codes adopted by professionals for dealing with the uncertainty and ignorance entailed in practice. It is beyond our scope to discuss all professions' methods for coping with uncertainty, so I shall try to provide a 'maximal contrast' pair of professions on the basis of the subjective/objective dichotomy.

At the risk of provoking howls of outrage from several quarters, let us take structural design in civil engineering and jurisprudence as our contrast pair. Not all engineering entails objectively discernible failure or malpractice, but structural failure (e.g., the collapse of a building or bridge) comes fairly close. On the other hand, matters of justice are inherently socially created and often not tied to physical outcomes or consequences.

First let us consider what these two professional spheres have in common. One shared feature is an overriding conservativism towards speculation and trial-and-error methods of inquiry. Unlike the scientist (or at least a Popperian), neither the judge nor the structural engineer wish their operational rules and procedures to be falsified. They are interested in avoiding failure, and maximizing certainty or knowledge is merely one strategy among many for accomplishing this goal. Accordingly, they also share what might be called a tendency to err on the side of caution. In structural mechanics, this may entail constructions that are deliberately made much stronger than theory says they need be; in the lawcourts this proclivity is embodied in standards of proof and the injunction to presume innocence until guilt is proven. Thus structural designers and judges have similar goals and cautionary tendencies. Their strategies for avoiding failure, however, are markedly different.

Structural engineers usually encounter two kinds of design problems: One-off and mass-produced products. Usually, a product that is to be mass-produced usually is designed and constructed initially in prototype form, and the prototypes are then subjected to destructive testing (or 'proving'). Standards of proof for the prototypes usually are set at some level well above the wear or stress expected during normal service. One-off designs, on the other hand, usually cannot be tested in prototype form either because they are genuinely unique or such testing would be uneconomical. Instead, the engineer may destructively test prototype components, nondestructively test the entire structure (rarely done), or make theoretical calculations of maximal stresses and multiply those by safety factors to provide the design criteria.

The difference between these two design problems leads to quite distinct orientations towards ignorance on the part of the engineer. As Blockley (1980: 23) points out, in a one-off job no theory or prior information about similar structures is strictly applicable, and while the mass production designer can resolve considerable uncertainty during the prototype testing phase, the structural engineer cannot. Structural engineers therefore tend to be more conservative and less empirical than mass production engineers, and often rely more on old-style rules of thumb.

Traditional engineering criteria and rules for ensuring structural safety in the absence of empirical tests are based on two related strategies. The first is to estimate a 'worst case' load (or stress) scenario and apply it to theoretical computations of the capacity of the proposed structure, and the second is to use safety factors to ensure that the 'worst case' calculations exceed the likely actual loads and stresses by an appropriate margin of safety. Both strategies have been employed for some time. Straub (1949) recounts the procedures used in assessing damage to the dome of St. Peter's in Rome and the repairs required in 1742, in which stress and load calculations were made by a team of mathematicians and then adjusted using a safety factor of 2.

The precise nature of the models used to estimate worst-case loads or stresses and the magnitude of the safety factors employed are mutually intertwined. The simplest and oldest method for estimating stress uses elastic theory and arrives at a 'critical stress' estimate, which in turn is multiplied by a single safety factor to arrive at a 'permissible stress' criterion. Elastic theory provides rather liberal estimates of how much loading a structure can withstand, however, and its efficacy therefore depends on conservatively large safety factors being employed. The resulting structures tend to be overbuilt and expensive, but reducing the safety factors is clearly a risky undertaking.

A similar method is called the 'load factor method', and involves the use of two estimates of load: a 'working' load of the magnitude that could reasonably be expected during service, and a 'collapse' load which is the load under which the structure fails. The safety factor may then be calculated by dividing the collapse load by the working load. It is then up to arbitrators of standards to decide what an acceptable safety factor is. Like the permissible stress approach, this method's success hinges on the conservativism of the theoretical model that is used to calculate the collapse load, since that estimate often is not empirically based for one-off design problems. The most common conventional method of assessing how much tolerance for error a given theoretical estimate contains is to systematically vary the theoretical assumptions or relevant parameters, thereby revealing how sensitive the final estimates are to those variations. The engineer may then use this sensitivity analysis to provide lower and upper bounds on his or her estimates, or alternatively to demonstrate that the variability is too small to worry about.

More recent methods (such as limit state design as outlined in Blockley 1980: Ch. 4) are essentially combinations and generalizations of the permissible stress and load factor methods--- a sophisticated armamentarium of partial and componential safety factors along with complex methods for estimating stress and load-bearing capacities. One important addition has been made to these methods, however, and that is the overt use of both subjective and empirically based probability estimates of load magnitudes. A probability distribution of loads is usually generated either empirically or theoretically, and this is used to arrive at a design value that has a very low (even if unknown) probability of being exceeded.

In summary, engineers and others involved in construction arrive at a quantified consensus on safety margins and methods of estimating load-bearing capacities and stresses. The main counterbalancing consideration in construction is cost, hence the motivation to find more sophisticated structural theories that will permit smaller safety factors. Only very recently have they overtly used probability, but even when it is utterly notional or subjective, engineers seem relatively comfortable with quantitative guesses. The nature of that consensus differs across nations and even among engineering subfields, but the overwhelming emphasis is on attempting to measure 'how near to truth' they are rather than what truth itself is (Blockley 1980: 51).

Judges, like structural engineers, must operate most of the time with less than perfect information, witnesses who are unreliable, and 'facts' of doubtful standing. Often they have at best partial knowledge of how the information presented to them was obtained. Unlike the engineers, however, the judges may also have to contend with dishonesty on the part of witnesses and/or lawyers, as well as legitimate attempts by the disputants to manipulate the judge's perceptions or feelings. Furthermore, ultimately the judicial system does not either stand or fall, as does a physical structure, on the accuracy with which experts have plumbed natural laws, but instead on social and political criteria. These criteria are changeable, and the law itself is vulnerable to multiple interpretations and revisions.

Every judicial decision is guided by a standard of proof analogous to the engineer's testing criteria. In civil cases guilt must be decided 'on a balance of probabilities', while of course criminal cases require the tougher standard of proof 'beyond a reasonable doubt'. Unlike the engineering criteria and despite the probabilistic language, these phrases have seldom been made more specific. Eggleston (1978: 102), in his study of judicial probability, comments that judges appear to avoid explaining or elaborating on these traditional phrases for fear that their attempts to improve them will be found wanting by their peers. One is reminded of Nagel's (1961) anecdotal quotation of Lord Mansfield's advice to an appointed governor of a colony to decide judicial cases intuitively but to "...never give your reasons, for your judgement will probably be right, but your reasons will certainly be wrong."

The vagueness and long history of those verbal standards may also helpfully conceal a potential lack of consensus should their precise meanings to judges, jury, and litigants ever be communicated to one another. Where new verbal formulations (e.g., 'leaving only a remote possibility') have been tried, judges have been unable to agree on their meaning (Cohen and Christie 1970: 60). Nor does quantification of subjective probabilities help. Simon and Mahan (1971) asked mock jurors to provide numerical probability levels that corresponded in their opinions with the phrase 'beyond reasonable doubt' and obtained levels ranging from 0.7 to 0.9; but the judges from whom they requested the same kind of estimates required stricter levels. Furthermore, they found that by asking jurors to quantify their estimates of probable guilt, the percentage of attributions of guilt was lowered.

In addition to the standards of proof, many Western judiciary systems also impose a greater burden of proof on the prosecution than on the defense, in line with the adage that the defendant be assumed innocent until proven guilty. However, again the nature of this burden is unspecified for the most part, although there is a widely held craft-knowledge of what kinds of evidence will or will not convict, and lawyers and judges alike may refer to volumes of case precedents or even tomes on the subject of evidence itself.

As with standards of proof, judges and lawyers have avoided formalization or quantification of the 'weighing' of evidence; as Eggleston observes, the "legal profession as a whole has been notably suspicious of the learning of mathematicians and actuaries..." (1978: 2-3). Suspicions notwithstanding, there have been many attempts to apply probability theory to legal problems. Indeed, Leibniz' early formulations of probability theory were motivated by problems of legal inference. Famous failures include probabilists of high rank: Laplace, Poisson, Cournot, Boole, and Keynes. To the extent that legal systems employ formal and quantitative rules of evidence, they greatly resemble the older engineering conventions. In fingerprinting, for example, experts in various Western countries require from 8 to more than 16 matching characteristics and no unexplained points of difference to risk the claim that two prints could have come from the same person (dare we call them rules of thumb?).

More often, rules of evidence are qualitative and broad. Traditional criteria for ascertaining the credibility of witness testimony include: (1) internal consistency of the account; (2) consistency and consensus with other credible witnesses; (3) evidence of mental soundness, competence, and credibility of the witness; (4) observable cues concerning witness mendacity, truthfulness, and confidence; and (5) the apparent (im)plausibility of the account. The first two criteria concern logical coherence and agreement, while the remaining three require considerable assumptions about human character and behavior. There is a large and rather inconsistent literature on how judges may ascertain whether perjury is being committed, and attempts to provide a 'scientific' basis for such inferences have provoked widespread controversy (e.g., the use of the polygraph in the U.S., U.K., and other countries. See the 1986 Report from the British Psychological Society).

Unlike the criteria for proof or evidence however, the legal systems in most Western nations are quite specific about the limits imposed on what kind of evidence may be admitted or heard, and for how long. These limits amount to (ir)relevancy rules which distinguish the matters that must be attended to by the judge and jury from those that may (or should) be ignored. A piece of evidence is considered relevant primarily if it may be used to prove the case. There are at least two ways in which this might happen: (1) If the evidence increases the credibility or believability of act(s) the court wishes to (dis)confirm; or (2) if the evidence is inconsistent with a relevant fact. Probativity is the dominant criterion for relevance insofar as it determines whether a piece of evidence has any bearing on the case, but veracity, credibility, adherence to proper procedures for obtaining evidence, and prejudiciality all define 'taboos' by which material may be banned from the court as inadmissible. For example, a defendant's prior criminal record, although verifiable and perhaps relevant, may be ruled inadmissible because it would prejudice the jury against the defendant.

The English system may be used as a model for explicating more specific points. Like most Western systems, it has two sets of rules pertaining to admissibility: Rules of exclusion and rules of extension. The former, of course, banish certain kinds of evidence from the proceedings. The latter permit the admission of evidence that does not appear to bear directly on the probativity of the case. Exclusion rules bar similar facts, hearsay, prejudicial material, privileged information, and evidence of character. Extension rules permit evidence of competence or credibility, surrounding detail or context, and facts 'deemed to be relevant' by the judge. Similar facts, hearsay, and rules of extension will be discussed in turn.

Similar facts refer to transactions or circumstances other than those which are the subject of the proceedings, and are inadmissible even if they might pertain to the case. For instance, the question of whether the defendant was speeding on a particular road cannot be resolved by undertaking a survey and finding that 95% of the motorists traveling that road also exceed the speed limit; nor is evidence that the defendant has or has not exceeded speed limits on other occasions allowable. Major exceptions to this rule involve expert witness testimony and, under some conditions, evidence of the defendant's behavior under similar circumstances to those resulting in the trial.

Hearsay distinguishes secondhand reports and attributions from primary evidence. But the hearsay rule is so exclusionary that it creates problems. For example, if taken literally it would eliminate a considerable portion of eyewitness testimony. Consequently this rule has many exceptions, including confessions, admissions, statements in public documents, and statements contained in the 'res gestae' (the things done). A similar juggling of exceptions may be found in the rule excluding evidence of character. The obvious exception is the defamation case, but oddly witnesses are not protected by this rule, and quite often the accused is permitted to present evidence of good character.

The rules of extension may override the exclusionary rules, mainly at the discretion of the judge. Thus witnesses may be cross-examined for indications of their competence, qualifications, or credibility. Accounts of surrounding detail may be sought as a check on the internal consistency or reliability of witness testimony. And judges may deem relevant any fact that could be used as evidence for or against another probative fact.

What does the comparison between the strategies adopted by judges and structural engineers for coping with ignorance reveal? A crucial insight is that these two professions differ profoundly in how they manage the two main kinds of ignorance in our taxonomy (i.e., error and irrelevancy). More specifically, the engineers have explicit criteria for dealing with error while the judges are explicit on the matter of irrelevancy. The engineers' relevancy criteria are implicit while their framework for dealing with incompleteness and distortion is explicit and quantitative. Nor are those implicit relevancy criteria beyond questioning. After all, the conventional approaches to preventing structural failure under uncertainty ignore systemic uncertainty (as distinct from uncertainty about particular parameters and individual components), human errors, and consequences of structural failure. Any of these could be deemed relevant.

The judges, on the other hand, have explicit and fairly extensive relevancy criteria encompassing outright taboos, but implicit, qualitative, and somewhat privatized ways of dealing with incompleteness and distortion in evidence. The sources of ignorance in jurisprudence are more social and less grounded in apparently physical reality than in structural engineering, and the differences between their brands of normative pragmatism may be explicable by that alone. The explicit, public aspects of ignorance control in these two professions seem to correspond to those that are manageable and consensual.

By the same token, the irrelevancy rules employed by these pragmatists banish unmanageable or intractable varieties of ignorance. For judges, the salient problem is to limit and control the debate over evidence, and their primary resources for doing this are relevancy rules that form a part of 'due process'. Control over courtroom proceedings is essential if a judge is to preserve his or her professional standing. Engineers find it easier to agree on their subject matter if they tacitly eschew the psychological, social, or moral realms and restrict discourse to physical problems. For them, the salient problem is how to avoid physical evidence of structural failure which could threaten their professional reputations.

The foregoing discussion, then, illustrates a central thesis of this book: Not only is ignorance socially constructed, but so are the normative frameworks adopted by professionals or intellectuals for dealing with it. If we want to understand how and why mainstream orientations towards ignorance in several fields have changed during the last 20 years, then it will not suffice to restrict considerations to the philosophical or even the psychological. We will surely have to search for explanations that take social, political, and cultural factors into account as well.

2.3 The Ultimate Mental Health Project

I have hinted that modern trends in ignorance management are strategic responses to a widely perceived increase in the variety and extent of ignorance as well as a breakdown in the older consensus on absolute knowledge. I also have indicated that some fields have undergone a crisis that could be termed a 'loss of certainty', and their practitioners' coping strategies may tell us something about what is happening in other areas now. I believe this kind of crisis, and the strategies that evolved as a result, are nowhere more clearly visible than in modern mathematics. Accordingly, this Chapter ends with an account of developments during the crisis in pure mathematics. These developments centered around a challenge thrown down to his fellow mathematicians by David Hilbert, who described a series of problems whose solutions would provide absolute foundations for mathematical knowledge and inference.

The Hilbert Program in mathematics and its offshoots were direct responses to the blows against mathematical absolute certainty, and therefore full belief, that befell mathematicians in the 19th century. Extending Wittgenstein's metaphor slightly, some of the best mathematicians in the late 19th and early 20th centuries found themselves embarking on an ultimate 'mental health' project whose goal was to establish the criteria for mathematical sanity and thence to rehabilitate both the field and its practitioners. The difficulties that mathematicians encountered in pursuing complete certainty resemble, at least in surface features, many of the problems that have arisen in connection with ignorance in other intellectual fields. Although this chapter in mathematical history is well known, it remains remarkable that irreducible ignorance should have arisen and persisted in what many people regarded (and some still regard) as the knowledge paradigm least susceptible to chronic ignorance. Perhaps more than in any other field, mathematicians would seem to be well equipped to eliminate ignorance. There is some value, then, in employing hindsight to understand the strategies of ignorance management adopted in mathematics, the problems they raised, and the current situation.

First, however, we must appreciate some of the difficulties pursuing mathematicians in the 19th century. The truly halcyon days for mathematics arrived with the birth of

Newtonian mechanics and a methodological shift in several of the physical sciences from emphasizing qualitative physical explanation to quantitative mathematical description and prediction. Galileo and Bacon were the precursors of this shift, heralded by Galileo's famous pronouncement in *The Assayer* of 1610 that the book of nature is written in mathematical language. While Newton, of course, was the main progenitor, the monumental contributions of such scholars as Euler, the Bernoullis, Laplace, Lagrange, Hamilton, and Jacobi not only raised the status of determinism to considerable heights but firmly installed mathematics as its terms of reference. By the end of the 18th century, many mathematicians and scientists had adopted the position later encapsulated by Kant's mildly tardy dictum that there is only as much real science as there is mathematics. Perhaps the most outstanding brief articulation of the full belief in the certitudes of mathematics during this time is the declaration by Laplace that

> "An intellect which at any given moment knew all the forces that animate nature and the mutual positions of the beings that compose it...could condense into a single formula the movement of the greatest bodies of the universe and that of the lightest atom: for such an intellect nothing could be uncertain..."

As Morris Kline (1980: 67) put it, natural law was mathematical law. Mathematical laws, therefore, were invested with universality and absolute certainty.

By the mid-19th century, however, several of the foremost mathematicians came to doubt that mathematical truths were either universal or absolute. As is well known, the seeds of this doubt germinated in the very mathematical fields held by many to be the most impermeable, namely geometry and algebra. The story of modern geometry is the more dramatic of the two, the prime mover being the pursuit of a justification of Euclid's so-called 'fifth postulate' (or 'parallel axiom'). The Jesuit priest Saccheri published an attempted proof by contradiction in 1733, in which he assumed a contrary postulate and claimed that it led to a contradiction in the resulting geometric system. However, by the mid 1700's at least a few mathematicians (e.g., Klugel, Lambert, and Kastner) went on record as recognizing that Saccheri's geometry was self-consistent and therefore non-Euclidean geometries were logically possible and constructible. Nonetheless, none of them pursued this realization to its conclusion. By 1813 Gauss had privately begun work on his non-Euclidean geometry which he apparently refrained from publishing for fear of the outcry it would raise. Lobachevsky and Bolyai, of course, received the majority of the credit for publicly articulating non-Euclidean geometries, and Riemann for the most famous case of a physically relevant nonstandard geometry.

Gauss reacted to the loss of absolutism in geometry by reposing his faith in arithmetic. Geometry, he claimed, was akin to mechanics, and its foundations ultimately were experiential. Arithmetic, on the other hand, was purely a priori and based on obviously certain numerical truths. Ironically, the foundations of algebra itself were threatened only a short time later by several developments which resembled those in geometry. Hamilton's quaternions were motivated by problems in physics, but their practical success came with a price: Quaternion multiplication is not commutative and thereby violates the 'absolute' properties of the algebra shared by all real and complex numbers. Other useful but noncommutative algebras soon followed (e.g., Cayley's matrix algebra and Grassmann's family of such algebras). Worse still, physicists such as Helmholtz pointed out that our most basic numerical and arithmetic concepts are grounded in experience and not a priori truths.

At about the same time as these unsettling developments were taking place, many excellent mathematicians were becoming uneasy about the lack of properly logical underpinnings for the calculus, particularly for the difficult concepts of limits and series which involved infinities and infinitesimals. The immense practical successes of the calculus promoted its application despite its unsound foundations, and those applications led increasingly both to exceptions to seemingly 'universal' theorems and to monsters in the form of bizarre logical paradoxes. By the latter half of the 19th century, many first-rank mathematicians were calling for an end to the confusion and chaotic developments that characterized the preceding 150 years. What strategies did they adopt for this return to sanity?

The classical strategies may be divided, roughly speaking, into three groups: Reductionism, banishment, and pragmatism. To use a metaphor that I shall extend later on, reductionism is essentially a rehabilitative or corrective impulse, while banishment is an attempt to bar problematic elements from mathematics. Reductionism is therefore an inclusionary form of control, while banishment is exclusionary. One of the most frequently used forms of banishment is an irrelevance rule, which dictates the boundaries of what 'proper' science or mathematics is and places unwanted uncertainties in the domain of irrelevant phenomena. Pragmatism, on the other hand, amounts to an abandonment of attempts to deal with the difficulties insofar as it entails evaluating mathematical concepts in terms of their utility for solving practical problems in other domains. Most sophisticated strategies for developing a general mathematical overview include elements of all three strategies. Nonetheless, some of them are identifiable by their emphasis on one or another such strategy.

Reductionists attempt to bring rigor to mathematics by reducing it to another system widely felt to be rigorous. This strategy has an impressive mathematical pedigree, and from ancient times manifested itself in various attempts to declare one mathematical field supreme over all the others. From the time of classical Greek civilization to the era of Descartes and Pascal, geometry held the crown. By the time Newton and his contemporaries had completed their work on the calculus, however, algebra increasingly gained favour, and by the Second International Congress of Mathematicians in 1900, Poincare could declare that mathematics had been arithmetized (Kline 1980: 182).

The logicists were the dominant reductionists in the modern crisis, claiming that any mathematical intuition was suspect until confirmed by analysis in terms of classical first-order (and sometimes second-order) logic. The most famous logicists included Frege, Whitehead, and Russell. Russell is the most widely quoted of the three, and probably the most accessible to the nonmathematician. For some twenty years he propounded logicism and defended it against its many critics. His recollection of his motivations in those days leaves no doubt of his agenda: "I wanted certainty in the kind of way in which people want religious faith. I thought that certainty is more likely to be found in mathematics than elsewhere... if certainty were indeed discoverable in mathematics, it would be in a new field of mathematics, with more solid foundations than those that had hitherto been thought secure" (Russell 1958). Here, then, is the ultimate mental health project incarnate.

Russell and his colleagues argued not only that mathematics is ultimately reducible to logic, but that the two are coextensive. In his 1919 *Introduction to Mathematical Philosophy*, Russell claimed that no clearcut distinction could be drawn between logical and mathematical axioms, propositions, and methods of deduction. To fully appreciate how radical this thesis seemed to many at the time, we must realize that by the mid-19th century the notion of mathematical proof was so disreputable that some quite able mathematicians hardly bothered with proofs at all. To them, an appeal to logic as the cornerstone of mathematics was an irrelevancy. However, it is noteworthy that not all mathematicians understood the same thing by the term 'logic'. Until the mid-1800s, Aristotelian logic dominated both philosophical and mathematical discourse on logical matters. But the logicists' program was based on and enabled by the modern development of symbolic logic, initiated by Boole and De Morgan, and further developed by Peirce. This new brand of logic, in fact, was derived in good part from ideas in algebra and therefore could be said to be mathematized logic.

Not all of the logicists' strategies were reductionistic, however. Even Whitehead and Russell found it necessary and sometimes convenient to use banishment tactics. The most famous of these is the Theory of Types, in which individual objects belong to a type 0 level, sets of objects belong to type 1, sets of sets belong to type 2, etc. Propositions that cross the boundaries are not allowed, so it becomes impermissible to speak of a set that belongs to itself.

Banishment is an even older strategy than reductionism, with roots in ancient mathematics. The primary objective is to find a way of avoiding paradoxes or monsters by constructing definitions and criteria that bar them in principle. Diophantus and his contemporary Arithmeticians declared negative numbers 'impossible' and 'absurd' (Newman 1956, V.1: 115), pointing out the apparent incomprehensibility of subtracting a larger quantity from a smaller quantity. These numbers were barred by the Arithmeticians from proper mathematical inquiry. Similar reactions may be observed as mathematicians encountered irrational, imaginary, and complex numbers. Their very names evoke their disrepute.

The modern banishers are the intuitionists, who directly oppose reductionism of any kind, and particularly logicism. The initiators of intuitionism included Baire, Borel, Kronecker, Lebesgue, and Poincare, all of whom raised objections to the axiomatics of the late 19th century and the logicists' programs in the early 20th. The fully-fledged intuitionist perspective, however, was expounded by Brouwer and elaborated by Weyl. Brouwer declared that mathematics is an autonomous mental activity, divorced from the world outside the mind, and independent of language. It is synthetic rather than derivative, and therefore cannot be based on some other system of thought or language such as logic. Intuition determines the validity of fundamental ideas, not logic. Certain mathematical ideas are intuitively clear but others are not and therefore cannot be accepted as part of true mathematics, whatever their historical roles or logical status.

It is at this point that intuitionism reveals itself as a banishment strategy. Thus, Brouwer claimed that natural numbers, addition, multiplication, and mathematical induction are intuitively sound. The latter permits infinite sets, but only those that are 'potentially' and constructibly infinite. Cantor's transfinites, Zermelo's axiom of choice, the continuum hypothesis, and other 'nonintuitive' consequences of logical or formal manipulations are banned. Moreover, logic is held by intuitionists to belong to language, not to mathematical thought. Not all principles in logic are acceptable to intuition, and so the existence of paradoxes indicates an undisciplined or improper application of logic, not a fault in mathematics. Brouwer and Weyl both castigated logicians for abstracting the Law of the Excluded Middle beyond its initial intuitively justifiable application to finite sets, and thereupon banished the Law of the Excluded Middle from arguments involving infinities. Partly as a consequence of this, the intuitionists also disallowed most reductio ad absurdum proofs (proof by contradiction) and many nonconstructive existence proofs (which establish the existence of a mathematical entity without specifying how to construct or observe it). A number of famous and widely used theorems were disqualified by the intuitionists.

Neither the reductionist or the banishment strategies provided satisfactory solutions to the problems raised in the 19th century. Logicism was widely attacked, both for specific axioms or arguments that opponents founds questionable, and more generally by those who rejected either its somewhat Kantian a priorism or the primacy of logic over mathematics. Though criticised on various specific counts, logicism's chief shortcoming is its inability to guarantee security against not only the paradoxes and monsters it was designed to handle, but any unforseen difficulties of the same kind.

The major complaints against the intuitionist school of thought probably were identical to those facing the Diophantines, namely that they were throwing the baby out with the bathwater. Intuitionism, if taken seriously, dismantled so much classical and modern mathematics that the field would be devastatingly impoverished. Although the intuitionists themselves have devoted considerable effort to reconstructing mathematics in accordance with their philosophy, both the slow pace and cumbersome complexity of that reconstruction have earned them many criticisms among practicing mathematicians. Hilbert spoke for many of them in his well-known 1927 declaration: "For, compared with the immense expanse of modern mathematics, what would the wretched remnants mean, the few isolated results incomplete and unrelated, that the intuitionists have obtained."

Furthermore, there is the awkward fact that the intuitionists themselves do not always agree on what is intuitive or not and how to establish this. In recent times the "father of fractals", Mandelbrot (1983), has argued that many so-called "monstrous" mathematical creations (such as the Peano curve that covers a square) are intuitively obvious and clearly manifested in natural forms, contrary to mainstream mathematical commentators such as Hahn (1956) and others before him. In fact, Mandelbrot's own position embodies the postmodern rejection of *both* classical strategies for handling mathematical creative uncertainty: "In any event, the typical mathematician's view of what is intuitive is wholly unreliable; it is impossible to permit it to serve as a guide in model making; mathematics is too important to be abandoned to fanatic logicians" (1983: 150).

Some modern classicists have attempted to combine the banishment and reductionist strategies. One such approach, which still enjoys some popularity among mathematicians, was constructed by Zermelo, Fraenkel, and their fellow workers in set theory. The Zermelo-Fraenkel revision of 'naive' set theory amounts to a compromise between the reductionist and banishment strategies, which may partly explain its appeal. Another appealing aspect is that some of the most galling monsters and paradoxes were initially generated by Cantor's version of set theory, which itself was an attempt to provide a foundation for the calculus. The Zermelo-Fraenkel axiomatization of set theory avoids the all-inclusive set and hence the paradoxes associated with that concept, but retains a sufficiently rich set theory (e.g., permitting transfinite cardinals and certain other concepts that go beyond experience and/or intuition) to satisfy the requirements of classical analysis.

The main criticisms raised against the Zermelo-Fraenkel scheme were technical, and concerned specific axioms. The axiom of choice, for instance, created considerable controversy but most of the initial arguments questioned its technical feasibility or intuitive plausibility rather than its logical or formal potential for generating paradoxes and monsters. Eventually the controversy was settled in a manner of speaking by proofs that the axiom of choice is both consistent relative to Cantorian set theory (Godel) and independent of the other axioms (P. Cohen). These proofs ensured that the axiom of choice could be added to or dropped from the Cantorian collection of axioms without damaging the formal system (Kramer 1970: 595), thereby rendering its inclusion in set theory a matter of choice indeed. The price for this is, however, that the axiom of choice itself is not decidably true or false within the Zermelo-Fraenkel system, and mathematicians are therefore bereft of any clearcut reasons for choosing to include it or not.

Reductionism and banishment are examples of traditional Western ways of dealing with ignorance, namely by outright reduction or elimination. Furthermore, they are not managerial strategies, insofar as neither of them approach the problem of ignorance in an anticipatory or generalized fashion. Instead they are crisis-oriented attempts to solve here-and-now problems posed by specific paradoxes and mathematical monsters. A more modern managerial approach had to await an attack on foundational problems from David Hilbert.

Hilbert not only was the first mathematician to coherently delineate several of the major problems facing the foundations of mathematics, but he also dismissed both classical logicism and intuitionism and proposed his own strategy, as he put it, "to establish once and for all the certitude of mathematical methods" (a quotation from his 1925 paper "On the Infinite", as cited in Kline 1980: 246). Hilbert's approach is often referred to as the "formalist" school of mathematical thought. He began with the assertion that a proper mathematical science must include axioms and concepts of both mathematics and logic, and all mathematical or logical statements must be expressed in a symbolic form that frees them of all ambiguity or vagueness. The symbols themselves need not have any meaning or intuitive sensibility. They must, however, be manipulated in a logically consistent fashion.

The kind of certainty sought by Hilbert differed in one important respect from that of the logicians and intuitionists, in that he wished to construct proofs of absolute consistency for mathematical systems. Indeed, he had already shown that Euclidean geometry is consistent if arithmetic is consistent. A proof of the consistency of arithmetic became the formalists' Holy Grail during the 1920's. The second goal of Hilbert and the formalists was an absolute proof of 'completeness'. A mathematical theory is consistent if it never generates contradictions, but it is complete if every possible meaningful statement generated by it is decidably true or false.

The crucial difference between this strategy and its classical forebears lies in the level of analysis involved. The classical approaches to establishing certainty operate at the *first-order* level of mathematics or logic itself. Hilbert's program required a *second-order* approach, that is, a metamathematical methodology. Logicism could not guarantee that new paradoxes or undecidable propositions would not arise someday in the future, nor could intuitionism guarantee that no unforeseen monsters would pop up inside the enclosure they had built. If Hilbert's program could be achieved, however, then the unknown in mathematics would literally be tamed.

The ultimate mental health program ended when Godel published his famous 1931 paper. The two most devastating results dashed the hopes Hilbert and his followers had of establishing absolute certainty in mathematics on terms acceptable to mathematicians of the time. First, Godel demonstrated that the consistency of any mathematical system as extensive as whole-number arithmetic cannot be proved by the logical principles adopted in any of the foundational schools of thought. Secondly, and more fundamentally, he established that if any formal system that encompasses whole-number theory is consistent, it must be incomplete. This result applies to both first-order and second-order predicate logic. Although some of the undecidable propositions generated by a self-consistent arithmetic could be decidable by arguments using logical principles that transcend that arithmetic, the transcendent system itself might not be consistent.

The problem of incompleteness was driven home in an even more profound form by the Lowenheim-Skolem results, which implied that any axiomatic system is vulnerable to unintended interpretations or models. One of the primary aims of axiomatic methods is to fully describe and define a particular 'species' of mathematical entities (e.g., the positive whole numbers). Lowenheim and Skolem's theorems proved that for any system of axioms there are other unanticipated 'species' that fulfill the axioms and yet were not intended to be included in the original class encompassed by the axioms themselves. A major source of these new mathematical monsters is the inevitable inclusion in any axiomatic system of some undefined terms. The upshot of these results is that mathematical reality cannot be described unambiguously by axiomatic methods.

In short, ignorance in mathematics is here to stay. Mathematicians now labour under what appears to be unavoidable twin spectres of doom: The Scylla of inconsistency and the Charybdis of incompleteness. Postmodern mathematicians have adopted various identifiable strategies for living with irreducible ignorance, and these echo recent trends in Western intellectual culture at large while providing particularly instructive examples of those trends.

One of the most noticeable trends is the relativization and consequent privatization of mathematical truth (cf. Wilder 1981: 39-41), a phenomenon remarkably similar to the modern privatization of religion. This tendency has been aided by the rapidly escalating specialization among mathematicians since the Second World War. The result is a subdivision and resubdivision of territory which, for want of a better term, I shall call "suburbanization". Specialists form homogeneous epistemic subcommunities within suburbs and either outright ignore or agree to disagree about the nature of mathematical truth with mathematicians from other suburbs. These practices amount to more than mere specialization, since they also involve the erection of boundaries.

Suburbanization not only involves dividing up mathematical subject matter and domains, but also selecting specific approaches to fundamentals. Thus today we find more than a dozen major versions of set theory, encompassing more than 50 axioms of various kinds. And with suburbanization has come the mathematical equivalent of urban planners, imperialists of a different sort than the would-be hegemonists of yesteryear. These system-builders engage in meta-axiomatics, which amounts to specifying which axioms may be added to or dropped from various versions of set theory without damaging the system. Having determined the menu, the choice of axioms is then left to the user.

A fairly typical example of this brand of theorizing is Vopenka's (1979) "alternative set theory" (AST). As one proponent of AST puts it, "...even the property 'to be a book' is not quite precisely defined... the decision whether a collection of objects... will be formalized as a set or as a proper semiset depends on our standpoint" (Sochor 1984: 174). Sochor then exemplifies the grander scheme of mathematical suburbia according to AST: "...Cantor's set theory becomes a study of one particular class of a certain type of extended universe studied by AST" (Sochor 1984: 186).

What criteria are used by the suburbanites for selecting axioms, guiding creative activity, and deciding the legitimacy of truth-claims? The postmodern choices of criteria include, of course, the classical schools amongst the options. But for the nonclassicists, the determining criteria tend to be either utility or aesthetics. If we cannot attain absolute certainty, then at least mathematics should be useful in applications (the 'applied' point of view) or beautiful to behold (the 'pure' viewpoint). To absolutists, these criteria beg the question. For instance, even granting the much remarked and possibly arbitrary success mathematics has had in modeling the physical universe, mathematicians (and other scientists) still are unable to answer the question of why that success should be the case and whether it will continue. Heaviside, on the other hand, spoke for most applied mathematicians and scientists when he asked whether he should refuse his dinner because he did not understand the digestive process.

The most important strategy for handling irreducible ignorance in mathematics to emerge in recent times, however, is to apply mathematical methods to the study of ignorance itself. Wilder (1981: 82) has suggested that pure mathematicians are less threatened by paradox or undeciability in formal systems than their predecessors, which if true is a reflection of a more managerial approach to mathematical ignorance. Thus, the specific nature of paradoxes, the types of mathematical propositions that are undecidable within particular systems, and the formalization and quantitative measurement of vagueness, ambiguity, and complexity in mathematics have all become topics of study in their own right (e.g., Skala et al. 1984).

These developments are, like Hilbert's program, a 'second-order' strategy for dealing with uncertainty. Unlike Hilbert's attempt to eliminate ignorance completely, however, this new second-order strategy primarily attempts to impose some order on ignorance by understanding and adapting to it. Therein lies its importance, as an exemplar of a major shift in Western intellectual culture away from the pursuit of absolute certainty. I shall not attempt to explain these developments in the recent history of mathematics here, since their primary utility at this point is illustration and suggestion. The sequential unfolding of strategies by mathematicians for dealing with ignorance suggests a crude but plausible model that will serve as a template later on.

The sequence of strategies may be briefly summarized in four stages. First outright banishment is tried, by way of directly eliminating uncertainty and clearly demarcating the boundary between the true and the false. If banishment eliminates some useful techniques or concepts, however, then pragmatists usually argue for inclusionary modes of control. The second stage therefore involves attempts to rehabilitate the 'monsters', and the most popular strategy is reductionism.

When reductionist or other inclusionary efforts generate their own monsters, a third stage begins with a leap to second-order strategies for taming first-order ignorance and uncertainty. For the mathematicians, this strategy entailed the search for proofs of absolute consistency and completeness. More generally, such strategies involve the first serious attempts to describe, classify, and explain ignorance itself. The failure of the second-order strategies leads to the fourth stage, which continues the study of ignorance and uncertainty but also entails suburbanization, relativistic coexistence or epistemic pluralism, and the abandonment of the pursuit of certainty.

The explanation of this history poses a fascinating problem in the sociology of knowledge (see Chapter 6). As the material in section 2.2 demonstrated, and has sociologists of science have argued for some years, what passes for knowledge and ignorance in mathematics is socially constructed by the community of mathematicians. That construction cannot be explained entirely by referring to the strategies mathematicians have employed to deal with ignorance. Fuller explanations must await an investigation into what accounts the human sciences offer of how people deal with ignorance.

That investigation will be taken up in the second part of this book.

Meanwhile, this section has provided a stage-model of strategies for coping with ignorance. I will employ this model in Chapters 3 and 4 to provide an account of the developments stemming from probability theory and its offshoots. We will find that the probabilists have used the classical first-order strategies of banishment and reduction in their attempts to tame uncertainty. The 'new look' in the mathematics of uncertainty, with its nonprobabilistic formalisms and appeals to nonstandard logics, represents the leap into second-order strategies and the prospect of irreducible uncertainty.

Chapter 3:
Probability and the Cultivation of Uncertainty

"If anything is to be probable, then something must be certain."
C.I. Lewis.
"Ex nihilo, nihil." Leslie Ellis.

3.1 Probabilists as Strategists

If there is any approach to ignorance that bears a creditable claim to generalizability and rationality simultaneously, it is probability. Virtually all modern accounts of uncertainty refer to the concept and theory of probability as a benchmark. However, it is crucial to realize that probability theory actually consists of a cluster of competing approaches, each of which claims either exclusive or universal status as the one true theory. The differences among these theories are considered fundamental by many probabilists, especially the division between the so-called 'objective' and 'subjective' approaches.

I do not intend this chapter to be an exhaustive scholarly, historical, philosophical, or mathematical treatment of probability theory. After all, this subject occupies entire books on its own. Instead, the goals and strategies of various approaches to probability will be examined, with an emphasis on the controversies that have arisen in each of them. The more hard-nosed probabilists discount some of these concerns as irrelevant, especially if they focus on apparently historical or social and psychological factors. Nevertheless, there are several reasons for this kind of treatment.

First, probability theories are mental and social creations, despite even the most ardent realist's exhortations on behalf of objectivity or rationality. Even the initial definitions in a given theory often are justificatory fortifications constructed in anticipation of particular criticisms. Few serious modern theorists write without sustained comparative and critical reference to other theorists and their work. As various controversies have become elaborated and argued during recent decades, probabilists have become increasingly concerned with incorporating viable theories into their own approach and excluding rival nonviable approaches.

Furthermore, debates about and applications of probability theories occur within communities comprised of both probabilists and nonprobabilists, and the theories themselves are shaped by the discourses within those communities. There is even a long-standing controversy about the relationship that should be fashioned between probability theory and its various applications. A strict realist (e.g., von Mises) would strive to divorce probability theory from its ordinary or even scientific precursors. Modern soft realist philosophers such as Weatherford (1982: 4) allow that a theory of probability, while prescriptive, nevertheless should either incorporate or at least account for as many pretheoretical usages of the term in either scientific or ordinary discourse. In fact, outright subjectivists (i.e., the Bayesians) argue that the main guidance provided by probability theory merely ensures some coherency in people's own beliefs about probability. I.J. Good (1962: 319-320) is one of the few probabilists who has tried incorporating the communicative aspect of probability itself into an overview of the topic, albeit in a way largely uninformed by modern behavioral scientific knowledge.

Finally, and most pragmatically, probability theories are created for purposes of prescriptively (normatively) dealing with uncertainty or ignorance. Wilkins explicitly announced this intention in 1675, and Joseph Butler's 1736 aphorism "Probability is the very guide in life" made it famous. In the Butlerian tradition, a theory of probability usually is intended as a guide in situations characterized by less than perfect certainty, confirmation, or determinacy. The most popular approaches to probability tend to be those that 'work', despite the lack of apparently solid foundations for them. Indeed, one of the clues to unraveling both major controversies and the riddle of modern pluralism among users of probability emerges from asking where each approach appears to work best, and to what ends. In doing so, we must also inquire how people decide 'what works', and what personal and social norms shape those judgments.

Accordingly, let us begin by examining attempts to talk about probability, or even about theories of probability. Talking about probability requires that we both define it and discuss others' definitions of it. We need at least a working classification of probability theories. A number of taxonomies have been proposed for various purposes (cf. Nagel 1939, von Wright 1941, Carnap 1950, Good 1950, Kyburg and Smokler 1964, Black 1967, Barnett 1973, Fine 1973, Weatherford 1982), but the most popular in recent years seems to be a threefold division of modern approaches into Relative Frequency Theory,

Logical (A Priori) Theory, and Subjective (Bayesian) Theory, with Classical Theory occupying a fourth, mainly historical category. I will adopt that scheme, since it enables the simplest and clearest translations among various metatheoretical discussions.

3.2 Definitions of Cultivatable Uncertainty

Prior to modern probability theory, uncertainty was largely uncultivated territory. The ancient strategies for dealing with it marked the earlier meanings of the very words that are etymologically related to the term 'probability'. Hacking (1975: Ch.3) makes a case that probability at one time meant something like 'worthy of approbation', and in support of this cites related modern words such as 'approbation', 'approval', and 'probity' as well as quotations in which 'probability' itself seems to have been used in that fashion. But it is also worth remembering that 'probability' may share etymological roots with quasi-legal terms like 'probative', 'probe', and 'provable' which stem from an alternative source whose principal meaning is 'to test'. Either meaning carries some explanatory power in accounting for the immense popularity of a strategy for taming uncertainty that appears to rest on both mathematical authority and empirical testability.

The basic strategy behind the construction of a probability theory has two components. The first is a 'bracketing' operation in the phenomenological sense of the term. Every approach to probability begins by excluding various kinds of ignorance and even uncertainty from consideration and even in some cases by banishing particular interpretations of probabilistic uncertainty. Ignorance is subdivided into the cultivatable and the intractably wild.

The most popular cultivating strategy is the quantification of uncertainty and the imposition of a logical calculus for manipulating those quantities. Furthermore, this strategy appears to have been applied by the same thinkers to two rather different sets of problems. As Hacking (1975: 12) points out, the earliest attempts at probabilty theory are premised on a clear duality between 'physical' propensity and 'subjective' degree of belief. Huygens, Hudde, Graunt, and de Witt developed their systems out of the realization that the statistical propensity of a physical event is countable, and its proportion of occurrences out of the total number of events is a quantity capable of mathematical manipulation. On the other hand Leibniz, the Port Royal Logic group, and to some extent Pascal (in his Great Decision argument) began with the idea of establishing a metric for degrees of belief or credibility and utilizing logical criteria to rationalize them.

This duality seems to have an ineradicable nature to it, since it has resurfaced many times in the history and philosophy of probability theory and remains a current great theoretical divide. Hacking's terms are 'aleatory' for physical or objective, and 'epistemic' for subjective or personal probability. Hacking also poses some pertinent questions about this apparent duality. If it is so fundamental, then why did Classical Theory arrive more or less in one piece? Why not historically and conceptually separate developments for both kinds of probability? Why do so many contemporary practitioners ignore this distinction and operate as if there is only one kind of probability? And why have definitional distinctions between the two kinds failed to cool the controversies boiling among probabilists for nearly three centuries?

Hacking raises the possibility that probability and related concepts have a life of their own that defies human abilities to impose strict order. A Wittgensteinian 'family resemblance' argument might point out that many unitary names (as in the famous example of 'game') conceal multiple prototypes that do not share any common properties or features. Perhaps probability is inherently multiple, despite wishful thinking to the contrary. Finally, ethnomethodologists have borrowed the concept of 'indexicality' from philosophers (cf. Garfinkel 1967) to describe the infinitely revisable character of everyday concepts. Indexical concepts, by virtue of being continually reconstructed and negotiated in everyday practice, are impossible to exhaustively describe or define. This could be the case with ordinary or even scientific meanings of 'probability'.

I cannot accept Hacking's ideationist answers to the questions he poses (nor does he ask anyone to do so). The family resemblance argument might turn out to be descriptively true, but still begs the question of whether all casual usages of the probability should be assigned equal legitimacy. Indeed, a bone of contention between subjectivists and objectivists consists of exactly that issue. Likewise, while probability may or may not be indexical, neither outcome obviates the need for pragmatic, working, or official definitions of the concept.

Instead, I argue that the struggles between objective and subjective schools of probability have hinged on a strategic dilemma in the definition and application of the concept itself. This dilemma is similar to that facing the metamathematicians engaged in Hilbert's undertaking during the early part of this century. On the one hand, the attractive domains of application for probability include kinds of uncertainty and ignorance that confound attempts at total rehabilitation. On the other, the strategies of either banishing those kinds of uncertainty or reducing probability to a purely mathematical or logical system overly restrict the domain of application. Like the metamathematicians, probabilists confront a mutinous boundary between knowledge and ignorance, at least as far as their classical strategic options are concerned.

Before surveying the gambits and problems in the major schools of probability, let us briefly consider the conventional approaches to defining a concept like probability. A great deal hinges on these definitional matters, including the scope and justification of one's viewpoint. It is notoriously difficult to avoid presuming the entire basis for a theory in the very definitions, as witness Laplace's well-known verbal "trick" (according to Good 1983: 68) or "vicious circle" (according to Reichenbach 1949: 353) of defining equiprobability in terms of equipossibility. At least three strategies of definition are available: (1) Positive or direct; (2) by reduction to some other concept(s); and (3) by negation or exclusion. Examples of direct definitions are the classical notion that probability is the proportion of 'favorable' out of all 'possible' events, or Good's proposal (1952) that probability theory is the logic of degrees of belief. A popular reductionist definition is the pure mathematician's characterization of probability theory in terms of mathematical measure theory. Purely negative definitions of probability are rare but exclusionary asides often are attached to definitions, usually in reference to some undesired usage of the term in common language or by a rival school. Direct definitions share the advantage of not explicitly describing the limits of applicability, and thereby tend to make the concept appear universalistic. However, they impose the greatest justificatory burden on the theorist. They are the most easily attacked (and also perhaps the most likely to be misinterpreted). Reductionist definitions usually are invoked when the framework to which probability is being reduced appears to have agreed-upon foundations. Reductionism, if successful, obviates the problem of justification and application; one need only follow the precepts of an established framework. The attempt to reduce probability to a purely logical basis or to mathematical measure theory appeals to universalists who wish probability theory to attain the status of mathematics, and pragmatists who hope to avoid the limits on application imposed by substantive definitions of probability. Definitional exclusion or negation defends against pollution by association with undesired versions of probability, or against a takeover by a rival claimant to universality. Exclusionary riders on direct definitions often are intended to restrict the scope of application to those situations where the definition will work or seem reasonable.

The orientation of a definition may well depend on any or a combination of the following considerations: (1) The intended domains of application; (2) the kind(s) of justification on which to base the theory; and (3) its intended relationship to competing theories. While other classificatory schemes have been proposed (e.g., Fine 1973: 3-10), these concerns appear to arise most frequently in the literature on probability.

The domain of application issue has become central in recent debates, not only between different schools of probability, but also between probabilists and proponents of nonprobabilistic approaches to ignorance. Historically, the issue arose in conjunction with attempts to repair classical probability theory against its critics by defining its domain more restrictively. These restrictions paved the way for alternative and/or more universal theories, with the result that the field may be characterized in terms of four approaches to the question of domain. The most xenophobic response is to declare a limited domain for oneself, and then to banish all rival theories to an undifferentiated hinterland of illegitimacy. An imperialist approach is to claim all conceivable territory and other theories for one's own by reductionist or equivalence arguments. A suburban solution is to carefully subdivide domains so that each theory possesses its own sphere of applicability. Finally, there is relativistic coexistence, in which domains may be shared by more than one theory, depending on the preferences of the user.

Justifications in probability theory usually concern either logical internal consistency (occasionally completeness as well) or the apparent sensibility of conclusions derived from applications of the theory. A somewhat related concern is whether the theory in question performs as well as its rivals in domains where they are successful, and/or better in domains where they are unsatisfactory. This last concern is at the heart of Bayesian claims that their approach accomplishes all that the classical, logical or frequentist approaches accomplish as well as performing rationally in domains requiring subjective probabilistic judgments, and is echoed by Shafer's (1976) claim that his theory of evidence is a generalization of Bayesian methods.

Relationships with other theories are important mainly for understanding who is arguing with whom, and on what grounds. Various probability theorists have claimed to annex other theories, contradict or correct them, or talk past them. We even find occasional claims, as we shall see later on, of incommensurability between one approach and others.

As I mentioned above, the dilemma facing probabilists concerns a tension between justification and domain of application. This tension has its source in Classical Theory, and its essential form in that approach has remained with us. Accordingly, our exploration of this dilemma will begin with a review of the Classical approach and its problems.

3.3 Classical Theory and the Unifying Impulse

There is some debate over who qualifies as a Classicist, but most writers in the area seem to agree that this approach publicly surfaced in the decade of 1660 with the works of Pascal, Fermat, Huygens, the Port Royal Logic, and the actuarial efforts of de Witt, Hudde, and Graunt. It received further development primarily at the hands of the Bernoullis, Montmort, De Moivre, Euler, Lagrange, and Bayes, but most accounts agree that its final mold was constructed by Laplace. The historical roots of the Classical approach have been ably traced by several historians; I am more concerned here with the strategic aspects of Classical Theory as a normative paradigm for handling uncertainty. Most of the key elements of this approach may be considered as proposed solutions to specific kinds of problems. The discussion that follows will therefore contain only relevant references to historical developments, and Laplace will be used in most instances as the spokesperson for Classical Theory.

A dominant type of problem in aleatory probability is exemplified in games of chance, where 'fair' gaming devices such as dice, coins, and cards are used for bets on specific outcomes out of a class of possible outcomes. For setups involving the physical propensity of an event to occur out of a finite set of possible events, probability is defined as the ratio of the number of 'favorable' cases to the number of all 'equipossible' cases. The computation of a physical probability of this kind, then, requires an enumeration of all possible outcomes in terms of a fundamental set of equipossible events and their possible combinations. Out of this definition the classicists developed the probabilistic calculus that is still widely and successfully used today (but see later material in this chapter for dissenters and revisionists regarding even the basic calculus).

This initial definition posed the problem of how one can know whether two or more fundamental events are equiprobable. At least three solutions to this problem were proposed early on. One was outright empiricism, although this option did not appear to be widely considered until much later. Perhaps one reason for its lack of popularity initially was its nongeneralizability. The empiricist solution must be retested in every new situation.

Far more popular was an idealized a priorism in the form of a fair die or balanced coin. This solution provides a seemingly generalizable rule covering setups that could be argued on a logical basis to constitute a fundamental probability set. Usually the plausibility of this argument hinges on appeals to a sense of the physical and/or logical symmetry of outcomes (e.g., geometric symmetry in dice, or by weight in coins). However, the a priori approach still requires an assumption that each die or coin encountered is fair. Furthermore, many problems lack an apparent basis for arguing on logical or physical grounds that one event should be as likely to occur as another.

These difficulties motivated James Bernoulli, followed by Thomas Bayes and Laplace, to extend the equipossibility definition to cover any situation characterized by a lack of prior knowledge about the propensities of fundamental events. Their justification is now usually known either as the Principle of Insufficient Reason or (after Keynes) the Principle of Indifference, which essentially holds that alternatives are considered equiprobable if there is no reason to expect or prefer one over another.

Two kinds of objections have been raised against the Principle of Indifference: On grounds of insufficient justification, and excessive restrictions placed by the Principle on the domain of applications. The most obvious justificatory problem is its apparent circularity. Probability is defined in terms of equipossible (read "equiprobable") cases. The identification of equiprobability with equipossibility has been traced back to Leibniz, but it is noteworthy that the two concepts remained essentially confounded until quite recently. Probabilists did not solve the problems posed by the Principle of Indifference through a disentanglement of possibility from probability. We do not see the emergence of a distinct theory of possibility until the mid 1970's. It is also worth bearing in mind that not all modern probability theorists agreed that definitions could or even should be noncircular (a famous naysayer on this count was Borel, in 1909).

A related problem is that the Principle of Indifference confounds at least two different senses of prior ignorance. In one sense, we know in advance (by some means or other) that all events are equally likely, while in another sense we do not know how likely any of them are. A famous scenario has Jones ready to throw a die, and Smith whispers to us "It's loaded!" According to the Principle of Indifference, we should still regard the faces as equiprobable because we don't know in what way the die is loaded. It is also easy to generate paradoxes in which acceptable transformations of measurement scales yield different and equally plausible equiprobable sets of alternatives. For instance, if x is uniformly distributed over some domain, it is not possible for $1/x$, x^2, and other simple nonlinear transformed versions of x to be uniformly distributed in the same domain (cf. Fine 1973: 169-170).

Modern theorists have responded in two ways to this problem. The Relative Frequentists banish it from consideration by saying that probabilities have no meaning under such circumstances, while Subjectivists and A Priorists have attempted to rehabilitate concepts of prior ignorance to permit people to proceed 'rationally' in such cases. We shall find that virtually none of these proposals has proved entirely satisfactory, and that further attempts to solve this problem are still in the works. A principal source of difficulty may well be the assumption that a theory of probability can handle incomplete information (e.g., "It's loaded") as well as complete information about uncertainty (e.g., "the probability of getting a five with this die is 1/2"). We shall return to this issue in discussions about the controversy over whether probability is capable of representing different kinds of prior ignorance.

The insistence on knowing in advance that fundamental events are equiprobable restricts Classical Theory to a narrow range of applications indeed, if it is taken seriously. Laplace's works are filled with computations concerning cases where events are not equiprobable, but they pertain only to situations where prior probabilities of fundamental events are known. Furthermore, the Classical definition of probability as a ratio of integers cannot be applied to irrational numbers as probabilities, despite the fact that they arise in many applications.

Equipossibility and the Principle of Indifference have been shot down by so many theorists that the objections raised against them are almost less interesting than the question of how they enjoyed such success for so long. Hacking (1967 and 1975) claims that possibility itself has the same kind of duality that probability does, as reflected in various linguistic conventions (e.g., the different meanings ascribed to "possible for" and "possible that" in English). This very ambiguity in the concept of possibility thereby became a virtue in its utility as a definitional basis for probability. It could simultaneously paper over the split in the probability concept, while justifying its use in both statistical (physical) and judgmental (subjective) tasks.

The Principle of Indifference also was useful for obscuring some potentially troublesome distinctions among kinds of prior ignorance, as implied in the comments above and in my initial taxonomy of ignorance. Its utility may have stemmed simply from the fact that for some applications it was the only possible justification for using the probability calculus, and no other means to quantitatively and logically handling ignorance was available. The only alternative to the Principle of Indifference was to do nothing or act on sheer intuition.

Applications of Classical Theory, understandably, were not hampered by the official definition. It is easy to fall into the trap of believing that all members of a school or a particular epoch thought and acted identically. In the earliest statistical work, actuarial tables were used to provide initial probabilities along the lines of relative frequency concepts. Furthermore, Relative Frequentists such as von Mises were quick to exploit this loose end by arguing that it demonstrated the need for a frequency-based theory of probability: "... where something must be said about the probability of death, they have forgotten that all their laws and theorems are based on a definition of probability founded only on equally likely cases...no one has succeeded in developing a complete theory of probability without, sooner or later, introducing probability by means of the relative frequencies in long sequences" (1957: 70). As Weatherford points out (1982: 31), it is clear that Classicists were working pragmatically and normatively with a considerably broader concept of probability than their official definition encompassed. In fact, it could be argued that the most practical applications of probability were precisely those that strayed from their definition.

Classical Theory has no difficulty either with the probability of a single event or repeated events, primarily because its definition is couched in terms of the number of ways an event *can happen*, not 'has happened' or 'will happen in the long run.' Repeated events are dealt with by machinery devoted to the problem of inferring probabilities from frequency runs (so-called inverse probabilities): Inverse Bernoulli's Theorem, Bayes' Theorem, and Laplace's Law of Succession. Of these, Bernoulli's and Bayes' concepts have enjoyed the greatest success and longevity. All, however, have been criticised on grounds of insufficient justification. As we shall see in the next section, Relative Frequentists have pointed out that at least two assumptions about the behavior of long-run frequencies are required in order to logically accept the Law of Large Numbers (Bernoulli's Theorem).

Two final, rather all-encompassing criticisms of Classical Theory are that it is not based on the world of real experience; nor is it capable of learning from experience. It is important to realize in this connection, however, that most of the Classicists (Laplace especially) were determinists of the Newtonian variety. For them there was no absolute probability; a Laplacian demon (omniscient being) would know the future with complete certainty. Probability in this sense is a rational way of assessing our own state of imperfect knowledge. The Classicists, in short, identified *uncertainty* with *incompleteness*, which underscores their failure to distinguish incomplete from merely uncertain prior knowledge. Modern theories of probability tend either towards a belief in the existence of real probabilities or a disbelief in the possibility of Laplacian Demons. However, these approaches still often entangle uncertainty with concepts such as ambiguity, vagueness, or other varieties of incomplete knowledge.

What of the Classical Theory escaped widespread criticism? The calculus itself received few serious attacks. Virtually all major schools of probability still use the Classical methods of combining and manipulating known probabilities. The chief debates have centered on the preconditions necessary for using the calculus. Likewise, the notion that candidate values for an unknown probability could be rejected or retained on the basis of accumulated long-run frequencies has been adopted in all mainstream theories. Objections even to this claim nevertheless exist (see, for instance, Fine 1973: Ch.4), and we shall investigate them in the next section.

3.4 Relative Frequency Theory: Banishing the Monsters
The dominant modern paradigm of probability in many fields often is referred to as "frequentist" (Relative Frequency Theory for our purposes), and begins with the premise that probability "...is concerned with things we can count. In so far as things, persons, are unique or ill defined, statistics are meaningless and statisticians silenced..." (Bartlett 1962: 11). Although frequentist ideas pervaded Classical Theory, they often entered through an actuarial back door as servants and supports for the general theory in practical applications. Despite the restrictive, even prohibitionist, rhetoric of the Relative Frequency school, its motivation and current appeal both rest on its vast practicality. We find it in vogue throughout most sciences and professions that deal with uncertainty. Even where it is not used explicitly or directly, when probability is invoked in such fields it is most often the Relative Frequency version.

Venn often is credited with the first overt attempt at an unabashedly frequentist definition of probability, in 1866. Von Mises is usually cited as its chief founding spokesperson, although at times Reichenbach's ideas have been ushered into this role. Von Mises' views have the double virtue of being clear and popular, so in this discussion von Mises will be used as the principal source of doctrine unless otherwise indicated.

Von Mises begins his case for a Relative Frequency basis for probability by identifying some of the faults in the Classical approach outlined in the previous section. First, he argues against the viability of the Classical definition itself, both in terms of its circularity and the impossibility of founding a definition on hypothetically equiprobable cases. He echoes Venn and others in calling for a definition that is both objective and measurable. In fact, his criteria for a definition correspond to the then highly influential operationalist version of scientific positivism which insists that all theoretical concepts be reducible to concrete operational terms. Operationalism is a bracketing strategy akin to that used by the Intuitionist mathematicians, wherein probability is pruned of its difficult and paradoxical cases by declaring them meaningless or illegitimate, and therefore irrelevant or even taboo.

Operationalism also is an attempt to shake off the chains imposed by ordinary language. Unlike some of the Classicists (or, later on, the Subjectivists), operationalists like von Mises are not concerned to capture commonsensical notions within their definitions. His goals are purely prescriptive: "... the content of a concept is not derived from the meaning popularly given to a word, and is therefore independent of current usage... the value of a concept is not gauged by its correspondence with some usual group of notions, but only by its usefulness for further scientific development..." (von Mises 1957: 4).

Like the Intuitionists, von Mises has an operational definition in mind. In particular, he focuses on the fact that whenever the Classicists required any statistical application of probability theory, they turned to relative frequencies based on long series of repeated events. Obviously one cannot define probability simply as the relative frequency of an event in any finite series of trials, since there is no guarantee that any such finite series has converged at the 'true' probability of that event. However, Bernoulli's Strong Law of Large Numbers and other similar arguments appear to demonstrate that under repeated independent trials, relative frequency converges to a limit as the number of trials tends toward infinity. The convergence limit of relative frequency under such conditions becomes a natural candidate for an operational definition since (1) the limit is guaranteed even if unattainable in practical terms, and (2) there is a clear mandate to gather as much data (as long a run) as possible to ensure the greatest accuracy.

This definition has become by far the most widespread among users of probability and statistics. It is still the flagship of textbook concepts. Despite an apparently firm foundation, its acceptance exhibits a strange characteristic of several prominent statistical canons. As Poincare (1913) remarked concerning the assumption of a Gaussian distribution for measurement errors, the mathematicians believe it to have been widely confirmed by empirical observation, while the empirical scientists believe it to be mathematically proven. I have in mind the introduction to the textbook from my first course in probability: "This observed regularity, that the frequency of appearance of any random event oscillates about some fixed number when the number of experiements is large, is the basis of the notion of probability" (Fisz 1963: 5). Weatherford observes (1982: 189) that von Mises' sole reference to empirical data in support of his approach occurs in his attempt to show that the assumption of the existence of a limit is reasonable.

Von Mises adds one more important condition to his definition in an attempt to avoid a simple pitfall. Obviously systematic sequences such as turn-taking in doling out candies to N children exhibit a limiting relative frequency if continued indefinitely; the proportion of deals in which a given child receives a candy is 1/N. But the series is perfectly cyclical. If we select every 4th deal, then the probability that one child gets a candy becomes 1 and the others become 0. Von Mises invoked the concept of a "random collective" in which the limiting value "must remain the same in all partial sequences which may be selected from the original one in any arbitrary way" (1957: 24-25; see also pg. 29).

Many Relative Frequentists have disagreed with von Mises' second condition (this aside from its frequent disregard by practitioners). Reichenbach, for instance, claims that the axiom of randomness is unnecessary and restrictive (1949: 132). However, he is careful to distinguish among the probabilities of different pairs of trials in a sequence of trials in order to avoid the pitfall of cyclical or otherwise regular collectives. A more severe criticism from within the Relative Frequency school itself is that one can always come up with some selection procedure that alters success rates (in fact, pseudo-random number generators are an inverse counterexample to von Mises' criterion). Fine (1973: 99) gives examples of "unfair" collectives in which the lead never changes but which nevertheless pass von Mises' test. Nevertheless, this criterion and the concept of randomicity itself have spurred many efforts to rehabilitate it. The 1970's saw a renewed interest in the definition of 'computational complexity' as a surrogate, but this candidate faces some embarrassing competition from the concept of deterministic chaos that has emerged lately from the mathematics of strange attractors in dynamic systems.

Because its strategy is to force orderliness by restriction and banishment, the Relative Frequentist definition immediately poses several problems for the scope of application. First, there is a bootstrap difficulty in restricting the meaning of probability to an a posteriori sense. Where are initial probabilities to come from? Although von Mises is not entirely consistent on this matter, most Relative Frequentists take an inductive position. One establishes initial probabilities by using relative frequency from (finite) runs of observations. The underlying assumption, of course, is that past observed propensities for events to occur will continue on into the future.

A second issue concerns the distinction between the probability of a single event and that of a class of events. In keeping with his banishment of the Classicists' definitions, von Mises' injunction was there is no such thing as the probability of a single event and all that probability can refer to is a class of objects or events. Ordinary practitioners frequently disobey this injunction, and in some practical estimation problems it is almost impossible to avoid doing so. Actuaries, for instance, must ignore the philosophical quandary of estimating the probability that a given individual will die within a certain period of time. But their practical difficulties in obtaining this estimate illustrate the underlying problem of specifying the reference class. Should the actuary take into account the person's gender only? Or is the relevant collective those people of a similar age as well as of the same sex? What about the person's occupation, hobbies, income bracket, or eating and drinking habits, all of which surely bear on his or her life expectancy? The Relative Frequentist paradigm provides no guidance here.

Moreover, in the strict sense we are barred not only from defining probabilities of individual events, but of any finite collective. Thus, while we may refer to the probability of rolling a six from a die in the long run, we cannot define such a probability for any single roll, nor for any composite event consisting of however many rolls of the die we like.

A related kind of confusion arises in the Neyman-Pearson school of statistical inference concerning probability statements about decisions versus hypotheses. In the human sciences, for instance, there is a common confusion in significance testing between the probability of being wrong in rejecting the null hypothesis and the probability that the null hypothesis is true. The Neyman-Pearson significance test informs us about he former, but not the latter. Likewise, practitioners sometimes equate confidence intervals with the assurance at a certain probability that the true population parameter value lies inside a given interval rather than the probability that with repeated sampling the interval constructed will contain the true population value of the parameter. Once again, it is worth remarking that the 'wrong' probabilities in these cases are precisely those the hapless researcher really wants to know.

The impulse to stretch the definition to fit individual cases is understandable, and Reichenbach self-consciously attempted it with his "posit" that we behave as if the most probable alternative is true. Combining this idea with the logic of induction, he elaborated a multi-valued probabilistic logic in which probability plays the role of "truth value".

Finally, there is the issue of epistemological status. Are probabilities 'real' in some absolute or necessary sense? Or is von Mises' positivist abhorrence of metaphysics and his resolute inductivism tenable? Owing to the well known difficulties in pure operationalism and inductivism, even Relative Frequentists are forced to admit a few unsubstantiatable (and unfalsifiable) assumptions. Chief among these, of course, is the assumption of convergence to a limiting relative frequency. Many Relative Frequentists are happy to admit this is an assumption and even that it is unfalsifiable. However, they refuse to move from there to grant any a prioristic status to probability since that would seem to land them back in either the Classicists' or Subjectivists' camps. The obvious alternative is the reductionist strategy: Probability theory is purely analytic and metaphysically empty. It reduces ultimately to a mathematical system that is axiomatizable.

Some of the major criticisms of Relative Frequency Theory have already been alluded to, including those which correspond to some of the standard attacks on positivism generally and operationalism in particular. Probabilities of the Relative Frequentist kind can never be truly known, nor can they even be known to exist. There is no guideline for how long the 'long run' has to be in practical estimation problems, and attempts to provide one suffer problems of infinite regress. There is no litmus-test for randomicity. The hypothesis of a limiting relative frequency is unfalsifiable within the Relative Frequentist paradigm.

Two major criticisms remain, however, which will repay attention later on. These also are special cases of anti-positivist critiques. The first one is that the very nature of the criteria used in defining a 'random collective' imply that all sequences involving such collectives will have convergent relative frequencies. Von Mises foreshadowed this, in a sense, by saying that apparent nonconvergence in a long run would not falsify the initial assumption of a limiting frequency. Instead, it would only show that this sequence "does not satisfy the conditions of a collective" (1957: 142). As N.R. Hanson would have it, seeing is a theory-laden undertaking.

But Fine (1973: 91-94) demonstrated that even adopting the more recent computational complexity approaches results in a logical impasse rather than von Mises' dogmatism. First, he demonstrates that an appropriate definition of sequence complexity passes classical tests for randomicity. Then he proves that high complexity guarantees narrow convergence. He concludes from this that scientists and statisticians who select collectives on the basis of their *apparent* randomicity or complexity erroneously attribute their convergent properties to underlying natural processes or the empirical validity of the convergence principle. "It is the scientist's selectivity in using the relative-frequency argument that accounts for his success, rather than some fortuitous law of nature" (1973: 93).

The second criticism anticipates some issues to be discussed in the section on subjective probability. The nub of the criticism is that Relative Frequentists must engage in subjectivism in order to have a theory of probability at all, but this criticism takes several forms. One that was recognized by von Mises and others as a genuine problem is that of specifying the reference classes for probability statements. Probability values may change, after all, with the nature and size of the referent class. Therefore, some prior judgment or knowledge must be brought to bear on the selection of the class in the first place.

Furthermore, in most of the statistical applications for which the Relative Frequentist paradigm became popular, decisions are required (e.g., rejecting a null hypothesis or measuring degree of fit for a model) that involve at least implicit use of subjective probabilities and utilities (see Good 1976 for an extended argument along this line). The point of this dispute is whether it is possible for the Relative Frequentist to fulfill Kerridge's requirement that the statistician must never decide for other people, only inform the decision maker (Barnett 1973: 15). Relative Frequentists attempt to avoid subjectivism by imposing severe limits on what constitutes 'relevant' information in assessing probabilities. Critics assert that subjectivism inevitably enters into the selection of background assumptions, inferential and estimation methods, and the styles of data summarization and conclusion reporting. In the human sciences one of the best commentaries on these matters has been provided by Oakes (1986).

What accounts for the considerable popularity of the Relative Frequency Theory? A simple and popular explanation is "it works". Unpacking this explanation somewhat, we find several components. First, the Relative Frequency approach, with some modifications and rule bendings, seems easily applied to many scientific and practical problems. There are at least two reasons for this. One is that its heart lies in the realm of statistics rather than a priori or logical probabilities, and many practical problems are heavily empirical. Another is that the Relative Frequency approach is compatible with traditional scientific positivism, and thereby offers security to positivistic practicioners.

These assertions are debatable, but let's grant that part of the explanation for the moment because it immediately suggests another important component. Relative Frequency Theory is widely acclaimed as *successful*. In the face of all its defects, how can Relative Frequency work? If its epistemological basis is doubtful, its assumptions and choices arbitrary, and if it is unable to learn from experience, why does it succeed? One obvious rejoinder is that its crucial tenets are unfalsifiable anyway, but this dodges the issue to some extent. Another is that it is more useful than successful, which is perhaps closer to the mark. These issues will be taken up at the end of this chapter.

3.5 Subjective Probability and the New Imperialists

Subjective (or Bayesian, as they are popularly called) probabilists, considered quite disreputable among many statisticians in the second quarter of this century, have since climbed to a relatively high status. There are spirited defenders of subjective probability against some of the newer normative paradigms for dealing with uncertainty (see Chapter 4) and even predictions of complete ascendancy, e.g., that we shall all be Bayesian by the year 2020 (de Finetti 1974: 2). Even the most soberly conservative commentary presently puts the Subjective school second only to Relative Frequency.

Subjective probability also is multifarious, perhaps more so than other schools, with numerous sectarian divisions among its advocates. Early disputes between the Subjective and Relative Frequentist schools tend to obscure these subtleties. Most agree that the founders of the Subjective approach were Ramsey, followed by de Finetti (1974), Good (1950), and Savage (1954), but varations have proliferated over the past 30 years. As an antidote Good (1971) mischievously presents an eleven-faceted typology, claiming a resultant 46656 varieties of Bayesians. Even so, I will refer later to some of his facets in discussing the more nontrivial differences among Subjectivists.

A Subjective Theory of probability defines probability in terms of degrees of belief, which is something of a descendant from the older philosophical notion of credibilities (but see Shafer's "theory of evidence" in Chapter 4). The motivations for this definition are threefold. First, the claim often is made that any practical use of probability theory requires subjective judgments both a priori and a posteriori, but especially the former. Secondly, it is arguably possible to represent degrees of belief in terms of subjective preferences with sufficiently strong conditions to imply the use of quantitative probabilities (and hence the probability calculus). Third, Subjectivists claim that their approach ensures that decision makers behave rationally in accordance with their subjective preferences. Let us briefly consider each of these motivating arguments.

The first claim actually has at least two significantly different versions. One is the kind argued by Good (1976) which says that Subjective Theory simply systematically deals with the inevitably subjective elements of statistical analysis and decision making that Relative Frequentists sweep under the carpet. One of the more compelling arguments of this kind strikes at the weak point of Relative Frequency Theory, namely sources of initial probabilities. Subjectivists argue that this source is inevitably either entirely personal or based on incomplete prior evidence with some subjective judgments. In the former case, one speaks of 'prior ignorance', while the latter denotes either 'vague' or 'substantial' prior knowledge.

The second version of this claim is exemplified in Savage's (1954) argument that prior subjective judgments ought to influence subsequent analysis and decisions. He presents three studies which have the same formal structure in classical terms, and for which the Relative Frequentists would arrive at the same conclusions. In Study 1, the notorious tea-drinking lady of Fisherian fame claims to know if milk or tea was poured first into the cup, and correctly diagnoses on 10 out of 10 trials. In Study 2, a music expert correctly distinguishes a page of Haydn from one of Mozart for 10 out of 10 pairs of pages. An inebriated friend at a party, for Study 3, correctly predicts the outcome of a spun coin in 10 out of 10 spins. The significance level is 2^{-10} for a one-tailed test in each of these studies. But Savage says we should react quite differently to these three results. Some folk beliefs have some degree of truth to them; the tea lady may be correct in her assertions. But a music expert should be able to tell Haydn from Mozart; we should not be surprised by this accomplishment in Study 2. Yet again, we might not be impressed by our friend's extreme good luck in his drunken predictions during Study 3.

In essence, the Subjectivist contention against the Relative Frequentists on this point concerns rules of relevance (first version) and legitimacy (second version). The Subjectivist wishes to declare some kinds of prior information relevant or legitimate that are ignored or jettisoned ex cathedra by the Relative Frequentist. This relevant information may include not only prior subjective probability judgments but also preferences for different possible outcomes.

The second major claim in the Subjectivist position is that personal probabilities can be measured and quantified according to the usual probabilistic scale and calculus. In order to motivate a subjective account of probability, one must not only make a relevancy claim on subjective judgments but also an argument that quantification of those judgments is possible and justifiable. It is rare these days for anyone to argue that these judgments are entirely rational, but appeals to concepts such as 'coherence' and 'reasonableness' abound in the foundational literature. These appeals comprise the third claim mentioned above, that a theory of probability (even a subjective one) should provide decision makers and judges with rational guidelines. The mainstream approaches begin with a behavioral anchor and then rationalize it by introducing constraints that lead to quantitative probabilities as degrees of belief.

De Finetti's version of this justification emphasizes betting behavior. He identifies personal probability with the conditions under which the person would be disposed to bet on the event in question. An individual who would be willing to pay $1 per toss in order to be rewarded with $2 whenever a heads occurs, according to de Finetti, is expressing a personal probability of heads as 1/2. This definition gets us as far as numbers between 0 and 1. To move from there to probabilities, de Finetti requires the person be 'coherent' in her or his betting behavior. Thus, the avoidance of a so-called Dutch Book (odds so arranged that the bettor inevitably loses money) requires that the probabilities sum to 1 across all alternatives. Likewise, the individual is required to be 'consistent' in the ordering of probability judgments; if E_1 is more probable than E_2 and E_2 more than E_3, then E_1 must be judged more probable than E_3. In place of physical event independence, de Finetti argues for the subjectively based concept of 'exchangeability'. Events in a sequence are said to be exchangeable if their reordering does not influence the subjective probabilities assigned to them.

Not all systems of well-ordered degrees of belief can be expressed in betting terms, however. An early but currently fashionable axiomatization of subjective probability is Cox's (1946) conditions for a probabilistic calculus of a belief function. For Cox, as for many probabilists, even subjective probabilities are never primary, but always conditional on some known (or assumed) evidence. He refers to these as likelihoods, wherein likelihood is a relation between a proposition and some hypothesized or known state of affairs. The characterization below combines Cox's treatment with commentary and interpretations provided in Horvitz et. al. (1986).

1. (*Clarity*) Any proposition can be defined clearly enough that an ommniscient being could determine its truth or falsehood. In fact, Cox's primary basis for his development of probability is Boolean algebra, and this condition amounts to a version of the Law of the Excluded Middle.

2. (*Unidimensionality*) A single real number on a continuum between truth and falsehood is both necessary and sufficient to represent a degree of belief.

3. (*Completeness*) A degree of belief can be assigned to any proposition that is sufficiently clearly stated.

4. (*Conditional Likelihood*) Degree of belief or credibility in a proposition Q is conditional on some other proposition or evidence E, and may be denoted by Bel(Q:E).

5. (*Complementarity*) If Bel(Q:E) denotes degree of belief in Q, then Bel(notQ:E) = h(Bel(Q:E)), where h is a continuous monotonically decreasing function of Bel(Q:E).

6. (*Conjunction*) Belief in the conjunction of two propositions Q and R is Bel(Q.R:E) = g(Bel(Q:R.E),Bel(R:E)), where g is a continuous function monotonically increasing in either argument when the other is constant.

7. (*Consistency*) There will be equal belief in propositions that have the same truth value.

Cox is able to prove that these properties are necessary and sufficient to motivate the usual probabilistic calculus, including finite additivity and the multiplication rule for conjunction. This proof forms the basis for his (and other Subjectivists') assertion that "the rules of probable inference are credited by common sense with a wider validity than can be established by deducing them from the frequency definition of probability... they were derived without reference to this definition, from rather elementary postulates" (1946: 9). While de Finetti's concept is operational insofar as it contains a crude theory of measurment, Cox's characterization is purely normative with no measurement basis at all. Systematic attempts have been made to provide a measurement basis for subjective probability, however (cf. Krantz et al. 1971), and we shall return to the issue of measurement shortly.

A somewhat more philosophically oriented debate over subjective probability as a normative basis for dealing with uncertainty has fermented for some time in the literature on induction and confirmation. To some extent the debate owes its impetus to a confusion in terminology perpetrated by Carnap (1950), who used 'degree of confirmation' to emphasize the difference between subjective and relative frequency concepts of probability but then went on to use the same phrase to denote changes in belief (although he did clear this matter up in the 1962 edition of his classic work). Popper was motivated, however, to refute theories of induction which identified degree of confirmation with subjective probability, and went on to propose his famous desiderata for a measure of 'corroboration' (his term for a well-defined measure of change in belief on the basis of incoming evidence). Other philosophers continued to confuse confirmation with absolute probability and to debate the paradoxes generated thereby. This problem has resurfaced in the literature on uncertainty in artificial intelligence (see Chapter 4). As a byproduct, however, Popper argued for retaining subjective probability theory to quantify change in belief, and Good (1960) demonstrated that a modified version of Popper's axioms for such a measure is indeed satisfied by certain probabilistic quantities.

Not all Subjectivists agree with formulations like de Finetti's and Cox's. Some argue against the requirement that propositions be completely ordered in terms of degree of belief. Others take issue with the restricted expression of belief in terms of a single number. Koopman (1940) and Good (1950) were early proponents of approaches that allowed for partially ordered probabilities; Good in particular argues that quantification is no more than a mathematical convenience. Good also claims that in principle subjective probabilities really are inequalities (intervals) rather than point estimates, but again simplicity favors using points instead. C.A.B. Smith (1961) provides a cogent and axiomatic theory that embraces intervalic probabilities which has garnered respect but few applications thus far. The vast majority of Subjective probabilists operate within the strictures laid down by axiomatizations similar to Cox's or de Finetti's.

Once one has admitted the usual probability calculus into a Subjective Theory, then people are able to correct their initial subjective probabilities on the basis of incoming evidence by repeated uses of Bayes' Theorem (Lindley 1965: xi). As a consequence, most Subjectivists have no problem with either the probability of a single event or a repeated event. Let H_i denote hypotheses (for $i=1,...,N$), $P(H_i)$ be our subjective prior probability for each hypothesis, and let E be new observed evidence. Then we may update our prior probability, $P(H_i)$ to an a posteriori probability, $P(H_i:E)$ by Bayes' formula:

$$P(H_i:E) = \frac{P(E:H_i)P(H_i)}{\sum_{j=1}^{N} P(E:H_j)P(H_j)} \tag{3.1}$$

Subjectivists differ, however, on the metaphysical status of personal and physical probabilities. Some are comfortable with a division of labor between Relative Frequency and Subjective Theories, with physical probabilities being accorded real existence and subjective probabilities referring to underlying belief states. Others do not commit themselves on the existence of physical probabilities, but utilize approaches that induce them to act as if they do under certain conditions. Extreme subjectivists like de Finetti argue that probability does not exist, and the only meaningful way to speak of any kind of probability is in personal terms. They are akin to nominalists but closer to rationalists, since the entire basis for their theory of probability becomes rational. Let's examine the criteria for rationality more closely.

In some Subjective Theories rationality is virtually tied to a mathematical or logical system, and those approaches resemble the extreme reductionists among the Relative Frequency school or (as we shall see) the logical emphasis in the A Priori camp. The objections to this style are similar to those raised against the reductionist strategies in the Relative Frequency school. Despite its apparently wide applicability, appeals to logic alone often do not convince people with practical concerns.

Cox is an exemplar of this approach to rationality. Although Cox may argue for a wider domain of application, de Finetti's prescriptions for rationality have the advantage of seeming less arbitrarily imposed because they are linked to outcomes that are (un)desirable in terms of the decision maker. As he puts it, "...such conditions, although *normative*, are not (as some critics seem to think) unjustified impositions of a criterion which their promoters consider 'reasonable': they merely assert that 'you must avoid this if you do not want...' (and there follows the specification of something which is obviously undesirable)" (1974: 85). De Finetti's claim is tied to his characterization of subjective probability in terms of losses and gains through betting, and other Subjectivists have followed his lead by proposing 'utilities' that may be extended to situations not involving betting.

Conventional utility theory begins with the assumption that people are able to completely order their preferences concerning possible outcomes, and then in a similar fashion to the construction of subjective probability, proceeds to place restrictive criteria of coherence and consistency on those preferences. In many such approaches, preferences are quantified to enable the usual definition of expected utility, which is the probability of an outcome multiplied by its utility for the decision maker. Decision rules are then proposed on the basis of expected utility. If the utilities and probabilities of all possible outcomes are known in advance, then the obvious choice is to maximize expected utility. The most popular rule for doing so is to find the decision that maximizes the mean expected utility over all possible consequences. If the probabilities are uncertain or we are not willing to entertain a prior assignment of probabilities to outcomes, then the most popular fallback is the 'minimax' rule which stipulates that we avoid the worst possible outcome (the one with the least utility).

Although it is possible for a Subjective Theory of probability to be developed without reference to utility, many of its applications have involved decision theoretic machinery. But there is a more fundamental link between this school and utilitarian concepts, through the foundational arguments themselves. Ramsey's, de Finetti's, and Savage's formulations all have utilitarian bases. Bayesian inference often uses loss functions directly related to the error of the estimate or model being tested.

A final major issue for the Subjective Theory is the specification of prior distributions themselves. The earliest formulations resuscitated the Classical Principle of Indifference, so that if the balance of prior evidence does not favor any alternative over another then the prior is uniform across alternatives. This prescription caused immense controversy (see criticisms below), despite several attempts to place it on firmer foundations than the Classical arguments. Some Subjectivists have gone along with Jaynes' (1968) principle of maximizing entropy by the choice of a prior distribution. This conservative principle requires that we select the least 'informative' element of the family of distributions consistent with our prior knowledge (where informativeness is defined and measured according to Shannon's information theory). Good was one of the earliest proponents of using 'quasiutilities' for the purpose of assigning priors or designing experiments. Among quasiutilities, Good (1971b) includes weight of evidence (in the formal sense proposed by Turing), information in either the Fisherian or Shannon senses where appropriate, strong explanatory power, an inverse of estimation error, and the inverse of computational time and/or effort.

Let us turn now to the criticisms raised against the Subjective approaches. Many early criticisms focused on the use of inverse probability as a legitimate tool for statistical inference, and on the issue of subjectivity itself. The first criticism we may ignore since few probabilists object to the use of Bayes' Theorem in situations where the prior frequency distribution rests on a Frequentist foundation. Strong criticisms still are raised against the 'degree of belief' concept, however. After all, if personal probabilities need only avoid the Dutch Book problem and otherwise satisfy personal utilities, then why should one idiosyncratic prior distribution be preferred over another?

The dilemma here is that the stronger the prior opinion the tighter and therefore the more influential the prior distribution on later results. What if the prior distribution is badly mistaken? As Oakes (1986: 137) observes, it would be highly embarrassing for Subjectivists to concede that strong prior opinion could not be trusted and should be discounted. Yet that is exactly what some of the leading proponents of this school suggest. Lindley (1971: 436) advises practitioners that if a prior distribution leads to an "unacceptable posterior" then they may modify it to cohere with desirable properties. Likewise, Good (in Savage, et. al. 1962: 77) says that one may imagine possible final results of an experiment and use Bayes' theorem in reverse to find out what one's prior distribution "must be for the sake of consistency."

A variation on the subjectivism criticism accuses Subjectivists of reifying personal probabilities into absolute probabilities by treating posterior distributions with greater scientific status than they merit. A somewhat more thoughtful version points out that even when the Subjectivist doesn't do this laypeople may. Recent examples of this outcome include publications on the likelihood of nuclear reactor failure and the pre-Challenger space shuttle reports, in which subjective probabilities given by scientists and engineers were treated as if they were real probabilities of events that had never occurred. In some such cases, linguistic probabilities were converted by social fiat into quantities with no attention to the questions of measurement and translatability raised by such conversions. And yet, Bayesians themselves (e.g., Phillips 1973: 78) like to claim that personal probability is saved from being utterly subjective by the fact that given enough evidence, eventually diverse opinions converge through the repeated application of Bayes' theorem.

The primacy of preference and utility over probability as a means to rationalizing personal belief also has been criticized. Von Wright (1962), for instance, has suggested that probability is logically prior to preference and that only after we have assessed probabilities do we assign preferences. Fine (1973: 236) makes the ironic comment that the argument against the existence of a best rational procedure of decision making partially justifies the utilitarian approach, since the Subjectivist may argue from there that the only alternative left is to consistently maximize one's self-satisfaction according to one's personal utilities.

Subjectivists' responses to these criticisms have been fourfold. First, some of them argue that any other approach to probability requires an appeal to statistical ad hoc principles with subjective elements, so subjectivism is inevitable and might as well be rationalistic. Secondly, they invoke the principle of 'precise measurement' which posits that under reasonable conditions the prior will be swamped by incoming information anyway, so its particulars are not crucial to the inferential enterprise. A third response has been to propose rules for combining personal priors from various disagreeing parties (section 3.8 on in this Chapter). Fourth, particularly recently, some Subjectivists have proposed foundations for measuring subjective probabilities in an attempt to at least rationalize the procedures by which these quantities are generated.

A second stream of criticism is directed at the nature of the prior distributions and their capacity to represent prior knowledge or ignorance. The question of when one may justifiably claim to know the form (or shape) of the prior has been widely debated, principally in connection with criticisms of the Subjectivist version of the Principle of Indifference. This problem is linked with the worry that personal probabilities might be neither systematic nor stable, hence the nature of the prior distribution could change arbitrarily even for the same person.

A more fundamental question has been raised concerning the viability of point-estimates of prior probabilities. I have already mentioned that some Subjectivists propose subjective probabilistic intervals rather than points, but the vast majority still use points. This practice does not enable the Subjectivist to distinguish between, say, a uniform prior distribution that rests on considerable accumulated empirical evidence and one generated in blind ignorance by the Principle of Indifference. Shafer (1976), in motivating the foundations for his theory of evidence, points out that a truly ignorant person might wish to accord 0 degrees of belief to both a proposition and its negation. The probabilistic requirement that P(not A) = 1-P(A) precludes this, and in fact requires the ignoramus to assign equal nonzero degrees of belief, 0.5, to A and its negation.

A related argument whose roots reach back to the original criticisms of the Classical definition of probability is that Subjectivists ignore the issues of relevance judgments and the partitioning of cases. Again, Shafer provides an example illustrating the impact of these issues on the nature of the prior distribution. Suppose an individual wishes to declare ignorance of whether there is life in the Sirian system. She may decide the relevant possibilities are (A) there is life, or (B) there is no life, and accordingly assign P(A) = P(B) = 1/2. Another individual, however, decides the relevant possibilities are (C) there is life, (D) there are planets but no life, and (E) there are no planets. He will assign P(C) = P(D) = P(E) = 1/3. But his propositions D and E are equivalent to her B, so the partitioning of possibilities alone is responsible for a disagreement between their prior distributions.

Finally, Relative Frequentists in particular point out that Bayesian inference fails to take into account the sampling space or procedure by which data are obtained. There is no sampling framework in conventional Subjective Theory, and therefore no use of concepts that are tied to probability samples (e.g., unbiasedness in estimation).

Although Subjectivists themselves and many practitioners seem convinced that their paradigm has secure foundations, it is undoubtedly the case that much of the drift towards subjectivism among engineers, scientists, and other users of probability and statistics stems from its universalism rather than a reasoned conversion. Many of these professionals are forced to apply probability under conditions that do not admit a Frequentist interpretation, and rather than concede the battle, they incline toward cut-and-try methods that seem to produce results. Not only do they apply such methods in the absence of sound theoretical rationale, often they seem unaware that they have strayed beyond the Frequentist's prescribed domain. Ironically, some of the defenders of subjective probability in fields such as knowledge engineering and artificial intelligence against newer nonprobabilistic alternatives claim that computer scientists and engineers have eagerly adopted the new techniques because they think probability is limited to the Relative Frequentist approach and are unfamiliar with the Subjective paradigm (cf. Cheeseman 1985).

3.6 Neoclassicism and the Logical Probabilists

Subjective theory may be currently the strongest claimant to a theory of probability that encompasses common meanings disallowed by the Relative Frequency school, but it is not alone in this claim. The Logical, or A Priori, school also has made its bid for universalism, and it has many linkages with the Subjective theory. A purely a prioristic approach to probability has not been fashionable for some time now. The quantity of literature sharply declines after the early 60's, and a recent advocate (Benenson 1984:1) says that Logical Theory as a universal account of probability has not been taken seriously since Carnap's efforts during the late 40's. I will not deal at length with it here, except to survey points about the Logical approach that distinguish it from the other two.

However, even if we discount the Logical school as a dying gasp on the part of Classical theory, it still provides insight into contemporary debates about probability and uncertainty generally. Its impressive pedigree (Keynes, Jeffreys, and Carnap, among others) testifies to the depth of the ideas and issues addressed by this school. It receives its name in part from the application of probability concepts to the problem of induction and partly from the utilization of quantitative inductive logics in formulating a priori probabilities themselves. The primary focus, then, is on a priori probabilities and rational methods for constructing them.

In a famous passage, Keynes (1962: 3-4) outlines his definition of probability as a logical relation between a proposition and a corpus of evidence. He proposed that while propositions are ultimately either true or false, we express them as being probable in relation to our current knowledge. Conversely, it is meaningless to call a proposition probable without specifying the evidence on which such a judgment is based. Unlike the Subjectivists, however, Keynes holds that this relationship between knowledge and a proposition is objective. A proposition is probable (or not) with respect to a given body of evidence regardless of whether anyone thinks so.

As to where the degrees of probability are to come from, Keynes provides a recycled version of the Principle of Indifference. Not all Logicists agree. Carnap, for instance, permitted weights to be assigned to the various possible events or states so that they need not be equiprobable. Furthermore, he set out an elaborate technical machinery to enable the logical implications of any weighting scheme to be followed through. While the Keynesian approach emphasizes intuition and the Principle of Indifference, Carnap's system focuses on logical calculation as the primary source of initial probabilities. One of the advantages claimed by Carnap for his approach is that in his version the probabilist not only can learn from experience but also is able to deal with 'unfair' partitions of event-spaces. However, neither Carnap nor Keynes are invulnerable to all the attacks mounted against the Classical Principle of Indifference, since even Carnap fails to provide a rule for choosing one set of weights over another under all conditions.

Another notable feature of the Logical approach is its attempt to construct a viable probabilistic logic of induction. Classical inductive logic itself foundered in many ways, but the temptation to find a way out of the snares by permitting a probabilistic basis for uncertainty has proved very tempting. Even critics of Carnap's whole enterprise have not been able to resist trying to improve on it. Indeed, as Weatherford (1982: 143) points out, if we accept a particular partition and define our probabilities in an a priori fashion, inductive logic becomes possible. The main limitations in Carnap's system reflect the primitiveness of the languages he constructed rather than an inherent flaw in the approach itself. Several philosophers have attempted to provide richer languages that can handle numerical measurements and other common scientific observations. Likewise, recent work in expert systems utilizing 'certainty factors' and related measures of belief updates are more or less direct descendants of the Logical as well as the Subjective school.

3.7 Nonquantitative Probability

Although the major theories of probability attempt to justify the application of the usual calculus for quantitative probabilities, some critics of these approaches have suggested that for many important applications this calculus cannot be used. Where not nihilistic for probability theory, the critics' alternatives mainly land in one of two camps: Those who advocate an interval-based approach to probability estimates, and those who recommend a nonquantitative pointwise approach (I am not aware of anyone who has provided a nonquantitative intervalic approach). I have mentioned the concept of probability intervals before, but this approach is best deferred until the next chapter since it amounts to a 'second-order' uncertainty framework which properly belongs with other frameworks of its kind. Instead, I shall discuss nonquantitative theories of probability, primarily in terms of their normative strategies for dealing with probabilistic uncertainty in the face of other kinds of informational ignorance.

The usual case made for nonquantitative probabilities is the kind made by Fine (1973: Ch.2), who points out that where there is insufficient (or incomplete) prior information to estimate quantitative probabilities there may still be enough information to enable us to make statements like 'A is at least as probable as B'. A 'comparative' probability built out of a weak ordering of events would have wider applicability than a strictly quantitative theory. Where most authors focus on the means by which nonquantitative probability may be 'upgraded' to quantitative (cf. Kraft et al. 1959, Savage 1954, Luce 1967), Fine devotes some effort to investigating the extent to which the tools of quantitative probability may be reconstructed using material from a nonquantitative approach.

Fine's system begins with five axioms:

1. The universal set of all events, U, is not the empty set, ϕ.

2. For any events A and B, A is at least as probable as B or B is at least as probable as A: $P(A) \geq P(B)$ or $P(B) \geq P(A)$.

3. If $P(A) \geq P(B)$ and $P(B) \geq P(C)$, then $P(A) \geq P(C)$ (*transitivity*).

4. What is impossible must be improbable: $P(A) \geq P(\phi)$.

5. If A and $(B \cup C)$ are mutually exclusive, then $P(B) \geq P(C)$ implies $P(A \cup B) \geq P(A \cup C)$.

Axioms (2) and (3) permit two events to be equiprobable (which happens when $P(A) \geq P(B)$ and $P(B) \geq P(A)$), but they require that the ordering among events be complete. Such a condition permits a similarly complete ordering of the complements of events, thereby obviating the need for a functional definition of P(not A) in terms of P(A). Not all theorists agree with this criterion for a nonquantitative probability; some of them would weaken this to allow partial orderings. Even so, as Fine takes pains to show, these axioms are sufficiently flexible to allow orderings that are incompatible with the finite additivity of ordinary probability theory. The immediate consequence of this is that the usual calculus cannot be applied to nonquantitative probability.

To understand this, and since such an understanding aids further discussions in this book, let us follow Fine's argument on this matter in some detail (1973: 22). Once we have defined a probability scale P that is compatible with Fine's axioms (including one additional axiom that need not concern us here), then there exists a two-place function G(x,y) such that for mutually exclusive (nonintersecting) events A and B,

$$P(A \cup B) = G(P(A), P(B)) \qquad (3.2)$$

which is symmetric, increasing in x, associative, and has the property $G(x,0) = x$. One might expect that some suitable increasing transformation of P could be found for which a corresponding function $G'(x,y) = x+y$. But Fine presents a counterexample from Kraft et al. (1959), consisting of the following ordering of events A,B,C,D,E and their unions:

0<a<b<c<ab<ac<d<ad<bc<e<abc<bd<cd<ae<abd<be<acd<ce<bcd< abe<ace<de<abcd<ade<bce<abce<bde<cde<abde<acde<bcde<abcde (3.3)

where a = P(A), b = P(B), ab = G(a,b), abc = G(a,G(b,c)), etc. If we assume that P is additive, then from (3.3) we obtain a+c < d, a+d < b+c, and c+d < a+e. Adding these inequalities together and subtracting repeated terms gives us a+c+d < b+e, which violates (3.3). Hence the assumption of additivity cannot agree with the ordering in (3.3).

Fine also goes on to demonstrate that his system for comparative probability does not always lead to comparative conditional probability, although the converse does hold; a suitably defined quaternary comparative conditional probability will also yield a simple comparative probability system. In nonquantitative probability, simple probability is more fundamental than conditional probability. He is also able to construct a concept of pairwise event independence in the context of comparative theory, and shows that independence does not automatically lead to the conventional combination rule for joint probability. Combining independence with conditional comparative probability, he generates definitions for expectation and rationality in decision making. These constructions apparently are intended to make the point that a nonquantitative probability theory may be capable of reasonably powerful applications in situations where quantitative probability cannot be used. Fine's is essentially a universalistic strategy for managing uncertainty, with some degree of bracketing.

A more strategically radical philosopher advocating nonquantitative probability is L.J. Cohen (1977), who argues that conventional probability (which he calls "Pascalian") cannot sensibly form the basis for judicial probability. He then constructs a nonquantitative probability theory that fulfills his criteria for judicial probability and demonstrates that it is incompatible with finite additivity (or associativity in the combination rule for independent events), thereby separating it effectively from conventional probability theories. Cohen is a suburbanite among probability theorists; he proposes subdividing the frontier of applications and establishing inviolable territories for various probability theories.

Nor is his claimed subdivision a small one by probabilists' standards. The earliest concerns of probability theorists included forensic probability. As Hacking (1975: 86) observes, a coherent theory of subjective probability requires a distinction between what causes something to happen and what tells us that it has happened. One of the few professions that has held fast to this distinction is civil law, which distinguishes between testimony and circumstance. Leibniz, for instance, initiated his work in probability by considering problems of partial implication and proof in law. He began with conditional probability as a basis for probability rather than the combinatorial basis adopted by Pascal and others whose inquiries arose in the context of gambling problems. Like other probabilists, however, Leibniz assumed that gradations of certainty were quantifiable and he had no quarrel with the Pascalian combinatorial rules.

Cohen also notes the etymological and philsophical connections between probability and judicial concepts of probativeness or provability. He argues for a theory of probability designed around the concept of probability as degrees of judicial provability. Clearly such an approach must also take account of the traditional judicial concerns in civil cases with the 'balance of evidence' and in criminal cases with 'reasonable doubt'.

As noted in the earlier discussion of how the legal profession handles uncertainty and ignorance, attempts have been made for some time to incorporate probability into legal deliberations. The very concept of 'standard of proof' may be "capable of being expounded in terms of probabilities, though lawyers prefer more ambiguous expressions" (Eggleston 1978: 5). Eggleston's caveat is heralded by a bald statement that the "legal profession as a whole has been notably suspicious of the learning of mathematicians and actuaries..." (1978: 2-3), and he hazards that lawyers and judges both find the ambiguities and vaguenesses of judicial standards for evidence and proof strategically useful.

Putting that possibility aside for the moment, the legal profession's encounters with probability have been less than salutary. An infamous outcome was the American 1967 trial of The People vs. Collins, in which a number of erroneous appeals to probability theory were utilized by the prosecution. Bystander evidence concerning a robbery indicated a couple, one a male negro with a beard and mustache and the other a blonde caucasian woman with hair worn in a ponytail, escaped after the deed in a yellow automobile. Police arrested the Collinses, a couple fitting this description. The prosecution pulled estimates of the probabilities of several of these characteristics out of thin air, multiplied them together under the assumption that they were all independent, and obtained a probability of 1/12,000,000 of finding one such couple possessing all these characteristics. Their conviction was eventually overturned, but many commentators have since cited this case as evidence of a real need for members of the legal profession to avoid the use of probability.

The legal profession, however, has mounted some telling criticisms of probability theory itself. They have objected to the requirement of quantifying uncertainty into a single numerical index on the grounds that for most judges, lawyers, and juries, ignorance is multifaceted. Furthermore, although probability theory makes no distinction between past but unknown facts and future predictions, the law does. In cases that hinge on establishing past facts, once judge and jury have made their (probabilistic?) assessments, the matter is treated as a hard-and-fast decision, as if the facts have been established. The court takes a different view, however, on payments for prospective losses in the future. If there is uncertainty about future loss, then the court may make allowance for this by multiplying the award by some numerical measure of that uncertainty.

Cohen's system incorporates the incompleteness of information by not requiring that evidence favor either a proposition or its opposite. This is an important and deliberate consideration on his part, since by his own account he is a Laplacean determinist (1977: 223-224) and he attributes this position to judges and juries in the courts as well. If all facts were truly known, then the courts would make infallible dispositions. But only under the assumption of complete information, according to Cohen, should we expect that P(not A) = 1 - P(A) if by P we mean probability in the sense of 'provability'. In incomplete systems there are two candidates for a lower limit on the scale of provability: Disprovability and nonprovability. This distinction is similar to the distinction between 'internal' and 'external' negation in modal logic, wherein 'proving (not A)' is not identical to 'not proving A'.

Furthermore, Cohen finds that the judicial combination of evidence for inferential purposes does not share certain properties of the multiplication rule. In order for the multiplication rule to yield a high probability of a compound event the elementary events must be established at high probabilities and must be few in number. For juries and judges, the believability of a story does not appear to decrease as a function of the number of components in that story.

Cohen raises doubts about whether judicial probabilities can be quantified after all. Nor is he the first to raise this issue. James (1941) argued that while most juries would not accept the syllogism "people who have fixed designs to kill are likely to kill", they would accept one that said "people who have fixed designs to kill are more likely to kill than those who do not." The latter is, of course, a comparative probability statement of the kind catered for in Fine's nonquantitative framework.

There is also the problem of prior guilt. Western judicial systems require that evidence be admitted only if it is 'probative', that is, if it has an impact on the probability of guilt. Furthermore, classical judicial principles require the judge and jury to decide guilt only on the basis of evidence presented to the court, and to effectively presume innocence until guilt is proven. In conventional probabilistic terms, according to Cohen, this requires assigning a prior probability of guilt of 0, which implies that no amount of subsequent evidence to the contrary can alter that probability (presuming that the court uses a Bayesian approach to correcting a prior probability on the basis of new evidence). The only alternative is to begin with some nonzero prior probability of guilt, but what should it be: 1/2?

Cohen has raised some of the problems suggested by critics of Bayesian methods for establishing prior probabilities in that he is pointing out the inability of standard probability to distinguish incomplete knowledge from uncertainty. But his critique attacks a deeper issue: the combinatory calculus itself. Although he is not the first to do this, Cohen's critique is unusual for its appeals to normative criteria from a specific domain of practical applications.

His proposed solution for these difficulties is a measure of support for a given proposition A given evidence E, P(A:E), which is conceptually akin to subjective probability but abides by a nonstandard calculus. He draws upon inductive logical arguments from scientific inference, arriving at different results than Popper or Good. According to Cohen's conclusions, the combinatorial rule for judicial probabilities should be

$$P(A \text{ and } B:E) = \min(P(A:E), P(B:E)). \tag{3.4}$$

A chain of propositions presented before the court should be no more (but no less) believable than the probability accorded the weakest link in the chain. This seems to solve the problems raised concerning the product rule for joint probability, although it raises other questions.

Cohen's principle of negation is if P(A:E)>0 and P(not E) = 0, then

$$P(\text{not } A:E) = 0. \tag{3.5}$$

Thus, the acquisition of inductive support for a proposition and its negation is a nonzero sum game. This definition is crucial for Cohen's interpretation of the 'balance of probability', which is that P(A:E)>P(not A:E). It also permits the judge and jury to begin with probabilities of 0 for both guilt and innocence, since no inductive evidence has been applied to either proposition. Likewise, Cohen argues that 'beyond reasonable doubt' is equivalent to the attainment of maximum possible support for the guilt proposition relative to the number of relevant considerations (although he leaves the issue of relevancy decisions aside).

Although his foundational arguments are intuitive, Cohen's demonstration that this scheme results in a nonquantitative probability measure is more than that. He demonstrates that the underlying logic of his 'inductive probability' is not Boolean (unlike the usual quantitative probability) but a modal logic (S4, specifically). The noncomplementary negation and combination rules imply that these probabilities may be mapped onto numerals but not real numbers; they are inherently unquantifiable (1977: pp. 227-229). However, clearly his scheme does not amount to a generalized nonquantitative theory of probability, since it is easy to construct examples of orderings that do not admit his combinatorial rule.

Despite its incompletely specified nature, Cohen's approach is strategically instructive in at least two respects. First, it demonstrates that a domain-specific theory of probability can be constructed independently of the usual calculus. Secondly, it indicates there may be room even in probability theory for alternative norms for rationality. Cohen does attempt to tie his approach to some respectable forbears, insofar as he likens his sense of probability to Keynes' 'weight of evidence'. It should not be surprising that Cohen's claims sparked some counterattacks from Subjectivists in particular, and one of the more famous exchanges between Cohen and the Bayesians will be examined in our discussion of the psychological debates over probabilistic thinking.

No simple theory of probability that focuses entirely on a single person making decisions or choices on her or his own can address another problem that has been widely debated in the legal profession, and that is how to combine evidence from two or more witnesses. A cognate problem is how to rationally arrive at decisions based on assessments of probabilities and utilities by a group of people. These issues have occupied considerable space in the probability literature and claimed some famous victims early on (Condorcet, Laplace, Poisson, Cournot, and Boole are among the notable failures in this arena). Accordingly, the next section examines normative proposals for achieving consensus under uncertainty.

3.8 Combining Probability Judgments

While there is a huge literature on group decision making as a function of group members' preferences, until recently relatively little attention has been paid to problems of group-level judgments of uncertainty per se. The Bayesian literature occasionally refers to this problem (e.g., Savage's 1954 mention of the "jury" situation), and early treatments are provided in Raiffa 1968 and Winkler 1968. However, it is only within the last 15 years that we find a substantial literature developing and comparing methods for combining probability distributions arising from individual judgments. This recency may well reflect the fact that Bayesian (and other Subjectivist) views themselves have become popular only recently, and nonsubjective views presume all decision makers have access to the same evidence so that 'rational' individuals invariable arrive at the same probability estimates.

For Subjectivists, however, even if the Bayesian paradigm is entirely satisfactory for guiding the individual judge, it leaves groups without an equivalent to the classical concept of objectivity or even coherence. Subjectivists therefore are eager to construct norms for probabilistic intersubjectivity that are as defensible as those for individual judges. I will discuss only the issues and methods pertinent to combining individuals' probability judgments, leaving aside the problems of aggregating utilities or preferences. I will also focus on the attempts to provide normative guidelines for aggregating probabilities, leaving aside psychological and group-level behavioral considerations until Part II of this book. This section owes much to the technical reviews provided in French 1985, Berenstein et al. 1986, and Genest and Zidek 1986; as well as the more psychologically oriented treatments by Beach 1975 and Hogarth 1975.

The concept of a normative consensus on probability estimates may take one of several forms, depending on the meanings ascribed to terms such as 'consensus' and 'aggregation'. Distinguishing among these meanings is not merely a pedantic exercise, since real groups and practical situations may require quite different kinds of aggregation and group-level information processing. Some of the literature distinguishes between 'epistemological' and 'operational' consensus, which roughly parallels the notions of 'cognitive' and 'behavioral' consensus in the human sciences. Epistemological consensus involves a unanimous commitment to a specific epistemology, which becomes the standard for rationality. An operational consensus, on the other hand, may require only the maximization of individuals' utilities or a procedural agreement for arriving at a final judgment.

A somewhat more practical distinction, perhaps, is between intersubjectivity and compromise. Intersubjective methods require that a final probability distribution be arrived at which is subscribed to by all group members, while compromise methods involve pooling individual estimates to arrive at some value which represents the central tendency of those estimates. Intersubjectivity has been proposed by more than one writer as a replacement for the older notion of objectivity. It makes a good starting point, since some aspects of intersubjectivity are necessary for opinions to be combined at all. First, virtually all aggregation methods assume that every individual has expressed her or his opinion as a quantitative point-estimate probability. Most such methods further assume that these individuals are coherent in the Bayesian sense. If they are not, then their estimates are incommensurable with the others in the group.

Secondly, some writers address the issue of 'calibration'. Mere coherence does not guarantee that two individuals' probability scales are directly comparable. One such person might, for instance, differ in his assessment from another by a simple monotonic transformation that preserves all required properties of a probability distribution (e.g., transformation by a scalar exponent). Ordinary Bayesian 'scoring rules' (cf. Lindley 1982) cannot prevent this, and recent attempts to provide normative calibration guidelines (e.g., Kadane et al. 1980 and Dawid 1982b) have not yet been widely accepted.

Given the difficulties inherent in the attempt to be rid of scale differences by forcing everyone onto the same scale, a natural alternative is to consider treating probability distributions as unique only up to some set of transformations. Thus, two individuals whose probability judgments over a range of alternatives differed merely by some permissible transformation would be considered identical in their judgments. To my knowledge, this possibility has not been actively pursued in the aggregation literature, although it is fundamental to psychological measurement theory.

A third criterion that often enters into establishing intersubjectivity as well as methods of achieving compromise aggregations is dependence among judges. It is implied in Weerahandi and Zidek's (1981) and Dawid's (1982a) definitions of intersubjectivity, which requires that the same conclusions from an experiment be reached by a succession of individuals faced with the same results. In real world situations, of course, experts often communicate opinions to one another and are accountable professionally to one another for their judgments. They may also base their judgments on overlapping data, using overlapping methods. The very nature of the peer review process among scientists and other professionals generates dependency among experts.

Insofar as the probability aggregation literature deals with dependency at all, it speaks mainly to experts communicating opinions. Very seldom is the issue of 'common method variance' raised, although this concept has been discussed in several substantive contexts (e.g., psychology and sociology) for some time. The primary consideration here is that two similar methods arriving at the same conclusion are less convincing than two different methods accomplishing the same feat. Morris (1986: 141) declares that dependency is the most important issue in practical applications, because it significantly affects the amount of uncertainty or disagreement that can reasonably be attributed to the group. But it is only rarely addressed in practice, and although some attempts have been made to cope with dependency (e.g., Winkler 1981) nearly all normative proposals for intersubjectivity and pooling make a default assumption of independence among opinions.

The dependency issue is also confounded with the actual methods recommended by many writers for achieving a genuine consensus. The Delphi technique, for instance, while not permitting open discussion, does require feedback (which may not be anonymous) of either group or individual estimates. Individuals then reassess their probability distributions in the hope that at some point in the process opinions will converge (cf. Pill 1971 and De Groot 1974). Bargaining theory, as pointed out by Weerahandi and Zidek (1983), suffers even more from this problem because the feedback is less constrained and motivational factors are permitted to influence the process.

Most of the probability aggregation literature proposes various pooling, or compromise, models. These models implicitly assume that the group is in a state at which no subsequent discussion among group members is possible or, if possible, that it would not change anyone's opinions. However, the synthetic judgment resulting from the pooling recipe need not represent consensus on the part of the group itself. The pooling recipes themselves may be broadly classed (following Genest and Zidek 1986) in the linear, nonlinear, and supraBayesian categories.

The linear opinion pool is perhaps the most natural pooling definition, and quite likely the most popular in practice since a simple unweighted arithmetic average of probability estimates from judges is its simplest special case. If $P_i(E)$ is the estimate given by the ith expert for the probability of event E, then we may assign nonnegative weights w_i to each expert such that they sum to 1 across all experts, and express a weighted linear opinion pool L(E) by

$$L(E) = \sum_{i=1}^{n} w_i P_i(E). \tag{3.6}$$

McConway (1981) and Wagner (1982) have shown that this is the only form L(E) can take if the 'strong setwise function property' is fulfilled, which stipulates that

$$L(E) = f(P_1(E),...,P_n(E)). \tag{3.7}$$

However, Dalkey (1972) demonstrated that if this requirement is extended to conditional event probabilities (i.e., L(E:A)), then only dictatorships in the weights will do. A similar requirement is the 'independence preservation property', such that L(A and B) = L(A)L(B) whenever P_i(A and B) = P_i(A)P_i(B). But Lehrer and Wagner (1983) prove that only dictatorial linear opinion pools can satisfy this property.

A weaker version of (3.7) is given in Bordley and Wolff (1981) and further discussed by McConway (1981). They permit the function f to depend on the event (or event set) E:

$$L(E) = f_E(P_1(E),...,P_n(E)). \tag{3.8}$$

It turns out that this condition permits weights in the interval [-1,1] with conditions imposed so that L(E) remains a probability measure. This pool still cannot have the independence preservation property without becoming necessarily dictatorial in the weights.

The difficulties encountered with finding linear opinion pools that preserve various aspects of the probability calculus while permitting flexibility in the weights has led mathematicians to suggest nonlinear opinion pools. The most popular has been the logarithmic opinion pool (cf. Bacharach 1972):

$$L(E) = g(\prod_{i=1}^{n} P_i(E)^{w_i}), \tag{3.9}$$

where g is a normalizing function (usually the integral of the argument in (3.9) over all E. This pool is invariant under certain kinds of rescaling, and it tends to be less dispersed than a linear opinion pool. However, it suffers some disadvantages. Any 0's among the judges become vetoes, and great emphasis is placed on small subjective probabilities. An extreme optimist or pessimist could unduly sway this function.

Genest and Zidek (1986) claim the most compelling reason to use the logarithmic pool is that it is 'externally

Bayesian', by which they mean that L(E) itself behaves as a single Bayesian judge would by virtue of being uninfluenced by the order in which pooling and updating of individual opinions is done. Unfortunately, none of these criteria shed light on how weights are to be assigned in the first place. The linear opinion pools turn out to be relatively insensitive to the choice of weights, but that could be an advantage or a drawback.

An important special case of the compromise methods is the 'supra' decision maker problem, wherein a final arbiter must aggregate opinions from a number of judges. In this case, the pooling operators discussed so far are not sensible, since they do not allow this arbiter to take her or his own prior distribution and subsequent information into account. Keeney and Raiffa (1976) named this arbiter the 'supra Bayesian' in an obvious appeal to the use of Bayesian norms to determine how such an arbiter should act. In fact, writers such as Winkler (1968) and Morris (1977) claim that in cases where the arbiter actually exists, the pooling problem itself vanishes. She or he can treat the experts' opinions as data and update her/his prior distribution by Bayes' theorem.

If the supra Bayesian is fictitious, on the other hand, the choice of an appropriate prior distribution falls to the group itself. This situation has caused the greatest disagreements among researchers in this paradigm. Lindley (1985), on the one hand, argues that the pooling problem should be handled by introducing supra Bayesian criteria from the start. Genest and Zidek (1986: 120) dissent from this on the grounds that normative prescriptions must permit questions about their axiomatic basis. The main motivation for considering a fictitious supra Bayesian appears to be that seemingly innocuous axiomatic properties like independence preservation lead to dictatorships in intuitively appealing opinion pools.

While the supra Bayesian criterion offers a way out, it exacts a price. The entire likelihood functions of all experts must be elicited, including covariances among expert judgments. Not only are many of these parameters bound to be subjective, they very likely have little psychological meaning for the experts themselves. Genest and Schervish (1985) attempt to avoid this tradeoff by finding minimal conditions under which simple pooling operators like the linear one satisfy supra Bayesian criteria, thereby foregoing expensive and possibly invalid elicitation requirements.

Virtually all the compromise and intersubjective approaches discussed presume the object of their methods is a single point-estimate probability measure that represents 'group opinion' in some sense. While some writers have argued that an individual's uncertainty cannot be represented after all by a single number, many more conventional arguments have been proposed to the effect that the uncertainty of a group cannot be so represented. Baird (1985) raises most of the relevant arguments in the context of pooling probability estimates, pointing out that even conventional statistical summaries lead to using a measure of dispersion and some measure of central tendency to encompass group opinion and variation within the group. West (1984) and Baird (1984) both argue, from quite different standpoints, that group belief may not be representable as a probability *per se*. Others (e.g., Walley 1981 and Shafer 1976, 1986) argue that individual beliefs should be permitted to be bracketed by intervals, and group opinion should be similarly represented. The issue of group-level belief has arguably motivated at least some probabilists to reconsider the fundamental strategies and prescriptions of probability theory for dealing with uncertainty.

An area of application where probability and group level decision making coexist is risk assessment. Understandably, the techniques mentioned in this section are little used in risk assessment. When more than one expert is making a judgment about the probability of an event, assessors usually either vote, take a simple arithmetic mean, or impose what might be called 'componential dictatorships' (wherein each expert takes sole responsibility for the probability estimate concerning one particular component of the overall system). In view of the heavily normative discussions among probabilists, it is instructive to examine the mixture of normative and pragmatic approaches taken by professionals applying probability concepts to the assessment of uncertainty in complex systems.

3.9 Probability and Risk Assessment

Risk assessors have employed probability frameworks for some time in evaluating risk, so much so that Probabilistic Risk Assessment has its own widely used acronym (PRA). PRA owes most of its conceptual basis to statistical decision and game theory (for a popular introduction, see Morgan 1981). Understandably, most of the literature in this tradition does not closely investigate the philosophical basis for assigning probabilities to undesirable events, but instead focuses on the assessment of utilities, loss functions, welfare functions, and other evaluative measures. The last several years, however, have seen an increase in the number of articles dealing with the assessment of probabilities themselves, and debates over the suitability of probability theory as a means for representing uncertainty.

The traditional PRA approach sits squarely in the Relative Frequentist or Logical schools of probability theory whenever possible, although practical demands and political pressures often force technical experts into a Subjectivist position. Furthermore, unlike many professional arenas in which probability is applied, the PRA experts are fully cognizant that differing schools of probability exist. The British Royal Society Study Group on risk assessment adopted a tripartite definition of probability that is virtually identical to the typology we have been using here. Likewise, many risk assessors probably would agree with Winkler's statement that there are three "potentially valuable" sources of probabilities: Data, output from models, and expert judgments (1982: 353).

Nevertheless, the broad faith in probability theory as the sole means by which to rationally and quantitatively assess uncertainty is reflected among risk assessment experts by statements such as Lord Ashby's exclamation that "those of us who are familiar with the concepts of probability find its conclusions so persuasive that we are surprised how unconvincing they are to many people" (1982: 5). Whatever people Asbhy may have had in mind, many policymakers seem prepared to go along with probability theory, even when they have reservations about the capacity of probabilistic techniques to yield exact answers. The German Federal Minister of the Interior, for instance, in laying down safety criteria for nuclear reactor design and operation, declared that "the reliability of safety related systems and main components should be evaluated using probabilistic methods as far as this is possible with the techniques in their present state of development" (cited in J.C. Consultancy 1986: 15).

The majority of risk assessors seem to favor a quantitative framework as well. Advantages claimed for this framework include the capacity to locate specific weaknesses in a system, to calculate potential financial liability for compensation in respect of damage caused by system failure, and to directly assess reliability, efficiency, and acceptability. Some of these claims are made even while recognizing that PRA may yield rather imprecise quantitative results (e.g., J.C. Consultancy 1986: 66).

The naysayers tend to argue against quantitative PRA mainly on grounds of infeasibility or high cost. J.C. Consultancy (1986: 73-74) cites an interesting statement from 1975 by NASA, in which explanations were given for why NASA did not employ quantitative risk assessment methods. The reasons given were that the data base was too small, that reliability parameters tend to improve with time rather than stabilizing over time, the reliability of new systems is difficult to predict because of the numerous factors involved, and that 'absolute' reliability numbers under these conditions are misleading. However, events following the Challenger Shuttle disaster revealed that even NASA engineers and managers did not resist converting verbal subjective estimates of probability into numerical ones.

When are risk assessors likely to be able to use the Relative Frequentist and Logical approaches? The primary requirement for the Relative Frequentist approach, of course, is a long run of empirical data on a process whose stability is known, from which frequency estimates of probabilities may be reliably obtained. With some exceptions, those ideal conditions tend to occur only for relatively high-probability, low cost, low consequence hazards, which are sufficiently regulated or predetermined that the processes generating those hazards may be considered stable in the long term. Such requirements also tend to limit the motivations for risk assessment to a purely evaluative function unless no changes in the strategies for coping with the hazard have occurred for some considerable time.

The requirement of considerable empirical data for the Relative Frequency theory induces a paradox for PRA whose form is identical to the paradox about the protection of innocence. Just as the protector of the innocent must be tainted or even fallen with knowledge of sin, the guarantee of a risk-free environment via Relative Frequentist assessment methods requires a large number of beings to have been exposed and sacrificed to risk.

Logical probabilistic concepts may be invoked when risk assessors have a complete, credible model for the risk-generating process. The most popular kind of model for artificial systems is 'fault-tree' analysis, in which component failures are linked together by treelike branching structures that in turn lead to specific kinds of system failure. The probabilities of various kinds of failure may then be computed using the conventional calculus for combining probabilities. The requirements of such a model include prior probabilities of component failures, known or assumed (in)dependency relationships among failures of particular components, and deterministic links between component failures and system failures. Clearly, these requirements favour relatively simple systems with either deterministic or highly predictable linkages among components.

The widespread preference for the Relative Frequentist and Logical approaches understandably results in a tendency for risk assessors to base their projections and assessments on well-known and simplified systems or models, and to attempt to reduce less well-known or more complicated systems to those terms. When reduction by simple analogy is not being used, assessors tend to introduce simplifying assumptions. The three most common kinds in the Logical camp are to treat specific components in the system as equivalent to components in the model, to define component events as nonoverlapping, and to assume independence among the failures of various components. Consider a model proposed by Chicken (1978) for the failure probability P(F) of a component:

$$P(F) = P(I)[P(D)+P(M)+P(C)+P(O)], \qquad (3.10)$$

where I refers to inspection techniques failing to reveal a fault, D is a design fault, M is a material fault, C is a fault in construction, and O is a fault in operation. Chicken assumes from the outset that inspection techniques are equally likely to detect a fault regardless of its source or cause, and that the various kinds of faults are mutually exclusive. For at least some kinds of components, these two assumptions are debatable.

Assumptions facilitating reduction of an unknown system to one for which assessors have empirical data include presupposing stability over time or homogeneity (equivalence) among various specific systems. One consequence of these assumptions is that even crude models or poor quality empirical data may be given more credibility than highly informed opinion. Consider the criticisms raised by Glasstone and Jordan (1980) against the Rassmussen Report (1975) on nuclear reactor safety, which included the use of "inadequate data bases", and the inability to quantify common-mode failures and human responsivity during accidents (and hence the absence of such considerations from the model). In a criticism of the same report, Brzustowski (1982: 7-8) notes that in practice the exact nature of the data base from which probabilities were estimated is unclear. Does an estimated gasket failure rate of 5×10^{-5} per year stem from the observation that for every 100,000 gaskets in service 5 failed per year, or the observation of many times 20,000 gaskets in operation for several years in identical environments?

Finally, any probabilistic model of uncertainty requires either a full enumeration of all relevant events and possible system outcomes or the use of a 'miscellaneous' category for unforeseen outcomes. Turner (1978: 169) points out that many practical decision making contexts require decision makers to specify the range of foreseeable outcomes, and that imposing the further requirement that the probabilities of these outcomes sum to 1 implicitly assumes that the probability of unforeseen outcomes is 0. Davis (1979) summarizes the criticisms of a risk assessment of a Liquid Natural Gas terminal by Science Applications Inc., which focus on the fact that the assessors either ignored a large number of plausible initiating events or declared their probabilities to be negligible.

Some of the more sophisticated PRA experts acknowledge this problem but almost invariably place it in the 'too hard' basket. In an article on the estimation of probabilities for risk assessment, Winkler expresses the conventional tendency to bracket this concern by declaring it irrelevant: "Although the identification of posible outcomes is an important problem, the concern in this paper is with the probabilistic aspects of risk assessment" (1982: 352).

A growing literature in the risk assessment field has taken PRA to task for ignoring what are increasingly regarded as crucial (mainly human) factors in system failures and outcomes. The relevancy requirements of PRA are being questioned. Thus, J.C. Consultancy's critique of 7 mainstream methods of risk assessment claims that 6 of 7 ignore or exclude socio-political and economic factors (1986: 91). Their assessment of existing methods for taking such factors into account points not only to rather drastic limitations in the relevance and effectiveness of those techniques, but also the sheer lack of explanatory knowledge of how people perceive and respond to risks.

Even when conditions are reasonably satisfactory for using the Relative Frequency approach for estimation and inference, PRA is plagued with the fact that many of their events of interest are low-probability events. The lower the probability to be estimated, the more data are required to estimate it confidently (cf. Kalbfleisch, et al. 1982: 20-22). Suppose, for instance, there have been 0 reactor failures in 1000 reactor-years of operation. Then if a Poisson model is assumed (such that reactor accidents are randomly distributed through space and time), a 95% confidence interval for the probability of an accident occuring is (0, 0.14). In the wake of Chernobyl, it is sobering to realize that a 1980 study in West Germany of risks due to nuclear reactor accidents broadly concluded that for a population of 25 nuclear power plants, the risk of a core meltdown is of magnitude 10^{-3} per year, the risk of one fatality is about 10^{-5} per year, and of 2,000 fatalities is 10^{-7} per year (for details on the methods of assessment used, see Birkhofer 1980).

The assessment of low probabilities arises in several contexts. One is in determining whether risks are at 'acceptable' levels. In process industry related legislation on toxic safety levels, the definition of 'virtually safe' often is associated with probabilities of harmful effects on the order of 10^{-6} or lower. Likewise, a study by Chicken (1975) indicates that the British trade unions' views concerning acceptable risk of a nuclear reactor accident ranged from no risk to probabilities of one accident of 10^{-6} to 10^{-7} per year. Reported event probabilities considered by British, French, and German authorities in relation to various nuclear reactor accidents range in order of magnitude from 10^{-2} per year downward (see J.C. Consultancy 1986: 17-29). Probabilities of these minute sizes would require sample sizes of several million to reliably establish by frequentist methods.

In process industries, a pragmatically empirical compromise is made by testing several hundred animals at several large dosage levels and obtaining a 'dosage response' curve by using the data to fit a curve of preconceived form (often S-shaped). This curve is then extrapolated (beyond the range of the data, of course) to estimate the 'virtually safe' dosage level for the test animals. The assumptions usually made in this extrapolation include not only the curve's shape but also homoscedasticity (homogeneity in the response variation about the curve with respect to dosage level).

Epidemiological studies lack the control of experiments but often have somewhat larger sample sizes. Here, methodologists have recommended a somewhat different compromise. The primary concern in most epidemiological studies is to compare the probability of getting a disease D given exposure to some 'risk factor' R with the probability of getting the same disease when not exposed to R. Let us denote these probabilities by P(D:R) and P(D:R'), respectively. Since P(D:R) and P(D:R') often are small, epidemiologists focus instead on what they call 'relative risk':

$$r = P(D:R)/P(D:R'), \qquad (3.11)$$

or its odds ratio version as defined by

$$a = \frac{P(D:R)/(1-P(D:R))}{P(D:R')/(1-P(D:R'))} \qquad (3.12)$$

If the two probabilities are small, then a and r approximate one another.

Cornfield (1951) suggests using Bayes' Theorem to estimate relative risk from retrospective data by reformulating it in terms of probabilities that are easier to confidently estimate. First, he observes that

$$P(D:R) = \frac{P(R:D)P(D)}{P(R)} = \frac{P(R:D)P(D)}{P(R:D)P(D) + P(R:D')(1-P(D))} \qquad (3.13)$$

and

$$P(D:R') = \frac{P(R':D)P(D)}{P(R)} = \frac{(1-P(R:D))P(D)}{(1-P(R:D))P(D) + (1-P(R:D'))(1-P(D))}$$

so that

$$a = \frac{P(R:D)/(1-P(R:D))}{P(R:D')/(1-P(R:D'))} \qquad (3.14)$$

The point is that P(R:D) and P(R:D') usually are much larger than P(D:R) and P(D':R) and so are easier to estimate on the basis of limited data. Furthermore, there is a hidden appeal here to the concept of likelihood, wherein the likelihood function L(D:R) is proportional to P(D:R). Adherents to the Likelihood school of statistical inference claim that by considering the ratio of likelihoods to one another, one has all the required information concerning the relative merits of the diagnoses D and D'.

The proliferation of uncertainty through a system also has concerned some risk analysts. An order of magnitude uncertainty about the individual probabilities in Chicken's model (3.10) of component failure, for instance, implies 2 orders of magnitude uncertainty in the final probability P(F) of component failure. In general, the assumption of independence maximizes the proliferation of uncertainty, mainly as a function of the number of component failure probabilities that are to be combined with a Boolean 'and' (hence, multiplication). Suppose there are N necessary but not sufficient components to fail before the system fails, and assume the failures of those components are independent events. Let u_i be the order of magnitude of uncertainty associated with the ith component. Then the order of magnitude of uncertainty associated with the probability of system failure is the sum of the u_i.

In view of the problems of estimating low probabilities and the proliferation of uncertainty throughout complex systems, one might anticipate that risk analysts would use intervalic probability statements of the kind favoured by some subjectivists. In fact, they rarely do. Instead, the preferred strategy for dealing with such problems is to introduce engineering style 'safety factors' into practices based on risk assessments. These factors are not confined to structural or material failure. In epidemiological studies, once the 'virtually safe' dosage has been estimated on the basis of animal data, a further inferential leap to humans is made by introducing a safety factor, $c<1$, such that if d is the estimated safe dosage for animals then cd is the level presumed safe for humans. The magnitude of c usually is established subjectively and such that cd does not fall below the minimal dosage required for the substance to be effective.

The practices of insurance underwriters are particularly indicative of the potential acceptability of this practice. As Lord Ashby (1981) comments, underwriters remain solvent by realistic risk assessments, but they must also appeal to perceived risk to attract customers in the first place. While this position might strike some as placing a naive faith in the honesty of insurance companies and the critical faculties of their customers, even a cynic must admit that the indubitably successful strategies of these companies may be worth considering. Where individual cases are members of a homogeneous class, they may be fairly charged the same pure premium rate, R. But some classes of risk are not entirely homogeneous, and individual risk claims must be taken into account. In that case, the premium often is based on a convex weighted sum of the premium based on individual experience, R_i, and the value of R associated with the class:

$$R_e = CR_i + (1-C)R, \qquad (3.15)$$

where C is a 'credibility weight' such that $0 \leq C \leq 1$, assigned on the basis of underwriting judgment (cf. Houston 1964). In general, C increases with the extent of risk exposure experience. Interestingly, C may not be a probability of any kind.

The traditional PRA paradigm reveals an uneasy normative consensus on the centrality and usefulness of quantitative probability theory. The principal divisions of opinion seem to focus on when and whether to use the Relative Frequentist and/or Logical approaches, the extent to which Subjective probabilities may be permitted, and the viability of nonprobabilistic safety or credibility factors. There is, of course, much more controversy among risk analysts about how to assess costs, benefits, and utilities. Furthermore, contemporary writings on risk assessment frequently address the rifts between experts and nonexperts in this regard (for a good summary of the major battle-lines, see Covello 1984: 232). In this context it is intriguing that until recently probabilistic methods for classifying and quantifying uncertainty have been so universally adopted. Even in the emerging literature comparing expert and nonexpert perceptions of risk, the normative primacy of probability goes largely unchallenged (as will become evident in Part II of this book). To a great extent, quantitative probability assessments still form the benchmark against which human judges must measure up.

Chapter 4:
Beyond Probability?
New Normative Paradigms

"It is this sudden confrontation with the depth and scope of ignorance that represents the most significant contribution of twentieth-century science to the human intellect... Because of this, these are hard times for the human intellect." Lewis Thomas.

"...everybody is ignorant, only on different subjects." Will Rogers.

4.1 The Modern Managerial Approach to Ignorance

The modern orientation towards ignorance contrasts starkly with traditional approaches. In the older view there is no room even for irreducible uncertainty, and all intractable forms of ignorance are banished from analysis. The modern view, on the other hand, in the words of Renee Fox (1980: 9) sees "errors and mistakes, as well as uncertainty and chance, as perennial parts of the... human condition." Nor is this view consensual or stationary. At the very least, the past two decades have seen a dramatic increase in public awareness of uncertainty and fundamental challenges to traditional methods for coping with it. Concurrently, applied mathematicians, scientists, engineers, and philosophers have questioned the exalted position of probability theory as the dominant formalism for analyzing uncertainty.

It seems untenable to dismiss these developments as mere faddism. Fox claims that "the degree and kinds of ferment over error, risk, hazard, and the like that are now occurring in our society may be indicators that we are in the midst of questioning and altering some of our fundamental, cultural ways of thinking about, and dealing with, uncertainty" (1980:11). Although it is too soon to tell whether Fox is right, a number of new normative perspectives on uncertainty itself have emerged in recent years, and the research literature on these matters has burgeoned to an unmanageable size.

The modern impulse with regard to uncertainty (if not all other kinds of ignorance) is managerial rather than exclusionary, but only up to a point. Much of the contemporary debate over uncertainty concerns how far the managerial impulse should or can be extended and what other kinds of ignorance must still be excluded from normative theories.

Despite their disagreements, the various schools of probability are agreed in excluding even various kinds of uncertainty from consideration. A quick appreciation of this may be gained by reexamining Figure 1.1. Probability theory handles only certain special kinds of uncertainty while banishing vagueness, ambiguity, sheer absence of information, distortion, and irrelevance.

The new normative perspectives are, for the most part, more inclusive. Several extensions and/or generalizations of probability theory permit analysis despite the absence of some information that makes point-estimates of probabilities impossible, fuzzy set theory attempts to quantify and analyze certain kinds of vagueness and ambiguity, while surprise theory and possibility theory deal with looser kinds of indeterminacy than probability. Correspondingly, widespread interest has arisen concerning second-order relations (e.g., uncertainty about uncertainty), resulting in intervalic approaches to belief, probability, and fuzziness.

It is noteworthy, however, that even these trends are limited to including more aspects of *uncertainty* rather than ignorance in general. If modern Western culture is undergoing a transformation of its methods for dealing with ignorance, then uncertainty and risk are the foci of that transformation, at least in the intellectual sphere. In terms of the four stages of strategic shifts that described developments in meta-mathematics, normative theories about ignorance are well into the third stage, characterized by second-order strategies for taming first-order ignorance and attempts at grand consensual syntheses, although some current work hints at a move to the fourth stage of suburbanization and even the possibility of a Godelian debacle.

This chapter provides an exegesis on the emerging normative perspectives and debates indicated above. Accordingly, the tone in this part of the book again must be more discursive and even didactic than analytical, critical, or explanatory. I shall begin with a review of the objections raised against probability in psychology, computer engineering and artificial intelligence, and risk analysis. These arguments motivate the introduction of new normative theories of uncertainty, which comprise the main body of the chapter. The concluding section returns to a consideration of the nature and magnitude of the normative revolution.

4.2 Normative and Pragmatic Objections to Probability

Several problems with probability theory in general have been pointed out already in Chapter 3. This discussion extends and systematizes those examples to incorporate the mainstream objections that have arisen in several fields. A rough but fairly functional classification of objections includes the following: (1) Phenomenological, philosophical, and semantic arguments for the existence of nonprobabilistic uncertainties; (2) Evidence that people are not probabilistic in their judgments under uncertainty; and (3) Meta-rationalist claims that probability theory is untenable.

The claim that nonprobabilistic uncertainty exists certainly is not new, and numerous attempts have been made to classify or define various kinds of uncertainty. One of the most famous is Max Black's classic (1937) article in which he distinguishes among vagueness, generality (or nonspecificity), and ambiguity. Most arguments against using probability theory to handle these kinds of uncertainty simply appeal to either definitional issues or to some kind of phenomenological account. A well-known case in point is the debate in the fuzzy set theoretic literature over the relationship (if any) between probability and fuzziness. Fuzzy set apologists claim that one may have certainty about a fuzzy event (e.g., "right now we have light to moderate drizzle" as an approximate answer to the question "is it raining?") or a graded attribute (e.g., "John is tall" or "this carpet is reddish-brown"). The nub of their argument for distinguishing fuzziness from probability is that it seems ungrammatical and/or counter-intuitive to describe the vagueness inherent in such events or attributes by subjective probabilities or even degrees of belief (cf. Zadeh 1980, 1983).

In Zadeh's terms, the objection is not to the axioms of probability theory itself but instead to its limited capacities for expression. Thus, he claims that probability theory is unable to represent the meanings of propositions involving fuzzy predicates (e.g, small, young, safe, soon), fuzzy quantifiers (some, most, several, often), fuzzy probabilities (likely, long odds), fuzzy truth values (very true, not untrue, almost true), and modifiers (very, rather, slightly, somewhat). As a consequence, probability theory cannot make inferences from fuzzy premises and so must abandon problems that have fuzzy components (Zadeh 1986).

A stronger version of this argument goes one step further and makes a case that probability theory is inadequate to reproduce or represent known or reasonable characteristics of a given uncertainty because any attempt to do so introduces a paradox. Perhaps the best illustration of such an argument is Ellsberg's paradox. We are to imagine two urns, each containing black and red balls. We are told that each urn contains 100 balls, and we are further informed that Urn 2 contains 50 red and 50 black balls. If asked to bet on the color of a ball being drawn from one of these urns, most people are indifferent as to which color they would choose whether told the ball will be taken from Urn 1 or Urn 2. A probabilist would infer from this that the subjective probabilities people are assigning to either color are the same (namely $P(R_1) = P(R_2) = P(B_1) = P(B_2) = 1/2$). But Ellsberg finds that if people are asked which urn they would prefer to use for betting on either color, they favor Urn 2. According to probability theory and the usual assumptions about utility, this second finding implies $P(R_2) > P(R_1)$ and $P(B_2) > P(B_1)$.

Not only do the two findings contradict one another, but the second one poses an insurmountable representation problem for probability theory, since it implies that either Urn 2 has probabilities that sum to more than 1 (superadditivity) or Urn 1 has probabilities that sum to less than one (subadditivity). The latter would seem the more likely choice, since it is Urn 1 for which our knowledge is the most incomplete. The point is that the nature of the incompleteness of our information about Urn 1 cannot be represented using probability theory without generating the paradox just described. The only justified conclusion appears to be that some component of our uncertainty about the contents of Urn 1 is nonprobabilistic in nature.

Ellsberg calls this nonprobabilistic component "ambiguity" which, in the terminology proposed in this book, is not quite correct--- the probabilities for Urn 1 are *vague*. But he heralds the developments to come when he declares that "it should be possible to identify 'objectively' some situations likely to present high ambiguity, by noting situations where available information is scanty or obviously unreliable or highly conflicting; or where expressed expectations of different individuals differ widely; or where expressed confidence in estimates tends to be low" (1961: 660). Clearly Ellsberg proposes that we not only classify such situations, but eventually analyze them as well.

Ellsberg's paradox is intimately related to L.J. Cohen's (1977) arguments about guilt versus innocence and nonprovability versus disprovability, Shafer's (1976) concept of the 'vacuous' belief state, and I. Levi's (1977) four kinds of ignorance. All of these involve disputes with the usual additivity requirement in probability theory that P(not A) = 1 - P(A). Specifically, these authors favor a representation of subjective degrees of belief which is subadditive under a wide variety of conditions. The major proponent of a superadditive calculus of uncertainty is Zadeh (1978), who maintains that because possibility is a looser concept than probability possibility must always be greater than or equal to probability. To abandon the additivity requirement means to abandon the standard probability calculus altogether, since it also destroys the de Morgan relations between P(A and B) and P(A or B).

It is also a short step to abandoning the probabilist's injunction that we restrict the representation of subjective belief to a single number between 0 and 1. Again, Cohen's and Shafer's arguments are among the most intuitively plausible, given that they both present cases for using two numbers to define an interval that bounds the individual's degree of belief. Of course, they are not the first to argue that case, and indeed the earliest such arguments came from probabilists themselves who were justifying the use of upper and lower probabilities (which will be discussed in the next section). Likewise the idea enjoyed some currency among philosophers of science concerned with establishing criteria for the confirmation and disconfirmation of hypotheses. Harre (1970), for instance, pointed out that evidence favoring a particular hypothesis to some degree does not simultaneously disconfirm it to any degree, since it may not given any support to the contrary. But the main point is that the motivating arguments against standard probability theory and those in favor of second-order uncertainty formalisms are linked.

A second major source of objections to probability theory stems from studies of or practical experiences with human judgment and behavior. The majority of these arguments in recent years have stemmed from cognitive psychology but some also have come from knowledge engineering and even risk assessment, particularly in the area of expert systems. The cognitive psychologists, for the most part, provide evidence and theories indicating that people do not treat uncertainty as probability theory dictates. Whether this is good or bad depends on the ideology of the experimenters, but the field is largely agreed on the nonprobabilistic nature of human judgment. The knowledge engineers, on the other hand, point out that experts and even laypeople do make apparently functional judgments with neither the probabilistic machinery nor sufficient information to invoke it. Since the engineers would like to build systems that accomplish similar feats, they are motivated to seek something other than probability for handling uncertainty and ignorance.

The inferences psychologists draw from the evidence that people do not adhere to probability theory depend mainly on the stance taken about probability as a normative model of rational judgment. At a time when many applied scientists, engineers, and applied mathematicians have been questioning the utility and rationality of probability theory, cognitive psychology does not seem to have seriously questioned its appropriateness. As Jungermann (1983) observes, studies of judgment and decision making under uncertainty have almost all been conducted using probability theory as the standard of reference, unlike most other areas of cognitive and social psychology (even those areas for which similar normative yardsticks are available). As we will see in Chapter 5, other branches of cognitive psychology either have a physical basis for establishing such standards, or they adopt a more critical stance towards them.

One result of this comparatively uncritical stance towards probability theory is that only a minority of cognitive researchers are prepared to defend human heuristics for reasoning under uncertainty (and indeed, only a minority of studies are directed to understanding in depth what those heuristics are). As we will observe in Chapter 5, the mainstream researchers in this area (e.g., Tversky and Kahneman 1974) are relatively strident in their pronouncements on the subrationality of human (nonBayesian) heuristics, while the tone of the defenders (e.g., Cohen 1979, or Einhorn and Hogarth 1981) is rather muted on the subject of human rationality per se. Slovic (1972) could be a spokesperson for this camp when he admonishes researchers "not too easily to adopt a crude view of human rationality."

One statement which could probably elicit widespread agreement from these researchers is that probability theory certainly does not model human judgment processes under a variety of conditions. If we are to account for or model those processes, then we require a different theory. To date, some limited claims have been made by proponents of alternative normative models of uncertainty that those models come closer to human heuristics than does probability theory (e.g., Shafer 1976: 27 concerning belief functions and Zadeh and his colleagues on numerous occasions concerning fuzzy set theory). Fuzzy set theory has even garnered some support for its claims through empirical studies (see Smithson 1987, Ch.2 for a review). But there is no systematic normative approach at present that demonstrably accounts for what people do and at the same time provides widely accepted standards for some kind of rationality.

One of the most trenchant points made by the defenders of human heuristics accords with a type of criticism raised in several quarters against applications of probability theory to real-world problems. These criticisms amount to an argument for 'meta-rationality', whereby one considers the costs of implementing a so-called rational solution. Meta-rationality arguments against probability refer primarily to its computational complexity and the amount of information required, especially in the context of applying Bayes' Rule. Some psychological researchers claim that people actually employ meta-rational considerations (cf. Beach and Mitchell 1978) while others (Miller and Starr 1967, Janis and Mann 1977, and Payne 1982) are more cautious and only suggest that this might characterize human rationality under certain conditions.

The computational complexity and required information criticisms have been well articulated in Shortliffe and Buchanan (1975). They use it as a motivating argument for the development of their Certainty Factors method for modeling uncertainty in medical expert diagnosis (more on this in the next section). They consider the situation where Bayes' Rule (as expressed in 3.1) has been applied to update prior subjective probabilities $P(H_i)$ that a patient is suffering from the ith disease in some finite list of possible diseases (for i=1,...,N). In the usual diagnostic interview, symptoms or evidence are presented sequentially, and according to Bayes' Rule we should update our $P(H_i)$ by using the following version:

$$P(H_i|E\&e_k) = \frac{P(e_k|H_i\&E)P(H_i|E)}{\sum_{j=1}^{N} P(e_k|H_j\&E)P(H_j|E)} \qquad (4.1)$$

where E is all the evidence to date, and e_k is the new piece of evidence. To implement this formula requires not only the $P(e_k|H_j)$ for each of the various e_k, but also the interdependencies of the ek within each H_j. Furthermore, the computational requirements become quite sizeable as the number of diseases and pieces of evidence grow. Finally, in actually constructing a knowledge base, one must frequently add or delete individual pieces of knowledge. Clearly this cannot be done without also revising the dependency information on all relevant subgroups of evidence.

Similar points have been raised about the availability of extensive statistical information of the kind required for a Bayesian analysis. Edwards (1972) quotes doctors as claiming that even sophisticated hospital records cannot provide that kind of associative information, but worse still, diseases and the medical profession's classification of them change, as do treatment modalities. The same arguments have been made about crime, suicide, and accident statistics (cf. Douglas 1967 and Atkinson 1968). Note that the informational unavailability criticism has two distinct components: One concerning the sheer absence of such information, and the second pertaining to its changeable, unreliable, or irreducibly contextual nature.

Having briefly reviewed the major contentions against probability as either the sole or even dominant normative perspective on uncertainty, let us turn to the counter-proposals. I will permit the probabilists the right of reply in section 4.6.

4.3 The New Uncertainties
4.3.1 The Certainty Factors Debate

In much of the work on expert systems during the 70s, criticisms concerning the practicality and desirability of probability theory motivated a search for alternative heuristics for handling uncertainty in expert knowledge and inference. Perhaps the best known of these early systems are MYCIN, from the Stanford Heuristic Programming Project (cf. Buchanan and Shortliffe 1984) and PROSPECTOR, a geological expert system whose developmental history appears to be unpublished but is often referred to in AI writings on uncertainty (cf. Duda, Hart, and Nilsson 1976, see also the acknowledgement in Grosof 1986:163). I will introduce both systems here, although they are familiar to expert systems researchers, since they both effectively raise crucial issues in the construction of alternative formalisms for handling uncertainty. Of the two, MYCIN's system has been the more widely debated, and the rest of this section will focus on that debate, again because of the issues raised therein.

Shortliffe and Buchanan (1975) constructed a "certainty factor" (CF) approach to be implemented in a medical diagnosis program called MYCIN. Their aim was to provide an alternative to a fully fledged Bayesian approach that would overcome the objections to that approach while still retaining certain desirable features. They based their requirements on the philosophical discussions of confirmation and induction stemming from Carnap's (1950) work. Specifically, the desiderata they selected included the following:

(1) Evidence confirming H_j in some degree should not also disconfirm H_j (that is, confirm $\sim H_j$) in a compensatory degree, as is required in probability theory.

(2) The CFs should behave 'modularly', which means that evidence may be added to or deleted from the knowledge base without affecting dependency relationships among the CFs for various H_j (although of course those additions and deletions will affect the values of the CFs for those H_j to which the evidence pertains).

(3) The CFs should be combinable in a way that is independent of the order in which evidence is presented (commutativity).

The first requirement comes from the philosophical literature on confirmation. As Carnap (1950) and several others recognized, a confirmation measure $C(H_j,e)$ does not equal $1 - C(\sim H_j, e)$: "...to confirm something to ever so slight a degree is not to disconfirm it at all, since the favourable evidence for some hypothesis gives no support whatever to the contrary supposition in many cases" (Harre 1970). The second, of course, overcomes a practical objection to the Bayesian framework and amounts to restricting experts to dealing with first-order updates, $P(H_j|e_k)$, where e_k is the evidence introduced on the kth step in the diagnostic interview, say. The third is a property of the probabilistic rule for combining evidence that is generally considered desirable, just as it is for premodern algebras.

Shortliffe and Buchanan implemented their desiderata by adopting independent measures of 'belief' and 'disbelief' as the starting point for their CF approach. That the two measures must be independent follows from (1), whereby confirmation and disconfirmation of a hypothesis are separate undertakings. They define $MB(H_j,e)$ as the degree of increased belief in hypothesis H_j given evidence e, and $MD(H_j,e)$ as the degree of increased disbelief in H_j given e. If $P(H_j|e)$ is greater than $P(H_j)$, then a judge's belief increases by $(P(H_j|e)-P(H_j))/(1-P(H_j))$. On the other hand, if $P(H_j|e)$ is less than $P(H_j)$ a judge's disbelief increases by $(P(H_j)-P(H_j|e))/P(H_j)$.

Somewhat more formally, MB and MD are defined by

$$MB(H_j,e) = \frac{\max(P(H_j|e),P(H_j))-P(H_j)}{1-P(H_j)} \text{ if } P(H_j) < 1,$$
$$= 1 \text{ if } P(H_j) = 1, \qquad (4.2)$$

and

$$MD(H_j,e) = \frac{\min(P(H_j|e),P(H_j))-P(H_j)}{1-P(H_j)} \text{ if } P(H_j) > 0,$$
$$= 1 \text{ if } P(H_j) = 0. \qquad (4.3)$$

Clearly both MB and MD are in the [0,1] interval, and when one of them is nonzero the other is set to 0. Shortliffe and Buchanan then introduce a third measure which combines belief and disbelief into a single number after all, by taking their difference. This is their 'certainty factor':

$$CF(H_j,e) = MB(H_j,e)-MD(H_j,e) \qquad (4.4)$$

and its range is the [-1,1] interval. $CF(H_j,e)$ is 1 when $P(H_j|e)$ is 1; it is -1 when $P(\sim H_j|e)$ is 1.

The CF system requires, of course, rules for combining evidence and hypotheses. Shortliffe and Buchanan provide four of these.

(1) Aggregation of evidence: (4.5)
$MB(H_j,e_1\&e_2) = 0$ if $MD(H_j,e_1\&e_2) = 1$,
$\qquad = MB(H_j,e_1)+MB(H_j,e_2)(1-MB(H_j,e_1))$
otherwise.
$MD(H_j,e_1\&e_2) = 0$ if $MB(H_j,e_1\&e_2) = 1$,
$\qquad = MD(H_j,e_1)+MD(H_j,e_2)(1-MD(H_j,e_1))$
otherwise.

(2) Conjunction of hypotheses: (4.6)
$MB(H_1 \text{ and } H_2,e) = \min(MB(H_1,e),MB(H_2,e))$
$MD(H_1 \text{ and } H_2,e) = \max(MD(H_1,e),MD(H_2,e))$

(3) Disjunction of hypotheses: (4.7)
$MB(H_1 \text{ or } H_2, e) = \max(MB(H_1,e), MB(H_2,e))$
$MD(H_1 \text{ or } H_2, e) = \min(MD(H_1,e), MD(H_2,e))$

(4) Uncertain evidence: (4.8)
If some evidence e_k is not known with certainty and if a $CF(e_k,e)$ based on prior evidence e is given for e_k, and if $MB^*(H_j,e_k)$ and $MD^*(H_j,e_k)$ are the measures of belief and disbelief that would be assigned to H_j were e_k known with certainty, then the adjusted degrees of belief in H_j given e_k are given by
$MB(H_j,e_k) = MB^*(H_j,e_k)\max(0,CF(e_k,e))$ and
$MD(H_j,e_k) = MD^*(H_j,e_k)\max(0,CF(e_k,e))$.

Clearly $CF(H_j,e) + CF(\sim H_j,e) \neq 1$ and in fact it equals 0, since $MB(H_j,e) = MD(\sim H_j,e)$. The CFs therefore satisfy the first criterion listed above. They are also modular and commutative, so the order in which evidence is encountered is irrelevant. At least three other important properties follow from the above definitions. First, although the CFs are not absolutely sum-constrained (ipsative) as are probabilities, their sums do have normative upper and lower bounds according to Shortliffe and Buchanan. Given K mutually exclusive hypotheses,

$$-K \leq \sum_{j=1}^{K} CF(H_j,e) \leq 1. \quad (4.9)$$

Secondly, MB or MD each increases towards 1 as confirming or disconfirming evidence is found, respectively. The implicit assumption is that the conjunction of two, say, confirming pieces of evidence always provides stronger confirmation for the hypothesis involved. Third, (4.5) and (4.8) imply that missing data are casewise ignored.

Finally, the authors provide a rationale for CFs in terms of their suitability for modeling real expert belief estimates. These arguments are crucial because Shortliffe and Buchanan begin with a claim that elicited quantitative estimates of belief from experts are not like probabilities, particularly with regard to the complementarity requirement of standard probability theory. Even more importantly, they base their implementation of the CF approach in MYCIN on elicted CF estimates from experts, and they assume that these quantities are close approximations to the CFs that would be obtained from the MBs and MDs if the relevant probabilities were known. If CFs are not an appropriate model of expert belief and disbelief, then of course the entire enterprise fails. As in the other foundational aspects of the CF framework, the authors openly declare that the basis for their claims are intuitive rather than evidential or logical. They have, however, a rather persuasive piece of circumstantial evidence in their favor: MYCIN performs very well.

PROSPECTOR contains another essentially ad hoc scoring system for updating uncertainty measures on the basis of sequentially acquired evidence. In contrast with the CF methods, however, PROSPECTOR makes use of a modified Bayesian approach employing strong assumptions about conditional independence which amount to the same thing intended by the concept of modularity in the CF framework. Although an account that is faithful to the original format of PROSPECTOR is available in Duda et. al. 1976, I shall follow Grosof's (1986) format to enable some direct comparisons between PROSPECTOR and revisions of CF later on.

The basis for the updating system in PROSPECTOR is a log-likelihood form of Bayes' Rule:

$$\frac{P(H_j|e)/P(\sim H_j|e)}{P(H_j)/P(\sim H_j)} = \frac{P(e|H_j)}{P(e|\sim H_j)}, \text{ or}$$

$$O(H_j|e)/O(H_j) = L(H_j,e) \qquad (4.10)$$

where the O stands for 'odds' and L denotes 'likelihood ratio' (this latter often is written as 'lambda' in the Bayesian literature). The main use of (4.10) in updating is to evaluate $O(H_j|e)$ on the basis of $L(H_j,e)$ and $O(H_j)$. Given the assumption that an update due to evidence e depends only on the summary measure $O(H_j)$ used to indicate the odds of the hypothesis given all previous evidence (conditional independence, hence modularity), the combination rule for two updates is simply multiplication:

$$L(H_j, e_1 \& e_2) = L(H_j, e_1) L(H_j, e_2) \qquad (4.11)$$

This rule is, of course, commutative and associative. PROSPECTOR's system is equivalent to a simplified and restricted version of Bayesian inference, and yet appears to satisfy much the same criteria as the CF framework. Like MYCIN, PROSPECTOR appears to have performed quite well in its domain of application.

Now let us turn to the debate stimulated by the CF proposal. One of the first objections raised against the original definition was that a single piece of negative (or positive) evidence could overwhelm the effect of a number of pieces of positive (or negative) evidence. Consider, for instance, a series of 9 pieces of evidence supporting a particular hypothesis, whose asymptotic behaviour has resulted in a cumulative MB value of 0.999. Now suppose a single disconfirming piece of evidence were to yield a CF value of -0.8. Then the net support for the hypothesis would be CF = MB - MD = 0.999 - 0.8 = 0.199, which seems to many people a counterintuitive result. It also apparently led MYCIN into some mistaken conclusions.

By 1977 a modification of the definition of a CF and the rule for aggregating evidence had been proposed by Van Melle and incorporated into MYCIN and its relations (Buchanan and Shortliffe 1984: 216). Let $B = CF(H_j,e_1)$ and $C = CF(H_j,e_2)$. Then

$$CF = \frac{MB-MD}{1-\min(MB,MD)}, \text{ and} \qquad (4.12)$$

$CF(H_j,e_1 \& e_2) = B + C - BC$ when $B,C \geq 0$

$\qquad = \frac{B+C}{1-\min(|B|,|C|)}$ when B, C have opposite signs

$\qquad = B + C + BC$ when $B,C < 0$

This definition alters CFs only when two of opposite sign are combined. Thus our new value for a combined CF based on input values of 0.999 and -0.8 is 0.199/0.2 = 0.99.

A concurrent issue raised within the MYCIN group was sensitivity. To what extent could MYCIN's conclusions be altered by either the revised combination rule proposed in (4.12) or, for that matter, varying CF input values? The group conducted a sensitivity analysis (reported in Buchanan and Shortliffe 1984: Ch.10) and concluded that MYCIN was relatively insensitive to alterations in CF values or rules for two reasons: (1) Inference chains are frequently short; and (2) Conclusions often are reached by a single rule, thereby avoiding the use of the combination formula altogether in many instances. An additional contributing factor to MYCIN's 'success' apparently also came from the fact that the same recommended therapy often holds across a number of similar medical complaints, so that even where diagnosis varies the recommended treatment may not. In short, it is possible that the domain of application was 'solution rich', which tends to devalue MYCIN's performative competence as evidence for the suitability of CFs.

Recently two more fundamental problems with the CF formulations have been raised by critics. One of these is an apparent inconsistency between the definition of a CF and the combination rule for updating CFs on the basis of aggregated evidence. An interesting demonstration of this inconsistency is presented in Heckerman (1986). First, he rewrites the CF definition in (4.2)-(4.4) in the following form:

$$CF(H_j,e) = \frac{P(H_j|e)-P(H_j)}{1-P(H_j)} \text{ when } P(H_j|e) > P(H_j) \quad (4.13)$$

$$= \frac{P(H_j|e)-P(H_j)}{P(H_j)} \text{ when } P(H_j|e) < P(H_j).$$

Then he argues that since CFs are intended to represent measures of change in belief, the probabilities themselves in (4.13) should be interpreted as measures of belief which in turn must depend on one's current state of information E prior to encountering the new evidence e. He therefore extends the definition in (4.13) by conditioning each of the probabilities on the evidence E prior to finding out e:

$$CF(H_j,E,e) = \frac{P(H_j|eE)-P(H_j|E)}{1-P(H_j|E)} \text{ when } P(H_j|eE) > P(H_j|E) \quad (4.14)$$

$$= \frac{P(H_j|eE)-P(H_j|E)}{P(H_j|E)} \text{ when } P(H_j|eE) < P(H_j|E).$$

The nub of Heckerman's claim is that (4.14) contains an *implicit* combination rule which is noncommutative. Suppose we have $CF(H_j,E,e_1) = 0.9$, $CF(H_j,E,e_2) = -0.9$, and $P(H_j|E) = 0.5$, the latter of which is the degree of belief before knowing e_1 or e_2. By the modularity property, $CF(H_j,Ee_2,e_1) = CF(H_j,E,e_1)$ and likewise $CF(H_j,Ee_1,e_2) = CF(H_j,E,e_2)$. That is, the CF associated with e_1, for instance, should remain the same regardless of whether e_2 is known prior to encountering e_1 or not. Therefore, we may use (4.14) to update our belief in H_j given e_1 in the following way. We plug the value of $CF(H_j,E,e_1) = 0.9$ and $P(H_j|E) = 0.5$ into (4.14) and solve for $P(H_j|e_1E)$, obtaining 0.95. We then plug that value and $CF(H_j,Ee_1,e_2) = -0.9$ into (4.14) again and find $P(H_j|e_2e_1E) = 0.10$. If we reverse the order and update on the basis of e_2 first and then e_1, we find $P(H_j|e_1e_2E) = 0.90$, which immediately demonstrates that the order in which the evidence is considered affects the resultant update. Thus, Heckerman claims, the CF framework fails implicitly to satisfy a major criterion, namely commutativity.

The second problem in the CF framework is an apparent confusion between belief updates and measures of absolute belief. Horvitz and Heckerman (1986) point out that while CFs were constructed as measures of belief updates on the basis of incoming evidence, in practice they have often been elicted from experts as measures of absolute belief. They cite Shortliffe's (1976) elicitation frame "On a scale from 1 to 10, how much certainty do you affix to this conclusion?" as evidence, along with a justification provided in Shortliffe and Buchanan (1975) for treating CFs as absolute belief measures under certain conditions. Specifically, the 1975 paper observes that when $P(H_j)$ approaches 0 and if $P(H_j|e)$ is not too close to 0, then $CF(H_j,e)$ is approximately equal to $P(H_j|e)$. The problem with all this, according to Horvitz and Heckerman, is that the CF scheme overcounts prior beliefs.

Then these two authors note that the primary philosophical source of the CF creators' intuitive justifications, Carnap's 1950 confirmation theory, confused absolute belief and belief updating. Carnap initiated the confusion by an inconsistent application of terminology (which he eventually acknowledged), and this in turn provided some philosophers with the same motives as those given by Shortliffe and Buchanan for turning away from subjective probability as a framework for a formal theory of scientific confirmation. Popper, among others, pointed out that probability theory allows for belief update and therefore need not be abandoned, and I.J. Good (1960) proved that some probabilistic entities satisfied Popper's criteria in slightly modified form. This is the key to why PROSPECTOR's creators were able to use an essentially Bayesian framework to perform the same updating tasks as the CFs, and indeed poses the question of whether the CF desiderata might be susceptible to a probabilistic approach after all.

That is the conclusion reached by Grosof (1986) and Heckerman (1986), although a less thoroughgoing case had been made earlier by Adams (1976). Heckerman uses the same odds version of Bayes' Rule as (4.10) to define a version of CFs that differs from the original only by an extra factor in the denominators of (4.13). Furthermore, Heckerman's CFs are simple functions of the likelihood ratio $L(H_j,e)$ and he demonstrates that the multiplication rule for combining two likelihood ratios is exactly the parallel combination rule used in the original CF framework. Both Heckerman and Grosof provide strong formal arguments that any CF-type framework that satisfies Popperian style desiderata and also has a probabilistic interpretation must be a simple transformation of $L(H_j,e)$. These arguments have historical roots in the work of Good (1960) and Cox (1946).

Finally, several authors have criticized the modularity assumption. Horvitz and Heckerman (1986) are among those who have pointed out that it implies a strong form of conditional independence such that beliefs in pieces of evidence e_k do not depend on the knowledge that any prior evidence is true when H_j is definitely true or definitely false. That is, the e_k are conditionally independent given H_j and $\sim H_j$. The main thrust of the criticism against this consequence is that while it ensures computational ease and efficiency, it fails to model most real-world knowledge and evidence domains. But R. Johnson (1986) raises an additional telling point on this matter, when he demonstrates that multiple updating of a hypothesis is impossible when there are more than two mutually exclusive and exhaustive hypotheses, if the evidence is conditionally independent given each hypothesis and its negation.

The debates over CFs are instructive on several counts. First, they demonstrate how difficult it is to construct a formalism for handling uncertainty that is truly distinct from probability theory while being both self-consistent and intuitively plausible. Secondly, despite their severalfold formal problems and shaky philosophical grounding, CFs have performed well in several applications. When expert systems such as MYCIN, PROSPECTOR, or INTERNIST are modified in their uncertainty calculi or inputs, the final output often remains surprisingly unchanged. It is quite possible that more complex domains will prove more sensitive to the formalists' objections, but so far the evidence does not contradict the conviction expressed by at least one of the MYCIN 'gang' that "...MYCIN's rule set is not 'fine-tuned' and does not need to be...there are few deliberate (or necessary) interactions in the choice of CFs" (Buchanan and Shortliffe 1984: 219).

Third, it is intriguing to observe how little of the debate is devoted to the question of how well or poorly the CFs (and their likelihood ratio counterparts) really conform to human experts' heuristics for belief updating and reasoning under uncertainty. There are few references in this literature to empirical studies or cognitive psychological theories about human information processing, and the closest the expert systems builders appear to come to conducting empirical studies of their own is to compare the final output of MYCIN with conclusions reached by medical doctors. The vast majority of the arguments in these debates are normative, with a few pragmatically based recommendations thrown in.

This normative emphasis contrasts vividly with the oft-stated ambitions of AI and expert systems researchers to duplicate at least some of the feats of human intelligence, and also seems to reflect the same uneasy ambivalence about human thought processes that was highlighted at the beginning of this chapter in connection with cognitive psychological studies of humans' probabilistic judgments. That same ambivalence surfaces repeatedly throughout the normative literature on alternatives to probability, and may well reflect an underlying conflict over how much ignorance really is manageable after all.

These concerns and conflicts have loomed larger than life in the polemics generated by proponents and critics of fuzzy set theory and its spinoffs. Like Certainty Factors, fuzzy set theory was born in a computer science environment. Unlike the CF framework, however, fuzzy set theory made a strong case early on for being distinct from and irreducible to probability. Understandably, fuzzy set theory has been repeatedly attacked by probabilists and vigorously defended by its proponents. Let us examine the case for fuzzy sets, its applications, and the kinds of ignorance it claims as its domain.

4.3.2 Fuzzy Sets, Vagueness, and Ambiguity

Fuzzy set theory, like several other formalisms for handling uncertainty, seems to have been discovered independently by more than one individual and anticipated in many ways by works in mathematics, logic, philosophy, and linguistics. However, Zadeh (1965) is generally accorded the status of 'founding father' because his classic article, unlike others (e.g. Klaua 1965), appealed to an applications-oriented audience and consequently became a 'citation classic' as fuzzy sets became more widely known. Although a number of fundamental articles on fuzzy sets appeared during the late 60's, the 70's saw an explosion in the number of publications and, more importantly, the first applications of the new theory. Like the CF framework and various other nonprobabilistic uncertainty formalisms, fuzzy sets entered into Western intellectual culture during the decade of the 70's.

As with the initial development of the CF framework, the first rationales provided for fuzzy set theory were intuitive and commonsensical rather than axiomatic or rationalistic. The oft-repeated opening gambit to motivate fuzzy sets is the observation that many natural categories and sets are not 'crisp' in the sense that an element either belongs to the set or is excluded from it. Instead where classical set theory permits only 'red' and 'not red', for instance, ordinary language allows elements to belong *partially* to the set of red things, as reflected by words such as 'reddish' and hedges (e.g., 'sort of red', 'very red').

If classical set theory cannot model this kind of blurriness, then one simple remedy is to construct a set theory which can, and this is precisely the tack that Zadeh's original formulation takes. Conventionally, membership of some element x in classical (crisp) set S is denoted by a membership function $m_S(x) = 1$, while exclusion from the set induces the value $m_S(x) = 0$. Fuzzy set theory models partial membership by permitting $m_S(x)$ to take values on the [0,1] interval, which in turn is called the 'valuation set'.

If membership in S is linked to a scalable attribute of x, then the scale $c_S(x)$ for that attribute is called the 'support set', and $m_S(x)$ becomes a membership *function*, a mapping from the domain of c_S to [0,1]. The most popular example used in the early fuzzy set literature was the set of 'tall' people. The support set is whatever scale one might use for measuring height (feet, centimeters, cubits, etc.), and the values for $m_S(x)$ are monotonically nondecreasing in $c_S(x)$. Thus, $c_S(x_1) = 5'6"$ might yield $m_S(x_1) = 0.4$, but $c_S(x_2) = 6'2"$ might result in $m_S(x_2) = 1.0$.

Fuzzy set theorists utilize these concepts to define such entities as fuzzy numbers (e.g., 'several', 'few'), fuzzy variables (variables whose scales are fuzzy numbers), fuzzy probabilities (where the probability scale is the support set, as in 'likely'), and fuzzy quantifiers (e.g., 'most', 'frequently'). A claim often made on behalf of fuzzy set theory is that phrases such as the news headline "Experts predict a big San Francisco earthquake unlikely soon" are decomposable into fuzzy predicates (Zadeh 1986: 104), each with their associated membership functions. So 'big', 'unlikely', and 'soon' all lie within the purview of fuzzy sets, as does the implicit notion of an earthquake 'near' enough to San Francisco to be called a 'San Francisco earthquake'.

Fuzzy set theorists also have proposed second- and higher-order fuzzy sets (called 'type k', with k denoting the order) that are analogous to concepts such as probabilities of probabilities or uncertainty about uncertainty. A type 2 fuzzy set is one whose membership values are themselves fuzzy numbers (e.g., 'about 0.5' or 'close to 1.0'). We shall have a closer look at second-order uncertainty formalisms in section 4.4.

Having defined membership functions, the next step in the construction of a theory of fuzzy sets is to define operations on and relations between those sets (e.g., negation, inclusion, union, and intersection). Again, the original formulations of fuzzy set theory use a combination of intuitive and semi-formal desiderata as guiding criteria, along with the requirement that fuzzy set theory always reduce to classical set theory when m_S values are restricted to 0 and 1. Negation is defined as in classical set theory:

$$m_{\sim S}(x) = 1 - m_S(x). \quad (4.15)$$

Fuzzy inclusion also follows the classical model closely. The intuitive argument is that if S includes T, then any element x should be at least as much a member of S as it is of T so that $m_S(x) \geq m_T(x)$ (e.g., if John is very tall then he also is definitely tall, but the converse does not necessarily hold). Fuzzy set intersection is based on the classical Boolean logical 'and', whereby x should be no more a member of the compound set 'S and T' (that is, the intersection of S with T) than it is of either S or T. Zadeh's original formulation, and still the most popular version of fuzzy set theory, strengthens the 'no more than' to equality. He also requires that the union of two fuzzy sets be related by the usual de Morgan law to intersection, and therefore

$$A(m_S, m_T) = \min(m_S, m_T) \text{ and} \quad (4.16)$$
$$O(m_S, m_T) = \max(m_S, m_T),$$

where A is the membership function for the intersection of S with T, and O is the membership function for the union of S with T.

These definitions comprise the kernel of fuzzy set theory in its earliest form. Since Zadeh's 1965 article, developments in the theory itself have taken three broad paths. One of them is the elaboration of fuzzy set theory itself via additional definitions of operations on fuzzy sets, which themselves are relatively independent of specific definitions and operations in the theoretical kernel. An example comes from Zadeh (1972) again (but for a thorough survey of such developments see Dubois and Prade 1980), in the form of a fuzzy set theory of linguistic hedges.

The hedges Zadeh is concerned with include phrases such as 'very', 'sort of', and 'more or less' which seem to directly modify fuzzy sets in an adjectival or adverbial manner. He argues that these may be modeled by simple monotonic transformations of membership functions, and he defines two classes of hedges: concentration and contrast intensification. Modifiers such as 'very' are concentrators since they result in a more exclusive fuzzy set; the set of 'very tall' men is more restricted than the set of 'tall' men. The hedge 'moderately', on the other hand, is a dilator since it defines a looser set ('moderately tall') than its unmodified counterpart. Zadeh's proposal for suitable membership function transformations is simply exponentiation since any number in the [0,1] interval raised to any power remains inside that interval. His concentrator and dilator transformations consist of

$$m_{ConS}(x) = (m_S(x))^2 \text{ and} \qquad (4.17)$$
$$m_{DilS}(x) = (m_S(x))^{1/2}$$

and his definitions for contrast intensification and diffusion follow similar lines. Other developments of this type involve relations between fuzzy sets such as measures of inclusion, likeness, similarity, difference, distance (surveys of these may be found in Dubois and Prade 1980), correlation (Murthy et al. 1985), and overlap (Smithson 1987).

The second major developmental path for fuzzy set theory is what might be termed the 'fuzzification' of those branches of mathematics and logic which are based on set theory per se (which of course is a very large domain indeed). Kaufmann's volumes from the mid-70's and Dubois and Prade (1980) provide the most exhaustive summaries of those extensions, including the agreed-upon principles for effecting them. One of the most widely known and used products of these developments is fuzzy logic, which is worth examining not only because it has been widely used but also for the clarity with which it highlights a number of issues and debates about fuzzy set theory itself.

Interest in fuzzy logic appears to have stemmed from at least two sources. One is the problem of inference using propositions containing fuzzy predicates. A second is the possibility of permitting the truth-value of a proposition to take intermediate states between complete truth or falsehood (e.g., 'somewhat true'). Zadeh's proposals for fuzzy logic (1973,1975a) used the min-max operators for logical 'and' and 'or', with the immediate result that the Law of the Excluded

Middle no longer holds (if t_q is a truth value in [0,1] associated with proposition q, then whenever $t_q < 1$ then $1-t_q > 0$, so $t_{q\&\sim q}$ = min($t_q, 1-t_q$) > 0).

The problem of defining fuzzy implication yielded several competing solutions. For example Zadeh's version, the so-called Maxmin Rule, defined the truth value of 'if p then q' by

$$t_{q|p} = O(A(t_p, t_q), 1-t_p), \quad (4.18)$$

while Lee (1972) used the classical equation of 'if p then q' with 'not p or q' to define the Arithmetic Rule:

$$t_{q|p} = O(1-t_p, t_q). \quad (4.19)$$

These proposals turned out to be linked to work that occurred much earlier (in the 20's and 30's in some cases) on multivalent and infinitely-valued logic, as well as stochastic logic. Gaines (1976) provides an excellent exegesis on those connections. They also pointed out an emerging fact of life about fuzzy set theory in the early 70's: Zadeh's original desiderata were not sufficiently precise or restrictive to uniquely define fuzzy set theory.

The third developmental path, therefore, has been the pluralization of fuzzy set theory. Rather than recounting the history of this development, I will merely summarize the major versions of fuzzy set theory that are currently in use and describe the criteria that have been proposed for preferring some over others.

At the heart of the impulse towards pluralism and the debates over variations on fuzzy set theory is the question of what kinds of uncertainty or ignorance fuzziness encompasses. Fuzzy set theorists themselves have not been entirely consistent on this issue. Instead, they have been rather more willing to say what fuzziness is not than what it is. In particular, they are anxious to make a case that fuzziness is distinct from probability since if it is not, then the fundamental definitions and operators used in fuzzy set theory disobey the norms laid down by the probability calculus. It is worth bearing in mind that although many aspects of probability have been disputed, the basic calculus has withstood almost all criticisms.

The case for distinguishing fuzziness from frequentist probability seems obvious to nearly everyone and has largely been conceded by probabilists anyway, so we need not dwell on that here. It is the subjectivist notion of probability as a degree of belief that is closest to and hence most dangerous for fuzzy set theory, and as we shall see later there are many probabilists who argue that the two either cannot or should not be differentiated. Zadeh's original definition of degree of membership differs from the usual axiomatic definitions (e.g., Cox's 1946 system) for probability in only two respects: One is the use of the min-max pair of operators rather than the product and probabilistic sum, and the other is the attribution of a nonprobabilistic meaning to the concept of a graded set. Zadeh and his allies argue that one may be entirely certain that a particular object is 'dark'; there is no probabilistic uncertainty associated with this proposition, the only source of ignorance is the graded nature of the concept of darkness.

This gradedness is said to be linked to *vagueness* rather than probability. Zadeh uses terms like 'vagueness', 'ambiguity', and 'nonspecificity' in rather nonstandard ways. He declares (1976,1979) that the proposition 'Ruth has dark skin and owns a red Porsche' contains fuzzy attributes ('dark' and 'red') because these are graded but not ambiguous. On the other hand, the utterance 'Ruth lives somewhere near Berkeley' is not merely fuzzy because 'somewhere near' is too nonspecific to precisely characterize Ruth's location. Fuzziness, then, is distinct from ambiguity and also is a specific (or precise) kind of vagueness. But Zadeh then defines vagueness itself as a combination of fuzziness and ambiguity.

The standardized usages of these terms in linguistic philosophy have their source in Max Black's classic (1937) paper on vagueness. He defines ambiguity as a condition in which a referent has several possible and distinct interpretations (e.g., 'hot' in the utterance 'the food is hot', which could mean either spicy or of high temperature). Generality, on the other hand, is Black's term for nonspecificity and he upbraids earlier philosophers for failing to distinguish it from vagueness. Vagueness, for Black, is akin to what Zadeh seems to mean by fuzziness and indeed Alston (1964) uses the term 'degree vagueness' in a manner uncannily like the usual characterizations of fuzziness. The upshot of all this appears to be that many fuzzy set theorists use fuzziness in a way that locates it as a special kind of vagueness, which intuitively seems like a qualitatively different kind of uncertainty than probability.

However, there are at least two other competing kinds of definitions. One of these is quite close to subjective probability, and was proposed by Sugeno (1977) and Giles (1976). For these theorists, the membership function is an expression of the uncertainty associated with the statement 'x belongs to S', where S is in reality a crisp set. Giles goes as far as to construct a de Finettian definition whereby $m_S(x)$ is the amount of money one would be willing to bet in a 'fair wager' against an opponent who has agreed to pay \$1 if x is not in S. This kind of definition makes it more difficult to argue that fuzziness is distinct from subjective probability. Giles and Sugeno do so by axiomatic methods in which they eschew certain axioms held by probabilists and substitute others which yield versions of fuzzy set theory (including Zadeh's).

Recently, however, a semantic case has been provided for a priori separating even this kind of fuzzy set theory from probability. Klir (1987) claims that Sugeno's 'guessing game' is founded on the assumption that, given the information at hand, x does not a priori belong to any crisp set S but might possibly belong to any of them in degree $m_S(x)$. Therefore, the state of knowledge concerning the true location of x is not a matter of vagueness as in Zadeh's formulation, but *ambiguity* in Black's sense. Klir's argument is that these two versions of fuzzy set theory actually address distinct nonprobabilistic kinds of uncertainty.

A third characterization of fuzzy set theory described in Dubois and Prade (1980) is based on a concept of possibility. They point out that a fuzzy concept like 'tall' conveys a possibility distribution over the set of all heights, so that $m_S(x)$ reflects the degree to which x might possibly belong to the set of tall people given his or her height. Possibility is a looser kind of uncertainty than probability, since anything that is impossible must be improbable but whatever is possible need not be probable. In fact, Zadeh (1978) founded a theory of possibility based on his version of fuzzy set theory, which we will investigate in the next subsection.

Finally, even within a particular approach to defining fuzziness there is room for differences over what operators should be used for fuzzy intersection, union, negation, and so on. Bellman and Giertz (1973) and more recently Trillas et al. (1982) have provided axiomatic rationales for limiting the theory to Zadeh's version, but these have not prevented a proliferation of alternatives. The dominant perspective in recent years, as embodied in the leading journal *Fuzzy Sets and Systems*, is that union and intersection operators should come from a class of triangular norms associated with certain modern algebras.

These triangular norms (t-norms) share the following properties: (1) $T(1,a) = a$ for any a in [0,1]; (2) Monotonicity; (3) Commutativity; and (4) Associativity. $T(a,b)$ is affiliated with the intersection operator. A co-norm $S(a,b)$, usually called an s-norm, is affiliated with the union operator, and its properties are identical to those of $T(a,b)$ with the exception of (1) which is replaced with $S(0,a) = a$ for any a in [0,1]. All $T(a,b) \leq \min(a,b)$ and $S(a,b) \geq \max(a,b)$. These are not only Zadeh's original fuzzy set operators but they also are the operators that would obtain in probability if a and b were maximally dependent (overlapped). Furthermore, $T(a,b) \geq \max(0,a+b-1)$ and $S(a,b) \leq \min(1,a+b)$, which are the operators that are called the 'bounded sum' pair in fuzzy set theory and also correspond in probability to conjunction and disjunction when two events are maximally exclusive.

The major criticisms of fuzzy set theory may be conveniently divided into three groups: (1) Normative, (2) Pragmatic, and (3) Explanatory. The first refers to philosophical or mathematical arguments against the theory, the second concerns its performance in applications, and the third pertains to how well it matches human judgment under uncertainty. I will only briefly summarize these here; a more detailed overview is provided in Smithson (1987:Ch.2).

Many of the normative arguments against fuzzy sets focus on the definitions of intersection, union, negation, and inclusion rather than the fundamental thesis of gradeness itself. The major exception to this, of course, is the probabilistic counter-attack, which we delay until section 4.6. There is a large cognitive psychological literature, on the other hand, which supports the gradient thesis (e.g. Rosch 1978, Rosch and Mervis 1975, Hersh and Caramazza 1976, Kempton 1978, Burgess et al. 1983). Fuzzy negation has been criticized only for not preserving the law of the excluded middle (e.g., Osherson and Smith 1981 and their example of an apple that is also not an apple). Zadeh's (1982) rejoinder naturally points out that fuzzy set theorists never considered the excluded middle to be a valid law in the first instance, although at least one variant of fuzzy set theory (that which uses the 'bounded sum' operators for intersection and union) does preserve a version of this law.

Intersection and union, on the other hand, have been roundly criticized from several sources. One argument points out that the membership function for the intersection of two sets may behave counterintuitively when direct membership judgments for a third set representing that intersection are also available. For instance, joint membership in 'red' and 'yellow' never approaches 1, but intuitively the peak of that membership function should so as to reflect membership in the intermediate category 'orange' (Kay and McDaniel 1975). Likewise, a guppy might be judged a better example of a 'pet fish' than either a 'pet' or a 'fish' (Osherson and Smith 1981). At least five proposals for explaining or dealing with this problem have been put forward, all of them ad hoc.

Another criticism of the fuzzy set intersection and union operators comes from the probabilists' camp, although often by way of the recent literature on expert systems. In the framework of probability theory, P(A&B) = min(P(A),P(B)) only when A and B are maximally dependent events. Probabilists therefore interpret the fuzzy set framework as making strong assumptions about dependency without any apparent basis for so doing. Even when the amount of dependency between A and B is unknown, they argue, assuming independence is a better compromise than assuming maximal dependence.

Perhaps the most persuasive rejoinder from fuzzy set theory is that their framework does not deal with probabilities, but with set membership. Definitions of intersection are therefore a matter of mathematical semantics and should take as their referent the semantics of 'and' and 'or' in logic and natural language, not probability. Boolean logic does not favor either the product or min operators. Natural language, on the other hand, may well entail that if 'big' is 'tall' and 'heavy', then John is no less 'big' than he is 'tall' and 'heavy' (which is what the min operator does). This defense, of course, is all the more persuasive if it can be shown that people actually use language in that way.

The empirical picture, however, is not clear. Several studies testing the fit between fuzzy set operators and human aggregation judgments have been conducted (see Smithson 1987: 62-65), and the main outcome is that people often do not seem to adhere to either a strict 'and' or 'or' when aggregating fuzzy attributes. However, convex combinations of 'and' and 'or' operators fit the data from some studies reasonably well (cf. Zimmermann and Zysno 1980 and Smithson 1984). As to which operators best fit, while Zimmermann and Zysno claim that the product is superior, a reanalysis of their data by Smithson (1984) using least-squares techniques finds no basis for that claim. More recent studies by Zwick, et al. (1988) suggest that while the max operator fits their subjects' data for 'or', a generalized mean best describes their use of 'and'. They employ Smithson's convex combinations of standard fuzzy 'and' and 'or' operators as well, but unlike Smithson's findings regarding the Zimmermann and Zysno data, Zwick et al. obtain results for about 20% of their subjects that are out of the acceptable range for the convexity parameter.

Inclusion also has been attacked, ironically for not being sufficiently fuzzy. Osherson and Smith (1981) raise the issue conceptually, but earlier Kempton (1978) provided empirical data to suggest that Zadeh's definition is inadequate and several reasonable propositions already had been published for fuzzifying inclusion (cf. Dubois and Prade 1980).

Finally, Zadeh's theory of hedges has been criticized on grounds of being mathematically arbitrary and philosophically untenable (Lakoff 1973) and not fitting human judgmental data well (Hersh and Caramazza 1976, Smithson 1987). One of its main drawbacks is that intuitively at least some common modifiers of the kind Zadeh models with his exponential concentrators and dilators are not even monotonic transformations of the original membership functions. However, fuzzy set theorists have not been forthcoming with counterproposals, possibly reflecting the fact that most interest currently is focused on the more fundamental aspects of the framework.

For some time the fuzzy set literature avoided the issue of performance evaluation in applications because applications were rare. That is not the case today, but the evaluative picture remains muddy. Some critics (e.g., Stallings 1977) have claimed that fuzzy set theory is outperformed by probability in specific applications, but an increasing number of applications have been found in which fuzzy sets appear to perform better than a conventional probabilistic approach. In the majority of instances, however, the same problem arises as in the CF debate: The performative data and/or criteria simply do not yield a clear verdict. Bonissone and Decker (1986) demonstrate this problem quite well in their sensitivity analysis comparing various t-norms under different levels of precision in the data. They find that many of the more popular t-norms perform indistinguishably when given realistically granular data. The majority of arguments for and against fuzziness rest primarily on appeals to axiomatics, meaningfulness, ease of expression and use, and semantic or philosophical distinctions. As a result, it is difficult to find any absolute justifications for adopting or abandoning this framework. Most attempts at justification boil down to pragmatic arguments or sheer conventionalism. Nonetheless, fuzzy set theory stands as one of the most successful of the new normative uncertainty paradigms, both in terms of staking a claim to new territory and inspiring widespread use in several fields.

4.3.3 Possibility and Belief

At least two other strong cases have been made for dissenting from the probabilistic hegemony in the past 10 years. One of these, possibility theory, is a spinoff from fuzzy set theory, although its roots go much further back into the history of probability itself. The other is a family of theories that purport to represent quantitative subjective belief in a nonprobabilistic context, which I shall call belief theories.

Zadeh (1978) has constructed a modern theory of possibility based on fuzzy sets. His concept of possibility hails from an earlier notion of a 'fuzzy restriction' (Zadeh 1975b) which is implicit in such aspects of fuzzy set theory as fuzzy numbers. The fuzzy number 'several', for instance, could be said to be a fuzzy restriction on the set of integers which assigns to each integer a grade of membership in the set referred to by the term 'several'. A slightly different way of looking at it is that those grades of membership reflect the degree to which a given integer could *possibly* be meant when the word 'several' is used. This viewpoint is quite similar to Dubois and Prade's (1980) alternative characterization of fuzzy set theory itself, in which a statement like 'John is tall' fuzzily restricts the range of heights John could possibly have.

Like probability, possibility is implicitly a dualistic concept, although its duality has been treated somewhat differently by various authors. Zadeh's semantic distinction is between compatibility, and feasibility or availability. Using the example above, 5'2" is not as compatible with the statement that John is tall as is 6'2", hence 6'2" is the more 'possible' height if John really is tall. Feasibility and availability do not apply to intransitive concepts like tallness, but instead to actions or events (e.g., 'it is almost impossible to find a needle in a haystack,' or 'it is entirely possible for Lisa to have cake').

Hacking (1975), on the other hand, begins with a distinction implied by the English phrases 'possible for' and 'possible that'. The former refers to feasibility or availability, while the second denotes a state of knowledge (as in 'it is possible that George has changed his mind'). Hacking calls the first kind 'de re' and considers it in the same vein as frequentist or aleatory probability. The second kind he calls 'de dicto' and likens it to a subjective possibility or even a state of belief. This distinction, as Hacking points out (1975: Ch.14), can be traced back to the origins of probability itself. As we have already observed in Chapter 3, probability was initially defined in terms of 'equipossibility' and grades of possibility. Not only did 17th and 18th century mathematicians such as Leibniz, Jacques Bernoulli, and Laplace hold such views, but so did modern classicists such as Borel and von Mises. Events were recognized as being possible in degree, and the frequentist von Mises argues in 1928 for grades of possibility using the same intellectual weapons Zadeh does 50 years later.

Zadeh's original framework (1978) emphasizes the de dicto, or subjective version of possibility. However, a straightforward frequentist case can be made for possibility as well. If 90% of the adult residents in suburb A own bicycles but only 45% of those in suburb B do, then legitimate transport by bicycle is arguably twice as possible for the residents of suburb A as for those in B. Furthermore, it seems reasonable to assign a grade of possibility 0.90 in the first instance and 0.45 in the second. Nor do these grades refer to probabilities, since anyone owning a bicycle may or may not choose to ride their bicycle on a particular occasion. The figure 0.90 puts an upper limit on the proportion of people in suburb A who could use their bicycles on any particular occasion, and therefore an upper limit on the probability that a resident randomly sampled from suburb A at 10 AM on a particular day would turn out to use a bike at that time. Zadeh calls this weak relationship between probability and possibility the 'consistency principle'.

As in L.J. Cohen's (1977) inductive probability and Shackle's (1952) surprise theory, there are two kinds of negation in possibility theory: the possibility of not driving a car, and the impossibility of driving a car (termed 'internal' and 'external' negation respectively). Denoting the possibility that x is A by $po_A(x)$, the impossibility of A is clearly $1-po_A(x)$, whereas the possibility of $\sim A$, $po_{\sim A}(x)$, refers to the degree to which x is 'free' not to be A.

Definitions for joint and conditional possibility follow the lines laid down by fuzzy set theory, although again independent arguments may be constructed to motivate them.

Thus, if the proportion of people who have access to option A is po_A and the proportion who have access to B is po_B, then the largest possible value for the proportion who have access to both A and B is $\min(po_A, po_B)$. This definition assumes, of course, that A and B are maximally dependent or overlapping, but if one holds strictly to the definition of possibility as the upper bound on probability then the definition is entirely appropriate.

Possibility theory has provoked considerable discussion in the fuzzy set literature, with the main issues consisting of how best to define conditionality, whether and when to use normalizing on possibility distributions, and the construction of a rational approach to assigning or eliciting grades of possibility in both subjective and empirical contexts. It has also enjoyed some popularity, particularly in computer science and expert systems applications (e.g., as the basis for Baldwin's 1986 support logic programming).

Independently and nearly simultaneously with the advent of possibility theory, Shafer (1976) elaborated a belief theory based on ideas formulated by Dempster (1966,1967) which also had their roots in probabilistic 'antiquity'. This belief theory has a number of elements in common with possibility theory and earlier antecedents (e.g., Shackle's 1952 theory of surprise, which he first elaborated in 1949). Shafer begins from the criticisms of Bayesian probability reviewed in both this and the preceding chapters. A central concern for him (1976: 26-28) is that a fullly fledged Bayesian approach requires an implicit assumption that evidence can always be expressed as a certainty and yet the prior Bayesian probability distribution, itself presumably based on evidence, usually expresses degrees of belief that fall short of certainty.

Traditionally, the two paths out of the difficulty are what distinguish 'logicist' from 'personalist' Bayesians. The problems with the either view are well known. Shafer proposes a framework in which a portion of belief can be committed to a proposition but need not entail any degree of commitment to its negation. He begins by defining a 'frame of discernment' Θ which may be partitioned into subsets that may or may not overlap. A portion of belief committed to one subset of the frame is also committed to any subset containing it. Thus, the portion of belief committed to any subset A of Θ will therefore have a component that is committed precisely to A itself, and another that may be committed to one or more proper subsets of A.

The portion of belief committed exactly to A is called a 'basic probability number', m(A). By definition, m(ϕ) = 0 and the sum of the m(A) over all subsets of the frame equals 1. Because m(A) represents only the belief committed exactly to A and not any of its subsets, it does not necessarily equal the total belief in A. To find the belief Bel(A), we require

$$\text{Bel}(A) = \sum_{B \text{ in } A} m(B). \qquad (4.20)$$

Clearly nonzero portions of belief may be uncommitted to any subset of the frame, so that Bel(Θ) > 0. Under that condition, the sum of beliefs over all subsets will be less than 1. The amount of uncommitted belief represents the extent to which the perceiver feels ignorant, and so, unlike the Bayesian approach, Shaferian belief theory is able to represent ignorance by the 'vacuous belief function' in which Bel(A) = 0 for all A in Θ while Bel(Θ) = 1. At the same time, it includes Bayesian probability as a special case, namely one in which Bel(A) + Bel($\sim$A) = 1 for all A in Θ.

Shafer offers an intuitive invocation of the kind of uncertainty represented by degrees of belief which sum to less than 1. If the elements of Θ are considered as points and each basic probability number a 'probability mass', then m(A) is the probability mass that is confined to A but not confined to any point in A. Hence m(A) is free to move to any point in A even though it is confined to A. The total probability mass that is free to move to any point of A (whether confined to A or not) is called the 'commonality number' for A, and is defined by

$$Q(A) = \sum_{A \text{ in } B} m(B) \qquad (4.21)$$

where the sum is taken over all B in Θ that contain A. As in possibility theory, Shaferian belief theory implies two kinds of nonbelief. Shafer calls Bel($\sim$A) the degree of doubt in A.

Up to this point Dempster-Shafer belief theory does not greatly differ from other belief theories (cf. Levi 1984). However, the central problem for these theories is the same one that faces the Bayesian and all its alternatives: The combination of evidence and updating of beliefs. On this issue

Dempster's rule of combination provides a unique, if controversial, prescription. Consider Bel_1 whose focal subsets are $A_1,...,A_k$, and Bel_2 whose focal subsets are $B_1,...,B_n$. For any pair of focal subsets A_i and B_j whose intersection is nonempty (let us call it C), the measure of belief committed to C due to the intersection of A_i and B_j is $m_1(A_i)m_2(B_j)$. However, C may also be the intersection of some other pair of focal subsets. So the amount of belief committed to C must be defined by

$$m(C) = \sum_{i \& j | C} m_1(A_i) m_2(B_j) \qquad (4.22)$$

where i and j run over all pairs of A_i and B_j whose intersection equals C.

Bel(C) could be constructed from the appropriate basic probability numbers resulting from applications of (4.22) to all supersets of C, except that this scheme commits some belief to the empty set so that the derived degree of belief in ϕ, $m_{12}(\phi) > 0$, whenever the intersection of A_i and B_j is empty. The remedy is to discard all such commitments and to renormalize m(C) by dividing (4.22) by $1-m_{12}(\phi)$. The combined belief function is called the 'orthogonal sum'. The orthogonal sum is commutative and associative, so that the order of combination does not matter.

Clearly if the none of the intersections between any pair of A_i and B_j is nonempty, then the orthogonal sum does not exist. Bel_1 and Bel_2 are said to be 'uncombinable'. The situation is equivalent to two utterly conflicting beliefs, and Shafer points out that the quantity $1-m_{12}(\phi)$ is a measure of the degree to which Bel_1 and Bel_2 conflict. Denoting this quantity by K_{12}, he defines the 'weight of conflict'

$$Con(Bel_1, Bel_2) = -\log(1-K_{12}). \qquad (4.23)$$

Its functional form and range (from 0 to infinity) identify Con as a weight-of-evidence measure in the sense employed by modern Bayesian probabilists, as introduced by Turing and elaborated by Good (1960). Indeed, Shafer goes on to define weights of evidence based on his $m(A_i)$ basic probability numbers which combine additively and include probabilistic weights as a special case.

Unlike any of the uncertainty formalisms we have discussed so far, Shaferian belief theory purports only to be an extension of subjective probability. One might anticipate that this framework should find readier acceptance than the others.

However, the normalization remedy for conflicting beliefs and its implications have caused some consternation over Shaferian belief theory and indeed appear to be its most controversial feature.

Some critics have taken issue with the apparently counterintuitive results that can be produced by normalization. They point out that if K_{12} is large enough it can 'swamp' the resultant combined belief function, thereby producing a large aggregated degree of belief in an hypothesis whose constituent degrees of belief are small. Zadeh (1984) has given the most widely cited example, but I shall reproduce an illustration given in M.S. Cohen (1986: 283-284).

Suppose we have two experts from whom we have obtained Bel_1 and Bel_2 on three focal hypotheses H_1, H_2, and H_3. The first expert's belief assignments are $Bel_1(H_1) = 0.99$, $Bel_1(H_2) = 0.01$, and $Bel_1(H_3) = 0$; but the second expert's beliefs are $Bel_2(H_1) = 0$, $Bel_2(H_2) = 0.01$, and $Bel_2(H_3) = 0.99$. Obviously these are highly conflicting assessments. Applying the normalized version of (4.22) we obtain $Bel_{12}(H_1) = 0$, $Bel_{12}(H_2) = 1$, and $Bel_{12}(H_3) = 0$. All the belief has been assigned to H_2, despite the fact that neither expert reposed more than 0.01 degrees of belief in it.

Table 4.1 Basic Probability Numbers for Two Experts

	m1	m2	m12
H1	0.9801	0	0.656
H2	0.0099	0.0098	0.013
H3	0	0.9702	0.325
Θ	0.01	0.02	0.007
	m1	m2	m12
H1	0.7029	0	0.4243
H2	0.0071	0.007	0.0085
H3	0	0.693	0.4044
Θ	0.29	0.30	0.1751

One of the points made in Shafer's (1984) rejoinder is that the argument is technically correct, but presupposes that both experts are perfectly reliable. Now suppose we discount the first expert's judgment by 0.01 and the second by 0.02, thereby making those amounts uncommitted probability masses. The top half of Table 4.1 shows the basic probability numbers resulting for each expert and the numbers resulting from their combination via Dempster's rule. This small alteration in the distribution of the probability masses yields a very large change in the allocation of combined belief. H_2 is given only 0.13 degrees of belief instead of 1, while H_1 receives more than twice the support that H_3 does.

The bottom half of Table 4.1 shows what happens if we discount the experts even more, and nearly equally. Support for H_1 and H_3 becomes more evenly distributed and support for H_2 remains low. The problem seems to be that in combining belief functions Dempster's rule does not permit any partial reconciliation of conflict on the basis of $Con(Bel_1, Bel_2)$ or any other indicator of the degree of conflict. Cohen (1986) claims that perhaps this is what ought to be allowed, and suggests a qualitative framework for so doing, in which individual assessments are jointly discounted as a function of apparent conflict. This suggestion amounts to disbelieving the belief allocations in the face of conflicting evidence, or, in Shaferian terms, rendering the individual belief functions more vacuous and less committed.

Levi (1984) makes a somewhat more general criticism of Shafer's system which focuses on the requirement, in the face of conflicting assessments, that we assign positive belief to whatever overlap or agreement is found between the assessors, no matter how trivial or small. Levi claims that this is not a rational requirement for pooling evidence, although he provides no alternative that fits into Shafer's framework. Instead his system involves assigning 'credal probabilities' and either accepting or rejecting various hypotheses through the use of criterion cutoff values for those probabilities.

Subadditive belief theories, while enabling representations of ignorance that are not within the scope of probability theory, also allow in 'untamed' varieties of ignorance such as conflict or dissonance. The construction of subadditive belief theories and superadditive possibility theories also invoke dual concepts of internal and external negation, which in turn suggest upper and lower bounds on belief or probability. Both of these issues have motivated a recent shift in emphasis from first-order (single number) uncertainty to second-order, intervalic formalisms. The idea of second-order uncertainties is not new, but it has seldom been treated seriously in the past. The shift to second-order uncertainty and ignorance may represent a significant development in normative approaches to these matters.

Accordingly, our survey of emerging paradigms moves up one level.

4.4 The Leap to Second-Order Relations
4.4.1 Probability of Probabilities, Saith the Preacher

The concept of higher-order probabilities has been raised a number of times in the probabilistic literature, mainly by the subjectivists (e.g., C.A.B. Smith 1961 and Good 1962, but see also Walley and Fine 1982 for a frequentist account) as a means for representing nonspecificity or imprecision about probabilities. One of the most succint cases for this is made in Good's 1962 article on types of probability, in which he points out that the very fact that a point-estimate of a probability has a finite number of digits after the decimal point implies that the estimate actually lies inside an interval. In his view, single-valued (that is, first-order) probabilities are a mere convenience rather than the axiomatically justified entities that either relative frequentists or subjectivists claim them to be.

The concept of intervalic bounds on probabilities leads quite naturally to an intervalic calculus for such probabilities. Suppose we decide that $d_1 \leq P(A_1) \leq u_1$ and $d_2 \leq P(A_2) \leq u_2$, where A_1 and A_2 are independent events. Then the appropriate interval for the conjunction of these two events is $d_1 d_2 \leq P(A_1$ and $A_2) \leq u_1 u_2$. Now suppose we do not know whether A_1 and A_2 are independent or not. The intervalic subjectivists find no difficulty here; they simply find the lowest value for the conjunction of two events (which is when they are exclusive) and the highest value (which is when they are dependent), and then incorporate the d_i and u_i accordingly to obtain $\max(0, d_1+d_2-1) \leq P(A_1$ and $A_2) \leq \min(u_1, u_2)$. Implicit in this exercise is the creation of intervals around the interval endpoints for $P(A_1$ and $A_2)$. That is, the lower bound lies in the interval $[\max(0, d_1+d_2-1), \min(d_1, d_2)]$, while the upper bound lies in the interval $[\max(0, u_1+u_2-1), \min(u_1, u_2)]$. A similar exercise for the disjunction of the two events yields $\max(d_1, d_2) \leq P(A_1$ or $A_2) \leq \min(1, u_1+u_2)$, along with corresponding intervals for the lower and upper bounds on this disjunction.

A simple generalization of intervals is to permit judgments about which values in the interval are more likely to be the 'true' probability than others, which induces a second-order probability distribution on the first-order probabilities. In short, the metaprobabilist claims to represent incompletely specified probabilistic uncertainty via uncertainty about uncertainty, which in turn is handled by second-order probabilities.

An immediate objection to upper and lower bounds for probabilities is that while it may seem to solve the problem of arbitrary precision for point-estimates, it raises an identical precision problem for the bounds themselves. This objection is raised even more strongly where probabilities of probabilities are used. The argument is that the metaprobabilist's rationale for second-order probabilities also entails having probabilities of the metaprobabilities, and so on. We thereby fall into an infinite regress of metaprobabilities. Subjectivists have not generally provided strong rejoinders to this criticism. Either they claim that the higher-order imprecision has little practical importance, or they argue that human judges do not usually resort to levels higher than two.

A second objection that numerous authors have pointed out is that if we assume a uniform p.d.f. over an interval-valued variable to express our ignorance of that p.d.f., then a nonlinear transformation of that variable yields a nonuniform p.d.f. over the transformed variable's interval. Thus, computing an intervalic mean of intervalic data is not troublesome in this regard, but computing an intervalic standard deviation is.

A more serious objection is that second-order uncertainty should not be taken to be probabilistic at all. A simple interval around a probability estimate could be taken to be a mere possibility distribution, for instance, or it might even reflect conflicting evidence from different sources about where the true value lies. This is a phenomenological argument, but it is noteworthy that the usual probabilistic phenomenological justifications are not compelling here. That is, there seems no compelling reason to conceive of second-order uncertainty in terms of betting on bets and avoiding a 'Dutch book'.

I shall close this section with an example given by Gardenfors and Sahlin (1982) which they used to demonstrate the limitations of first-order Bayesianism, but which also points out the importance for decision-makers of second-order information. We are asked to consider Miss Julie who has been invited to bet on the outcomes of three tennis matches. She knows that in Match A the players are very close in ability, that they have an even win-lose record against each other in previous meetings, and so forth, so she has good reasons to suppose the probability of one or the other player winning Match A is 0.5. She is not informed at all about the players for

Match B, however, and knowing nothing about them cannot predict who will win. In Match C she has overheard someone say that one of the players is very good but the other a rank amateur, so the contest is lopsided--- but she doesn't know which player is the better one.

A first-order probabilistic approach to the decision problem involved in these bets would make the same prescription for all of them, namely to represent each player's chance of winning with a probability of 0.5. Indeed, the contrast between Matches A and B is familiar to us from the aforementioned criticisms of the inability of first-order probabilities to represent ignorance. But the second-order information indicates three quite different betting situations. Match A connotes a tight second-order interval around 0.5, whereas Match B indicates an interval spanning [0,1]. Match C seems to correspond to two nonoverlapping (and therefore conflicting) intervals, one including 0 and the other including 1. The value of 0.5 would not be included in either of those intervals.

In Match A we have what Gardenfors and Sahlin would term an epistemically reliable probability of 0.5, while in Match B the probabilistic information is so unspecific as to be unreliable. Insofar as reliability is a utility, they claim any reasonable decision theory should take it into account when maximizing expected utility. Therefore, Miss Julie should prefer to bet on Match A rather than Matches B or C because she is more reliably informed about Match A. This, then, is their resolution of Ellsberg's paradox.

Although they do not mention it, Match C is not only problematic in terms of reliability, it also involves conflicting evidence of the kind that causes problems for the Dempster rule of combination. Though Gardenfors and Sahlin acknowledge that Miss Julie should be very suspicious of someone offering her a bet on Match C with that sort of information (even if she is allowed to decide which player to back), they do not indicate a normative ordering for Matches B and C. In general, metaprobability frameworks seldom acknowledge conflict and none provide a persuasive method for dealing with it.

4.4.2 Rough Sets, Shaferian Belief, and Possibility Theory

A rather large number of nonprobabilistic second-order uncertainty formalisms have been proposed during the last 15 years. They share several major features, and this introductory survey will briefly describe those formalisms before discussing the differences among them.

Perhaps the simplest, but in some respects most general, second-order system is one for crisp sets articulated by Pawlak (1982,1985). His 'rough set' theory is based on the assumption that we are faced with a collection of objects, X, which cannot be precisely characterized in terms of a set of attributes, while a lower and upper approximation of the collection (X_- and X^+ respectively) can be so characterized. He demonstrates that this system has the following properties:

(1) $X^+ = \sim X_-$,
(2) $(X \cup Y)^+ = X^+ \cup Y^+$,
(3) $(X \cap Y)_- = X_- \cap Y_-$,
(4) $(X \cup Y)_-$ includes $X_- \cup Y_-$, and
(5) $(X \cap Y)^+$ is included by $X^+ \cap Y^+$.

Pawlak also declares that these properties are not generally reproducible in ordinary (first order) fuzzy set theory. Dubois and Prade (1983,1987a) propose a rather similar scheme based on possibility theory which they call 'twofold' set theory. But their concern is to represent the situation where the *elements* are incompletely specified rather than the *set*.

Fuzzy set theory itself, as mentioned in the previous section, has been extended to a second-order account of metafuzziness, in which degrees of membership are themselves fuzzy numbers. Fuzzy arithmetic has been developed out of this concept, and it has enjoyed some applications.

However, the formalisms that are inherently second-order are the subadditive and superadditive alternatives to probability, including Dempster-Shafer belief and possibility theories. Their intervalic nature arises from the noncomplementarity of their negation rules. For belief theory, $1-Bel(A) = Pla(\sim A)$, which Shafer calls the 'plausibility' of $\sim A$. Clearly because belief functions are subadditive, $Bel(A) \leq Pla(A)$, with equality occurring only when complementary negation holds (the special case corresponding to ordinary probability). In possibility theory, $1-po_A = ne_{\sim A}$, which is called the 'necessity' of $\sim A$. Because possibility is superadditive, $po_A \geq$

ne_A. Given that possibility theory and belief theory share several features and differ only in their rules of combination, it might seem that they are formally related. Indeed they are, in that the combination rule in possibility theory can be shown to be a special case of Dempster's rule (Shafer 1985, Dubois and Prade 1986). Specifically, the two frameworks are identical when the focal subsets in the Shaferian perspective are nested (overlapped).

One of the key issues in the debates inspired by competing second-order perspectives is the aggregation of conflicting evidence. We have already seen the criticisms leveled at Dempster's rule for its renormalization procedure. Dubois and Prade (1986) also take issue with the assumption in Shaferian theory that $m(\phi) = 0$, claiming that some situations might be better represented by allowing it a positive value.

One approach that has been implemented in a fuzzy version of PROLOG (and also, more recently, the "fuzzy relational inference language" FRIL) utilizes the information about meta-uncertainty to influence the combined support for a proposition from two (or more) sources of evidence in a way that eschews normalization. Baldwin's (1986) 'support logic' begins with a possibilistic framework but departs from it in his construction of a combination rule. In his system, ne_A denotes the 'necessary' support given to A while po_A is the greatest possible support for A, which is akin to Shafer's notions of belief and plausibility. Baldwin labels the difference between po_A and ne_A the 'uncertainty' of A (that is, $po_A - ne_A = un_A$). Thus, for Baldwin any proposition has support favoring it (ne_A), support for its negation ($1-po_A$), and uncommitted or uncertain support (un_A), all of which sum to 1.

Now suppose we have two sources of information yielding ne_1, $1-po_1$, and un_1; and ne_2, $1-po_2$, and un_2. Baldwin's framework defines the combined support for the proposition by $ne_{12} = A(ne_1, ne_2) + A(ne_1, un_2) + A(ne_2, un_1)$ where A is an aggregation operator whose form varies with the dependency assumptions regarding the two sources. Assuming independence, for instance, entails that $A(a,b) = ab$. Combined support for the negation of the proposition is defined by $ne_{\sim 12} = A(1-po_1, 1-po_2) + A(1-po_1, un_2) + A(1-po_2, un_1)$.

Problems arise with the allocation of simultaneous support to the proposition and its negation, which takes the value $A(ne_2, 1-po_1) + A(ne_1, 1-po_2)$. Baldwin's initial suggestion is a Dempsterian renormalization, but his second proposal is to allocate this quantity to the 'uncertain' box and combine it with the joint support for uncertainty. In that event, we have $un_{12} = A(un_1, un_2) + A(ne_2, 1-po_1) + A(ne_1, 1-po_2)$.

Suppose we apply Baldwin's scheme to the examples used in the previous section to test Dempster's rule? Consider three hypotheses assessed by two sources of judgment. Let $ne_{(i)j}$ denote the ith expert's belief in the jth hypothesis. According to the first source $ne_{(1)1} = 0.99$, $ne_{(1)2} = 0.01$, and $ne_{(1)3} = 0$; but the second source yields $ne_{(2)1} = 0$, $ne_{(2)2} = 0.01$, and $ne_{(2)3} = 0.99$. Although highly conflicting, neither of the two judges allows any uncommitted support. Recall that Dempster's rule resulted in a belief of 1 being reposed onto the second hypothesis, despite the fact that each judge accorded it only 0.01 support. Baldwin's rule produces combined support $ne_{(12)1} = ne_{(12)3} = 0$ for the first and third hypotheses and $0 \leq ne_{(12)2} = A(0.01, 0.01) \leq 0.01$ for the second. The remainder of the support becomes uncommitted for the first and third hypotheses; we have $un_{(12)1} = un_{(12)3} \geq 0.99$. For the second, $0 \leq un_{(12)2} = 2A(0.01, 0.99) \leq 0.02$ and the support for its negation lies between 0.98 and 1.

These results seem more intuitively reasonable than those obtained using the Dempster rule. The contradictory judgments on the first and third hypotheses result in most of the support being thrown into limbo, while the combined support for the second hypothesis remains about as strong as for either judge with correspondingly strong support allocated to the negation of the second hypothesis. Applying Baldwin's rule to the modifications of this example given in Table 4.1 result in seemingly appropriate shifts in support values. The tiny discounting of the judges' beliefs provided in the upper half of the table yields, for example, $ne_{(12)1} = 0$, $0.9601 \leq un_{(12)1} \leq 0.99$, and $0.01 \leq ne_{(12)\sim 1} \leq 0.0399$. The most obvious drawback to this combination rule is similar to the original CF proposal; a low support value from a single source can swamp several high values from other sources.

Baldwin's identification of conflicting with uncommitted support raises some difficult issues for normative paradigms. However, his use of a measure of second-order uncertainty to direct a rule of combination is innovative and speaks to Cohen's (1986) criticism of the Shaferian perspective. If we differ with Baldwin and decide that conflict and uncommitted support are distinct, then we require separate measures of both. And what about Shafer's introduced notion of 'discounting' judges' beliefs on the basis of their supposed unreliability? These intervalic schemes yield several new varieties of ignorance, most of them second-order. As the reader should expect by now, there have been a multitude of proposals for second-order measures (and classifications of) uncertainty. Let us turn our attention to these.

4.4.3 Measures of Uncertainty

The classic measure of second-order uncertainty, of course, is Shannon's (1948) 'entropy' function:

$$H = -\sum_{i=1}^{n} p_i \log p_i, \qquad (4.24)$$

where the p_i are probabilities. Most of the new developments in measures of second-order uncertainty begin from information theory, with a number of them claiming Shannon's function as a special case. An earlier measure of information, also called an entropy function, was defined by Hartley in 1928:

$$I(N) = K \log_b N, \qquad (4.25)$$

where N is the number of alternatives and K and b (the base) are arbitrary. Renyi (1970) proved that a normalized version of I(N) in which K = 1 and b = 2 is the only function that equates a choice between two alternatives with one bit of information (that is, I(2) = 1) and also possesses additivity (i.e., I(NXM) = I(N)+I(M)) and monotonicity (i.e., I(N) $\leq$ I(N+1)).

The fact that the concept of entropy has been linked with both functions (and various other measures as well) has caused some confusion and controversy. Boltzmanian and Shannonian entropy have quite different meanings and one is not a limiting case of the other (see, e.g., Guiasu 1977). Furthermore, the Hartley and Shannon functions clearly measure different kinds of uncertainty. Hartley's measure pertains to what many would term ambiguity or latitude of choice. Shannon's measure has been given several interpretations, ranging from his original 'informativeness' concept to more recent reinterpretation as a measure of relative variation (Smithson 1982; Kim 1984 calls it a measure of "qualitative" variation). Of the two, Shannon's is known much more widely and has been used as a 'loss' and 'utility' function for programming problems and other applications (see, e.g., Jaynes 1979, for a review of the maximum entropy principle).

Recent developments in the measurement of second-order uncertainty have occurred primarily in fuzzy sets, possibility theory, and Shaferian belief theory. Because most of these innovations are quite new, both their conceptual bases and terminology tend to be somewhat confused and subject to debate. Accordingly, this review will be more of a tentative survey than a true synthesis or overview.

The problem of measuring fuzziness or vagueness was first systematically addressed in the mid-70's and has received sporadic attention since then. Although most investigators agree that a measure of fuzziness should equal 0 for crisp (nonfuzzy) sets and attain its maximum when a set is maximally fuzzy, they differ over the meaning of 'maximal fuzziness' and what other properties such measures should have. The most popular class of measures I have called 'centered' (Smithson 1987: 111) because they define maximal fuzziness as the case when all membership values equal 1/2. De Luca and Termini (1972) provided the best known among these measures, which they called a 'fuzzy entropy':

$$f(A) = -\sum_{i=1}^{n}[m_{Ai}\log_2 m_{Ai} + (1-m_{Ai})\log_2(1-m_{Ai})], \qquad (4.26)$$

where m_{Ai} is the membership of the ith element in fuzzy set A. Kaufmann's (1975) distance-based measure is another example, which concerns the Hamming distance between a fuzzy set A and its nearest crisp neighbor. Loo (1977) provides a class of centered measures and discusses their properties.

A related approach is given by Yager (1979) generalized by Higashi and Klir (1982), which takes as its point of departure the comparison of a fuzzy set with its complement. The concept of fuzziness in this approach is tied to the lack of distinction between a fuzzy set and its complement. Hence the specific form this kind of measure takes depends on how distinctiveness is measured and how fuzzy negation is defined. Higashi and Klir use a measure of strict similarity to measure distinctiveness, but others could be employed (cf. Dubois and Prade 1980 for a review of such measures).

Loo (1977) notes that his class of measures is not 'expansible' with respect to nonsense elements whose membership in a particular set is 0. That is, the value of the centered fuzziness measures may be made arbitrarily small simply by adding more 0-valued elements. The same is true for measures based on the Yager approach. Moreover, they vary with sample size.

Smithson (1982, 1987) provides a class of uncentered measures based on relative variation coefficients (e.g., the Gini coefficient, coefficient of relative variation, and Theil's (1967) information theoretic coefficient of variation). In his approach maximum fuzziness is defined as the condition where all membership values are equal (they need not equal 1/2). His class of measures includes several others proposed in the fuzzy clustering literature (Dunn 1976, Roubens 1978, Backer 1978, and Bezdek 1974). These measures do not vary with sample size, and are bounded from below when 0-valued elements are introduced.

Ambiguity has been accorded less attention than vagueness and fuzziness, perhaps because of the initial confusion of terminology in the fuzzy sets literature which conflated ambiguity with other kinds of uncertainty. Recently, however, ambiguity has surfaced as an independent concept in possibility theory and Shaferian belief theory. Possibilistic measures of ambiguity have been defined in at least two ways. The first is due to Yager (1982) and may be written in the following form due to Klir (1987):

$$A = 1 - \sum_{i=1}^{n} \frac{po_i - po_{i+1}}{i}, \quad (4.27)$$

where $po_i \geq po_{i+1}$ for all i, and by convention $po_{n+1} = 0$. The second is a generalization of Hartley's measure which permits degrees of possibility for the alternatives (Higashi and Klir 1983):

$$IP(n) = \sum_{i=1}^{n} (po_i - po_{i+1}) \log_2 i \quad (4.28)$$

This measure shares the properties enjoyed by Hartley's I(N), and Klir (1987) claims it to be better justified than Yager's proposal. Dubois and Prade (1985b) generalize (4.28) to the setting of Dempster-Shafer belief, and also point out that because of the correspondence between fuzzy sets and possibility theory, IP(n) could be used to measure the 'ambiguity' of a fuzzy set with respect to grades of membership defined by the po_i.

Several other kinds of ignorance have been investigated in the setting of the newer second-order uncertainty schemes, but their epistemological status remains unclear. Measures variously termed 'dissonance', 'entropy' and 'confusion' have been proposed which are extensions of the Shannon information function in (4.24) in the setting of Dempster-Shafer belief theory. Yager (1983) defined one such measure based on Shafer's coefficient of conflict between two beliefs, $Con(Bel_1, Bel_2)$:

$$E(m) = \sum_{A \text{ in } X} m(A) Con(Bel, Bel_A), \qquad (4.29)$$

where Bel denotes the belief function associated with the basic probability numbers $m(A)$ and Bel_A denotes a belief function associated with basic probability numbers for which $m(A) = 1$ and $m(B) = 0$ for all $B \neq A$. $E(m)$ therefore represents a weighted sum of the conflicts between the $m(A)$ assignments and total belief in A, hence its description as a measure of dissonance. $E(m)$ reduces to (4.24) when the $m(A)$ are probabilities. It shares the properties of Shannon's information function. Furthermore, given a plausibility measure Pl associated with the $m(A)$, the $Con(Bel, Bel_A)$ term may be replaced with $\log_2 Pl(A)$.

It is natural to consider an analogous function with $Pl(A)$ replaced by $Bel(A)$, and such a function $C(m)$ has been introduced by Hohle (1982) and discussed by Dubois and Prade (1987b). They call it a measure of 'confusion', given the fact that $C(m)$ attains its maximum value when there are as many focal elements as possible such that none of them is included in any other and the weights of evidence are uniformly distributed among them. Dubois and Prade suggest several applications for $C(m)$, $E(m)$, and related measures, including utilizing a 'minimum specificity' principle in ways analogous to the application of the 'maximum entropy' principle.

Finally, there are some attempts to measure the extent to which bounded parameters are 'free' to vary inside their constraints. Baldwin's measure of 'uncertainty', $po_i - ne_i$, is an obvious example. Smithson (1988) generalizes this to n-option systems in which each option is assigned its own degrees of possibility and necessity, calling the resulting coefficient a measure of 'relative freedom'. This measure is proportional to the volume of the convex polytope resulting from cuts made to a unit simplex by the po_i and ne_i (see also Cohen and Hickey 1979). Smithson's treatment covers two conditions: Where the

Bel_i that lie between po_i and ne_i are required to sum to 1 (additivity), and where they are subadditive.

As mentioned before, the primary purpose of this section has been to give the reader a tentative survey of an emergent literature. Perhaps nowhere else in the welter of normative discussions about uncertainty is the recent proliferation of types and approaches more evident than in the debates over measures of second-order uncertainty. In terms of the typology presented in the first chapter, the normative formal paradigms have moved not only into vagueness, nonspecificity, and ambiguity, but also unreliability, undecidability, distortion, and conflict of opinion. It is understandable that the most recent articles in this area focus mainly on establishing normative criteria and justifications that uniquely prefer one measure over others in each category (usually the preference goes to a Shannon-type measure). Equally understandably, there is also a strong impulse towards unification in this literature, and indeed throughout the recent corpus of normative writings on nonprobabilistic uncertainties. Both impulses share their origins in a desire to construct an order that makes sense out of a rather bewildering pantheon of youthful, and contentious, gods and demons.

4.5 Imposing Order on the New Chaos: Banishers, Reductionists, and Synthesists

The ferment among mathematicians, philosophers, and applied scientists over alternatives to probability has provoked several attempts to (re)establish order amidst all these new uncertainties. This section briefly reviews those responses in preparation for the interpretive discussions toward the end of this book. As in the history of mathematics and several other disciplines, we find reactionaries who wish to banish the new uncertainty formalisms as untenable and/or improper; conservative reductionists who claim that the new frameworks really may be assimilated into older ones; and progressive synthesists who already wish to construct a new system that unifies the old with the new.

4.5.1 The Probabilists' Rejoinders

The new reactionaries are mainly probabilists (but see Scheffler 1979 for a searching philosophical critique of the most popular justifications via vagueness for abandoning or revising dualistic logic), and theirs is perhaps the most understandable response. In their terms, the new paradigms have nothing to offer and also are unsound. Though the probabilists' critiques have been scattered throughout this chapter, it is worth assembling them here.

The foremost argument for banishment is an axiomatic one: Any self-consistent system for quantifying degree of belief with a single number must be probabilistic. Cox's 1946 version of this argument is but one expression; it has been reiterated in terms of probability as the measure of a non-measurable set (Good 1962) and probability as the inevitable result of certain scoring rules (Lindley 1982). These are powerful normative arguments if one accepts the axioms that give rise to them. Since at least some of the alternativists begin with a departure from those axioms, some probabilists have replied to these apostasies.

One of the most frequently invoked motivations for formalisms such as possibility and Shaferian belief theory is that one number is insufficient to represent subjective belief, particularly in the face of what some writers call 'ignorance' (to distinguish it from mere probabilistic uncertainty). Probabilists reply that we need not invent a new theory to handle uncertainty about probabilities. Instead we may use meta-probabilities (cf. Fung and Chong 1986 for a comparison between second-order probabilities and the Dempster-Shafer approach). Even such apparently nonprobabilistic concepts as possibility may be so represented. A possibility distribution, for instance, is merely the subjective probabilities associated with possibilistic statements (Cheeseman 1985, 1986). Thus, the probabilist would replace the expression po('several' = 6) = 0.9 with P('several' is possibly 6) = 0.9. The so-called problem of representing ignorance, the probabilists claim, poses no difficulty for probability theory. One merely induces a second-order probability distribution over the first-order subjective probabilities.

A second common departure from the probabilistic axioms is the complementarity of negation. As we have seen, some probabilists claim that this has arisen from a confusion between measures of absolute belief with measures of belief updates. This is the nub of Horvitz and Heckerman's (1986) critique of MYCIN's certainty factors and Heckerman's (1986) and Grosof's (1986) probabilistic reformulations. Again, the probabilist's stand is backed up by axiomatics, wherein a modified version of Popper's desiderata for a measure of confirmation is satisfied by simple transformations of Bayes' Rule (Good 1960). Though no one seems to have reinterpreted L.J. Cohen's (1977) inductive probability from this standpoint, it might be possible to argue there as well that his version of judicial probability really is a qualitative description of belief updates using a Bayesian likelihood measure.

These axiomatics are all very well, but as the history of mathematics and its applications repeatedly demonstrates, formal reactionary protests often are no match for nonrigorous but practical progressivism. After all, the primary motivations for at least two major nonprobabilistic alternatives (certainty factors and fuzzy set theory) have been applications. If these novel uncertainty frameworks deliver the goods in applications, then they will continue to be used regardless of probabilistic naysaying unless the probabilists can show similar capabilities for their framework.

A less defensive line of argumentation, then, takes the stand that probability can do anything any of the alternative systems can do, and performs better because it is consistent and conservative. An early detractor of fuzzy set theory from a Bayesian standpoint (Stallings 1977) compared the performances of fuzzy set and Bayesian analysis in specific applications and claimed superior performance by the Bayesian approach. More recently Cheeseman (1986), Hisdal (1986), and Wise and Henrion (1986) have made similar illustrative comparisons between these two frameworks, while the Dempster-Shafer approach has been criticized for poor performance by some nonprobabilists (e.g., Zadeh 1984) and in comparison with metaprobability (Fung and Chong 1986).

The probabilistic counterattack indicates strong resistance to the ambitious territorial claims made by proponents of the new uncertainty formalisms. At the same time there is no doubt of the appeal that the alternativists have had in several fields. Fuzzy set theory alone currently generates several hundred publications a year, and the Dempster-Shafer framework has become something of a hot topic among researchers in artificial intelligence. Most of the probabilists' answers to their rivals involve proposals that those rivals have already rejected. It is clear that these matters are far from settled at present, even though there are some signs of moderation among several contributors to these debates (e.g., the probabilist Natvig 1983 and the pro-fuzzy Blockley et al. 1983).

4.5.2 Reductionists: Logic, Topoi, and Probability Again

Reductionistic responses to the new uncertainties appear to stem from two motivational camps. One of these is a response from the proponents of novel frameworks to accusations of philosophical or mathematical (hence normative) unsoundness, and the primary agenda item is the provision of better foundations for a given framework by demonstrating that it is a (heretofore unrecognized) special case of some well-respected paradigm. The second camp is a variant on the banishers, and their goal is a demonstration that the so-called new framework is nothing other than old wine in a new bottle. This demonstration may consist either in showing that the framework is well-justified only if it is made isomorphic with some traditional paradigm, or that it is a special case of that paradigm and provides no new features or insights. The differences between these two kinds of reductionism are perhaps best apprehended through illustration.

Since one of the most common criticisms of fuzzy set theory is that it lacks proper philosophical and formal justification, a number of fuzzy set theorists have attempted to provide those justifications. The main problem with a mathematical justification of fuzzy set theory is that some of the traditional justificatory paradigms (e.g., classical dualistic logic) seem inappropriate to most proponents of fuzzy sets (nevertheless, even they are not above embedding fuzzy set theory in a meta-level context that possesses ultimate certitudes of the old kind, as we will see in the next section). However, as indicated in Chapter 2's discussion of the meta-mathematical pursuit of certainty, the new varieties of meta-mathematics are relativized and suburbanized, and at least one kind has appealed to fuzzy set apologists. This is the theory of elementary topoi (or toposes).

Topoi theory unites classical logic and algebraic geometry, as well as providing a relativized foundation for mathematics that generalizes logic in the sense required by Intuitionists. That is, only a weakened version of the Law of the Excluded Middle holds (i.e., $\sim X$ and $X = 0$, although $\sim\sim X \neq X$). Furthermore, topoi provide an adequate axiomatization for ordinary set theory. But perhaps most importantly, topoi theory appears to provide an Intuitionist (that is, a constructivist) account of higher-order logic (Tierney 1972). Eytan (1981) first suggested that fuzzy sets form a topos, with Pitts (1982), Carrega (1983) and Ponasse (1983) dissenting. The issue appears to have been resolved among mathematicians (cf. Stout 1984 and Barr 1986), if not between mathematicians and practitioners (see the remarkable exchange between Johnstone 1988 and Graham 1988).

Without going into the nature of the controversy itself, let us examine its motivational basis. Graham's (1987) exegesis on the controversy best expresses the hopes reposed in topoi theory when he says "If it were known that the 'fuzzy world' were a topos then we would not only know that the crisp quantifiers lived in that world, but would have an explicit, essentially algebraic construction of them, and a guarantee of good behaviour" (pg.19), and later "Thus the need arises to search for a profound level of justification for fuzzy logic which will, at a minimum, allow fuzzy logic to enjoy the apotheosis of classical logic; i.e., its situation within the foundations of mathematics" (pg.31). Here we have the first kind of reductionist agenda in an ambitiously explicit form, whereby the annexation of crisp sets and dualistic logic as special cases of fuzzy sets and fuzzy logic is to be ad hoc or intuitive no longer, but instead christened by immersion in topoi theory.

Among the best examples of attempts to demonstrate that the new uncertainties are merely old ones in disguise is the probabilistic characterizations of fuzzy set theory by researchers such as Giles (1976, 1979, 1982) and Hisdal (1986). These have a similar flavor to the probabilistic counterattack on Certainty Factors. Like the banishers, these investigators begin by pointing out difficulties in the definitions of fuzzy set definitions, concepts, and operations. Giles, for instance, points out that it is insufficient to equate possibility with 'degree of ease' or 'compatibility' and then provide a few examples illustrating its use. He calls for an operational definition of degrees of possibility analogous to that of probability which would strictly guide the translation of a judge's opinions into numerical terms. His own proposals amount to identifying fuzzy concepts with upper and lower probabilities.

The second step in this reductionist exercise is to show that a recasting of standard probability theory can capture the essential properties of fuzzy set theory in a way that eliminates the difficulties. Hisdal's list of problems is one of the largest in the literature, encompassing some 17 main disputes with the original definitions and properties of fuzzy set theory. Her solution is an elaborate model based on dualistic logic and probability theory. For example, she claims the concept of membership grades is confused in fuzzy set theory, since it may be taken to mean either the possibility that 6'2" is in the set of 'tall' heights for people, or that a 'tall' person could have a height of 6'2". In our earlier notation, Hisdal asks us to compare the conditional probabilities $P(c_S(x) = 6'2''|'tall')$ and $P('tall'|c_S(x) = 6'2'')$, where $c_S(x)$ is the height scale, c_S, applied to person x. She identifies the concept of membership and also possibility with the latter and probability with the former. That is, $m_S(x) = P('tall'|c_S(x) = 6'2'')$. She then proceeds to argue that this approach reproduces the relationships claimed by Zadeh between possibility and probability, but without the operational and conceptual problems posed by possibility theory.

Similar battles have been joined over Shaferian belief theory, despite its close relationship with Bayesian probabilism. On the one hand we have claims that the Dempster-Shafer framework subsumes Bayesian probability as a special case (Shafer 1976, of course, initiated these claims at the first-order level, but they have been echoed and extended in recent years by, e.g., Grosof 1986). But there are numerous counter-claims. Some authors such as Lemmer (1986) and Fung and Chong (1986) argue that Bayesian meta-probabilities work better than the Dempster-Shafer approach on the same cases. Others (Levi 1980, Kyburg 1985, 1987, and Loui 1986) point out that there are several second-order generalizations of probability available which subsume the Dempster-Shafer theory by virtue of having weaker requirements for coherency.

One of the main issues underlying the reductionist debates is the meta-criteria for representing and managing uncertainty or ignorance. From Graham's hope for a guarantee of "good behavior" to Levi's discussions of "acceptance" standards, we find a common theme among reductionists concerning the one true or best set of justifications for an orientation towards ignorance. This theme evokes Good's notions of "type II" rationality and "quasi-utilities" as aids to deciding whether to deal with first or second-order probabilities. One class of discussants in the ferment over the new uncertainty frameworks could be called 'pluralists' or 'synthesists' insofar as they propose meta-frameworks which permit users to select a given uncertainty formalism depending on what attitude towards ignorance, error, or other (dis)utilities they wish to assume. I will briefly survey their suggestions before attempting a summary and evaluation of the recent developments in normative uncertainty formalisms.

4.5.3 Pluralists and Synthesists

Pluralists may be thought of as those who take the position that ignorance is multiple and that we therefore require distinct methodologies for managing its various manifestations. Even in a single framework such as probability or fuzzy set theory, several competing definitions and characterizations have sprung up. The 'suburban compromise' is to divide the territory into exclusive domains, over each of which one or another approach presides. The traditional distinctions drawn between 'objective and 'subjective' probability, and Klir's (1987) proposal to distinguish among varieties of fuzzy set theory by separating 'vagueness' from 'ambiguity', are examples of this kind of subdivision.

Once pluralism is established, synthesis of the various separate paradigms is relatively straightforward and consists of two types. The first arises when the domains are independent (or distinct) and therefore not mutually exclusive (or competitive). Then one may apply more than one method for dealing with ignorance at the same time, given an appropriate context for joint application. The main problem faced by users is how to represent the various kinds of ignorance appropriately and how best to manipulate the combined information. The second kind of synthesis arises when the domains are mutually exclusive or, in some cases, when a choice among distinguishable domains must be forced, so that the chief problem is deciding which framework or method applies. In this case the criteria for deciding this may amount to more than mere recognition and so some investigators have cast this situation into a decision theoretic mold. Unfortunately, the question of whether various formalisms may coexist independently or not often remains unanswered.

Perhaps the most widely known pluralism in the literature on new uncertainty paradigms involves probability and fuzziness. The point of departure for this position is that fuzziness and probability are distinct kinds of uncertainty, the former pertaining (usually) to vagueness and the latter concerning likelihood or propensity. This was Zadeh's (1968) original position, and it has been echoed and elaborated many times since (e.g., Gaines 1975, Zadeh 1980, Ralescu and Ralescu

1984, and Dutta 1985). Fuzzy set theorists adopting this attitude were quick to recognize the scope for aggregating these frameworks. Zadeh's initial paper on probability measures of fuzzy events and his later publications on linguistic variables (1975) generated a flurry of writing on combinations of the two frameworks. One stream continued Zadeh's suggestions concerning the probability of a fuzzy *event*, usually attempting to formulate axiomatic foundations for the concept (e.g., Smets 1982, Yager 1979b and 1984), and led to accounts of fuzzy random *variables*, namely random variables whose values are fuzzy numbers rather than ordinary real numbers (cf. Kwaakernak 1978). These schemes have since been elaborated into a theory of 'fuzzy statistics' (cf. Kruse and Meyer (1987). A sizeable literature on fuzzy decision theory also quickly evolved from this particular combination because the concepts of fuzzy utilities and fuzzy decision states could be readily imported into the standard decision theoretic framework (see Kickert 1978 for an early review of this development).

Some fuzzy set theorists argue that the vagueness or imprecision associated with subjective probabilities requires that they be treated either as linguistic variables (Zadeh 1975 and Nguyen 1977) or at least as fuzzy numbers. Others (e.g., Yager 1979b) have claimed that even the measurement of the probability of a fuzzy event should not consist of an ordinary real number, but instead should itself be a fuzzy set. But in Yager's model, statements of the form 'the probability of x having membership $m_S(x) = y$ in S is p' should be assigned a fuzzy truth value (between 0 and 1, presumably).

Somewhat ironically, what began as a pluralism has moved recently towards a genuine synthesis, but from an unexpected direction. If we replace the concept of a random *variable* with that of a random *set*, we have a mathematical formalism that not only neatly expresses many aspects of probability and sampling theory, but also generalizes classical set theory (see Kendall 1974 and Matheron 1975). This fact has been used by several writers to subsume fuzzy set and possibility theory under a general theory of random sets (Goodman 1982, Manes 1982, Nguyen 1978, 1984, and Goodman and Nguyen 1985), thereby combining the pluralist and reductionist standpoints.

The second kind of synthesis which requires a choice among competing domains or methods of handling uncertainty, has been approached by invoking second-order criteria in the manner of quasi-utilities for decision making. The literature on generalized intervalic probabilities, credal states, and subjective belief provides the best examples of this approach. The fundamental idea here is that the assignment of positive belief to some hypothesis H_j is itself a decision that may be made on the basis of some acceptance criterion. The classical example is Neyman-Pearson significance testing, wherein if a sample estimate q falls within a critical region we reject H_0 but otherwise reject nothing. In the notation of belief theory, the rule stipulates that if q falls in the critical region, then $Bel(\sim H_0) > 0$, otherwise $Bel(\sim H_0)$ and $Bel(H_0)$ both are 0. The second-order criteria here, of course, are the probabilities of making a Type I or Type II error. The lower the level permitted for a probability of Type I error, the more cautious the test. By the same token, the lower the level permitted for the probability of a Type II error, the more powerful (or, we might say, 'bolder') the test.

Levi (1980) provides a generalized discussion of this kind of criterion-driven acceptance rule and points out that although intervalists such as Dempster, Shafer, L.J. Cohen, and Shackle do not explicitly introduce these rules, they are tacitly committed to some version of them by virtue of the measures they introduce of conflict, dissonance, surprise, and the like. In Shafer's case, for instance, as both Levi and M.S. Cohen (1986) point out, the amount of conflict between two or more belief assignments ($CON(Bel_1, Bel_2)$) actually provides a parameter for assessing whether one should assign positive belief to any given hypothesis after all.

It seems intuitively reasonable to them that this information should be used *prior* to the application of Dempster's rule for combining Bel_1 and Bel_2. Zadeh (1984) has a similar idea but proposes a different measure of combinability than CON. Rather than automatically invoking Dempster's rule and renormalizing the resulting basic probability assignment, thereby producing the counter-intuitive belief values that Shafer's critics decry, these authors prescribe setting a level of acceptable conflict, k, such that whenever $CON(Bel_1, Bel_2) \leq k$, the H_j are permitted positive degrees of belief as assigned by Dempster's rule. Otherwise, the H_j must all be assigned 0 belief.

It is but a short step from this prescription to a similar method for selecting which of various available combination rules to use (in, say, intervalic systems where Dempster's rule is merely one of a family of such rules). One can invoke a quasi-(dis)utility such as CON and select that rule which is minimax, minimal regret, maximally beneficial, or satisfies some Hurwiczian pessimism-optimism ratio. This, then, is the basis for a synthesis and second-order decision making paradigm involving alternative methods of representing uncertainty. Its prime mover is merely a measure of caution about conflict, dissonance, or even (as in traditional significance testing) proneness to error.

Loui (1986), among others, has expressed dissatisfaction with this synthesis. For one thing, the choice of an orientation towards conflict or error (i.e., whether to be minimax, Hurwiczian, etc.) seems ad hoc. But more importantly, while avoiding conflict or error seems to solve the 'estimation' problem under partial ignorance, it throws any decision maker into a quandary. Wide confidence intervals, for instance, may immunize the estimator against making a mistake but ultimately they become vacuous: Why bother with belief intervals that are any narrower than [0,1]? The answer, of course, is that we at least implicitly want to compromise between caution in estimation, on the one hand, and sufficient power, informativeness, or boldness to enable us to make a decision where necessary.

Loui's own proposal for this compromise is instructive, whether justifiable or not. He asks that we rank-order a sequence of intervals (or they could be rules for combination, or uncertainty representation frameworks) in terms of our measure of caution and search through them until one of three states is attained: (1) The power-boldness criterion is met by a single option and we are able to make a decision on the spot; (2) The intervals have become points so that we are left with a standard decision under risk problem; or (3) The boldness level has become intolerably large so that no decision is possible. What he is proposing is in fact very similar to a standard optimization or multi-objective programming approach in which there are two loss functions that are somewhat negatively correlated.

M.S. Cohen (1986) dissents from these second-order schemes, claiming that we require a nonmonotonic probability framework for revising beliefs in the light of apparent conflict or dissonance. He has a prescription that begins with a procedure similar to Baldwin's alternative to Dempster's rule.

But instead of merely throwing all of the conflicting belief into a box labeled 'uncertain', Cohen proposes that the *sources* of the conflict be identified and that their respective initial belief assignments be revised by discounting them. That is, if k is our maximum acceptable level for $CON(Bel_1, Bel_2)$, we revise Bel_1 and/or Bel_2 until CON falls below k by selectively decreasing the basic probability assignments to various H_j. Thus, instead of altering or selecting an uncertainty representation framework, Cohen proposes that the belief assignments themselves be changed. His main criterion for selecting which H_j to discount and by how much is that the minimum possible amount of discounting should be done that is required to reduce CON to k or less. Once that is accomplished, Dempster's rule of combination may be applied.

Although it is far too early to evaluate most of these proposals (even normatively, let alone in pragmatic terms), clearly the pluralists and synthesists embody much that is genuinely novel in recent normative approaches to uncertainty and ignorance. At the very least, they have rejected both lines of retreat offered by banishment and reductionism. And even the more applications-oriented pluralists have begun to propose authentic syntheses of what they believe are distinguishable aspects of uncertainty, while those who claim that various systems are competitive alternatives to each other have attempted specify second-order decisional criteria for selecting among them. Nevertheless, it seems worthwhile to close this chapter and the first part of this book by addressing the question of how revolutionary these developments really are. Several claims have been made in previous chapters concerning the current shifts in paradigms dealing with ignorance, and uncertainty in particular. How much of the normative style in Western thinking on these matters is being overturned, and to what extent are recent developments merely extensions of traditional elements in that style?

4.6 A Normative Paradigm Shift?

There are a number of ways in which the current paradigmatic turmoil over uncertainty could be evaluated. It is too early, for instance, to attain an overview of the influence the emerging perspectives may exert on various professions and applied sciences, although an impressionistic survey of several fields would indicate that this influence may be relatively profound. I have chosen instead to attempt an assessment of what is conceptually innovative in the emerging normative perspectives, and what remains of the traditional Western orientations towards ignorance.

Perhaps the most strikingly novel characteristic of recent formal approaches is that they seem to accept varieties of uncertainty and even ignorance that heretofore were either ignored (both in the sense of being declared irrelevant and constituting part of our meta-ignorance) or eliminated. More than passively admitting those varieties into the picture, the new approaches prescribe methods for managing and understanding matters such as vagueness, ambiguity, nonspecificity, and possibility. The rhetoric produced by advocates of the new perspectives leaves no doubt of their hopes and ambitions for their progeny. For Zadeh (1975c: ix-x) the primary motivation behind the creation of fuzzy set theory is "...a rapprochement between the precision of classical mathematics and the pervasive imprecision of the real world... What we need is a... body of concepts and techniques in which fuzziness is accepted as an all pervasive reality of human existence... Such methods could open many new frontiers on psychology, sociology, political science, philosophy, physiology, economics, operations research, management science, and other fields, and providen a basis for the design of systems far superior in artificial intelligence to those we can conceive today."

Shafer (1976) also has an expansiveness to his program, only he wishes to liberate the concept of degrees of belief from the doctrine of chances. First he points out (pg.9) that for some time the idea of numerical degree of belief has been unified with the idea of chance. He then goes on to say (p.16-17) that "The chances governing an aleatory experiment may or may not coincide with our degrees of belief about the outcome of the experiment. If we know the chances, then we will surely adopt them as our degrees of belief. But if we do not know the chances, then it will be an extraordinary coincidence for our degrees of belief to be equal to them... Chances, then, must be conceived of as features of the world. They are not necessarily features of our knowledge or belief. And it would be quite untenable to claim that a chance is *merely* a feature of our knowledge or belief (emphasis in the original)." While the last remark is aimed at Laplacian determinists, his earlier statements are intended to persuade readers that rejecting Bayesianism need not entail rejecting the use of numerical grades of belief, nor the coherency offered by a mathematical framework.

A second novel feature common to most of the new perspectives is an explicit move to formalizing or modeling second-order uncertainty. While second-order schemes had been suggested prior to the 1970's both within probability (Koopman 1940, Kyburg 1961, Smith 1961) and outside it (Shackle 1952), none of these seem to have received widespread attention at the time, either in the theoretical literature or in applied realms. It is intriguing to contrast that state of affairs with the present-day situation, in which second-order schemes are widely discussed in most recent major publications on uncertainty, conferences are being held on intervalic mathematics, and debates over the implementation of second-order schemes routinely crop up in the expert systems and artificial intelligence literature. Additionally, the kinds of second-order uncertainty that are being investigated appear to be new. One of the best examples is the increasing discussion (again, both within and outside the probabilistic camp) of *conflict* or *dissonance* among separate belief structures, bodies of evidence, or expert judgments.

Despite the innovative nature of these developments, however, there are also important characteristics of the new perspectives on uncertainty that are wholly traditional and in some instances even atavistic. The most obvious is that much of the meta-level analytic machinery being deployed to handle the new uncertainties is taken straight out of the old dualistic logic and classical set theoretic frameworks. Fuzzy logicians do not prove their theorems in degree less than 1; in fact they frequently resort to proofs by contradiction and thereby utilize the strong version of the Law of the Excluded Middle.

This cross-level inconsistency may be interpreted as the old uncertainty reduction impulse being established at the second-order level: If we cannot have first-order certainty, let us attain certainty about uncertainty. Far from being the province of reductionists like Hisdal or Eytan, this impulse is spelt out by some of the most ardent revolutionaries. Kaufmann's (1975) introductory book on fuzzy sets in fact bears the title Theory of Fuzzy *Sub*sets, and he explains that third term in his prefatory remarks (pp.xiii-xiv): "...a fuzzy set will never be a concept proper to the present theory; the reference set will always be an ordinary set... I am not a stubborn and rigorous boubakiste,... but there are some definitions that must not be fuzzy." Goodman and Nguyen (1985: xvii), in their comprehensive mathematical treatise on uncertainty models, devote an entire 'prolog' to explicating this perspective. After noting the apparent inconsistency inherent in proving theorems on fuzzy or multivalent logic with the tools of standard dualistic logic, they go on to say: "This is consistent with the philosophy that on one hand describes both subjective and objective real-world phenomena (and related models) within a multiple-truth-valued context, yet at the same time, in order to prevent a possible infinite regress of nested multi-truth evaluations, the meta-level of description must be classical." In this tangle of language lies the central reason behind the imposition of meta-level certainty, and it amounts to something stronger than Good's 'computational convenience'. Goodman and Nguyen, along with others working in these areas, cannot quite tolerate unbounded meta-uncertainty. They end their confession on a wistful note: "As researchers in probability theory, information theory, possibility theory, and multi-valued logical systems in general, we all seem to be collectively stating... 'don't do as I do, do as I say!'"

A stronger statement of this position has been articulated by P.T. Johnstone (1988) in his recent 'Open Letter' to the Bulletin Editor of Fuzzy Sets and Systems. There, he states that the "truth values of mathematical assertions are crisp and not fuzzy" (pg. 248). Moreover, he argues that this position does not entail a Platonist philosophy of mathematics, since crispness does not equal immutability. He then provides one of the clearest statements of the current suburban solution by mathematicians for living in post-Godelian times. While acknowledging that the axioms and modes of reasoning adopted by mathematicians are social products and therefore mutable, Johnstone nevertheless asserts that the "proof of a given assertion, *within* a given system of logical assumptions, is a finite, surveyable thing: if two mathematicians look at it, they are bound to come to the same conclusion about its validity..." (pg. 248, emphasis in the original). Cultivation may yet proceed, so long as one remains within the walled garden.

A second way in which the new perspectives seem to be 'more of the same' is their emphasis on quantifying uncertainty and ignorance. Both the Certainty Factors and Dempster-Shafer belief frameworks seem to assume from the start that numerical representations of belief strength are feasible and reasonable. Fuzzy set and possibility theory both have exhibited an ambivalence towards quantification. The earliest writings on these frameworks heralded the development of fully-fledged linguistic versions of fuzzy set and possibility theory (e.g., Goguen's 1969 extension of fuzzy set theory to incorporate valuation sets that are lattices), and to be sure some quite interesting work has been done in that direction. However, in both theory and applications the bulk of the output from these perspectives has assumed and indeed required a numerical representation of fuzziness and possibility.

In artificial intelligence and expert systems research there is, however, a minority of investigators who advocate qualitative approaches to dealing with uncertainty and the revision of beliefs. One of the earlier examples is the rule-based system for diagnosing indigestion created by Fox, Barber, and Bardhan (1980) which marks hypotheses with strings like 'maybe' and 'very likely'. Most discussions of this kind of system compare the advantages of using linguistic or numerical labels. Nagy and Hoffman (1981), in a preliminary study comparing judgment performances of subjects using linguistic and numerical estimates of probability, found in favor of the linguistic estimates mainly because they apparently discouraged extremely inaccurate estimates. In the same vein Zimmer (1983) found that subjects could handle complex dependency information better verbally than they could numerically. On the other side of the debate, Beyth-Marom (1982) claimed superiority for numerical estimates of probability in forecasting, Behn and Vaupel (1982) made the same claim for decision-making in situations where effective communication is required, and Bryant and Norman (1980) and Nakao and Axelrod (1983) argued in favor of numbers in medical contexts.

More elaborate qualitative approaches have come from three concepts concerning mundane reasoning under uncertainty: Endorsements, defaults, and circumscription. P.R. Cohen (1985) and Cohen and Grinberg (1983) propose that the strength of a belief and the adequacy of its evidence may be represented by the reasons for believing and disbelieving them, which they term 'endorsements'. While endorsements may be ranked in terms of superior quality and even propagated across inferences, they resist quantification and their propagation requires sensitivity to the context in which they are being used. Their system contains rules for semantically combining evidence and handling conflicting beliefs. Perhaps most importantly, the concept of endorsement includes distortion and dishonesty as possible attributions, neither of which are mentioned in the quantitative approaches to ignorance.

Default reasoning has a relatively long history in artificial intelligence and the human sciences (see, for instance, Minsky 1975, and Schank and Abelson 1977). Reiter (1980) advocates the use of defaults for filling gaps in knowledge, and as such promotes defaults as a method for managing incompleteness. He claims that default rules are akin to meta-rules insofar as they instruct us in how to extend an incomplete theory or knowledge base so that we may act. In fact, Moore (1985) extends this concept to the notion of 'autoepistemic' reasoning, whereby beliefs are revised and uncertainty dealt with by reasoning about one's own belief system. McDermott and Doyle (1980) provide an alternative perspective on defaults by considering them in the framework of nonmonotonic modal logics. A closely related concept, circumscription, involves explicitly stating rules for eliminating possibilities on the basis of relevancy (McCarthy 1980).

Debates over whether uncertainty is best represented quantitatively or qualitatively are, of course, nothing new. But these proposals point out two limiting aspects of the new quantitative approaches which echo traditional constraints. One of these is the fact that virtually none of the quantitative frameworks deal effectively with irrelevancy or distortion (especially in the sense of dishonesty and falsehood). The second is the lack of contextualization, although in fairness it should be noted that quantitativists also castigate qualitative frameworks for their domain specificity.

In summary the developments of the past two decades in normative perspectives on uncertainty, although remarkable in the light of nearly 300 years' domination by probability theory, are perhaps not quite as revolutionary as some of their advocates claim. At most we have a cautious revolution, or a quantitative reformation. Nonetheless, that a paradigmatic shift is afoot seems undeniable, and there is immense potential in a dialogue between the normative reformists and their counterparts in the human sciences, whose investigations into ignorance also have largely taken place within the past 25 years. These two intellectual streams have run parallel, uncommunicative courses for the most part, and they are just beginning to merge.

Chapter 5:
Psychological Accounts:
Biases, Heuristics, and Control

"To be able to function and live amid uncertainty--- that's a clinical definition of sanity." E. Boulding

"Whoever denies he is certain of anything is neurotic. Whoever claims certainty of everything is foolish." Isaac Levi.

5.1 The Normative View from Psychology
5.1.1 Three Traditions

The field of psychology is ostensibly concerned with explaining human thought, emotion, and behavior, and we may properly turn to it when seeking explanations of how individuals perceive and respond to ignorance. However, traditionally psychology also has provided clear normative messages concerning ignorance (and especially uncertainty), of sufficient consistency across theories and schools of thought that they qualify as a dominant ideology. Those messages, in turn, are derived from frameworks which have crucially influenced psychological research and theories on these topics, as have debates over the relationship that should obtain between normative and explanatory frameworks. This chapter is an attempt to construct an overview of the interactions between the normative and the explanatory in the psychological literature which addresses ignorance. Once again, the reader should not expect an exhaustive treatment; preference has been given to the modern mainstreams of theory and research and a comparative synthesis.

There are, roughly speaking, three traditional and popular normative orientations towards ignorance in psychology, and each has its roots in a particular theoretical development. Perhaps the oldest is contained in the psychoanalytic canons for the well-adjusted individual, and has since been carried into personality and motivation research and indeed most branches of ego psychology. This is a view that favors the person who seeks novel information and experience, is open to full and honest communication, who can tolerate various kinds of uncertainty and even ignorance in the short run in order to gain knowledge, and who is not defensive about prior beliefs. This normative orientation will be termed the 'Knowledge Seeker' thesis. The second tradition concerns the debilitating consequences of incomplete information, uncertainty, unpredictability, and uncontrollability for the affective, cognitive, and physiological capabilities of the affected organism. Conversely, this normative orientation claims positive effects from exposure to fully informative, certain, predictable, and controllable stimuli or environments. Most of the impetus and evidence for this viewpoint have come from behaviorist research and theories concerning learning and adaptation, and it will be referred to as the 'Certainty Maximizer' thesis. The third normative orientation has its origins in psychophysics, perception, and cognitive psychology, and reflects the mainstream information processing models of cognition. It is primarily concerned with criteria for rationality in thought and behavior, and in research on uncertainty the dominant normative viewpoints have been Bayesian probability and the maximization of expected utility in decision making. I will call this perspective the 'Statistician' orientation.

These three normative viewpoints reflect most of the dominant cultural assumptions and orientations that have been referred to throughout this book. In fact, taken sequentially, they amount to a brief digest of traditional and modern Western prescriptions for dealing with ignorance. First, reduce ignorance as much as possible by gaining full information and understanding, and ignoring nothing that is relevant (the Knowledge Seeker). Secondly, attain as much control or predictability as possible by learning and responding appropriately to the environment (the Certainty Maximizer). Finally, wherever ignorance is irreducible, treat uncertainty probabilistically, ignore the other kinds of ignorance, and select that alternative which maximizes expected utility in the long run (the Statistician).

The link between each of these prescriptive orientations and psychological explanations of thought and behavior under uncertainty and ignorance lies in the supposed benefits that accrue to the well-adjusted, fast-learning, adaptively responsive, rational individual who follows their prescriptions. Conversely, the primary object of much research and a source of debate on these topics consists of apparent deviations by people from prescribed orientations and strategies. Accordingly, each of these prescriptions will form the focus for a review of relevant psychological research in pertinent subfields. The most penetrating review will focus on the 'Statistician' viewpoint, however, since it is the most modern and has provoked the sharpest debates on fundamental issues.

5.1.2 Coping, Defending, and Ignorance

In the psychoanalytic tradition and its descendants, a major source of anxiety for any individual is conflict, whether between the drives and the ego (and/or superego) as in Freudian theory, or from the incongruity of experience with one's self concept (as in Roger's framework). These anxieties lead to 'defensive' behavior involving, among other things, unawareness or distortion of those experiences and underlying conflicts. Ignorance in the sense of incomplete or vague self-knowledge arises from unconscious defenses such as denial, subception, or suppression. Distortion, on the other hand, involves reactions such as fantasy, rationalization, or projection.

Early psychoanalytic therapeutic interventions, then, were oriented partly towards breaking down these defenses and giving the patient full information and certainty about these heretofore inaccessible reaches of the self and experience. The adjusted, coping, nondefensive individual was equated with full self-knowledge and analysis. More importantly, the lack of full information about the self was believed to explain a number of maladaptive behaviors and symptoms.

The belief that personal orientation towards uncertainty and ignorance (or certainty and knowledge) is a crucial explanatory factor has thrived in theories of personality, motivation, individual differences, and other variants of ego psychology. It has consequently spawned many attempts to measure such orientations which in themselves are informative of psychologists' agendas concerning these topics.

The earliest measures of this kind appeared in attempts to explain right-wing conservativism in terms of personality traits or motivational orientations. Among the more famous of these are the authoritarian personality study by Adorno et al. (1950) and the open and closed mind theory by Rokeach (1960). In both research programs, orientations toward uncertainty and ignorance were tapped at least indirectly by subscales contained in the batteries of attitudinal measures.

The authoritarian personality according to Adorno et al. is indicated on their F-scale by, among other things, cognitive rigidity and defensiveness, lack of receptiveness to new and counterindicative ideas or information, and rigid categorical schemata. Authoritarians are intolerant of ambiguity, heterogeneity of ideas or beliefs, and deviations from religious or political dogma. Correspondingly, nonauthoritarian personalities possess the opposite cognitive traits. Similar claims have been made by researchers investigating creativity during the same period. Studies by Barron (1953) and MacKinnon (1962), for instance, sought to substantiate hypotheses that creative people prefer novelty, surprise, ambiguity, complexity, and other uncertainty-related attributes while uncreative people do not. Some writers such as Koestler (1964) have gone as far as to suggest that creative people also dislike secure, conventional, usual, clear-cut, simple, and obvious ideas.

Rokeach's trait theory has a similar flavor. He sets up a bipolar continuum with 'gestalt types' at one end and 'psychoanalytic types' at the other. These terms reflect the psychoanalytic origins of the constructs underlying this typology. The gestalt type possesses a need to know and understand, while the psychoanalytic type is characterized by a need to defend against threatening aspects of reality. The gestalt type is therefore 'open-minded' in the sense that she or he has a cognitive system that is receptive to new or unfamiliar information and ideas. The psychoanalytic type, on the other hand, resembles the cognitive part of the personality that has become fixated; she or he is defensive and prefers the familiar and predictable.

Like the research on authoritarianism and creativity, Rokeach's typology and his correlational studies leave no doubt about which type is normatively preferred. Open-minded gestalt types are less prejudiced, less authoritarian, less religiously dogmatic, more politically progressive, better at problem-solving, and more artistically appreciative than their psychoanalytic counterparts. The fact that these results and the research that produced them is partly driven by a normative orientation is underscored by Billig's (1982) account of the Nazi German social psychologist Jaensch's 'J-scale' which apparently was designed to measure personality characteristics desirable for citizens of the Third Reich. Jaensch's constructs heavily overlap with those of the F-scale for authoritarianism but the valence of the labels is reversed. Thus, 'cognitive rigidity' for Jaensch becomes evidence of one's strength of will and courage of convictions, and the 'defensive' avoidance of the unfamiliar or novel becomes an indication of suitable intellectual rigor.

More recently the motivational research and theory on need for achievement (originating with McClelland et al. 1953) has spun off measures of uncertainty orientation. The most widely known among these is Hofstede's (1980) Uncertainty Avoidance Index (UAI), which consists of three items which emerged from a factor analysis of a questionnaire used in a large-scale study of a multinational corporation's employees during the 70's. These items tap the individual's felt level of anxiety, need for fixed company rules, and need to continue with the company.

Although his study is primarily cross-national and oriented towards the social and organizational level (we shall return to it in Chapter 6), Hofstede reports a number of correlations between UAI and various attitudes, motivations, and beliefs. High UAI individuals are more anxious, stressed by their jobs, resistant to change, fearful of failure, and intolerant of ambiguity; they are less motivated to achieve, compete, take risks, compromise in the face of conflict, and trust co-workers; and they prefer to work in large organizations with clear rules and requirements, hierarchical structures, specialized work-roles, low conflict, and rewards accruing to loyalty and seniority.

More blatantly normatively-driven research in this vein comes from the research on the links between uncertainty and achievement motivation. Starting with the postulate that the uncertainty of outcomes in achievement-oriented situations is an important situational determinant (cf. Schneider and Posse 1982), several researchers (e.g., Weiner 1972 and Trope 1975, 1979) have argued that the difference between achievement motivated individuals and their opposites is not due to affective orientation but differences in information seeking style. The success-oriented person seeks more information about her or his abilities. These claims have sparked the development of a theory of 'uncertainty orientation' by Sorrentino and his colleagues (Sorrentino and Hewitt 1984, Sorrentino, Short, and Raynor 1984). They link uncertainty orientation with Rokeach's theory as well as the research on achievement motivation, and even claim that uncertainty orientation and achievement motivation have been confused in previous research (Sorrentino and Short 1986). Uncertainty oriented people are like Rokeach's gestalt types; they seek more diagnostic information about themselves in relation to their environments than do certainty oriented people. A similar measure has been proposed by Cacioppo and Petty (1982), which they term a 'need for cognition'.

Again there is no doubt about the normative orientations of these researchers. Sorrentino et. al. characterize the uncertainty oriented people as "normal", similar to "most of us in academia", and "rationally weighing all the pros and cons". By contrast, certainty oriented people are "bizarre" in their behavior, select options without any "rational or conscious deliberate planning", and are "at, or below, Kohlberg's (1976) level of moral authority" (all quotations taken from Sorrentino and Short 1986).

In accordance with the thesis I introduced above, the primary source of tension and debate in the 'uncertainty orientation' and UAI research is manifested in explanations of counter-normative attitudes, cognition, and behavior. Conformity with normative prescriptions are, by comparison, unproblematic. Sorrentino and his colleagues are once again exemplars of this tendency; they wish to draw attention to the fact that "many people", even "most adults" are certainty-oriented and that their behavior is puzzling in the light of standard achievement motivation theory. Hofstede is more ambivalent about uncertainty avoidance, although he does link it with the F-scale, Eysenck's 'tough-mindedness' scale, and intolerance of ambiguity. Nevertheless, he too finds it difficult to explain. He declares that the causes and origins of the UAI "syndrome" are much less clear than other constructs he investigates, and gestures briefly at broad-brush factors such as modernization, religion, population density in relation to wealth, and selected historical factors (Hofstede 1980: 185).

Finally, the ego psychology models have led some researchers to an interest in cognitive styles as a possible explanation for individual differences in automatic coping styles and effectiveness. Most of this research was carried out during the 50's and 60's, although it has seen some resurgence during the late 70's and early 80's. Pettigrew (1958, 1982) initiated a research program based on a measure of 'category width', which refers to a propensity to have broad, heterogeneous categories rather than narrow, exclusive and homogeneous ones. A similar concept emerged in the literature on cognitive complexity (e.g., Bieri and Blackner 1967) and cognitive controls (cf. the concept of 'leveling/sharpening' in Allport and Postman 1947). Budner (1962) followed up the suggestion by the authoritarianism literature that authoritarians are intolerant of ambiguity or unreality, and explored tolerance of ambiguity as a cognitive style. As would be expected, the normative orientations of these theorists are similar to those described above.

The conceptualization of ignorance in these research traditions is rather crude, both from a philosophical and measurement standpoint. None of the analytic distinctions are drawn among various kinds of ignorance that one finds in standard philosophical discussions. Ambiguity, incomplete information, probability, vagueness, and nonspecificity are lumped together or sometimes discussed as if they are synonymous (Hofstede is one of the few researchers in this tradition to explicitly acknowledge this, 1980: 155). Furthermore, the dominant philosophical orientation towards ignorance is a naive realism which takes ignorance as objectively given and located 'out there' in the environment. Little provision is made for the possibility that ignorance might be actively and socially constructed by the very people who must either cope with or defend against it. These assumptions and glosses are congruent with the older uncertainty-reducing and absorbing traditions of Western intellectual culture, but not with the modern managerial approaches which require a much finer-grained analysis.

5.1.3 The Control and Predictability Thesis

The major claims of the learning theoretic and S-R behaviorist streams of theory and research concerning ignorance pertain primarily to control and prediction. These two concepts and their attendant motivations also are central to the literature on stress and coping. In fact, much of the research from these streams does not disentangle predictability from controllability; only recently have researchers made systematic attempts to investigate the two separately. Broadly, the principal motivational tenets are quite simple. People are averse to uncontrollability, unpredictability, and risk. They suffer deleterious effects in the presence of such stimuli, and when given a choice they avoid those stimuli. Somewhat less attention has been paid to the converse, but a substantial body of theory and research nonetheless claims that people are control and predictability maximizers, and that they are rewarded by exposure to predictable and controllable stimuli.

Unlike the ego psychology research, predictability and controllability are quite narrowly and operationally defined in the behaviorist research tradition, although there are important variations in those operationalizations. An older style of research focuses on uncertainty in the form of 'conflicting' reinforcement. There are three major varieties: intermittency (e.g., randomicity) in reinforcement schedules (this was of course at the heart of Skinner's initial discovery of the importance of reinforcement schedules in his work with pigeons); 'conflicting' reinforcement whereby two or more incompatible behaviors are simultaneously reinforced, and 'inconsistent' reinforcement in which the same stimulus is the source of both reinforcement and aversion. The primary normative orientation here favors consistency or regularity across time and behavioral modes.

Each of these inconsistent types of reinforcement lead to problematic responses on the part of the organism. The intermittent reinforcement schedule is, of course, the classical paradigm for inducing unextinguishable behavior. Likewise, approach-approach (or avoidance-avoidance) conflict results from simultaneous attraction to (or fear of) two or more stimuli. Finally, approach-avoidance conflict stems from the inconsistent reinforcing-aversive stimulus.

A more modern research paradigm over the past 25 years has concentrated directly on controllability and predictability per se, with the most influential motif being the 'learned helplessness' thesis of Seligman and his colleagues (Overmier and Seligman 1967, Seligman and Maier 1967, Seligman 1975). Although the theoretical orientations of learned helplessness researchers over the years have vacillated between cognitivism and behaviorism, the main impetus of the research and theory has been behavioral.

Early findings pointed to behavioral and physiological changes in a variety of animal species as a consequence of exposure to uncontrollable aversive events (e.g., shocks or loud noises). The principal change is a reduction in the capacity to learn to escape or avoid such events later on. This reduction was hypothesized to result from the organism learning that it has no control over the events, leading to both an impaired ability to detect response-outcome contingencies and a decline in response initiation. Related 'adverse' effects found in various studies include increases in stress, impaired problem solving, increased depression, increased passivity, greater susceptibility to pain and disease, decreases in aggressiveness and competitiveness, and concomitant chemical changes. Nor are the negative effects limited to uncontrollable aversive stimuli. Uncontrollable appetitive stimuli lead to a decline in inclination to work for rewards and even crosses over to induce impairment of learning to escape aversive stimuli (cf. Overmier et al. 1980). Evidence for beneficial effects on humans of exposure to controllable stimuli focus mainly on so-called 'mastery' effects on learning in children (cf. Gunnar 1980, Dweck and Elliot 1983) and psychophysical benefits for institutionalized populations such as the elderly or hospital patients (Rodin and Langer 1977, Schulz and Hanusa 1980, Krantz and Schulz 1980).

Several attempts to improve upon the theory of learned helplessness have entailed some elaborations concerning the effects of uncertainty and unpredictability per se. The attribution theoretic reformulation of learned helplessness (Abramson, et al. 1978) casts the stability of the attribution of no control in a mediating role, such that internal, stable, and global attributions of no control are responsible for the most debilitating effects, whereas instability can ameliorate those effects. More important is the research that attempts to disentangle effects of unpredictability from those of uncontrollability. A substantial body of research indicates that individuals prefer signalled to unsignalled aversive events (cf. Badia et al. 1979), even when the signalled event is more aversive (Miller et al. 1983). Several researchers have argued that these results are explicable in terms of reduction in uncertainty (Imada and Nageishi 1982, D'Amato and Safarjan 1979).

An observable trend in this research over the past decade is an increasing ambivalence towards ignorance and uncertainty which has arisen primarily because the researchers' understanding of the role played by unpredictability and control in producing learning effects has increased in complexity. First, there has been growing recognition that learned helplessness is not the only possible response in the face of unpredictable or uncontrollable events. People may elect to retain an interpretive, vicarious, or even illusory sense of control (Rothbaum et al. 1982 refer to this as 'secondary control'). Furthermore, people's perceptions of controllability and predictability are not always accurate (Abramson and Alloy 1980, Alloy and Tabachnik 1984) and illusions of predictability or control may be cognitively adaptive (e.g. Taylor 1983).

Secondly, as Mineka and Henderson (1985) point out, the issue of whether predictability is reinforcing or unpredictability aversive remains contentious. Frustration research (Amsel and Stanton 1980) underscores a potential contradiction between the learned helplessness thesis and the older intermittent reinforcement paradigm by arguing that partial or intermittent success should be more reinforcing than the continuous success recommended by learned helplessness proponents. Another line of research suggests that the benefits of predictability may not accrue to individuals who use a 'distracting' or 'blunting' coping style for dealing with aversive experiences (Miller and Mangan 1983).

The learned helplessness research and related approaches present a somewhat less overtly ideological picture of uncertainty than does the ego psychology tradition. Nevertheless, even ignoring the problematic circularity in the definitions of aversiveness, reinforcement, and learning deficit, it is clear that the central normative thesis remains pro-certainty and control, and the implicit philosophical orientation toward uncertainty and ignorance is again naive realism coupled with a narrow operationalism. An interesting recent trend in behaviorial research on responses to uncertainty has equally narrowly operationalized it in terms of reinforcement delay rather than unpredictability or conflict. We will return to that research near the end of this chapter, since it is related to the more modern statistical perspective rather than the control maximization thesis. It is to the statistical perspective that we now turn, and it requires an in-depth treatment.

5.2 The Bayesian Inquisition
5.2.1 A Paradigm for Studying Statistical Intuitions

Unlike the mainstream ego psychological and behavioral learning literature, the cognitive studies of judgment under uncertainty neither openly nor implicitly adopt a negative view of uncertainty per se, aside from a general acknowledgement that both judgment and decision making are made more difficult by uncertainty. However, the literature on judgment and decision making is dominated by a statistical normative orientation that entails a strict Bayesian interpretation of perceived uncertainty and a prescription for decision makers to maximize subjective expected utility by employing the Bayesian probabilistic calculus. Although alternative normative perspectives on uncertainty are available, as we have seen in Chapter 4, none of them have guided cognitive and social psychologists in their studies of judgment. Edwards and von

Winterfeldt (1986) remark that few thinkers about decision making have questioned the normative appropriateness of maximizing subjective expected utility, although some have questioned its feasibility in real world situations. On another occasion, Edwards (1984: 7) remarks that no other normative principle deserves a moment's consideration, given that the appropriate numbers and task structure for a decision have been provided.

This remark and the implication Edwards derives from it, that the question of how to obtain the numbers and structure that constitute 'wise' decisions is at the heart of decision analysis, point to an important cornerstone in the statistical normative perspective behind judgment research: namely, the veneration of quantified degrees of belief. Quantitative subjective probabilities are required by Bayesian analysis and of course also for computing expected utility. There is little doubt that most judgment researchers favor numbers over words for representing uncertainty. Not only is the vast majority of studies couched in strictly numerical terms (presenting subjects with numerical probability problems and eliciting numerical responses), but a number of these researchers have published normative statements justifying their quantitative orientation.

It is intriguing to contrast such statements with some of the debates raised in the risk assessment literature over the usefulness of quantitative estimates of subjective probability. The textbook on thinking under uncertainty by Beyth-Marom et. al. (1985: 36-47) for secondary school students contains a clear elaboration of the quantitativist argument. Words are inappropriate for expressing uncertainty, according to the authors, because they are vague, ambiguous, and inconsistenly used. Their meanings vary with context. Verbal expressions can seldom even be rank-ordered by expressed degree of uncertainty, and even when ranked they cannot be quantified into a scale. Verbal expressions are insensitive to 'small but important' changes in degree of belief. Finally, some verbal expressions of uncertainty confound strength of belief with value or utility. Numbers, on the other hand, are "interpreted identically by all users" (1985: 44). They do not vary in meaning with context. They permit comparisons between degrees of belief, and are sensitive to small changes in degrees of belief. Furthermore, they separate uncertainty from value or utility, rendering uncertainty and utility independent.

This latter property is a crucial assumption in the statistical decision making perspective with far-reaching implications for the study of judgment. The 'Statistician' thesis requires the assessment of uncertainty to be independent of utility, or in other words value-free. Subjective probability estimates cannot justifiably vary with the goodness or badness of the consequences of uncertainty.

The normative orientation of most studies of judgment under uncertainty, then, assumes that the theory of probability is unproblematic. 'Rational' judges quantify their uncertainty about an event into a single numerical estimate and then manipulate those estimates according to the Bayesian calculus. Likewise, 'rational' decision makers order their preferences for outcomes, quantify those preferences into utility functions that are independent of their estimates of subjective probability, and then maximize subjective expected utility. It is small wonder, given this rather restrictive and elaborate set of norms, that most of the research on judgment under uncertainty has focused on judgmental errors, as more than one observer has noted (e.g., Kahneman and Tversky 1982a: 123, and Jungermann 1983).

Influential researchers such as Kahneman and Tversky justify the focus on errors by arguing that we may learn about our intellectual limitations from them, they may reveal to us the processes by which we make judgments and arrive at decisions, and they also may reveal which principles of statistics are nonintuitive. Others have argued that the study of errors is misguided (e.g., Phillips 1983, L.J. Cohen 1979, 1981, Macdonald 1986), or at least overemphasized at the expense of understanding judgment processes and heuristics (e.g., Einhorn and Hogarth 1981). We shall return to the debate over rationality later in this chapter, since it is central to any considerations of how the dialog between explanatory and normative frameworks on ignorance might be enhanced and developed. First, however, a review of the research on judgmental deviation from probabilism is in order. The literature on this topic is large and so is the list of 'errors' (Hogarth 1980 reviews some 27 distinguishable kinds of deviation from probability by human judges). Furthermore, there are a number of ways the material could be organized. Hogarth and Makridakis (1981), for instance, classify biases and heuristics according to whether they apply to the information acquisition, processing, output, or feedback stages of the judgment process. This review attempts to cover a representative sample, organized under the three primary directives of the dominant normative perspective: quantification of uncertainty, adherence to the probabilistic calculus, and the independence between uncertainty and utility or value.

5.2.2 From Words to Numbers and Vice-Versa

The human sciences are rife with quantitative-versus-qualitative debates in research methodology, and it is no surprise that a similar debate should occur over the representation of subjective uncertainty. Nearly all of that debate has focused on representing uncertainty probabilistically, and it has two discernible streams. One of them concerns the reliability and validity of measuring people's meanings and assessments of uncertainty by verbal phrases or numerical estimates. The other focuses on how well numerical subjective probability estimates are 'calibrated'.

The majority of researchers in this area ask subjects to make numerical estimates of the probability (or in some cases frequency) represented by selected phrases (e.g., 'quite likely', 'probable', 'almost impossible', 'often', 'rare'). Although several such studies report reasonable within-subject reliability (e.g., Lichtenstein and Newman 1967, Beyth-Marom 1982, Budescu and Wallsten 1985) and stability of group means over time (Simpson 1944, 1963), the most consistent findings have been considerable intersubject variation, overlap between phrases (Stone and Johnson 1959, Lichtenstein and Newman 1967, Hakel 1968, Beyth-Marom 1982, Budescu and Wallsten 1985), and, where investigated, variability with changes in context (e.g., Cohen, Dearnaley, and Hansel 1958c, Pepper and Prytulak 1974, and Wallsten, Fillenbaum, and Cox 1986; see Pepper 1981 for a review on quantifiers generally).

Both practical and methodological implications have been derived from these findings. The most popular is simply that people should use numbers rather than words to express probabilities whenever possible. Several authors have suggested that if people must use words, then they should restrict the vocabulary to a small set of relatively stable, nonoverlapping terms (Bass, et al. 1974, Beyth-Marom 1982). Budescu and Wallsten (1985) suggest that high intersubject variability implies the proper unit of analysis for studies of judgment under uncertainty is the individual and not the group.

A closer look at these findings and conclusions raises several questions. First, do the results indicate that verbal expressions for uncertainty change with time and context or that their meanings are ambiguous, vague, and/or nonspecific? There is no clear answer to be gained from the experimental evidence. Secondly, are those terms with large variances and substantial overlap vague because they are nonspecific about probability or because they in fact refer to some other (nonprobabilistic) kind of uncertainty? For instance, the term 'possible' was employed by Lichtenstein and Newman (1967), Beyth-Marom (1982), Budescu and Wallsten (1985), and Wallsten et al. (1986). In all but Beyth-Marom's study (which used a sample of professional forecasters), this term exhibited wide variation and considerable overlap with other terms (e.g., in Lichtenstein and Newman 1967, subjects assigned the term a probability anywhere from 0 to 0.99). But this is precisely what Zadeh's possibility theory predicts. A possible event is one whose probability may vary from 0 to 1, since possibility places an upper bound on probability. Wallsten et. al.'s first experiment contributes some circumstantial evidence for this interpretation via the finding that a great deal of intersubject variability occurs in assigning an upper limit of probability to the term 'possible'.

A third issue that is unaddressed by this research is the process by which people assess uncertainty in the first place. Zimmer (1983) claims that people process information for predictions in a way that is similar to putting forward arguments rather than estimating parameters. If so, then the entire task format in these studies is alien to most of the subjects and requires more mental effort than their usual mode. Furthermore, the estimates themselves will mistranslate the true meanings of the phrases being considered.

Zimmer (1984) and Wallsten et. al. (1986) both have utilized the fuzzy set theoretic concept of a membership function to model a translation from verbal to numerical expressions of probability. Its simplest version involves eliciting from subjects lower and upper bounds for a probability interval that would represent the meaning of a given phrase. Thus, a subject could claim that 'possible' can take a probability value from 0 to 1. The fully-fledged membership function, of course, permits a function shape other than a rectangle, and the shape itself may reveal semantic properties of the phrase to which it refers. These researchers argue that the membership function provides a superior quantitative representation of uncertainty which is sensitive to at least some differences in qualitative meanings.

The bulk of the work on translating between verbal and numerical phrases uses pointwise numerical estimates as the yardstick by which to measure the reliability, stability, and distinctiveness of verbal phrases. A few recent investigations have employed intervalic numerical estimates. I am unaware of any studies, however, which have elicited verbal phrases from subjects when presented with a numerical probability or probability interval. One plausible explanation for this asymmetry is that ordinary people use natural language to describe uncertainty more often than numbers or mathematics and so verbal phrases are the more salient object for investigation. Furthermore, the normative model from which yardsticks are obtained for assessing reliability and validity is mathematical and so of course numbers emerge as the most obvious yardstick. Finally, people seem to resist expressing their subjective feelings of uncertainty numerically, and so there is some concern over how accessible and understandable quantitative probability estimates are to them.

Nevertheless, we cannot assume that the reverse translation would yield the same kinds of results as has the previous research. Furthermore, an investigation into the numbers-into-words process would address the question of how people respond verbally to quantitative probabilities, which has not been dealt with in experimental studies to date. This question directly refers to a wide range of real occurrences and practical issues (e.g., the effects that numerical forecasts and risk estimates have on the public). Finally, the claim that verbal terms for uncertainty express aspects (e.g., vagueness, ambiguity, and generality) not captured by a single number that is tied to the concept of probability cannot be addressed by studies that judge those phrases solely in terms of numerical probability. It may turn out that when asked to translate numerical probabilities people prefer to use phrases that are explicitly probabilistic (e.g., 'highly probable') rather than phrases connoting other kinds of uncertainty (e.g., 'possible').

The calibration literature, by contrast, assumes the appropriateness of using numerical probability estimates and pays little attention to the elicitation format (an exception is Seaver, von Winterfeldt, and Edwards 1978, who investigated the effects of using fractile, odds, probability, and log-odds methods of elicitation on calibration). Instead, calibration research is concerned with whether people tend to overestimate or underestimate knowable probabilities of events. A variety of measures have been constructed to measure the magnitude and direction of miscalibration, but most of them fail to distinguish between two kinds identified by Lichtenstein et. al. 1982: Over(under)confidence in the truth of one's prediction, and over(under)confidence in one's ability to distinguish true from false predictions.

The most far-reaching calibration studies have been conducted on data from weather forecasters, particularly those whose organizations require them to make probability estimates (Winkler and Murphy 1968, Murphy and Winkler 1977). They find that weather forecasters are quite well-calibrated. However, early studies of children by Cohen and his colleagues found that their probabilities summed to less than 1 (Cohen, Dearnaley and Hansel 1956), and indicated overconfidence when the actual probability of an event was low and underconfidence when the true probability was high (cf. Cohen 1964).

A large number of studies of laypeople on general knowledge (e.g., Fischhoff et. al. 1977), difficult tasks (Fischhoff and Slovic 1980), performance tasks (Howell 1972), and various experts (Christensen-Szalanski and Bushyhead's 1981 study of physicians, Oskamp's 1962 investigation of clinical psychologists, and Stael von Holstein's 1971 study of weather forecasters' judgments of relatively unfamiliar events) show a general tendency towards overconfidence. Furthermore, overconfidence appears to increase with task difficulty (Nickerson and McGoldrick 1965, Pitz 1974) and varies inversely with knowledge. A second widespread finding in calibration research is that when people are asked to provide quantile or other distributional estimates involving the spread of probabilities, they tend to construct distributions that are too tight (e.g., Pickhardt and Wallace 1974; see Lichtenstein et. al. 1982 for a review). Furthermore, training does not seem to reduce this tendency (Lichtenstein and Fischhoff 1980).

The picture that emerges from these two research streams, then, effectively says that numbers are better than words for expressing subjective uncertainty, but that when people's numerical estimates can be checked against frequency data, they are found to be using the numbers incorrectly. Several competing explanations for the overconfidence and excessive distributional concentration effects have been proposed, and those will be discussed in the section on heuristics. However, it is worth observing here that none of those explanations consider whether the meanings that the estimation tasks have for the subjects correspond to the probabilistic framework assumed by the investigators themselves. Moreover, the relationship between the highly artificial estimation tasks in laboratory settings and their real-world counterparts is unclear.

5.2.3 Selective Attention to Evidence

At least three major kinds of bias in attention to and weighting of evidence have been unearthed by psychologists in the study of judgment under uncertainty. These include a tendency to ignore or discount negative and disconfirming evidence (the confirmation bias), conservative revisions of probability estimates as evidence is presented sequentially (the conservatism bias), ignoring base-rate information in the presence of other (perhaps worthless) evidence (the base-rate fallacy), and ignoring sample size in accounting for statistical variation (the sample size fallacy). I will discuss these in the order given.

The confirmation bias has long been noted by philosophers of science and it has an extensive literature in both cognitive and social psychology. Indeed, anyone teaching introductory scientific methods classes becomes familiar with it as students grapple with the concept that a disconfirming experiment is just as important as a confirming experiement. However, as Fischhoff and Beyth-Marom have noted (1983), accounts of the confirmation bias in psychological research encompass diverse phenomena. In studies of logical reasoning about simple binary hypotheses, Wason (1960, 1968) demonstrates that subjects are likely to ignore or fail to seek negative or disconfirming information, and if it is given to them they do not use it effectively (see also Johnson 1972). Studies of intuitive reasoning have documented that while lay judges are normative ('logical') in their use of modus ponens, they fail to be so when faced with a problem requiring modus tollens (Wason and Johnson-Laird 1972).

In a slightly different vein, several experimental studies show that in a 2x2 contingency table people tend to judge the strength of association between two dichotomous variables that are coded in terms of valency (e.g., 'yes-no' or 'present-absent') primarily on the basis of the 'positive hits' cell only (Jenkins and Ward 1965, Smedslund 1963 and 1966, Shweder 1977, Arkes and Harkness 1983). A similar kind of bias has been found in studies of how people assess covariation between two variables (Crocker 1981, Schustack and Sternberg 1981, Shaklee and Mims 1982).

Yet another line of investigation has found that people are inclined to ignore, discredit, or reinterpret information that is contrary to a belief they already hold (Nisbett and Ross 1980, Lord, Ross, and Lepper 1979). Furthermore, they tend to be biased in favor of 'positive tests' that are likely to reconfirm what they already believe in (Snyder, Tanke, and Berscheid 1977, Snyder and Swann 1978, Snyder 1982, Klayman and Ha 1987). The view that emerges from all this is that the 'confirmation bias' refers not only to selective attention but also to selective information seeking, discounting, interpretation, and testing.

In contrast, the possibly related bias towards 'conservatism' in the updating of subjective probabilities refers to a relatively self-contained series of studies using the Bayesian updating framework as the normative benchmark. Edwards' (1968) summary of this research claims that it takes from two to five observations for a typical subject to shift her or his subjective probabilities as far as one observation would do according to Bayes' theorem. Edwards' principal explanatory claim is that conservatism is due to people's tendency to misaggregate data (see also the conjunction fallacy below).

To a considerable extent, the early comparisons between ordinary people's probability judgments and the normative probabilistic framework were guardedly optimistic. The confirmation bias and its cousins could be explained as motivationally induced, while Edwards' assessments, the studies by Cohen and his colleagues during the 1950's, and Petersen and Beach's 1967 review of statistical intuitions suggested that human judgment differs from the normative model only in relatively minor fashion (e.g., by a scalar multiplier). The findings that people ignore base-rate and sample size, however, challenged this optimism by indicating that people differ more profoundly from the Bayesian model than conservatism implies, and that those differences may have a cognitive rather than a motivational basis.

The first experimental study investigating the impact of base-rate data was reported by Kahneman and Tversky (1973), although the neglect of base-rates by laypeople (e.g., Huff 1959) and experts (Meehl and Rosen 1955) was not a new claim. Kahneman and Tversky presented subjects with a brief personality description of an individual (the Tom W. problem) and then asked them to estimate the probability that this individual worked in a particular occupation or specialized in a given course of study. Subjects were also either implicitly or explicitly provided with base-rate information concerning the percentages of people in such occupations or study programs among the sample from which this individual was chosen, and with indications of the unreliability of either the description (it was based on a projective personality test) or the selection procedure (the individual was randomly selected). They found that the subjects' probability estimates did not correlate highly with base-rates, nor did altering the base-rates for different subject groups produce corresponding differences between those groups in probability estimates. They also reported evidence suggesting that subjects do attend to base-rates if the personality descriptions are uninformative, but ignore base-rates if the descriptions are informative, even when the information is normatively irrelevant (nondiagnostic).

Not all base-rate studies support the 'base-rate fallacy' findings. McCauley and Stitt (1978), for instance, report a correlation between judged base-rate of traits and the subjective probabilities of those traits given a particular nationality. Ginosar and Trope (1980) fail to replicate Kahneman and Tversky's findings of base-rate neglect in the face of specific but nondiagnostic information, but instead report a substantial impact from base-rates. Some other researchers have found that when the base-rate has a causal status that it exerts much greater influence on judgments than when it seems incidental (e.g., Azjen 1977). These findings have provoked a rash of studies in which both procedural and evidential variables are manipulated in attempts to understand the conditions under which base-rates are neglected or utilized (see Borigida and Brekke 1981 for a review).

Several issues are somewhat blurred by the fact that a number of variants on the base-rate problems used in these studies possess characteristics that may arguably influence the subjects' responses. The 'Tom W.' problem is an example of what Bar-Hillel (1983) calls the 'social judgment paradigm'. In this kind of problem, the probabilities are usually not given explicitly and the 'diagnostic' or indicative information contained in the personality descriptions invites comparisons between 'Tom W.' and the categories to which he might belong. The 'textbook paradigm' for a base-rate problem, on the other hand, presents the base-rate and diagnosticity probabilities precisely. A prototypical example of this kind of problem is the much-studied 'cab problem' from Tversky and Kahneman (1979). Subjects are informed that 85% of the cabs in a city are Green and 15% are Blue. A cab was involved in a hit-and-run accident in the city last night. An eyewitness identified the offending cab as Blue. When tested by the court for his ability to discriminate between Green and Blue cabs under appropriate visibility conditions, the witness made correct identifications in 80% of the cases from a sample composed of 50% Blue and 50% Green cabs.

The chief difference between this problem and the 'Tom W.' variety is that there is clearly a unique normative solution, assuming that the witness is equally accurate in identifying Green and Blue cabs. In the social comparison problems, on the other hand, the subject is not informed of the diagnosticity of the indicative information (e.g., the personality descriptions based on a projective test), so there is no unique normative solution, notwithstanding the fact that subjective probabilities should be affected by variations in the base-rate information. One possible source of vagueness even in the 'cab problem' is the required assumption that the eyewitness account is equally diagnostic regarding the hypotheses of Green or Blue cab. Many medical or industrial fault diagnostic tests, by contrast, have differing false positive and false negative rates (see, for instance, the study of Harvard Medical School staff and students' responses to a fairly realistic base-rate problem in Casscells, Schoenberger, and Grayboys 1978).

A bias that is related to the base-rate fallacy is insensitivity to sample size. Tversky and Kahneman (1974, also Kahneman and Tversky 1972) again provided the landmark studies of this phenomenon, although it had at least indirectly occurred in previous studies of 'conservatism' (Edwards 1968, Slovic and Lichtenstein 1971). A classic problem presented to subjects asks them to consider two hospitals one in which about 45 babies are born each day and another in which only about 15 babies are born daily. For a period of a year, each hospital has recorded the number of days on which more than 60% of the babies born were boys. Subjects are then asked which hospital recorded more such days. Most of the subjects respond that the hospitals will have recorded about an equal number of such days, whereas the normatively correct answer is that the smaller hospital is likely to have recorded a much larger number.

In an earlier article Tversky and Kahneman (1971) castigated psychological researchers for a 'belief in the law of small numbers', whereby researchers were found to be overly optimistic about the statistical power of their studies and replications. They also observed, in their 1972 study, similar neglect of sample size considerations in people's constructions of subjective probability distributions for proportions of gender, heartbeat types, and height intervals. The percentages or proportions assigned to gender categories, say, or height intervals did not vary with the sample size given to subjects, which Tversky and Kahneman interpreted to mean that this information was not being taken into account.

5.2.4 Aggregation and The Conjunction Effect

Early investigations by Cohen and his colleagues into how ordinary adults and children combine subjective probabilities led them to conclude that although adults produced answers tallying with the product rule for combining probabilities of independent events (Cohen, Dearnaley, and Hansel 1958a), children produced estimates that were larger than prescribed by probability theory (Cohen, Dearnaley, and Hansel 1958b). They also found that in gambles involving uncertainty due to a combination of skill and chance, although the product rule was used in combining uncertainties related to skill to determine subjective probabilities of success, uncertainties related to chance were combined additively (Cohen and Hansel 1959).

A second finding in the 1959 study echoed the tendency they had observed in children: adult subjects sometimes produced subjective probabilities of success in the presence of two sources of uncertainty that were greater than corresponding probability estimates in the presence of only one source of uncertainty. Furthermore, they found this effect was enhanced when the additional uncertainty had a larger subjective probability of success attached to it than was attached to the first source. Later, Cohen, Chesnick, and Haran

(1972) extended these findings on choices among gambles to conclude that people tend to overestimate the probability of conjunctive events and underestimate the probability of disjunctive events. These findings are perhaps the earliest experimental evidence pointing to the so-called 'conjunction fallacy', and other variants on the same phenomenon have been reported in Wyer (1970), Bar-Hillel (1973), Goldsmith (1978), and Beyth-Marom (1981).

However, the effect has been most directly studied and popularized, once again, in a series of studies by Tversky and Kahneman (1982, 1983). Their findings, termed "perhaps the most shocking example" of normative violations of logic by Edwards and von Winterfeldt (1986: 666), point to a consistent violation by both naive and well-trained subjects of the rule that $P(A\&B) \leq \min(P(A),P(B))$. Their first series of experiments were couched in social judgment terms similar to their investigations of base-rate fallacies. Subjects were presented with target persons and accompanying descriptions of those persons. They were then asked to judge the probability that the target possessed certain characteristics, some of which were simple and others compound. As in the Cohen et. al. findings, most of the subjects in these studies tended to judge $P(A\&B)$ as greater than $P(B)$, say, but less than $P(A)$ when $P(A)$ was considered quite high and $P(B)$ rather low. The effect did not emerge when both $P(A)$ and $P(B)$ were low, and was exaggerated when both were high.

The 1983 study presented university students with a more explicit problem. They were asked to consider a regular six-sided die with four green faces and two red faces. The die was rolled 20 times and the sequence of greens (G) and reds (R) recorded. The students were then asked to select which sequence they would prefer to bet on from the following three: (a) RGRRR, (b) GRGRRR, and (c) GRRRRR. Their instructions then went on to admonish them that sequence (a) could be obtained from (b) by eliminating the first G, and therefore by the conjunction rule (a) should be the more probable. However, sequence (b) appears to be more representative of the distribution of colors on the die faces, since it contains more G's than (a) does. Of the subjects playing for real payoffs ($25), 65% chose (b) and 62% of those playing hypothetically chose (b), instructions notwithstanding.

5.2.5 Randomicity and the Illusion of Control

An intriguing class of findings that purport to demonstrate normative violations refers to people's perceptions of and responses to randomicity and 'chance' or 'luck'. Unlike the previously discussed normative violations which involve deviations from the standard probability calculus, these are violations against normative criteria for defining and detecting randomicity itself. As clearly indicated by the material covered in Chapter 3, probabilists themselves are hard-pressed to agree on ultimate definitions and tests of randomicity, and some branches of applied mathematics (e.g., fractal theory, strange attractors, and algorithmic information theory) have recently distinguished 'deterministic chaos' and 'algorithmic randomness' from older definitions of random processes. Randomicity itself, then, is the weak underbelly of probability as a normative framework. Nonetheless, the bulk of psychological work in this area has unquestioningly accepted a frequentist perspective and compared human performance with frequentist prescriptions.

One of the oldest and best-known findings of this kind, the Gambler's Fallacy, is a special case of an apparent tendency for people to see patterns, predictability, and even causation in randomly generated data. Given a long run of heads in coin-tossing, the gambler who believes the coin is fair will also believe that a run of tails is now due because in the long run, the number of heads and tails should balance out. A related finding is that people tend to underestimate the extent to which 'patterns' in the form of long runs, clustering, and seemingly regular subsequences occur in randomly generated data. Feller's (1968: 160) well-known anecdote concerns the general belief by Londoners during WW II that German bombing patterns were not random because some areas of London were repeatedly hit while others remained unscathed. The fact that a distributional analysis of the bombing patterns fit a Poisson model quite well led Feller to conclude that untrained people see regularity and clustering in randomicity. Kates (1962) has made a similar claim about perceptions of disasters by threatened populations. Kahneman and Tversky (1972) presented subjects with sequences of heads and tails from 6 coin tossings, who in turn rated sequences like HTHTTH as more likely than 'regular'-looking but balanced sequences such as HHHTTT or unbalanced ones like HTTTTT. Furthermore, when asked to generate 'random' sequences of coin tossings, subjects tended to produce sequences which stayed closer to the proportion of 1/2 heads than the binomial distribution would predict, thereby overly replicating the long-run tendency in every subsequence and not permitting sufficiently long runs of heads or tails (Tune 1964).

The tendency to attribute causation to randomly generated events has been reported in several quarters. Furby (1973) and Kahneman and Tversky (1973) report that fluctuations in human performance are invariably attributed to the skill and effort of the performer while the regression effects predicted by the classical statistical framework is ignored. When rewards and punishments are being issued for good and bad performance, Kahneman and Tversky argue, the regression effect leads the would-be reinforcer to believe that punishment is more effective in improving performance than reward is in maintaining good performance. Similar violations of normative prescriptions have been observed among stock market analysts, who have overdetermined models for arguably random short-term fluctuations in share values and stock indexes (Fischhoff 1980).

Finally, several reseachers report tendencies for people to believe that they can exert control over random events. The most popular examples come from studies of gambling behavior (e.g., Henslin's 1967 study of dice players' beliefs and techniques). Experimental studies of this effect have been carried out by Strickland, Lewicki, and Katz (1966) and Langer (1975, 1977). Do people believe that chance can obtrude in tasks that require skill or effort? In the attribution literature, a widely reported self-serving bias is for successful task performers to attribute their outcome to skill or effort and unsuccessful performers to make external attributions for their failure, chance being among these (see Ross and Fletcher 1985 for a summary of this literature). A more important result, however, comes from a study by Elig and Frieze (1979), in which subjects were asked to explain their performances (on anagram problems) in their own words. A content analysis of their explanations revealed that ability and task difficulty were the most frequent kinds of attribution, with effort coming third. Heider's (1958) original formulation of attribution theory, however, posited four kinds of causes: Ability, effort, task difficulty, and luck. Luck was rarely mentioned by these subjects, thereby perhaps reflecting a lack of belief that luck could influence performance in a task requiring skill and/or effort.

5.2.6 Reframing Biases and the Independence Axiom

Our final class of normative violations involves the connections between judgments of probability and utility in decision making. While Jungermann (1983) correctly comments that biases and heuristics in utility judgments themselves remain largely unexplored, recent research in psychology and economics has focused on certain normative criteria specifying the 'rational' interface between probability and utility. This literature is voluminous and complex, so the coverage here must necessarily be confined to an outline of the main points.

The central claim of classical decision theory is that rational decision makers maximize expected utility, which in turn is constrained by the so-called independence axiom to be 'linear in the probabilities'. That is, given an exhaustive set $\{x_1, x_2, ..., x_n\}$ of possible outcomes, if $P = (p_1, p_2, ..., p_n)$ represents the probabilities of obtaining those outcomes and U_i denotes the utility assigned to outcome x_i, then the valuation function $V(P)$ is defined by

$$V(P) = \sum_{i=1}^{n} U_i p_i \qquad (5.1)$$

Given two probability distributions P and Q, P is preferred to Q if and only if $V(P) > V(Q)$, in which case the risky decision prospect that entails P is said to 'stochastically dominate' the prospect involving Q.

These assumptions do not appear constraining at first glance, since the utility function itself can account for general risk aversion or risk seeking. The classic essay by Daniel Bernoulli in 1738 argued that people's utility function for money is concave with respect to monetary value, reflecting a general aversion to risk that decreases with wealth. A typical illustration of this effect asks the subject to choose between a sure gain of $750 and a 75% chance of obtaining $1000. If monetary value is used to represent utility then the two prospects are equivalent in terms of expected utility, but most people prefer the sure gain.

However, the expected utility model's linearity assumption does impose a crucial constraint on normatively correct behavior, and to illustrate this I will use Machina's (1982, 1987) elegant representation. He considers a set of three outcomes such that $U_3>U_2>U_1$, and observes that since $p_2 = 1-p_1-p_3$, we may represent the domain of possible probabilities for the x_i by the triangle shown in Figure 5.1. Because x_3 is preferred to x_1, any increase in p_3 should increase preference (that is, any movement in a northwest direction within the triangle). Now we may represent the expected utility model by graphing 'indifference curves' inside the triangle. These curves must be lines because of the restriction imposed by (5.1) via the independence axiom. The lines one would expect if an individual is neither risk-averse nor risk-seeking (e.g., if their subjective utility regarding monetary outcomes was simply proportional to monetary value) would make a 45-degree angle with respect to the axes. Part A of the figure shows the steeper indifference curves of a risk-averse individual, while part B shows the flatter lines of a risk-seeking individual.

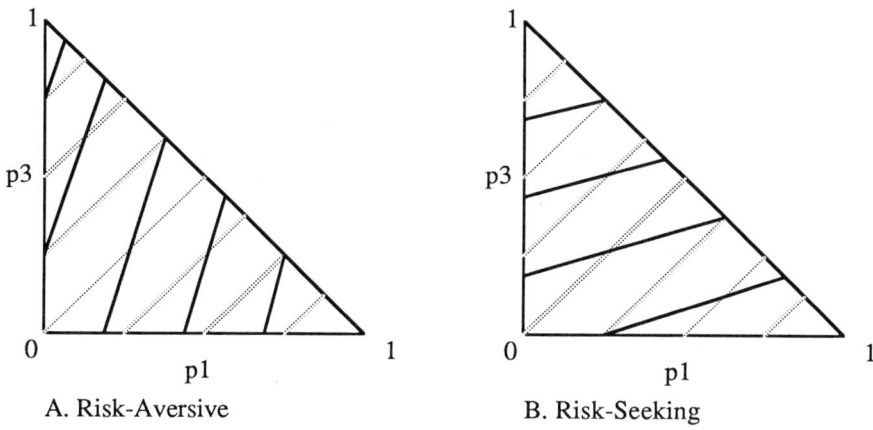

A. Risk-Aversive   B. Risk-Seeking

Figure 5.1. Indifference Curves

Both normative debates (e.g., Allais 1953) and empirical research reveal a large class of systematic violations of the independence axiom. An example of a special case in this class is the famous Allais paradox. The subject is asked to indicate her or his preferences for two pairs of gambles:
 (a) {1.00 chance of $1M} versus
 (b) {0.10 chance of $5M, 0.89 chance of $1M, and 0.01 chance of $0}; and
 (B) {0.10 chance of $5M and 0.90 chance of $0} versus
 (A) {0.11 chance of $1M and 0.89 chance of $0}.

Under the expected utility hypothesis and the independence axiom, rational individuals should prefer either the pairs {a,A} (indicating risk aversion) or {b,B} (indicating risk seeking). However, both intuitive evidence and empirical studies (Edwards 1961, Morrison 1967, Slovic and Tversky 1974, Moskowitz 1974) find that the modal individual favors the pair {a,B}. Allais managed to entrap several big-name decision theorists with this problem, including Savage himself. More generalized versions of this effect, called the 'common consequence' effect, have been empirically investigated and supported in MacCrimmon and Larsson (1979) and Chew and Waller (1986).

A second subclass of violations is the 'common ratio' effect. An example from Kahneman and Tversky (1979) is their finding that a majority of their subjects preferred a sure gain of $3,000 to an 80% chance of winning $4,000, but also preferred a 20% chance of winning $4,000 to a 25% chance of winning $4,000. The common ratio effect includes the 'certainty effect' described by Kahneman and Tversky (1979) and the 'Bergen paradox' in Hagen (1979). The common ratio effect involves pairs of gambles with the following form:
   (c) {p chance of $X and 1-p chance of $0} versus
   (d) {q chance of $Y and 1-q chance of $0};
   (C) {rp chance of $X and 1-rp chance of $0} versus
   (D) {rq chance of $Y and 1-rq chance of $0},
where p>q, 0<X<Y, and 0<r<1. The normative theory would prescribe the pairs of preferences {c,C} or {d,D}, but the dominant empirical finding yields the pair {c,D}.

Kahneman and Tversky (1979) have argued that Bernoulli was only half-correct when he argued that people are risk-averse. Their hypothesis is that people are risk-averse about possible gains, but risk-seeking when faced with possible losses. They then demonstrate that such an orientation leads to counter-normative choices among gambles. Unfortunately, even some of their own examples do not support this hypothesis. For example, they find (cf. Problems 8 and 8' in the 1979 article) that when probabilities are miniscule, people are risk-seeking in the face of possible gains and risk-aversive in the face of possible losses. They argue from this that people's decision weights are unstable in the region of low probabilities, and that they may even ignore such probabilities altogether.

Machina's (1982, 1987) account persuasively integrates these normative violations in terms of the representation provided in Figure 5.1. He argues that the common consequence, common ratio, and several other related violations of the independence axiom may be understood as symptomatic of a human tendency to be risk-aversive when there is a high probability of gaining the preferred outcome and risk-seeking when that probability is low. In Figure 5.2 (part B) this is represented by indifference curves which 'fan out' from the vicinity of the origin. Part A shows the points corresponding to the Allais paradox in the context of the expected utility indifference curves prescribed for a risk-aversive individual. If we move along the heavy line from b to a or from B to A we cross indifference curves in a positive direction, so the dominating pair is {a,A}. In part B, however, the same exercise demonstrates that the dominating pair is {a,B}, in accordance with most empirical findings.

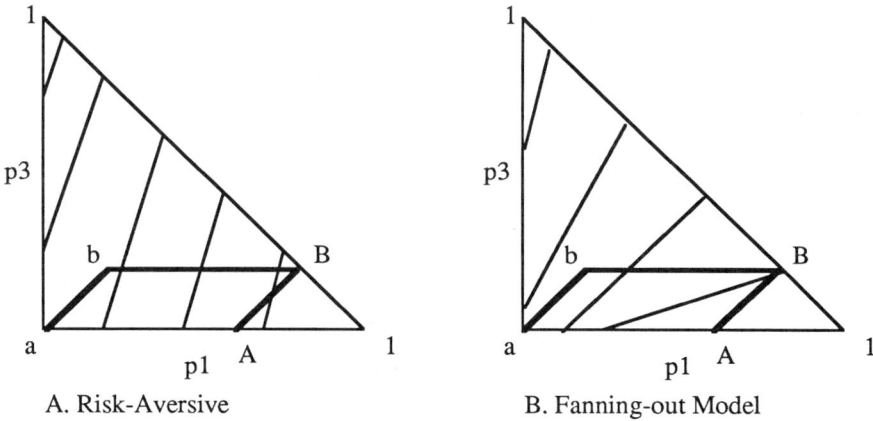

A. Risk-Aversive   B. Fanning-out Model

Figure 5.2. The Allais Paradox

If Machina's 'fanning out' indifference curve hypothesis holds true for a large percentage of humanity, then the maximization of expected utility hypothesis (as well as the independence axiom) fails to have descriptive as well as prescriptive validity. However, even more fundamental challenges confront it through two other kinds of normative violations that have been empirically demonstrated. One is the 'preference reversal' phenomenon, in which subjects' self-reported choices among pairs of gambles are normatively inconsistent with the rank order of their selling prices for those same gambles. An example from Lichtenstein and Slovic (1971) asks subjects to compare the following two bets: (E) .90 chance to win $4 and .10 chance to lose $2 versus (F) .30 to win $16 and .70 to lose $2. Most subjects choose E, but if asked to nominate a selling price for tickets on these two gambles they tend to ask for a higher price to sell F. This finding has proven quite robust, and even seems to hold for professional gamblers (cf. Grether and Plott 1979, Mowen and Gentry 1980).

The second challenge comes from framing effects, in which probabilistically equivalent but descriptively alternative statements of a decision problem yield different expressed choices. The most famous examples come from Tversky and Kahneman (1981), although earlier studies in the context of gambling had isolated this effect (e.g., Slovic 1969, Payne and Braunstein 1971). Subjects in one group are asked to choose between two programs to combat a deadly epidemic which threatens 600 people. Program A will save 200 lives, while Program B has a 1/3 probability of saving 600 lives and 2/3 probability of saving no lives. Most subjects in this group favor Program A. Subjects in another group are asked to choose between Program C which will result in 400 people dying, and Progam D which has a 1/3 probability that no one will die and a 2/3 probability that 600 will die. Most subjects choose Program D. Of course in terms of expected utility, Programs C and D are indistinguishable from A and B, respectively. Although some framing effects may be explained in terms of normative violations of expected utility previously discussed, not all of them can. Of the various classes of violations of expected utility, this one has so far proved the most impervious to explication in terms of basic underlying principles.

Finally, several researchers have addressed the question of whether valuation actually influences the perception of uncertainty itself. The most obvious hypothesis of this kind is the 'optimistic bias', that desirable events will be perceived as more likely and undesirable events as less likely. Milburn (1978) asked subjects to predict events up to four decades into the future and found evidence for this effect. Weinstein (1980) and Zakay (1983) also found support for the effect. A dissenting study was recently reported by Wright and Ayton (1987), but they concede that their results might be explained by the fact that the forecast period was short and the events relatively impersonal in comparison to the previous studies.

If an optimistic (or even pessimistic) bias is a widespread phenomenon, then it undermines classical decision theory even further by demonstrating that subjective probability judgments are influenced by the subjective utility accorded the outcomes whose probabilities are being judged. Subjective expected utility itself is then confounded by the interaction between its two components, resulting in 'double (dis)counting'.

5.2.7 Normative Violations: Is the Evidence Clearcut?

Before considering the various explanations of what people do when they deviate from probabilistic prescriptions, it is reasonable to inquire into the soundness of the evidence for those deviations. As previously indicated, the research into judgmental errors and biases has been extensively criticized by several writers. While some of the critics have argued against the primacy of Bayesian probability as a normative paradigm, many of them have either claimed that the problems themselves are open to more than one normatively correct interpretation or that the subjects' 'deviations' may be explained by some additional reasonable normative considerations. We may unpack these concerns slightly by posing the following questions:
1. Are the problems or tasks unambiguous?
    1.1. Is there a unique normative (probabilistic) solution?
    1.2. Are the normative criteria themselves clear?
2. Are the problems realistic and credible?
3. Do the subjects understand what the experimenters want?
4. Are apparent deviations by subjects from normative solutions explicable in terms of trivial adjustments (e.g., weighting)?
These questions will be discussed in turn.

As writers like Navon (1978) and Fischhoff and Beyth-Marom (1983) have pointed out, in order for the Bayesian framework to be applicable the problem must be structured in such a way that every competing hypothesis is well-formed, the hypotheses themselves constitute a partition of an exhaustive space of possibilities, and every piece of information is entirely reliable. Many of the base-rate problems (especially those in which the base-rate probabilities are not presented explicitly) fail on these counts, so much so that Scholz (1987: 21) argues that the terms 'normative solution' and 'base-rate fallacy' should be placed in semiquotes.

There are at least three kinds of ambiguities that have been located in base-rate problems by critics. The most common is the lack of specificity or explicitness about probabilities. Thus for instance the 'troops massing at the border' problem that has been employed in several studies specifies that the general knows 'from past experience' that the probability of the enemy invading whenever his troops are massed at the border is 0.75. The subject is asked to assume that whenever intelligence sources report that the troops are massing, they really are there. The questions posed for the subject are, first, to determine the probability of invasion given a report that the troops are massing, and second, to recommend a response by the general. What is missing is the probability of an intelligence report of troops at the border given that massing is occurring. In the absence of that information, there is no unique normative solution. Perhaps more importantly, it is possible that the unevenness of the available information in problems like this one make them seem unrealistic to the subjects. How can the general know from past experience that P(I:M) = 0.75 (where I = invasion and M = massing troops) when he doesn't know what P(I:R) is (where R = intelligence report of massing)? After all, where would the informational base for his past experience have come from? He must either have had exhaustive and reliable direct experience of M over time or exhaustive and reliable intelligence about M over time. The latter seems much more plausible, in which case P(R:M) = P(M:R) = 1 and 0.75 is the correct answer, as most subjects reply.

A second kind of ambiguity arises from alternative possible theories that differ in their implications for the relevancy of base-rates. Bar-Hillel (1983) point out a problem of this kind in the context of a doctor working in an emergency ward of a hospital. This doctor knows that during holiday weekends the ratio of stomach disorders to heart trouble is higher than on ordinary weekends. A patient arrives presenting symptoms that suggest the onset of heart trouble but that also sometimes are manifested by a stomach disorder. One theory accounts for the increase in stomach disorders during holiday weekends by the nonavailability of doctors on such weekends, but discounts this information by the supposition that anyone with this patient's symptoms would be referred to the emergency ward of the hospital anyway. Therefore the base-rate should be ignored. A second theory assumes that the increase in stomach disorders arises from different eating and drinking habits, and therefore the base-rate is relevant.

The third source of ambiguity arises from an incomplete or faulty normative solution specification. In base-rate and other similar probability problems requiring the subject to aggregate information, the Achilles heel is the often implicit assumption that diagnosticity is independent of base-rates. For example, Birnbaum (1983) argues persuasively that even textbook-paradigm base-rate problems such as the famous 'cab' example contain an unrealistic assumption that eyewitness false-alarm rates are independent of the base-rates of cab colors. According to evidence from the signal detection literature, the false-alarm rate depends on the base-rate, and a failure to specify the nature of that dependence leads to an unspecifiable normative solution to such problems.

A related difficulty arises from the interpretation of the sample space involved in the problem statement. Recalling from Chapter 3 that professional risk assessors can misunderstand one another's statements about sample spaces, it does not seem implausible that experimental subjects could do so too. Scholz (1987:18) provides an example of two possible interpretations of the statement "tests over several years have shown that 75% of the tubes are okay and 25% are defective" in his TV problem. And in the problems involving diagnostic tests for which 'false positive' rates of, say, 5% are given, naive subjects could either assume this to mean that 5% of those who test positive do not have the indicated condition, *or* that 5% of those who do not have the indicated condition test positive. These alternative interpretations yield dramatically different normative solutions.

A more serious kind of problem for researchers claiming to have found evidence of subrationality arises when the normative criteria in probability theory itself are unclear. This is not the same as the argument against the normative dominance of probability theory, which will be discussed later, but concerns the extent to which probability theory achieves consensus on particular issues. As the discussions in Chapter 3 tried to make clear, probability theory is by no stretch of the imagination consensual on all fundamental issues. The probability calculus itself is perhaps the source of highest agreement among the competing schools, and coincidentally that is where much of the research on judgment biases has focused attention. The group of studies most threatened by the lack of consensus among probabilists is those reporting unreasonable perceptions and concepts of randomicity by naive subjects.

As Lopes (1982) argues, these studies all implicitly assume that there is a completely specified and agreed-upon normative perspective on criteria for defining and detecting randomicity. In Chapter 3 we saw that this is not the case even for the relative frequentist school of thought. The difficulty lies in the proposition that all subsequences of a random sequence must also have randomicity. Lopes points out that in the production of random-number tables, sequences produced by presumably random physical processes often have failed mathematical tests of randomicity for exactly the same reasons that sequences produced by human subjects do. Spencer-Brown (1957) and Popper (1959) both present philosophically sophisticated arguments (and in Spencer-Brown's case, amusing paradoxical counter-examples) against the von Mises type of approach to defining randomicity, and in favor of criteria for randomicity that resemble the naive individual's criteria. Lopes brings the issue home to psychology by observing that the most damaging evidence in the case against Cyril Burt was that his data were too consistent to be believable, a judgment based on criteria for inter-study variation that are identical in principle to the criteria naive subjects use for judging inter-trial variation in games of chance.

The credibility of the source and type of information provided in problems can justifiably affect judgments based on that information, although this is not taken into account by any conventional probabilistic approach. The importance of this point for studies of judgment stems from the potential of credibility to generate artifactual findings or even confound experimental manipulations. As an example, consider once again the Asian disease problem from Tversky and Kahneman (1981), described in the previous section. In one condition, subjects are presented with the choice between Program A in which 200 people will be saved and Program B which has 1/3 probability of saving 600 people and 2/3 probability of saving none. In the second condition they are presented with the choice of Program C in which 400 people will die and Program D which has a 1/3 probability of no one dying and 2/3 probability of 600 people will die. First, as R.R. Macdonald (1986) observes, it is unnatural to predict an exact number of deaths in this kind of forecast. Most subjects would expect some approximate number instead, and may well interpret these exact numbers as approximations. Furthermore, Programs B and D sound quite strange because either a fixed number of people will be saved (die) or none will be saved (die). In short, the formulation of these estimates is unbelievable, and it is quite possible that B and D are less credible than A and C. If true, then that confounds the effect sought by the researchers. In the 'troops massing at the border' problem reviewed above, the modal subject's response may simply be an artifact of the subject's attempt to impose additional assumptions on the problem description in order to make it credible (i.e., that if the general knows from past experience that $P(I:M) = 0.75$ then $P(R:M)$ and $P(M:R)$ must both equal 1).

These considerations lead us to the issue of whether subjects and experimenters share the same definitions of the problems and tasks employed in these studies. After all, as the literature on scaling verbal expressions of probability indicates and as several writers have observed (e.g., Macdonald 1986), the concept of probability itself is both technically controversial and pluralistic in everyday usage. Furthermore, the inherent uncertainty of even simple problems is hardly limited to probability. Berkeley and Humphreys (1982) argue that there are seven kinds of uncertainty entailed in such problems that must be resolved before one can conclude that subject and experimenter viewpoints coincide, only four of which are explicitly addressed by normative decision theory. They further point out that there are two ways of establishing this common understanding. One is through process tracing methodology, whereby the experimenter uncovers the definitions and procedures actually used by the subjects while working through the problems. The second, and much more common, approach is to rely on task instructions to specify the problem sufficiently to foreclose any alternative interpretations. The major point in Berkeley and Humphreys is that while the second approach is inexpensive and quick, it is extremely difficult to achieve problem closure.

The process tracing approach has seldom been applied in judgmental bias research (exceptions include Huber 1983, and Montgomery and Svenson 1983). Huber, for instance, argues that some subjects relate the information provided in the problem to previously stored knowledge, thereby starting from a different understanding of the problem than that intended by the experimenter. A rare and thorough study of subjects' 'think-aloud' protocols while working through base-rate problems has been provided in Scholz' recent book (1987). Scholz finds some support for Bar-Hillel's (1984) suggestion that some subjects may invert the conditional probability requested in the problem. His evidence also suggests that subjects differ markedly in whether they employ quantitative versus nonquantitative and analytic versus intuitive strategies.

Finally, we may ask whether the nonnormative responses of subjects can be trivially explained by minor adjustments to the normative paradigm. Again, the base-rate problem has received the most attention by writers who have considered this possibility. Birnbaum's (1983) signal detection theoretic formulation is perhaps the most serious candidate, since it potentially explains a wide variety of responses. Another proposal is that subjects differentially weight the importance or relevance of information components, and if these weights are taken into account then the subjects' adjusted probabilities may be normative after all. Both post-hoc (Birnbaum and Mellers 1983) and empirically grounded (Scholz 1987) versions of this approach have been tried, but neither have proved convincing.

The overall picture that emerges from these critiques is that, with the possible exception of the studies of randomicity judgments, the nonnormative responses of subjects in judgment under uncertainty require explanation. Even after one accounts for the lack of specification in many of the problems and tasks, possible confounding or artifactual factors, and the potential for nonshared understandings between subject and experimenter, there is still a sufficient residuum of evidence to suggest strongly that Bayesian (or any other kind of) probability does not fit what many people do when making judgments and decisions under uncertainty. The dominant trend in the literature has been to account for subjects' judgments by fairly simple 'heuristics'; a minority trend is to propose an alternative normative framework that fits the data better than probability theory. I will review the mainstream explanations first, since much writing on alternative normative perspectives in the psychological literature is a reaction to the mainstream proposals.

5.3 Explaining Human Heuristics
5.3.1 Availability, Relevance, and Specificity

There are at least three major explanations referring to selective attention or retrieval for the seemingly counter-normative judgments people make when assessing probabilities. The most popular of these is the availability heuristic, which Tversky and Kahneman (1973) characterize as the tendency to assess likelihood by the ease of the relevant mental operations for information retrieval, construction, or association. They call this kind of judgment an assessment of 'associative distance'. People find that more likely occurrences or cases are easier to recall or imagine than unlikely ones, and so the argument is that they make use of this apparent association to judge probability. Tversky and Kahneman claim that this particular heuristic is most likely to occur when individual instances (rather than the generic features) of a category are made salient by the problem formulation.

A related heuristic (Kahneman and Tversky 1982c) is the 'simulation' heuristic, whereby people construct multiple possible scenarios in attempting to assess such things as predictions, counterfactuality, causation, and conditioned probabilities. They claim the simulation heuristic is biased in favor of events for which one readily imaginable plausible scenario can be found, and against those that could occur in a variety of unlikely ways.

Bar-Hillel (1980), by contrast, asks what cues might make some information more available or salient. Her broad thesis is that the relevancy of the information influences whether people will attend to or recall it or not. Her explanation accounts for a variety of findings suggesting that a causal link between base rate and target event increases subjects' attention to base rates (Hammerton 1973, Lyon and Slovic 1976, Carroll and Siegler 1977), as well as some widely reported ego-enhancing biases in causal attribution (cf. Ross and Sicoly 1979). Indeed, so apparent is causality as a criterion cue that a number of authors have proposed hypotheses about the impact of causal attribution on the assessment of probabilities.

Tversky and Kahneman's (1980) primary claim on this topic is that causal information has a greater impact on the judged probability of a consequence than diagnostic information on the judged probability of a cause. Other researchers have studied the impact of cues to causality such as apparent covariation, temporal order, similarity of cause and effect, and contiguity or copresence in time (see, e.g., Einhorn and Hogarth 1986 for a review). Still others have found differential effects on assessed likelihood depending on the order in which subjects are asked to consider reasons for and against the occurrence of an event (e.g., Hoch 1984).

An attempt to synthesize these various hypotheses into a coherent decision theoretic framework has been mounted recently by Pennington and Hastie (1986, 1987). They term their perspective a theory of 'explanation-based' decision making, whose distinctive assumption is that decision makers construct an intermediate representation of the evidence and impose a causal organizing structure onto it, and this representation then becomes the basis for the decision rather than the raw evidence. The representation often is domain-specific in its structure and resembles a story about the evidence.

In the first stage of their three-phase decision model, the decision maker constructs the representation, which is essentially a causal or explanatory interpretation of the evidence. The second stage involves learning or generating a set of alternatives from which a choice must be made. The final decision making stage has the decision maker match the representational schema to an alternative from the choice set. Pennington and Hastie claim that three influences from this process determine the confidence of the decision maker in her or his choice: (1) the coherency (i.e., completeness, consistency, and plausibility) of the representation, (2) its uniqueness, and (3) the apparent goodness-of-fit between the representation and the chosen alternative.

5.3.2 Representativeness, Anchoring, and Adjustment

In addition to availability or causal relevance, several explanations of judgments under uncertainty refer to the use of what might be called 'similarity' heuristics, or 'connotative distance'. Kahneman and Tversky (1972) proposed the most widely studied of these heuristics, which they term 'representativeness'. The representativeness of an event or sample is "the degree to which it: (i) is similar in essential characteristics to its parent population; and (ii) reflects the salient features of the process by which it is generated" (1972: 431). They provide evidence that the judged probability that an event or object is in a given category is judged by the extent to which that event or object seems similar to or representative of that category.

Although the definition seems plain enough, it has been criticized for being too vague and even unfalsifiable. Some investigators (e.g., Olson 1976 and Scholz 1987) present evidence from subjects' judgments that appears contrary to predictions based on the representativeness heuristic. In an attempt to clarify and defend their position, Tversky and Kahneman (1982: 84) distinguish two hypotheses involving representativeness: (1) "people expect samples to be highly similar to their parent population and also to represent the randomness of the sampling process" and (2) "people often use the representativeness heuristic to make predictions and judge probabilities." They call the first judgments *of* representativeness and the second judgments *by* representativeness.

Judgments of representativeness have been invoked to explain the gambler's fallacy, the exaggerated belief in the stability of results from small random samples, the belief that chance processes are self-correcting in the short-run, and several related counter-normative perceptions of randomicity itself. However, most of the empirical work on this matter has been confined to the context of judgments about random samples. Judgments by representativeness have received much wider attention by researchers. Although there has been some tendency to interpret the original thesis as meaning that judgments of subjective probability are entirely determined by representativeness, Tversky and Kahneman (1973, 1982: 88-89) have consistently taken a more moderate line. First, they claim that while intuitive predictions and probability judgments are sensitive to representativeness, they are not entirely determined by it. Secondly, they argue that representativeness is most likely to be used when the generic features of the problem (rather than specific instances) are made salient.

Their explanation for the widespread impact of representativeness stems from research on judgments of categorization and prototypicality (e.g., Rosch 1978, Rosch and Mervis 1975) which indicates that people find it easier to assess representativeness than subjective probability and that it often correlates with probability. Representativeness is most likely to generate counter-normative probability judgments when the evidence is fallible, the target event or object is highly specific, and/or when conditional and compound probabilities are requested.

Several authors have investigated the possible causes for conservatism and related phenomena. Edwards' classic (1968) paper proposed that the bulk of evidence at the time indicated a 'misaggregation' of information presented sequentially, such that subjects failed to sufficiently appreciate the diagnositicity of evidence subsquent to the initial conditions they were given. Related investigations have yielded an account called the 'anchoring and adjustment' strategy (Poulton 1968, Slovic and Lichtenstein 1971, Tversky and Kahneman 1974).

According to this account, people make sequential probability judgments by starting from an initial value (suggested either in the problem formulation itself or through some initial piece of data), and adjust that on the basis of incoming information to produce their final answer. Usually they adjust insufficiently so that the final answer is biased towards the initial estimate, an effect which resembles the well-known 'primacy effect' in studies of information recall and processing. Evidence for anchoring effects have been found in estimates of fatality rates (Lichtenstein 1978), lethality (Fischhoff and MacGregor 1980) as well as in abstract probability estimation or arithmetic computational tasks. It has also been incorporated into the basis of at least one nonstandard decision theory (Einhorn and Hogarth 1985, on which more later).

5.3.3 Nonstandard decision theories

The findings by decision researchers that the Savage independence axiom is frequently violated by decision makers quickly stimulated the development of several alternatives to subjective expected utility (SEU) decision theory. I will review some of them here, primarily for their accounts of the relationships between utility and probability. The earliest nonstandard theories replaced subjective probabilities with 'decision weights' (Edwards 1962, Fellner 1961) in an attempt to explain risk aversion; these were also investigated empirically (e.g., Anderson and Shanteau 1970). More recently there have been a variety of attempts to replace the independence axiom by some more generalized constraint on preferences and to substitute some nonlinear functional form for the bilinear preference function (e.g., Fishburn 1984, Yaari 1987).

An interesting dissenter from all this is Machina (1982, 1987) who argues that nonlinear forms are not required and that the SEU version of decision theory can be salvaged without the independence axiom and its accompanying linearity restriction. He proposes instead a weaker criterion of 'smoothness' for preferences, whereby decision makers are 'locally linear' in the probabilities and that in these vanishingly small neighborhoods of a probability distribution they maximize locally expected utility. He is able to argue this by using two familiar results from calculus. The first is that any differentiable function may be approximated by a Taylor expansion that is locally linear, so that one need not be restricted to considering a particular nonlinear functional form. Secondly, many differentiable functions exhibit a property globally if and only if that property is exhibited by its linear approximation at every point. He utilizes this last result to prove that for a wide class of preference functions global risk aversion is equivalent to local linear risk aversion in the usual sense of a locally concave utility function.

The most well-known nonstandard decision framework among psychologists is Kahneman and Tversky's (1979) 'prospect theory', whose major hypotheses rest on two points. The first is that decision makers' choices must be understood in terms of change rather than simple absolute values, and the second is the 'reflection hypothesis', which holds that people are risk-seeking when faced with the prospect of a loss and symmetrically risk-averse when faced with the prospect of a gain (other earlier theorists also suggested reflection effects: e.g., Markowitz 1952). Kahneman and Tversky provide a systematic exposition of this theory and demonstrate that despite its intuitively reasonable starting point, it violates the normative canons of transitivity, dominance, and invariance.

Prospect theory has had a mixed reception. On the one hand it has stimulated considerable debate. On the other, empirical support for it actually has been rather weak. The evidence cited by Kahneman and Tversky themselves deals only with cross-subject reflection effects, or in other words treating reflections in percentages or rates of preferences as if they referred to reversals by the individuals themselves. In an empirical investigation which sought actual reversals by individuals as well as attempting to replicate the cross-subject results from Kahneman and Tversky's studies, Hershey and Schoemaker (1980) found little support for the lawfulness of the cross-subject reflection effect and only a weak trend in the expected direction for individual reversals. Furthermore, they found no evidence that subjects' reasoning about their choices was reflected symmetrically when the signs were reversed. More recently, M. Cohen, Jaffray, and Said (1987) report a study in which about as many subjects demonstrate consistent risk aversion (seeking) as display a reflection effect. Even more strikingly, they find that risk aversion/neutrality/seeking on the loss side is statistically independent of the subject's orientation on the gain side. Keren and Wagenaar (1987) have looked for evidence of the certainty and possibility effects reported by Kahneman and Tversky in repeated gambles, noting that all previous empirical evidence referred to unique gambles. They find that those effects do not generalize readily to repeated gambles.

Finally, the preference reversal phenomenon has received theoretical attention from Goldstein and Einhorn (1987), who present 'expression theory' to account for it. Initial attempts to account for preference reversal focused initially on the possibility that it was an artifact of the different methods of eliciting judgments and choice (Lichtenstein and Slovic 1973, Grether and Plott 1979). Grether and Plott, in particular, experimentally investigated some 13 possible theoretical and methodological explanations for the effect, eventually discarding all of them except for the elicitation method explanation. They concluded that the reversal effect applies to areas relevant to economic theories of choice and that it lacks a satisfactory explanation. Goldstein and Einhorn have proposed that in addition to the usual two stages of judgment and choice (first stimuli are encoded; then the encoded stimuli are evaluated), a third stage exists in which the individual expresses the evaluated stimuli on some kind of response scale.

The essence of expression theory is that when subjects are asked to rate their preference for one of two gambles they do so on a subjective utility scale, whereas when they are asked to set a minimum selling price for each of the gambles they do so on a subjective monetary scale in which they attempt to interpolate some monetary value that corresponds to their sense of the subjective utility of the gamble. They make the two scales comparable by equating their endpoints and matching a proportional change in the utility scale with a proportional change in the monetary scale. Reversals are therefore accounted for simply by the fact that the difference between the ranges of the utility scale and monetary scale for one gamble may be greater or less than the difference between the ranges of the utility and monetary scales for another. The authors are able to systematically relate those differential range differences to six types of preference reversals in a predictive framework. Furthermore, their theory is consistent with the reflection effect posited in prospect theory, insofar as expression theory predicts that reflection reverses the reversals.

5.3.4 Criticisms of the Explanations

The major criticisms of the mainstream psychological theories on uncertainty may be roughly lumped into three categories: counterproposals invoking rival explanations, criticisms of the theories themselves, and critiques of the supporting evidence for the theories. I will delay discussing counterproposals until the final section of this chapter, since they are crucially linked with the debates in psychology over rationality. This section will address the second and third kinds of criticism.

Few psychological researchers would care to vouch now for the position taken 20 years ago by Peterson and Beach (1967) that probability theory or the SEU framework accounts for most people's judgments and decisions under uncertainty. It is one thing, however, to demonstrate that at least some people systematically violate the predictions of a normative theory, and quite another to construct a theoretically cogent and well-supported explanation of what they really are doing. The latter is considerably more difficult, and not aided by the fact that the bulk of psychological research in this field has concentrated on the end-products of judgment and choice processes rather than the processes themselves. In view of these difficulties, criticisms should be tempered by recognition of the fact that these are early days for serious descriptive theories of human judgment and decision making under uncertainty.

The most thoroughly criticized explanations are the representativeness, availability, causal schema, and specificity heuristics and their offshoots. Most of the criticisms against them (aside from debates over how well-supported they are by the evidence) boil down to two claims. One is that they are too vague and insufficiently specified to be truly testable (e.g., Olson 1976, La Breque 1980, Wallsten 1983, Groner et al.

1983). More specifically, none of these heuristics is well grounded in an account of human information processing, mental representations, or the usual concomitants of cognitive psychological theories. They also have an uncomfortably ad-hoc flavor to them insofar as their proper domains of application are not specified. As La Breque's examples illustrate, the representativeness heuristic can be invoked to 'explain' conflicting responses to the same problem.

These accounts also, ironically, are not explicitly quantifiable. There have been few attempts to genuinely scale the properties (or the experimental manipulations thereof) of the stimuli themselves, let alone the judgments being made about them (Fischhoff and Bar-Hillel 1984). One consequence of the qualitative nature of the heuristics is that no testable framework is available which permits a researcher to assess the extent to which judgments are being influenced by, say, representativeness to the exclusion of other competing influences. Conversely, assessing the degree of impact base-rate information has is easy because the base-rates themselves are quantified, as is the judgment output itself. But it is noteworthy that a number of researchers do not even make use of the available information in their own studies to quantitatively assess effects. Most of them eschew measures of the strength of effects or relationships in favor of tests of significance (often misapplied to nonrandom samples). Scholz' otherwise exemplary study of base-rate effects, for instance, merely reports whether there was a 'significant' effect from the base-rate manipulation. In fact, simple computations using his ANOVA tables (1987: 45) reveal that the percentage of variance in his dependent measures explained by base-rate effects vary considerably among his three problems (from 2.7% to 22.5%).

A related criticism which applies to virtually all the explanations discussed so far is that the various heuristics are isolated from one another in a kind of theoretical archipelago. No bridging framework has been provided that would link them or specify their interdependence. Such a framework would require that at least some of the forementioned problems be solved, especially empirical falsifiability and the specification of domains of application. Even in those cases where competing heuristics actually provide contradictory predictions (e.g., prospect theory's reflection hypothesis versus the more traditional hypothesis of generalized risk aversion) little progress has been made towards an empirical or theoretical rapprochement.

Furthermore, the term 'heuristic' has been applied in a rather slipshod fashion to processes that may occur at quite different stages of the judgment process (e.g., encoding versus evaluation versus expression), and that may also reflect distinct kinds of knowledge (e.g., procedural versus factual). Some recognition of this problem has been forthcoming in the recent literature (cf. Hogarth and Makridakis 1981, Goldstein and Einhorn 1987, Scholz 1987), but a conceptual framework that would fulfill well-known criteria for any theory of knowledge and thinking has yet to be proposed. In particular, little attention has been paid to the well-known distinction in psycholinguistics and cognitive psychology between a 'performance' model and a 'competence' model. These two kinds of model appear conflated in the judgmental heuristics literature.

The other line of criticism attacks the empirical support for the heuristics. Most of the studies claiming quite general support for a particular theory may be taken to task for their inattention to sampling issues. Not only are the vast majority of subject samples nonrandomly selected (and yet nearly all of the researchers conduct Neyman-Pearson significance tests on the data), but they are taken from unrepresentative or even unspecified populations (they tend to be the usual convenience sample of university students in psychology). The major exception to this, of course, is the sizeable plurality of studies dealing with particular kinds of expert judges. On the whole, however, we still know very little about how the majority of ordinary human beings make judgments about uncertainty. Furthermore, a large number of experiments base their conclusions on very small samples, so that even if they were well-selected their statistical power and the stability of their estimates would be quite low.

We also know comparatively little about how widely any account applies to the various kinds of judgment or decision tasks that face laypeople (or even experts), since the tasks utilized in the bulk of judgment and decision research are very stylized examples from a restricted domain of problems (cf. Macdonald 1986). Although some researchers have pointed to the likely possibility that motivational and cognitive predispositions may differ dramatically from one context to another (e.g., Lopes 1982), no systematic attempt seems to have been made to address the issue of task realism or representativeness. These problems lead commentators such as Edwards (1983: 503) to conclude that the judgment "studies are grossly unrepresentative both of tasks and of subjects who might perform these tasks."

A second methodological criticism which applies to many studies is their inattention to individual differences and indeed to the crucial distinction between cross-subject and inter-subject design issues and effects. As Scholz (1987: 26) remarks, these studies give us some knowledge of what the median or modal subject is doing but little or no account of what the rest of them are doing, to say nothing of an explanation for individual differences. Since it is theoretically possible that the median (the empirical focus in several of the Tversky and Kahneman type of experiments) score refers to no subject, and because even the mode often accounts for only a minority of them, this criticism is a reasonable one. It is especially justified for those theories that refer to individual-level judgmental phenomena (rather than to the central tendency of a population or group), as indeed most of them do.

Scholz' survey of base-rate effect studies finds that between 50% and 80% of the subjects do not display the kinds of 'fallacious' judgments that the authors of those studies focus on (e.g., the so-called diagnosticity response). But more importantly, in his own experiments few subjects respond to base-rate problems in ways that could be explained by heuristics such as representativeness or availability. In one such study where he asked subjects to provide written accounts of their problem-solving strategies, Scholz reports that even those subjects who gave the 'diagnosticity' response did not generally use a representativeness heuristic (see also Pollatsek et al. 1984). Instead, many of them seem to have approached the problem through 'blind calculation' "as if a conditioned response pattern is elicited, roughly taking the form: probability---mathematics---pick up the numbers---produce a calculation---perhaps it fits" (1987: 91).

Scholz' findings and the repeated calls for more attention to the judgmental process rather than its end-product (Payne et al. 1978, Phillips 1983) point to another criticism of the evidence for the heuristics. This is simply that it is difficult to dismiss rival inferences from the output back to the judgmental process that produced it. Until much more research has been conducted on what subjects actually think (or report that they think) about while performing experimental tasks, explanations of their output by appeals to vaguely specified heuristics will always rest on insecure foundations.

5.4 New Directions and New Uncertainties

Psychology and its cognate disciplines have not eagerly embraced the recent developments in normative paradigms on ignorance and, particularly, uncertainty. Most behavioral scientists during the 70's and 80's have based their research on a limited and rather uncritical familiarity with standard probability theory alone. Even within those confines, little has been made of the diversity and outright controversy that still distinguishes one school of probability theory from another. With few exceptions, psychologists have eschewed testing or applying nonprobabilistic theories of uncertainty, vagueness, or ambiguity and in this respect they display greater methodological and conceptual inhibitions than, say, their colleagues in artificial intelligence and expert systems.

Those exceptions have taken three forms. First, there are empirical investigations and critiques of nonstandard normative theories. The most popular among these so far has been fuzzy set theory, although there is some evidence that Cohen's Baconian probability framework, Dempster-Shafer belief theory, and other related frameworks may attract empirical research programs in the near future. I briefly reviewed the empirical work on fuzzy set theory in Chapter 4, but it is worth further consideration for the unusual relationship between the 'normative' version of fuzzy set theory and both the empirical evidence and explanations of that evidence.

A second key development in psychology is a recent surge of interest in ambiguity and vagueness, prompted mainly by the Ellsberg type of 'paradox' and the difficulties encountered in phrasing probability problems unambiguously. This interest also has received some impetus from the decision research findings on violations of the independence axiom. A potential spinoff of this development may be empirical research on second-order uncertainty and its relationship to utility.

Finally, probability theory has been imbued with new life from unexpected quarters. Both psychologists and behavioral economists have nearly simultaneously proposed behavioral frameworks for explaining the human responses to uncertainty.

The claims for this approach are ambitious indeed: they amount to nothing less than subsuming the entire corpus of cognitive explanations for 'heuristics' and extending its reach. Likewise, again from both psychologists and economists, a neoevolutionary account of probability theory has been proposed in which cognitive rationality is replaced with a biological rationality. These parallel developments constitute new reductionist strategies, whereby the behaviorist account reduces the explanatory corpus and the evolutionary reduces the probabilistic normative framework. An inquiry into each of them sheds new light on the rationality debate.

5.4.1 Fuzzy Set Theory as a Behavioral Science?

The empirical and normative criticisms of fuzzy set theory contrast interestingly with the mainstream judgment research in psychology. The attitudes adopted by the fuzzy set researchers towards the normative fuzzy set framework are considerably more relaxed and more critical of the framework itself than the almost unremittingly pro-probability attitude displayed by most of the psychologists. One consequence of this difference is that the relationship between the normative and explanatory approaches to fuzzy sets tends to be more mutual and interactive, with a potential for modifications to both kinds of paradigm and hence perhaps an eventual rapprochement between the two.

Consider, for instance, the research on fuzzy connectives ('and' and 'or') in comparison with the judgment research on probabilistic conjunctions and disjunctions. The main findings in the judgment research boil down to the following points: (1) people overestimate the conjunctive probability of two events, sometimes to the point of exceeding their estimates of the probability of one of those events; (2) they tend to underestimate the probability of the disjunction (that is, an inclusive 'or') of two events; and (3) these trends (but particularly the first one) are most marked when heuristics such as representativeness, specificity, anchoring and adjustment, or plausibility are invoked by the stimuli or problem frame. Point (3) provides the primary explanation for (1) and (2), while the normative validity of probability and standard logic are unquestioned. Tversky and Kahneman (1982: 98) summarize this orientation: "...we cannot defy the laws of probability, because they capture important truths about the world. Like it or not, A cannot be less probable than

(A&B), and a belief to the contrary is fallacious." Many of their colleagues who study intuitive logic hold a similar attitude about standard logic.

Research into and empirical critiques of fuzzy set aggregators have been conducted chiefly in the domains of prototypicality and category membership judgments rather than judgments of probability or logic. These domains do not appear to have been dominated by normative considerations or a concern with 'erroneous' information processing. Indeed, the extensive work on information integration by Anderson (1981, 1982) and his colleagues includes a consistent finding that subjects translate 'and' by an averaging operator when they aggregate, but nowhere is this labeled by these researchers as a fallacy or logical error.

Fuzzy set theory itself has been presented at times as a normative theory and on other occasions as a candidate for descriptive modeling, so its normative status is less specified than that of probability theory. In this less normatively constrained environment, it is no surprise that researchers such as Osherson and Smith (1981) and Kay and McDaniel (1975) can reject fuzzy set theory out of hand for its failure to model what Tversky, Kahneman, and their colleagues describe as the 'conjunction fallacy'. To Osherson and Smith, the fact that a guppy receives a higher membership value in the category of 'pet fish' than it does in either the category of 'pets' or 'fish' is no logical fallacy at all but merely an unproblematic feature of natural language. Kay and McDaniel, along with later contributions to color research by Burgess et al. (1983), go even further and argue that linguistic evolution functionally selects for useful aggregated categories and the conjunction effect (e.g., for a color category such as 'orange' which is an amalgam of 'red' and 'yellow') is evidence of such evolution having taken place.

Nor are the fuzzy set theorists themselves reluctant to modify their connectives in the face of empirical disconfirmation or even pragmatic directives. Zadeh (e.g., 1976) has consistently declared that applications of fuzzy set theory should entail a choice of connective to fit the purpose at hand. In his view, any search for the 'best' operators for defining 'and' and 'or' is a mistaken enterprise. As mentioned in Chapter 4, a recent normative consensus has emerged that the choice of such operators should be restricted to a family of 't-norms', which are pairs of 'and' and 'or' operators satisfying certain algebraic properties. A number of authors have proposed generalized connectives which vary from a strict 'and' to a strict 'or' by means of one or more free parameters, and where empirical research has been conducted some limited support has been found for these models (e.g., Zimmermann and Zysno 1980 and Smithson 1984). But even when faced with the prospect of other studies which do not support those models (e.g., Zwick et al. 1988) at least some proponents of fuzzy set theory have not hesitated to suggest abandoning t-norms (cf. Dubois and Prade 1985).

The Zwick et al. (1988) study is important in another respect, since it investigates models of vague probability statements. Not only is the traditional normative inhibition about the conjunction effect notably absent from their writing, but they appear equally comfortable with both fuzziness and probability. Zwick and his colleagues have undertaken a series of studies combining fuzzy set and probability theories via the fuzzy random set framework, thereby initiating the descriptive/explanatory counterpart to the normative move towards synthesis of these two approaches. Their program is also an exemplar of the recent normative expansion into areas of ignorance and uncertainty previously designated as off-limits. Zwick and Wallsten (1988), for instance, test four models for combining fuzziness and probability in a probability elicitation task, under conditions where neither pointwise nor intervalic probabilities may be sensibly applied.

It is important to recognize that such research represents not only a shift in explanatory framework, but a normative shift as well. Zwick and Wallsten begin their paper by taking the normative stance that the insistence on numerical expressions for probability in decision making is misguided. They point out that numbers convey an illusory precision and that a linguistic mode of analysis is the rule among decision makers; and they dispute (via their research) the widespread belief that the meanings of vague linguistic probability expressions cannot be effectively measured and agreed upon. Their findings also reflect this orientation, since the data support the linguistic rather than the numerical models, which they interpret to mean that people cope with complexity not only by blurring (in the sense of Kochen 1979), but also by using 'multiple-crisp' representations of vague or inexact concepts. Unlike the judgment researchers' explanatory view of human heuristics, these researchers propose to explicitly model those heuristics in a testable fashion via a linguistic variable approach.

5.4.2 Vagueness and Ambiguity

A scattered but growing literature has emerged in recent years on the impact of ambiguity, missing information, and vagueness on judgment and decision making. Most of this research parallels the mainstream judgment research paradigm, but there are signs of some important departures from that framework. For one thing, studies of ambiguity in particular have been partly motivated by such Bayesian bogeymen as Ellsberg's paradox and related arguments concerning second-order uncertainty about probability values. Indeed, much of this research points towards an increasing concern among psychologists with second-order ignorance and uncertainty which is moving into the territory that has begun to be mapped by the newer normative approaches discussed in Chapter 4.

The empirical literature on the effect of missing data to date echoes mainstream themes. Thus, F. H. Barron (1987) reports that in multiattribute decision problems, missing attributes can and do result in the selection of inferior options. In an early study, Yates et al. (1978) suggested that partially described decision alternatives are devalued relative to their completely described counterparts. Furthermore, they found a saliency effect, such that the decision maker's attention is drawn to options containing attributes that are incompletely described, even when in other options they are fully described. Finally, Levin and his colleagues (1985, 1986) reproduced some of the framing effects made famous in studies involving judgment under probabilistic uncertainty for judgment tasks where some of the information required for a judgment or decision is missing.

The study of ambiguity has so far been restricted mainly to ambiguity about probabilities, for reasons already indicated. Many writers have posited or found support for the claim that people negatively evaluate and try to avoid ambiguity in decision making or gambling. Einhorn and Hogarth (1985) provide what they term a "descriptive model" of how people make probability judgments and choices under ambiguity. They posit an anchoring and adjustment process that is influenced by the initial probability assessment, p, amount of perceived ambiguity (which they represent with a parameter, a, such that $0 \leq a \leq 1$) and the individual's attitude towards ambiguity. They characterize the latter in terms of another parameter, b, which reflects the individual's tendency to weight probability values that are greater or smaller than the initial estimate p. Their judgment model is the equation $S(p) = p + a(1-p-p^b)$, where $S(p)$ is the final subjective probability assessment after adjustment for ambiguity.

They report limited support for their model, and a fair amount of support for several of its qualitative features. In particular, they find that people are generally subadditive in their probabilities under ambiguity, thereby supporting the claim that they tend to be ambiguity-averse. Einhorn and Hogarth also note the connection between this finding and certain normative paradigms which combine ambiguity with measures of subjective belief, as in Dempster-Shafer theory or L.J. Cohen's intuitive probability. They also find that in the presence of ambiguity, desirability of outcome and probability are not independent because the b parameter varies with outcome desirability. This constitutes a novel explanation for violations of the independence axiom in SEU decision theory.

In a related study, Curley et al. (1986) have investigated the psychological reasons behind ambiguity aversiveness. They review the literature on avoidance of ambiguity and isolate six candidate explanations:

(1) People attribute greater likelihood to bad outcomes (or less likelihood to positive ones) under ambiguity.

(2) People believe that preferring the ambiguous situation would be less justifiable to others than selecting the unambiguous option.

(3) They make a similar attribution about justifying their own choice to themselves in the future.

(4) The less ambiguous option is selected only when all other considerations are equal, owing to the forced-choice nature of the task.

(5) Avoidance of ambiguity is merely another instance of avoidance of uncertainty generally.

(6) Ambiguity aversiveness is a systematic mistake or bias of some kind, which knowledgeable people would correct if it was brought to their attention.

Their major claim, based on a series of experiments, is that only the other-evaluation hypothesis (2) is supported empirically. However, one should view this claim with some caution for at least two reasons. First, compared to the other-evaluation study, the sample sizes for the rest of the experiments are small so their statistical power is undoubtedly low. Secondly, although a 'significant' main effect for other-evaluation is reported, the effect looks to be small. From their incomplete ANOVA information, it is clear that the percentage of variance due to the other-evaluation effect is less than 5.3%, and the mean percentage of subjects selecting ambiguity-avoiding choices in the low other-evaluative condition is still more than 70%. Thus, these authors have hardly uncovered an explanation for the existence of ambiguity avoidance; at best they have found a mild influence on it.

The psychological study of second-order uncertainty or ignorance is clearly still in infancy, but its immediate future probably is bright. Not only is there an emerging set of interesting normative accounts addressing ambiguity, conflict, confusion, vagueness, and fuzziness, but psychologists and economists have started to come to terms with the realization that ignorance and uncertainty are multiple and mutli-level concepts. A confluence between these two streams is not far ahead. However, as in the normative arena, there are outstanding controversies concerning rationality and the relationship between the normative and descriptive that will need to be dealt with before such research can make much headway. These issues will be reviewed at the end of this chapter as well as comprising the focus of the final chapter in this book.

Before moving to discussions about rationality, we must complete our survey of new directions in psychological research on uncertainty with a review of recent reductionist strategies. In Chapter 4 a case was argued that the new reductionists constitute a counterattack on the emerging normative paradigms by the probabilists (mainly Bayesians). The human sciences have their reductionist impulses also. While mathematicians like to reduce mathematical systems to small collections of widely accepted axioms (such as those of set theory or standard logic), behavioral scientists have long preferred reducing sociological explanations to psychological, and psychological to either biological or behaviorist. Recent proposals for reducing psychological accounts of judgment under uncertainty are no exception to this historical tendency.

5.4.3 The New Reductionists

Understandably, since the normative basis for judgment and choice under uncertainty has been accounts of cognitive rationality, the dominant paradigm for research on those topics has been cognitive psychology. However, this was not always the case in the past. In the 30's and 40's, for instance, some of the leading figures in behavioral psychology argued that animal behavior studies of choice would eventually provide the evidence and insights for generalizable laws of human choice behavior. During this decade there has been a resurgence of interest in this claim, or at least a modified version of it, from three related streams of research: (1) neobehaviorist models of choice under uncertainty, (2) studies of animal behavior in stochastically uncertain environments, and (3) recent developments in theories of evolution. The agendas motivating at least some of this research are also threefold: (1) an empirical assessment of parallels between human and animal predispositions under uncertainty; (2) a (possibly reductive) translation of cognitive theories of judgment and choice into behavioral and/or ecological terms; and (3) the provision of a biological rationale for probability theory and related normative concepts.

Animal and human behavioral studies of choice under uncertainty have operationalized uncertainty primarily in two ways: reward variability and reward delay. The linkage between reward variability and probabilistic uncertainty is straightforward because it is based on a frequentist account of probability. Accordingly, studies have been conducted on risk aversion and preference for a variety of foraging animals (see Real and Caraco 1986, Stephens and Krebs 1986 for recent reviews), revealing striking similarities between animal and human preferences (e.g., mixed evidence for and against prospect theory's certainty effect, with the majority of evidence against favoring a view of foragers as risk-averse).

A similar argument has been made for equating a frequentist definition of probability with a rate of reward over a series of trials. The operationalization of probability as the inverse of reward delay, however, is considered primarily as a link between a behavioral operationalization and subjective probability, and so requires a justifying argument. Rotter (1954) provided early arguments and empirical evidence that the two are subjectively equivalent. He proposed that the sooner a reward is promised the higher the subjective probability that it will actually be delivered. More recently, several experimenters (e.g., Rachlin et al. 1986 and Navarick 1987) have provided additional evidence for their equivalence. By replacing certainties with immediacies and uncertainty with delay, Rachlin and his colleagues (1986, 1987) make a case for linking findings that people are risk averse with positive outcomes with another body of empirical findings that people and animals discount positive delayed reinforcement, and correspondingly the findings that people are risk seeking with negative outcomes with the findings that people and animals prefer delayed to immediate aversive stimuli.

More specifically, Rachlin et al. (1986) provide a translation between prospect theory and Herrnstein's (1961) model of animal choice. They begin with the so-called 'matching law' of animal choice, which has the following functional form (abbreviated from Baum 1974):

$$B_1/B_2 = (A_1/A_2)^a (D_2/D_1)^d, \qquad (5.1)$$

where B_1 and B_2 are the rates at which the subject chooses options 1 and 2, A_1 and A_2 are the amounts of reward offered by those options, and D_1 and D_2 are the delays; while the exponents express individuals' sensitivities to reward and delay. They then translate delay into probability by the following equation:

$$D_i = (t+c)/p_i - t, \qquad (5.2)$$

where c is the time taken for each trial, t is the intertrial delay, and p_i is the probability to be equated with the delay D_i.

They utilize (5.1) to account for the certainty effect from prospect theory by pointing out that if d is significantly greater than 0 then near-immediacies (values of D_1 or D_2 close to 0) will dominate B_1/B_2 regardless of A_1/A_2, thereby reproducing the same failure of dominance that prospect theory explains through the certainty effect. The possibility effect, on the other hand, is accounted for by arguing that adding a sufficiently large amount of time to D_1 and D_2 in (5.1) will cause B_1/B_2 to become dominated by A_1/A_2, which can account for both the tendency in behavioral experiments for long delays to result in a change from short-term gratification to commitment and the possibility effect findings. In a similar fashion, the authors go on to translate various other 'effects' from the cognitive paradigm into behavioral terms by inferring certain value ranges for either the sensitivity exponents or delays in the (5.1) matching law.

That translation works in the following manner for the certainty effect. If we let $c_1=c_2=c$, $t_1=t_2=1$, and $A_1=A_2$, for instance, and substitute these values into (5.2) and thence into (5.1) we may rewrite (5.1) in the form

$$\frac{B_1}{B_2} = \left(\frac{p_1(1+c-p_2)}{p_2(1+c-p_1)}\right)^d \qquad (5.3)$$

and if in turn we set $p_2=1$, as Rachlin et al. do for an in-principle argument about the certainty effect (1986:39-40), we find from (5.3) that $B_1/B_2 = cp_1/(1+c-p_1) \leq 1$. The immediate implication is that option 1 will never be clearly preferred over option 2 unless we remove the restriction that $A_1=A_2$.

Rachlin and his colleagues claim four advantages for a behavioral over a cognitive account of choice under uncertainty. First, it is parsimonious insofar as a single equation accounts for a large catalog of disconnected 'effects' in the cognitive theories. Second, behavioral research techniques provide direct confirmation of equation (5.1) for individual subjects whereas most cognitive research can provide only indirect confirmation of cognitive hypotheses for groups of subjects. Third, the behavioral paradigm translates choice-making into a temporal perspective rather than the atemporal one of cognitivism. And fourth, the authors prefer a behavioral account of rationality to a cognitive account. A behavioral criterion for rationality of behavior refers to how much temporal integration is required to explain it. Rachlin et al.'s translation is not a complete one, which they recognize, but it has intriguing possibilities. Its strongest points are its explicitly testable and quantifiable form, and its theoretical unity. Its weaknesses are more difficult to judge fairly because it is a new approach, but perhaps the most urgent issues at hand are its generalizability to situations where uncertainty is not obviously operationalizable in terms of delay, and whether it can handle other nonprobabilistic kinds of uncertainty and ignorance.

Another important question concerns the uniqueness of Rachlin et al.'s particular translation. For instance, Mazur (1984) provides an alternative approach which is algebraically equivalent to equating delay with inverse odds rather than probability (although Mazur himself does not make this link), and it is not difficult to reformulate (5.2) in an odds form so that delay is inversely proportional to odds: $D_i = t/p_i - t = t/o_i$, where $o_i = p_i/(1-p_i)$. The objections raised by Rachlin and his colleagues over unbounded odds can be dealt with easily by assuming that no subject is a perfect detector of a probability (including a sure thing) so that the subjective p_i are always bounded slightly below 1 and slightly above 0. No matter how slightly those bounds differ from 0 and 1, the o_i also are bounded and a version of (5.3) results which is bounded from above by 1 when $p_2 = 1$ and p_1 is allowed to vary, thereby reproducing the certainty effect. Whatever the current status of these attempts at translating between the cognitive and behaviorist approaches to modeling choice under uncertainty, this line of investigation is intriguing and there is undoubtedly ample scope for further work here.

The third relevant reductionist development is a spate of arguments that the rationality of choice behavior may be explained in evolutionary or biological adaptational terms. One kind of argument is exhibited frequently in the economically oriented animal behavior and ecological literature, which holds that risk aversion and/or reflection effects are explicable in terms of pro-survival foraging strategies. The controversies over this type of argument are well-known in the sociobiological literature and will not be dealt with here. However, the fact that they are advanced at all is interesting because they contrast with the dominant verdict of 'irrationality' for the same proclivities among human decision makers in the psychological literature. It would seem that if the subjects are animals and the rewards are food, then ipso facto whatever risk aversion or seeking they display has survival value. But if they are humans and the rewards are money or something less tangible, then the same tendencies are suboptimal.

Another kind of biological rationality argument attempts to derive the axioms of some formal system (e.g., Savage's axioms) from some biological population process model or a related argument. An example from a professional statistician is Durbin's (1985, 1988) philosophical explanation for the development of the human capacity to do mathematics (and therefore statistics). He attempts to ground his explanation in an argument that a certain kind of logical thought had survival value for gatherer-hunters.

But the most ambitious attempt to argue the reduction of formal decision theory to a strictly biological basis is Cooper's (1987) recent theoretical article. Cooper recognizes the difficulties of this undertaking from the start: classical expected utility decision theory has more than two dozen versions, and evolutionary theory is likewise not monolithic either. However, he surmounts this by simplifying his population model via several assumptions and by limiting his derivation to the Savage axioms (with the exception of the partitionability of the event space). The heart of his derivation is the characterization of an 'evolutionarily stable strategy' (ESS) as an uninvadable one. That is, an ESS is capable of eliminating any mutations that adopt another strategy from the population in the long run. Thus, he is able to define a choice function that is an ESS if it maximizes fitness, as measured by the simple population rate of increase. He then proves a general theorem that if a choice function is an ESS, then there is a binary relation $\geq$ over the set of all acts that satisfies Savage's axioms.

At first glance, arguments such as this might seem a direct contradiction with those put forward by the behaviorists and ecologists about the rationality of counter-normative choice behavior. While individualized arguments such as Durbin's do run a danger of constituting such a contradiction, Cooper avoids this trap by pointing out that his reductionist program sheds no light on the question of whether individual organisms maximize subjective expected utility in short-run decisions or not. Instead, his framework pertains only to long-run (that is, evolutionary time-spans) stabilized processes at the population level.

5.5 The Rationality Debate
5.5.1 The Bayesians and Rival Rationalists

The debates over rationality in the psychological literature (and, for that matter, the literature of cognate disciplines) are many and varied, but certain themes are distinguishable. First, there are queries about whether the research on judgment under uncertainty has been 'fair' to the subjects themselves. Some of these queries have already been discussed, with mixed findings. We have seen that in at least some studies the subjects have not clearly understood what the experimenters were asking of them, and conversely the very problem statements themselves did not always yield unique normative solutions. A related concern has recently been raised by Christensen-Szalanski and Beach (1984) over whether the citation of these findings in the literature has been lopsided. They provide both empirical and circumstantial evidence that most surveys of the literature are biased towards emphasizing and citing those studies that demonstrate poor performance by subjects, a bias which is then reproduced in the attitudes of those who read that literature.

However, two more fundamental objections to the Bayesian hegemony have appeared increasingly frequently during the past decade. One is that probability theory itself does not constitute a monolithic normative consensus, nor a well-founded generalized normative framework. The other is that alternative normative perspectives exist which are as defensible in their own domains as probability may explain what people are doing. Both of these themes indicate a possible normative shift among psychologists, and in view of the starting-point of this chapter, deserve at least a brief survey (they will also receive further attention in Chapter 7). Specifically, both the 'Certainty Maximizer' and 'Statistician' normative orientations have been severely criticized and may be undergoing modification.

Recognition of the diversity of views on probability has come late to psychology, beginning with the introduction of Bayesian and related subjective probability theories to the field by Edwards and his co-workers during the 50's and 60's. However, it is in the 70's and 80's that we see an explicit realization by psychological researchers that probability and decision theories are multiple (e.g., Scholz 1987: 19, 58), relatively recent cultural and historical products that are still changing (e.g., Phillips 1983), and culture-bound (cf. Wright and Phillips 1980, or again Scholz 1987: 3-4). In related fields such as operations research and decision making, there are parallel criticisms that the simplifying and restructuring assumptions required by dominant normative perspectives both distort the original problem beyond recognition and reduce the chances of a useful solution (cf. Vari and Vecsenyi 1984, McArthur 1980, and the debate in the Edwards, et. al. 1984 forum).

To a lesser but detectable degree, there is an emerging realization that at least some components in the foundations of probability theories are not firmly established. Lopes' (1982) case against the widespread assumption by researchers that probability theory unambigiously specifies the nature of randomicity is an outstanding example. But many of the philosophical quandaries concerning (for example) the ultimate foundations for the 'long run' arguments for objective probability, the primacy of conditional versus simple probability, the conditions under which probability may or may not be quantifiable, or the problem of representing ignorance versus probabilistic uncertainty are put aside by most psychological researchers. These are assumed away both in the normative frameworks used for comparison with the performances of their subjects and in the very analysis of their own data.

An increasingly vocal group of researchers and commentators have advanced arguments for alternative rationalisms to the dominant Bayesian and SEU frameworks. Most of these arguments seem to take two forms. A 'strong' form is that there are coherent normative alternatives to Bayesianism or SEU and that at least some of the subjects in various studies can be shown to be acting rationally within those alternative frameworks. Many of these arguments in fact hinge on a claim that there is more to decisional uncertainty than mere pointwise probability, and that truly rational decision makers must take this into account. The most famous alternative is the Baconian formalism by L.J. Cohen (1970, 1977) which we reviewed in Chapter 3. Cohen has made strong claims for his system in regard to the Kahneman-Tversky results (1979, 1981), and despite sharp criticisms of him by those authors and others in the field, at least one has recently championed his case (Macdonald 1986). An intriguing example in the literature on SEU is a recent formal argument by M. Cohen and Jaffray (1988) that rational decision making under conditions of complete or partial ignorance (not merely probabilistic uncertainty) compels violations of Savage's independence axiom in order to comply with other rationality requirements.

The weaker version of the normative alternative argument holds that subjects often cannot be shown to be committing fallacies and in fact may be operating sensibly within some restricted notion of rationality. This claim has most persuasively been advanced for conditions in which the probabilities are subjective or epistemic rather than physical or aleatory, where choices or judgments must be made in the short term rather than the long run, or where the environment is stochastically turbulent rather than stable. The arguments over vague priors in Bayesian updating, the problems of specifying a subjective probability problem so there is a unique normative solution, and other related issues regarding the assessment of subjective probabilities have been surveyed at several points in this book and will not be reviewed again here. Many of them boil down to disputes over alternative structurings or representations of the problems (cf. Berkeley and Humphreys 1982).

The most cogent arguments differentiating short-term from long-run decision making have been advanced by Lopes (1981) and Einhorn and Hogarth (1981). They echo the meta-rationality concepts involving quasi-utilities referred to in Chapter 3 in connection with subjective probability theories. One mainstream argument is that if the costs of being 'rational' are overly large, then people may elect for a suboptimal method or choice. In one of the few attempts to compare the actual outcomes of choices made by supposedly subrational judgment and decision heuristics with normative prescriptions, Thorngate (1980) conducted a computer simulation within various decisional environments, varying number of outcomes, probabilities, and payoffs. He found that several such heuristics select optimal or near-optimal options considerably more often than predicted by chance, and that of the two most efficient heuristics one of them utterly ignores probability information (by assuming all outcomes are equally likely) and the other uses only a simple dichotomous scale for probability (probable versus not probable). He concluded that people may use these or similar other heuristics because they cost less effort and yield nearly as good results. Although one needs to be cautious about Thorngate's operational definition of 'efficiency', these are provocative findings. A behaviorist might add that Thorngate's results indicate that people are reinforced by the environment for using simple but relatively rewarding heuristics.

Related arguments for the utility of being 'nonrational' have been forthcoming from other psychological considerations. Taylor's (1983) model of how people with a terminal illness cope effectively virtually requires the use of so-called cognitive illusions concerning mastery and controllability. In an extended polemic against the 'Certainty Maximizer' and 'Knowledge Seeker' normative viewpoints, Reser and Smithson (1988) propose that the maintenance of ignorance and uncertainty serve several crucial psycho-social functions in enabling humanity to cope with the omnipresent threat of nuclear war. They further suggest that the usual prescriptions for greater knowledge and more education fail to provide similar functions for coping.

Yet another argument against the primacy of SEU is that if the decision maker knows she or he is not involved in a long-run series of choices or gambles, then long-run considerations may not apply (one of the more popularized accounts of this consideration is provided by Weaver 1963). Keren and Wagenaar (1987) provide some empirical support for Lopes' claims about how people might choose among unique versus repeated gambles, thereby strengthening their descriptive status. The heart of Lopes' (1981) argument is that even the SEU model contains an implicit long-run assumption that may be questioned, and in fact this problem is fairly well-known in the economic and decision science literature. Although Tversky and Bar-Hillel (1983) argue against Lopes on standard normative grounds, they no longer represent a normative consensus.

Finally, Hogarth (1981) points out an implicit bias in the judgment and decision literature which is towards treating gambles and other problems of choice as if they are discrete, isolated entities in a stable world. They point out that there may be good normative reasons for eschewing the standard Bayesian and SEU criteria for rationality if the decision maker believes the environment is unstable or if the decision itself is part of a nonstationary process. He also echoes (and anticipates) some of the ecological studies of animal behavior in stochastic environments by noting that many choices are made in the context of environmental competition, and that when competitive adaptation and survival are considered some of the seemingly nonrational heuristics people use may be highly adaptive.

The rationality debate has also been characterized by an increasingly liberal orientation by psychologists towards the legitimacy of second-order and nonprobabilistic uncertainty. While research into ambiguity, intervalic probability, fuzziness, and vagueness are still fringe-dwellers on the frontiers of the field, their status has become decidedly enhanced in recent years. Impetus for this development has come from at least two sources. One is obviously the recognition that at least some kinds of nonprobabilistic uncertainty play important roles in even very simple judgment or decision tasks. But no less crucial is the construction of new normative and descriptive models for those uncertainties, since in their absence research yields little more than dustbowl empiricism or, as Hogarth (1986) recently put it, "data without models".

5.5.2 The Normative and the Descriptive

The recent trends in judgment and decisional research have altered not only the normative perspectives that have informed and guided that research, but also the very relationship between normative and descriptive paradigms. One consequence of this is that psychologists are starting to raise the meta-normative issue of what that relationship ought to be. Jungermann (1983) considers this question in the light of the rationality debates, and argues that research on judgment and decision making under uncertainty has been unusually constrained by uncritical conformity with a normative yardstick. He asks, in effect, why researchers in this field are so hung-up on rationality.

In the psychology of perception, for instance, he claims that physical scales of loudness, wavelength, and so on are used primarily for eliciting responses rather than as a standard against which human responses are compared for evidence of deviation or 'inconsistency'. He also argues that the role of formal logic as a normative yardstick for the study of human reasoning is comparatively minor and that standard logic itself is increasingly criticized for its inadequacies. A closer look, however, reveals that Jungermann's claims are only partly true.

The early work in perception and psychophysics actually did concentrate on finding out how human senses and perceptions differ from physical reality, and some of the most famous work in both those fields in fact is that which demonstrates various perceptual 'illusions'. Likewise, some of the most widely exported results from that research are those which establish that humans are not linear in their perceptions of loudness, color, brightness, and so on with respect to corresponding physical scales. But the main difference in the relationship between the normative and descriptive in these fields and that in the psychology of judgment is that the normative consensus is much more secure. Psychophysics and perception researchers take the normative scales and their underlying physical theories almost for granted as being well-founded and beyond serious debate. Physicists have solved those problems to their satisfaction long ago, and there is no 'rationality debate'.

Such is not the case for logic and probability, on the other hand. In both the study of reasoning and the psychology of judgment and decision, there is scope for a debate over rationality. But the lack of normative consensus in logic is more widely known to psychologists than that in probability. Students of human reasoning have long been familiar with the endless debates among philosophers and logicians over what the best logic(s) is (are). As Braine (1978) points out, the early tendency for psychologists to uncritically use Aristotelean and later standard dualistic logic as the normative model was superseded some time ago. The relationship between logic and cognitive science therefore may be instructive for researchers on uncertainty, since its current tendencies could be harbingers.

In a climate where fundamental principles such as the Law of the Excluded Middle are debated and inconsistently applied, a simple dominance relationship between the normative and descriptive is impossible. No one can be entirely sure of which aspects of ordinary reasoning are normatively appropriate. In fact, more than one formalist (e.g., Dubois and Prade 1980) has treated descriptive theories of reasoning as normative criteria for selecting among different logics. Among mathematicians and logicians, as pointed out in earlier discussions in this book, there is a widespread recognition of the *social nature* of any normative consensus. The distinction between the normative and explanatory or descriptive is thereby blurred and changeable.

Braine's own proposals amount to the same kind of enterprise being undertaken by current decision theorists from Kahneman and Tversky (1979) to Machina (1982), namely an attempt to salvage some aspects of normative theories while constructing a descriptively adequate and insightful account of judgment and choice under less than perfect knowledge. Phillips' (1984) approach to a 'requisite decision model' openly embraces the concept of representing the social reality of the (un)shared understandings by 'problem owners'. His goal is to enable decision makers to create a new reality by using a combination of the participants' unease about current models and their own understandings and an adverserial process guided by the decision specialist. When no new intuitions arise, the model is considered requisite, or completed.

Finally, some of the differences between the normative-descriptive debates in this field and in others may be due to the social valuation of the skills involved and the (non)existence of a viable technology for matching or exceeding human performance. It is interesting to compare research on judgment with that on less highly valued skills such as perception or memory. Memory in particular is intriguing because there is historical evidence that it once was as highly valued in Western culture as reasoning or judgment. Today it has little value because books and computers are such good memory storage devices. A contemporary adage is that everyone complains about their poor memory but no one complains about their judgment. Some commentators have noted that spelling, logical, and computational skills also are becoming devalued as the technologies for duplicating them become superior to human performance.

The point is that research on memory is almost entirely explanation-oriented rather than being geared toward improving memory capacity or accuracy. Similarly we do not see substantial research programs on how to 'debias' people with respect to errors in spelling, arithmetic computation, or even the use of modus tollens. All of that goes under the name of 'education'. The absence of a viable, agreed-upon technology for making decisions under uncertainty simultaneously enhances the social value of human performance in this area and the importance ascribed to normative standards. It also links the fate and direction of psychological research in this area with the success or failure of work on expert systems and artificial intelligence, as well as a shifting social construction of rationality.

Chapter 6:
The Social Construction of Ignorance

"The true use of speech is not so much to express our wants as to conceal them." Oliver Goldsmith

"You can't fool me--- I'm too ignorant." Joe Penner

6.1 The View from Social Science

Ignorance has been a marginal and neglected topic in the social sciences, as is the case in cognate disciplines. Indeed, most of what mainstream social science says about ignorance is merely implicit in its outpourings about knowledge. As for direct statements about ignorance or even uncertainty themselves, at best, one could say there is a fragmentary literature that is loosely held together by common themes. This state of affairs might seem to justify ignoring altogether whatever contributions sociologists, social psychologists, and anthropologists may have made in this area. However, this literature has important redeeming features in that it discusses several aspects of ignorance that are not effectively covered in the perspectives we have reviewed so far. Moreover, at least some social scientists bring to their commentary philosophical perspectives that differ in crucial ways from those that inform either applied mathematicians or cognitive psychologists.

The most fundamental insight offered by social scientists into knowledge (and, by implication, ignorance) is that both they and the criteria by which they are established are socially constructed. What passes for knowledge in one culture may not do so in another, and today's ignoramus may well be tomorrow's visionary genius. The very criteria for deciding what is real or truthful have their origins in a negotiated social consensus, despite the undeniable reality of the physical world. This insight may be found in many parts of the social scientific corpus, but its original elaboration comes from the sociology of knowledge, whose first modern exponent was Karl Mannheim (1936).

Sociologists of knowledge have since attempted to follow the implications of this insight to its radical conclusions (e.g., Stark 1958, Berger and Luckmann 1967, Bloor 1976, and Barnes 1977). Perhaps the most obvious impact of the thesis that reality is socially constructed may be observed in the recent sociological studies of scientists and the production of scientific knowledge. The 'new look' in the sociology of science presents a striking contrast to the older view of science as the detached pursuit of objective knowledge by isolated intellects. Instead, the scientific community is found to be intensively social, engaged in constant negotiations over knowledge claims and the criteria for establishing them (see, e.g., Mulkay 1979 and Knorr-Cetina 1981). By a simple extension, as Merton (1987) has recently suggested, scientists also negotiate issues of 'usable' ignorance which provide the grounds for discovery and verification.

Few contemporary psychological theories of perception (including those on the perception of uncertainty, risk, and affiliated matters) treat social factors seriously. As Campbell (1975) argues, there is virtually a professional bias among psychologists against the notion that social factors influence perception, and this bias is even more strongly held if we pass on to discussions of 'rational' perception. Social scientists rebuke cognitive scientists for this bias by pointing out that social factors so fundamentally determine perception that to ignore them trivializes any would-be study of uncertainty or ignorance (e.g., Douglas 1985: 39).

One obvious reason for the reluctance of pro-rationalism analysts to embrace the sociology of knowledge thesis is that, at first glance, it appears to threaten any absolutist epistemology. If all knowledge originates in social interaction, then is it not also the case that there is no human knowledge which is based independently on accurate perceptions of absolute reality? The idea that knowledge is a social product appears initially to open the floodgates to rampant relativism, although in fact it is possible to construct a limited 'realist' philosophy which incorporates this thesis (see, for instance, Manicas and Rosenberg 1985). A starting point is the observation that the truth-value of knowledge is not inextricably linked with how that knowledge is discovered or constructed.

An unavoidable implication, however, is that it is impossible to refer intelligibly to either knowledge or ignorance without reference to the social context in which they are created and maintained. An additional spinoff is that if knowledge is a matter for social negotiation, then it may be possible to revise it or even its epistemological basis at any point in social interaction. This potential mutability has been referred to by sociologists as the 'revisable' character of knowledge.

Given that revisability is a 'built-in' feature of any body of knowledge, the interpretation of knowledge or any communication of it can never be completed in principle. No matter how much one knows about the context in which that knowledge or communication was generated, there is always a wider social context in which such interpretations may have to be changed. This property is called 'indexicality', in one of the more famous borrowings by sociology from philosophy. C.S. Peirce used the term to refer to pronouns, demonstratives, and tenses whose meanings can only be understood by knowing the speaker and the context in which they were used. Although the most radical proponents of this concept (e.g., Garfinkel 1967) stop short of reducing matters to a point where communication between people is rendered impossible, even a restricted version of the indexicality thesis imbues all communication and thought with a degree of inbuilt indeterminacy.

Alongside the fundamental claim that knowledge and ignorance are social creations stands the other pillar of social scientific wisdom in this area: If knowledge and ignorance are social products, then their production is driven by human motivations, values, goals, and interests. Ignorance does not simply exist because of the limitations on human knowledge and imagination; it is also created and maintained by people and the social structures within which they live (often deliberately so). Like their cognitive psychological colleagues, social psychologists and sociologists have elaborated the commonsense notion that knowledge is power. However, they have focused on social or interpersonal power and control rather than analytical or technological power. Sociologists, in particular, have made much of the link between the unequal distribution of knowledge (and therefore ignorance) across society and the genesis of political domination. Likewise, the social value of information has been connected with the extent to which others are ignorant of it, and so has its credibility. Social psychologists and anthropologists alike have explored personal and social norms concerning the control of information exchange, and there is a long-running commentary on secrecy and privacy as special kinds of ignorance arrangements.

Finally, a number of social scientists are fond of pointing out that social and institutional structures have a life of their own, so that the creation of knowledge and ignorance occurs at a structural as well as an interpersonal level. This version of the sociology of knowledge thesis has spun off an entire sub-literature in the sociology of organizations, whose accounts of decision making under uncertainty differ profoundly from the behavioral decision theorist's socially isolated individual mind. Conversely, sociologists have employed uncertainty itself as a causal influence in the analysis of organizational structures and processes.

This chapter, then, reviews contributions from the social sciences to the study of ignorance, developing the major themes in approximately the same order as they have been introduced in this section. We begin with the sociology of knowledge and its implicit 'theories' of ignorance. Discussion then focuses on the social micro-order and the strategic uses people make of ignorance in ordinary social interaction. Those concerns are further elaborated in a review of the literature on privacy, secrecy, and related ignorance arrangements, which in turn leads to a review of social psychological perspectives on ignorance-relevant norms, motives, values, goals, and interests. Finally, structural and political influences are considered in three related subfields: The sociology of organizations, the sociology of science, and the social psychology of risk perception.

These themes initially may seem rather far afield from the prevailing focus of this book. However, that impression is an artifact of two powerful tendencies in the literature reviewed thus far. One is the overly individualistic view of human thinkers, decision makers, and actors which is the legacy of such fields as cognitive psychology and behavioral decision theory. The second is the widespread assumption that ignorance is either a feature of the 'real world' or the result of limitations on our mental capabilities. The radical thesis from the social sciences is that everything that passes for ignorance is a social product. This is not to say that there is nothing we really do not know, nor is it an invitation to descend into solipcism. Instead, this thesis alerts us to the facts that ignorance can be understood only in reference to a social context, and that a great proportion of ignorance is created and maintained by people themselves, rather than being imposed on us by an intractable universe.

6.2 Ignorance and the Sociology of Knowledge

The central project of the sociology of knowledge has been to explain how knowledge is socially constructed and describe the social forces influencing the construction process.

The main goals of this project have, in fact, changed little over the past several decades (compare, for instance, Merton's (1957) 'paradigm' with Bloor's (1976) 'strong programme'), and in their general form they seem not much nearer to being achieved than they were in Mannheim's day. Furthermore, the field has been blighted by severe biases of its own which have considerably hampered its development. Nonetheless, the sociology of knowledge bears several important lessons for students of ignorance, some of which are necessary for any worthwhile progress in this area.

The classics in the sociology of knowledge are structural-functionalist theories which attribute the nature of knowledge to the social and psychological functions or 'interests' it serves. For our purposes, I shall include Marxist perspectives on knowledge in this category, since most of their arguments differ little in their formal properties from functionalism. Although they neglect the problem of explaining ignorance, most of these early theories contain an implicit account of ignorance and some of them attend to particular kinds of ignorance. As Weinstein and Weinstein (1978) observe, these theorists think of ignorance as either a distortion or neglect of 'true' knowledge. Moore and Tumin (1949) are explicit on this point. In both mainstream and Marxist functionalism, the largely unquestioned source of true knowledge is modern science (and of course these theorists are anxious to be counted as real scientists). An immediate implication of this position is that ignorance is defined in terms of deviations away from whatever is taken to be scientific truth (e.g., Parsons 1959: 295).

This implication immediately raises insurmountable problems for these theories. Foremost among them is the requirement that any idea or belief be judged and explained solely in terms of whether it neglects or distorts scientific knowledge, which in turn drastically prunes the original project in the sociology of knowledge. A related difficulty is the awkward fact that scientific truth itself changes, and is not free of controversy. Parsons recognized these problems, but counseled theorists to simply live with them; present-day scientific truth may be only an approximation to absolute knowledge but it is the best we have and the only sure guide to detecting ignorance. Less than a decade later, however, Berger and Luckmann (1967) argued that science should not be the sole arbiter of what passes for knowledge, and the sociology of knowledge project should be broadened to encompass everything that people call knowledge.

The early preoccupation with science as the standard-bearer for knowledge influenced sociologists' explanations of how knowledge is socially constructed. In particular, many of them fell into two traps, both of which have been alluded to before. The first of these is the tendency to posit a necessary connection between the origins of an idea and its validity. Stark (1958), another functionalist, baldly asserted that any idea that is influenced by social or psychological interests is thereby led astray from its true path. In one stroke, we have an explanation for all of 'ignorance' (in Stark's terms, deviations away from scientific truth). In a classic paper, Lavine (1962) coined the phrase 'genetic fallacy' to describe this philosophical error.

The second trap is known as the 'fallacy of dual residentialism' (cf. Dixon 1980), wherein a special group in society is assumed to be free of distortive social or psychological influences on their thought processes. For Mannheim and the functionalists after him, this group is the scientific intellectuals. For Marxists, it is Marxists and eventually, the revolutionary proletariat. The fallacy of dual residentialism, like the genetic fallacy, restricts sociologists' attention to a peculiar kind of ignorance, namely ideology.

For functionalists such as Parsons and Stark, ideology consists of erroneously selected and distorted interpretations of facts, which arise partly from the strains induced in the social system by the division of labor and also from individual value-systems. An isolated but identifiably functionalist account of ignorance by Moore and Tumin (1949), on the other hand, gives a more detailed picture of the social functions performed by ignorance in the sense of neglect (or absence of true knowledge). First, they claim that unequally distributed ignorance of this kind helps maintain the power of privileged individuals (e.g., experts). It also may be used by people in power to maintain a climate of insecurity which enables them to dominate their subordinates. Secondly, because isolation from alternative beliefs or values is itself a conservative force, these authors argue that neglect helps preserve traditional ideologies, thereby contributing to social stability. Likewise, ignorance reinforces stereotypes.

The Marxist concept of ideology has at least two senses. The first is similar to that of the functionalists, namely, a deviation from scientific thought. The second simply refers to ideology as thought that is influenced by sectional interests, but in which people have wrongly perceived their own 'true' interests (often this is also called 'false consciousness'). In so doing, the Marxists replace the problem of specifying an absolute epistemology with another: Identifying the true interests of each part of society. Their account of how ideology is generated combines a materialistic determinism with a conspiracy theory version of functionalism. The ruling class controls both the means of production, whose conditions influence thought, and the dissemination of ideas and information. It is in their interests to promulgate false consciousness among the proletariat in order to maintain their dominance. Thus, ignorance on the part of the proletariat is functional for the ruling class.

Both the classical functionalist and Marxist accounts of ignorance (and knowledge as well) have many obvious defects. First, they require a naive absolutist epistemology, with an idealized and static notion of modern science as its centerpiece. This epistemological stance virtually contradicts the fundamental insight of the sociology of knowledge (thereby committing the fallacy of dual residentialism). It also drastically curtails the explanatory power that any such theory can have, by begging the question of how scientific 'truth' becomes established in the first place, and by yielding only unsatisfactory teleological accounts of ideology. Most of the functionalists appear to take the scientific method as its own explanation as well as justficiation: Apply this method and you get truth. Parsons is somewhat more sophisticated, but concludes by positing a vaguely Darwinian 'steering' mechanism whereby the best ideas get naturally selected in the long run. Marxists posit that the processes leading to the overthrow of capitalist orders entail the proletariat waking up to what their real interests are, but they rely on mere exposure to the 'truth' about the ruling elites as a sufficient stimulus to trigger this transformation in consciousness.

The genetic fallacy which stands behind these difficulties is cut from the same cloth as the psychologist's and mathematician's reluctance to admit the influence of social factors into their accounts of uncertainty and ignorance. As Douglas (1985: 4) puts it, the lesson here is not simply that we must strive for greater objectivity. It is, in fact, a timid clinging to objectivity that has made even sociologists of knowledge so reluctant to embrace the implications of their own thesis. They and their psychological or mathematical colleagues assume at the start that it is impossible to construct an objective or rational framework that encompasses the relationships between mind and society. To give up the pursuit of an objective account of ignorance or knowledge because it cannot be complete is akin to giving up axiomatization and logic in mathematics because that can never be complete.

Modern perspectives in the sociology of knowledge have been more whole-hearted, although their accounts have revealed still more pitfalls and biases. Symbolic interactionists, for instance, have produced some of the most explicit and interesting discussions of ignorance in the sociological literature. Antecedent theories, by Cooley (1922) and Mead (1934), attempted to explain how social interaction is possible. The central tenet of their approach is that communication, and therefore a social stock of knowledge, is made possible by 'significant symbols' whose meanings are shared by their users in the sense that they elicit the same response from the receiver as they do from the sender. Some symbolic interactionists (e.g., Blumer 1969) have since modified this thesis to claim that reality is negotiated via social interactions in which participants share definitions of the premises on which those interactions are based. In either case, the result has been that the role occupied by scientific truth in the classical functionalist and Marxist perspectives is taken by social consensus in the symbolic interactionist approach.

Berger and Luckmann's (1967) monograph is perhaps the most famous articulation of a symbolic interactionist version of the sociology of knowledge. One of their primary claims is that through shared standardized meanings (significant symbols), subjective reality becomes consensually defined and eventually crystallized in institutional forms (pg. 28). Furthermore, it is this cognitive consensus which makes social interaction possible in all its cooperative forms; it is a fundamental underpinning of the social order. Understandably, most symbolic interactionists tend to view nonshared cognitive orientations (that is, ignorance) as socially divisive or even destructive. At the very least, ignorance poses a major problem for this perspective, since it lacks an account of how ignorance is even possible, to say nothing of its social nature.

Several theorists who are close to the symbolic interactionist position have attempted to provide accounts of ignorance. Lindesmith, Strauss and Denzin (1975), for instance, hold that cooperative group behavior requires shared cognitive perspectives but they also claim that nonshared perspectives help demarcate boundaries between social groups, and admit the possibility that deliberately manufactured ignorance can have prosocial consequences. Some researchers such as Braroe (1970) have found instances where the entire livelihood of social groups or even communities depends on sheer deception or even pluralistic ignorance. A somewhat more radical stance is taken by McCall and Simmons (1966), who claim that the social consensus is almost never complete and that the real basis for cooperative behavior is a variety of loosely negotiated 'workable agreements'. However, they never specify what the minimum requirements are for these agreements to function effectively. Their arguments are nevertheless symptomatic of a general trend among symbolic interactionists to decrease the scope of the social consensus argument and to admit ignorance into an increasingly central role in ordinary social life.

Perhaps the most radical dissident in the symbolic interactionist camp is Goffman (1959, 1963, 1974). Goffman not only allows that social interaction is possible when the participants do not share the same perceptions of the situation, he claims that for some kinds of social interaction it is necessary that the interactants be ignorant of one another's thoughts, plans, and intentions. Goffman borrows the analogy of the backstage from theatrical production, and explains a wide variety of social interaction in terms of people's strategic use of their own 'backstages' for impression management purposes. He even extends this metaphor to the organizational level, virtually claiming that within every nontrivial organization exists an informal secret society within which individual members contrive and collude to attain goals that cannot be reached through consensual action.

The limitations in the classical symbolic interactionist tradition are fairly clear. Although they are not scientistic in their absolutism, their account of ignorance still requires that the theorist or observer know what the 'real' state of affairs is (otherwise it is difficult to ascertain who is deceiving whom). They overcome their bias in favor of consensus only to the extent that they admit the possibility that some social interaction may be carried out via deception or, more rarely, in fortunate coincidences between unintentionally divergent understandings. The fundamental thesis remains essentially a functionalist one, namely that social agreements on what constitutes knowledge arise because they are needed in order for people to get on with social life. Lasting agreements of this kind become institutionalized and thereby take on a seemingly objective quality (which many sociologists refer to as 'reification').

Ethnomethodologists carry the implications of the sociology of knowledge thesis to a more radical extreme than the symbolic interactionists. First, they observe that much of what passes for shared cognitive orientations is simply manufactured as part of people's impression management strategies (Filmer 1972: 212). The fact is that social reality is very flimsy and transitory, and people could potentially renegotiate reality at any instant in social interaction (cf. the 'breaching' experiments in Garfinkel 1967). The fact that they do not is due to the work that people do to maintain an unbroken social fabric through the operation of mutual trust and normative as well as interpretive rules (Mehan and Wood 1975: 114).

Ethnomethodological writings utilize two ideas that bear directly on the topic of ignorance. One is the concept of 'indexicality' mentioned in the previous section. What shared meanings people obtain through social interaction are never complete; they always require the use of an 'et cetera' principle to get by. Ethnomethodologists are fond of pointing out how even the simplest instructions for activities like cooking or operating a machine are incomprehensible without invoking interpretive assumptions that are not explicit in the instructions themselves. The claim that extra interpretations are always required for the manufacture of apparent consensus and cooperative action is the second key element in ethnomethodology. This interpretive work is called 'filling in', and without it the most elementary kinds of social interactions would not be possible.

The ethnomethodological position appears to imply that ignorance is the unregulated state of affairs among people, and that shared knowledge is possible only to the limited degree permitted by the exercise of mutual trust and shared rules for interpretive 'filling in'. Thus, to the symbolic interactionist portrait of ignorance as a byproduct of improper or impaired negotiations, the ethomethodologists have added a new look: Ignorance as the stumbling-block in social reality which is defeated in the very process of social interaction. It is noteworthy that one of the elements in ethnomethodological procedures for demonstrating the existence of unsaid rules in social interaction is for the interactant to inquire 'too much' into the other person's meanings, thoughts or feelings. Their claim is that no social consensus, and hence no body of knowledge, can survive such inquiries.

Despite their highly effective critique of the older sociology of knowledge tradition for being too timid with its central thesis, the ethnomethodologists are far from having a complete social theory of ignorance. At best, they have pointed out the centrality of ignorance in everyday social life, and suggested that the micro-order of social interaction contains within it the potential for both strategic and unintended ignorance. They have not provided a usable typology of ignorance, nor have they explained how it arises from the constituents of human communication. They have virtually no account of how ignorance and knowledge function at the social structural level. No such general theory of ignorance has been forthcoming from any of the social sciences. Fortunately, however, other social scientists have accomplished some of these things within limited domains of social life, so we may turn to their accounts for some instructive examples.

6.3 Ignorance in the Micro-Order

It is not difficult to set the stage for some understanding of how ignorance and uncertainty are constructed in ordinary social interaction if we avoid the absolutist and pro-consensus biases of mainstream sociological and psychological theories. The definition and typology of ignorance developed in Chapter 1 were intended for that purpose, and they will serve us here. Ignorance therefore will always be understood to exist only in relation to some subjective and ultimately social point of view.

This section advances several arguments that ignorance is made possible by certain properties of natural language, social interaction processes, social norms, and agreements, all of which pervade everyday social life. These arguments utilize evidence from research on polite conversational norms and strategies, self-disclosure, question-asking, privacy, secrecy, selective attention. The main arguments are as follows. First, natural language is sufficiently flexible and incompletely specified as to permit partial communication or even outright misunderstanding. Secondly, those properties of natural language are used widely in, for example, polite conversation and self-disclosure to generate acceptable levels of ignorance. Specific strategies are used for creating particular kinds of ignorance, in fact. Moreover, there are pervasive social norms which encourage and legitimize less than fully informative communication (e.g., the maintenance of personal privacy). Finally, there are norms restricting information-seeking (e.g., constraints on reality-testing and question-asking).

6.3.1 Language, Politeness, and Ignorance

We may begin by firming up several of the ethnomethodological and symbolic interactionist insights into what makes ignorance not only possible but likely. By the same token we need some understanding of the conditions under which ignorance may be detected. Complete cognitive agreement between people requires at least three conditions: (1) A language that is sufficiently specific and unambiguous that meanings may be unconditionally agreed upon by its users; (2) Agreement on acceptable methods for information seeking, reality-testing, and verification; and (3) Equal and identical opportunities for people to engage in information seeking and reality-testing.

Mathematical and other context-free formal languages are perhaps the closest approximations we have to examples fulfilling the first condition. It is rather difficult to lie or dissemble in mathematics or PASCAL, for instance, given a sufficiently mathematically fluent victim. By the same token, it is comparatively easy for an expert in mathematics or PASCAL to detect errors. Thus, formal languages are restricted in their potential for generating ignorance. Indeed, they are designed to minimize incompleteness and distortion. However, the properties of those languages that make it difficult to fool another expert also make it easy to be found 'wrong' (and thereby ignorant in the passive sense).

Natural language, on the other hand, is notorious for its vagueness, ambiguity, nonspecificity, and indexicality. Those very properties make it relatively easy to lie, dissemble, obscure, and confuse so that even those who are fluent in a natural language are easily deceived. However, those same characteristics also make it more difficult to detect communicative or factual errors; there are too many let-outs for the communicator.

Criteria and norms for information seeking, reality-testing, and verification also are so constructed as to make ignorance not only possible but probable. Not only is there seldom anything like a consensus on these matters among interactants, but more importantly a variety of norms and rules inhibit both information seeking and reality-testing in most ordinary social situations. Moreover, those norms and rules institute unequal opportunities for the acquisition of knowledge. Some of the most complete accounts of ignorance in the micro-order of social interaction concern norms that militate against full knowledge or complete communication. Such norms exploit our capacity for linguistic distortion, vagueness, ambiguity, and context dependency in the service of politeness, avoidance of conflict, and the maintenance of social relations.

Consider the construction of ignorance in polite conversation. A considerable component of being polite involves the avoidance of matters or messages that might induce discomfort or social conflict. Thus, polite conversationalists must on the one hand find ways of ignoring such matters themselves, and on the other hand helping their co-participants to ignore them as well. It is probably no concidence that the Latin root for 'nice' is 'necius', which means ignorant. An early awareness of the connections between politeness and ignorance may have become expressed in language.

According to the taxonomy in Chapter 1, there are two ways in which polite communications can maintain ignorance about taboo areas: (1) Distortion, or (2) Incompleteness. Distortion requires outright disinformative communication (e.g., white lies), which is risky because of the demands it makes on the communicator. Under some conditions it is also socially unacceptable to be found out. Incompleteness, on the other hand, is not as risky and often more acceptable. It may consist of mere omission or absence of information (referential avoidance), or the use of vagueness, ambiguity, or nonspecificity to obscure the information (referential abbreviation).

Brown and Levinson (1978) propose an account of politeness strategies that distinguishes between two kinds of politeness which have different strategic requirements. Positive politeness, on the one hand, involves endorsement, approval, and camaraderie. Negative politeness, on the other, denotes deference and nonimposition. Brown and Levinson's list of strategies for positive politeness include the following:

(1) Claim common ground with the other person,
(2) Presuppose or assert common ground, and
(3) Convey cooperation with the other person.

Claiming or asserting common ground involves indicating positive interest or sympathy, using in-group slang, seeking agreement, and avoiding disagreement. These subgoals may be achieved either by disinformation (that is, promoting a false impression of common ground) or by referential abbreviation when discussing 'common ground' topics. Referential avoidance, of course, is used to screen out information that does not convey an impression of common ground.

Negative politeness strategies require the communication of deference and the avoidance of imposition. Brown and Levinson propose the following:

(1) Indirect speech acts,
(2) Hedging (qualifying or diffusing meanings),
(3) Avoiding coercion,
(4) Apologizing,
(5) Impersonalizing self or other,
(6) Stating a face-threatening behavior as a general rule,
(7) Nominalizing, and
(8) Conveying indebtedness.

Strategies 1, 2, 4, 5, and 6 use communicational devices. Hedging and apologies both rely on the use of referential abbreviation, particularly vagueness and ambiguity. Referential avoidance, on the other hand, is used for impersonalizing self and other, and also for stating coercive or other threatening messages as general rules (contrast, for instance 'No smoking is permitted in this room' with 'I will not permit you to smoke here').

6.3.2 Norms Inhibiting Communication

Strategies for polite communication are mediated by widely shared norms restricting what should be communicated. Social psychologists are fond of explicating such norms and following out their implications for social interaction. An example of a reasonably well-investigated norm is the reluctance of people to transmit bad news to a person who is directly affected by the contents of the message. Rosen and Tesser (1970) and Tesser, et al. (1972) provided the best experimental evidence for the operation of this norm in the absence of a variety of possible confounding factors (e.g., differences in status between the communicator and receiver).

They also found that the norms for actually transmitting bad news are vague in comparison with those for conveying good news. Good news tends to be communicated more frequently, more quickly, more fully, and more spontaneously (Tesser and Rosen 1975: 228).

This norm has several obvious implications and consequences. The person most likely to be badly affected by bad news is the least likely to receive it. This is probably enhanced when the potential recipient is of higher status than the communicator(s), which has immediate implications for the distribution of ignorance throughout organizational hierarchies. Furthermore, insofar as negative messages are more likely to motivate changes in behavior than positive messages, the reluctance to transmit bad news tends to act as a conservative force, favoring social and personal stability and retarding the rate of change.

A related concern in social psychology has been the study of self-disclosure. Initially, the main research topic in this area was the apparent norm of reciprocity in various properties of the information disclosed by interactants about themselves (cf. Derlega and Chaikin 1975, Lynn 1978, and Taylor 1979). However, violations of the reciprocity norm soon came to light and research interest shifted to them. Ironically, one of the most frequently noted exceptions to the reciprocity rule occurs in interactions between friends or intimates (Derlega et al. 1976). Davis (1977) observed that unless interactants communicate about how intimate to be with one another, one of them tends to exert unilateral control over intimacy levels in the conversation, with the other following or resisting. Researchers who brought status differences into their experiments found that subordinates tend to disclose more about themselves to superiors than vice versa (Taylor 1979).

A related stream of research has investigated factors that inhibit self-disclosure. Most of the factors explored are norms of some kind, although it is widely acknowledged that the norms governing the appropriateness of self-disclosure are likely to vary considerably across cultures and social contexts (cf. Barnlund 1975). In addition to the norm that inhibits the transmission of bad news about others, there is some evidence that revealing negative information about oneself is generally viewed as less appropriate than revealing positive information (Chaikin and Derlega 1974). More tellingly, some studies suggest that any self-evaluative information tends to be transmitted via third parties rather than by oneself when possible (e.g., Blumberg 1972).

A variety of studies have also indicated that lower-status individuals are expected to self-disclose more intimately than higher-status individuals, with most of the direct evidence for this coming from studies comparing men's and women's self-disclosing behavior (Jourard 1971). Caltabiano and Smithson (1982) construct a somewhat more complex picture of gender differences. Their findings suggest not only that women are expected to disclose more intimately than men, but that women are also more accepting of intimate self disclosure generally than are men. Indeed, women may be expected to receive more intimate information about another person. These findings raise the possibility that women are accorded less of a right to privacy than are men.

These studies indicate the existence of systematic, and widely shared, social norms limiting the exchange of information among people in everyday life. While researchers have barely scratched the surface of this phenomenon, their evidence is sufficient to demonstrate that a considerable portion of social interaction entails agreements among interactants on how much and what kind of information is to be shared. The micro-order contains even more direct constraints on knowledge than these agreements, however. Chief among these are privacy and secrecy, which involve norms against information-seeking and reality-testing.

6.3.3 Privacy, Secrecy, and the Valuation of Information

Privacy and secrecy both have received some attention from social scientists. Although early treatments of privacy tended to focus on the physical isolation of individuals from others (see Schwartz 1968 and Foddy and Finighan 1980 for reviews of competing definitions of privacy), an emerging position among social psychologists holds that privacy involves control over access by others to information, mainly about the self. Foddy and Finighan in particular argue that private information is concealed in order to protect its holder from others who they believe might use such information against them.

Warren and Laslett (1977) offer a viable distinction between privacy and secrecy, when they point out that privacy involves a consensual and essentially cooperative ignorance arrangement, while secrecy is a unilateral arrangement. Thus, while we often refer to a 'right to privacy', secrecy is almost never mentioned in that sense. Secrecy connotes the exercise of unilateral control over information. There is a parallel, then, between privacy and negative politeness (nonimposition in particular), and between secrecy and positive politeness (impression management). Privacy tends to be created and maintained via the same strategies used in negative politeness rituals, while secrecy is more likely to involve outright disinformation and referential avoidance.

Insofar as privacy requires cooperation from others and may be conferred upon people as a right or privilege, Warren and Laslett maintain that privacy is most likely to be granted to people with status and legitimacy. Secrecy, by contrast, is likely to be resorted to by those who are denied the right to privacy and is attained primarily by the exercise of power and cunning. These observations reinforce the earlier suggestion in the literature on self-disclosure that women are given less privacy than men.

Privacy and secrecy arrangements both are buttressed by a variety of norms that limit people's information seeking and reality-testing behavior, or in the words of Douglas (1985: 55), that promote the 'social control of curiosity'. Children everywhere quickly learn that there are private and secret things, privileged domains of knowledge into which they are not supposed to inquire. They also learn not to ask too many questions and to avoid certain kinds of questions altogether. Overinquisitiveness is widely censured in many societies (e.g., 'curiosity killed the cat') and is also denoted by pejorative terms (e.g., 'nosey' in America, 'sticky-beak' in Australia).

Two types of limitations on information seeking behavior have been discussed in the social science literature: Selective attention and question asking. Selective attention has long been the province of perception psychology (cf. the classic studies by Hebb 1949), but perception studies have consistently neglected the social influences on selection. Even most social psychological accounts of this phenomenon amount to little more than motivational explanations in terms of a self-serving avoidance of disagreeable, dissonant, or threatening information. These may be classed with the experiments on the use of modus tollens reviewed in Chapter 5, since they are really motivational studies of attentional 'bias'.

Perhaps the most interesting motivational perspectives are those which contain an economic element. These begin with the observation that while information is a multiplier positive-sum resource, attention is a zero-sum nonrenewable resource. If one person gives information to someone, that person still retains the information. For this reason, an 'information explosion' is possible to a degree that a 'monetary explosion' is not. However, a person is limited in the information to which she or he can simultaneously pay attention. In an information-dense world, therefore, most people cannot afford to indulge in an unselective consumption of information. A number of commentators have observed that ignorance is expanding faster than knowledge in two related senses. First, there is the well-known catchcry that as the social stock of knowledge increases, each individual is forced to become more specialized in her or his domain of comprehension and learning. A second, less popular claim has it that the very growth of knowledge throws up more unanswered questions than answered ones (e.g., Pirsig 1974, Ravetz 1987).

Simon (1978: 13) effectively portrays the immediate quasi-economic consequences of an information-rich society when he says "in a world where information is relatively scarce, and where problems for decision are few and simple, information is almost always a positive good. In a world where attention is a major scarce resource, information may be an expensive luxury... we cannot afford to attend to information simply because it is there." This line of argument clearly leads to an 'economics of attention' in which various information sources compete with one another to persuade people that attending to them is worth their time and effort (Derber 1979).

These observations contrast interestingly with the more widely known motivational account of 'reactance' (cf. Brehm and Brehm 1981), which posits that people react against any perceived threat to their freedom of action. Accordingly, people should respond to a proscription against information-seeking or a ban on certain information by trying to obtain that very information. Indeed, forbidden or scarce information is not only valued more highly than easily accessible information, but some writers argue that it is more likely to grab people's attention and is believed more strongly once obtained.

Cialdini (1984) provides empirical support for both these claims. He reports a study of mock-juries in which subjects not only failed to ignore evidence ruled by the judge as inadmissible (despite the judge's direction to them to disregard it), but actually gave it greater weight than those in a mock-trial where the same evidence was not ruled as inadmissible. He also finds in another study that consumers are more likely to respond to an advertisement of a special sale if they also believe that only a select group of customers know about the sale than if they think the information is widely available to the public.

Motivations and economics notwithstanding, it is obvious that a large portion of selective attention can only be explained in terms of social proscription. The relevant classical paradigm in the social sciences is the anthropological study of taboo, which is enforced ignorance, as indicated in Chapter 1. Douglas' (1966) landmark work remains one of the best articulations of a structural-functionalist perspective on taboo. According to Douglas, the larger patterns of taboos, privacy and secrecy arrangements, and social norms guiding selective attention all spring from a ubiquitous concern with the avoidance of danger and pollution. Furthermore, this concern is distributed in accordance with the boundaries and cohesions that are built into the social structure, so that societies possessed of either high cohesiveness or complex hierarchical structures will also have the most intricate and vigorously enforced proscriptions against information seeking, attention, or knowledge acquisition.

Related accounts have come from social anthropologists who have studied information-seeking behavior, and in particular question-asking. Goody (1978) and Martin (1981) both report that among the Gonja and certain groups of Australian Aborigines (respectively) informational question-asking is nearly totally forbidden in the training of children. They contrast this finding with the Western style of education, where constraints on question-asking are much looser and certain kinds of information-seeking questions play a central role. Goody provides a compelling argument that conventional explanations for this proscription (e.g., the children learn by example alone) do not account for the injunction against question-asking *per se*.

Instead, Goody observes that questions, like other speech acts, have both a locutionary and an illocutionary function. The latter exerts a kind of control over the conversational process, in the sense that a question demands some kind of answer from the person to whom it has been directed. But the controlling aspects of the illocutionary function can go beyond this, and Goody provides examples of 'control' questions of the kind that might be used in an interrogation on the one hand, and 'deference' questions that might be used in politeness rituals.

Goody's next point echoes Douglas' analysis of taboos, for she argues that societies with rigid and pervasive hierarchies inhibit even purely informational question-asking, because they are neither control nor deference questions. Their ambiguous status is problematic in interactions between subordinates and superiors. A subordinate will tend to interpret a request for information from a superior as an interrogation, while a superior will tend to interpret such a question from a subordinate as a challenge of authority. Perhaps the most available example for Westerners is to imagine the question 'What do you know about x?' being asked by a teacher of a student, and then by a student of a teacher. A more threatening question in that context, of course, is 'How do you know?'

Proscriptions against reality-testing are even more striking than those on information-seeking, perhaps because their functions are more obviously protective but also by virtue of the mass of untested propositions in which people are willing to place their faith. As sociologists have eloquently phrased it (e.g., C. W. Mills 1959, or Berger and Luckmann 1967), we live in second-hand worlds in which most of what we consider knowledge never gets tested or experienced in a first-hand manner. However, because reality-testing is instrumental and intrusive, it is highly circumscribed and usually reserved for high-status members of society with approved authority, qualifications, competence, and reliability. Reality-testing opportunities also are restricted in order to maintain privilege, secrecy, and advantage. Cases in point run the gamut from laws against the unauthorized repair or even investigation of technological devices and infrastructures to the restraints placed on educational activities in schools.

Wildavsky (1985) pursues the issue of normative restrictions on reality-testing to several interesting conclusions. He argues that Western society has become collectively more risk-averse in recent years, and there is a broadly held view that nothing new should be tried unless it is certain that it will do no damage (his phrase is 'trial without error'). He points out that without trial there is no error, and without error, there can be no learning. A similar point was made by the philosophically minded mathematician G. Spencer-Brown (1974: 110), when he claims that Western educational practices limit pupils' capacities for expanding their knowledge by reinforcing pride in what they already know and shame in respect of consciously acknowledged ignorance. Wildavsky recognizes that some sectors of society provide social environments that favor this norm more than others (e.g., bureaucratic institutions tend to be much more risk-averse in this sense than entrepreneurial organizations).

Across-the-board risk-aversion has several consequences for the creation and maintenance of ignorance, according to Wildavsky. First, increasingly strict standards for evidence that a proposed innovation or reality test will do no damage makes innovation more expensive and difficult. Increased expense and difficulty in turn retard the rate of innovation and discovery. A reduced capacity for innovation renders society more vulnerable to unanticipated changes or problems by decreasing its adaptability. In fact, its ultimate byproduct is what ecologists aptly term 'overspecialization'. Thus, sociological perspectives on the norms governing ignorance shed light on the debates about rationality in the face of uncertainty or ignorance, and their contributions will be considered in the next chapter.

Norms for the manufacture of apparent consensus and/or certainty can lead to inattention to, and bans on, information-seeking about the extent of 'real' ignorance. These practices form part of Janis' (1972) conditions for the development of 'groupthink'. In his provocative monograph on man-made disasters, Turner (1978: 194-196) dwells on the tendency for anomalies to be ignored or unattended to because they do not fit into the prevailing standardized norms for information-seeking and recording. Twenty years earlier, March and Simon (1958) and Cyert and March (1963) described the organizational norms that characterize a traditional mode of dealing with uncertainty by 'absorption'. Members of such organizations manufacture an atmosphere of certainty by curtailing the permissible ways of analyzing, anticipating, and communicating about the organization's problems. Instead, they construct a self-confirming environment from which inquiries or communications about anomalies or uncertainties are excluded.

Finally, at least one researcher has used norms about uncertainty as a crucial variable in a cross-cultural study of values, norms, and cultural practices. Hofstede's ambitious (1980) cross-national comparison of 'uncertainty avoidance' and related cultural and social factors suggests that high uncertainty-avoidance societies are distinguishable from their low uncertainty-avoidance counterparts on a number of counts. High uncertainty-avoidance countries share norms that promote hard work, strong regulation of individual action, an emphasis on the attainment of security, an emphasis on consensus, an avoidance of conflict, more aggression towards outsiders, absolutism, and a dependency on experts. These countries exhibit stronger nationalism, less tolerance for citizen protest, more elaborate legal systems, and greater specialization in their workforces. They are what Douglas would call 'high-grid, high-group' societies.

Taken seriously, these scattered research findings and theoretical concepts suggest that norms and social arrangements that promote ignorance are woven not only into the micro-order of social interaction, but also into higher-level cultural and social institutions. Clearly they perform a multitude of social functions, to such an extent that the traditional bias in the human sciences favoring total consensus and complete communication as the foundation of society is revealed for what it is: Less than half the story.

6.3.4 Problems and Limitations

While these more recent social scientific accounts of ignorance are not as problematic as their classical forebears, they still suffer from a number of defects. Chief among these is the absence of a clearly defined typology of ignorance, uncertainty, and related concepts. This lack may well be due to a residual tendency for most writers to treat ignorance implicitly as a simple opposite of 'true' knowledge. Nevertheless, this tendency considerably handicaps the development of research and theory in this area. While the typology in Chapter 1 may not be entirely rigorous, at least it avoids many of the pitfalls associated with using terms like 'ambiguity', 'uncertainty', and 'ignorance' as if they were synonyms.

Unlike the cognitive psychological literature, the social scientific accounts of ignorance are weak on operationalizations of their main concepts, and also on empirical evidence for their primary claims. Most of them are still suggestive perspectives rather than fully-fledged theories, and although the empirical evidence they do have is from the real world instead of laboratory studies, much of that evidence is little more than anecdotal. Hofstede's cross-national survey is an outstanding exception to this rule.

Another problem with this material is its (perhaps unintended) bias in covering various kinds of ignorance. The most prominent bias is an emphasis on ignorance in the active voice (that is, taboo and other kinds of irrelevancy rules) and intentionally created ignorance in the passive voice. The modern literature on the micro-order of social interaction has little to offer for those interested in unintentional ignorance. It is noteworthy that this bias is almost the opposite of that which is found in cognitive and decisional psychology as well as in normative/mathematical accounts. Those focus almost exclusively on unintended error or omission, and pay virtually no attention to intentionally created ignorance or irrelevancy rules.

These complementary biases may well reflect more than historical artifacts within each discipline. Unintended ignorance is rather difficult to observe, let alone explain, in social psychological terms. The ignorance-creating values, motivations, rules, and norms in any society are far more likely to refer to irrelevancy and deliberate incompleteness or distortion than they are to unconscious error; and the social processes by which ignorance is intentionally created are much more easily observed than those resulting in unintentional ignorance. Applied scientists, cognitive psychologists, and decision theorists on the other hand, are more inclined to view uncertainty as a property 'out there' in the real world and ignorance as a largely unintended product of improper thinking. These biases also have their *raisons d'être*. The notion that uncertainty is an environmental property implies that it is potentially knowable, or at least classifiable, and even measurable. It can thereby be managed with the correct approach. Likewise, if ignorance is an unintended byproduct of incorrect thinking, then it is potentially corrigible. Dishonest or deliberately ignorant people, on the other hand, may not be swayed by reason.

These last few points highlight the potential contributions that a social psychological or sociological view of ignorance could make to both normative and descriptive paradigms. First, this view reminds us that ignorance, like knowledge, is socially constructed and that social norms, rules, and other sociocultural concomitants exert an inevitable influence on both the existence of and the criteria for detecting ignorance of any kind. Second, it explodes the myth that uncertainty or any other kind of ignorance is a disembodied property of the environment. It is also a human creation, particularly when the environment involves human organizations. Third, it alerts analysts and psychologists alike to the fact that much ignorance is intentionally created and that it is essential to nearly all human interaction and organizational functioning. The 'honest system' implicitly assumed by the paradigms reviewed in Chapters 3, 4, and 5 often does not exist. Any psychological perspective on ignorance that fails to recognize this is socially and politically naive, and any normative perspective that does not consider it is condemned to be ineffectual in the management of human systems. Finally, it suggests that deliberately creating or maintaining ignorance is not always irrational or maladaptive, and may even be beneficial in some contexts.

6.4 Ignorance in Organizational Life

Some of the major contrasts between the individualistic and organizational views of decision making under uncertainty have been highlighted by several writers. Linnerooth (1984) points out that in organizations and institutions, decisions are seldom made by an individual (rational or not), but instead are negotiated sequentially by interest-groups consisting of public officials, representatives of public constituencies, and industrial representatives. Organizational and public decisions therefore tend to be 'stretched' over time rather than being 'made' at some definite point, in a process which Weiss (1980) calls "decision accretion".

Weiss also observes that knowledge of the kind that would be directly taken into account by an individual 'rational' decision maker often plays only a background role at the organizational level, with its influence increasing gradually at best (she terms this "knowledge creep"). Other writers (e.g., Downs 1966) claim that bureaucratic officials often attempt to minimize or even cover up the extent of ignorance in respect of a given problem to avoid complicated negotiations and controversies. To this Linnerooth (1984: 228) adds three commonly occurring political agendas (or needs) that militate against governmental regulatory agencies being open and honest about ignorance: (1) The need to maintain apparent control in order to legitimate and bolster authority; (2) the need to justify policy decisions with persuasive and apparently certainty-producing analyses; and (3) the need for narrow rather than comprehensive analyses. Agencies operating with these agendas tend to reframe problems so that uncertainties appear normal, ordinary, and manageable.

In some cases, on the other hand, organizations may be motivated to emphasize or even exaggerate ignorance. Campbell (1983) presents a case-study of a strategic use of certified scientific arguments to bolster the case for uncertainty in relation to a policy decision. Smithson (1980) makes a general claim that ignorance becomes strategically important when a case needs to be made for stopping some activity (because not enough is known about its effects) or against changing the status quo (the 'better the devil you know' argument). He presents an analysis of the early debates over the ozone-depleting effects of fluorocarbon propellants in aerosols in which appeals to scientific ignorance were made by both environmentalists and industrialists to support their respective demands that aerosol production be stopped or continued.

To some extent the fact that organizations negotiate decisions and take longer than individuals to arrive at them could be explained away by referring to the competitive relations among many organizations and their size or complexity. However, this begs many questions. Also, there is at least persuasive circumstantial evidence that organizational practices, structures, and their members' impromptu actions may collude in delaying decisions under some kinds of ignorance-prone conditions. Cohen, March, and Olsen (1972) argue this point in their 'garbage can' model of organizational choice. Hofstede (1980) points out several 'uncertainty avoiding rituals' in organizations that serve to slow down or stop time.

The mixed picture that emerges of how organizations use knowledge and ignorance is even more vexing. Clearly organizations are not always certainty maximizers, and even when they do maximize certainty it is not in the same way individuals do. Unfortunately, the organizational literature does not provide anything like a comprehensive overview of these matters. Instead, it contains a hodge-podge of studies, suggestions, and conceptual schemas in which ignorance (mainly uncertainty) appears fitfully and plays ambiguous roles. Moreover, as in much of the sociological literature, all nonknowledge is lumped together as if 'uncertainty' stands for any kind of ignorance. Curiously, this literature also treats ignorance in a rather unsociological fashion, not as the social product it is, but instead as an objective feature of organizational or physical environments. Despite these faults, the organizational literature does have some insights to offer that would serve other students of ignorance well.

6.4.1 Definitions of Uncertainty and Related Concepts

Let us begin by examining the definitions of ignorance that have been utilized in this literature. Most writers have used the term 'uncertainty' quite broadly, and for purposes of discussion I will follow their usage in this section. The classical approach to uncertainty grew out of the seminal works in the so-called open systems school, in part as a reaction against the traditional closed-systems perspectives which treated organizations as socially isolated. Part of the price paid by the open systems theorists for their greater realism was the admission of uncertainty into their frameworks on organizational structures and processes.

Perhaps because of the emphasis in the open systems perspective on relations between the organization and its environment and a tendency to view that environment in objective, physical terms, the classical view of uncertainty also was objective. Even so, there was sufficient diversity in definitions and operationalizations to cause considerable confusion. March and Simon (1958) defined uncertainty as a lack of internal control, which led some years later to researchers confusing objective with perceived uncertainty. Cyert and March (1963), on the other hand, focused on external uncertainty and considered the impact of environmental uncertainty on organizational equilibrium and control.

Burns and Stalker (1961) viewed uncertainty as synonymous with environmental instability and rapid change, which they termed 'turbulence'. Other analysts quickly followed (e.g., Chandler 1962, Emery and Trist 1965), with the consequence that turbulence became a widely used substitute for uncertainty. To this Terreberry (1968) added systemic or environmental complexity, thereby bringing some of March and Simon's original concept back in.

More recent work in the classical vein has consisted mainly of attempts to refine and quantify indices of environmental instability, change, or complexity. Thus, Tosi et al. (1973) define three indicators of 'volatility' (market, technological, and income), using normed coefficients of relative variation as measures of short-term variability (see also Bourgeois 1985). Osborn et al. (1980) argue that uncertainty has two main components, disparity (or heterogeneity) and volatility, the latter consisting of the rate, magnitude, and directional predictability of change.

The classical approaches may be criticized on several grounds. Like their sociological, psychological, and economic counterparts, these definitions tend to lump all of uncertainty together and utterly disregard several crucial kinds of ignorance *per se*. So while any kind of unpredictability, stochastic tendency, lack of complete information, and even merely rapid change is branded 'uncertainty', most crucial distinctions are not addressed and other kinds of ignorance (deliberate or unintentional distortion, irrelevance, and taboo) are disregarded. This practice has led Downey and Slocum (1975) to remark about this literature that many writers simply assume that everyone knows what is meant by 'uncertainty'.

A critical-minded social anthropologist such as Douglas might justifiably remark that this assumption and the lack of specific terminology are symptomatic of the taboo surrounding ignorance for organizations theorists of the time. The original open systems school, like the psychological traditions reviewed in Chapter 5, had a consistent normative message. The nearly universal assumption was that uncertainty upsets organizational equilibria and is dysfunctional for organizational performance and relations with the environment (cf. Jauch and Kraft 1986 for a review of these and related assumptions). The corresponding prescription from classical theorists was to eliminate uncertainty and/or buffer a technical core of structures and procedures by which to absorb uncertainty.

A related criticism is that the classical approach does not take organizational members' perceptions into account, nor does it recognize that the very concept of uncertainty is socially constructed. An underlying assumption in the classical view which supports the objective approach to operationalizing uncertainty is that the organization is unable to influence the surrounding environment and therefore must adapt to whatever the environment is or does through internal structural change. Unfortunately, many of the objective indicators of environmental uncertainty are arguably characteristics which may or may not result in uncertainty, rather than uncertainty itself. For example, as Lawrence and Lorsch (1973), Pfeffer and Salancik (1978), and Cameron et al. (1987) point out, uncertainty is better thought of as a likely consequence of environmental turbulence rather than a synonym. Bourgeois (1978) goes one step further and argues that in many cases measures of change such as the Tosi et al. indices simply detect systematic and predictable trends rather than uncertain fluctuations.

Controversy has persisted on these points. Subjectivists (as we shall see) have argued that uncertainty is in the eye of the beholder, and we may understand organizational responses to uncertainty only insofar as we know how uncertainty is perceived by organization members. Classicists respond by pointing out that managers may be overconfident of their knowledge or otherwise mis-estimate the degree of 'real' uncertainty. Therefore, if we wish to assess the impact of actual uncertainty on organizational performance, we require objective as well as subjective measures.

Serious attention was given to perceptions of uncertainty by Thompson (1962) in his early account of decision making strategies in the face of uncertainty. Thompson's goal was to link his typology of decision strategies with another typology of competition and cooperation, and in his scheme perceived uncertainty played a mediating role. Interestingly, Thompson began by recognizing that he blurred the distinction between risk and uncertainty (1962: 335), claiming that this distinction is important for selecting among the tools for decision making but not the strategies. Instead, he distinguished between uncertainty concerning causation and uncertainty about one's own preferences for different outcomes. The latter kind of uncertainty could arise, of course, from not knowing the costs or benefits associated with various possible outcomes.

Duncan's (1972) scale for quantifying perceived uncertainty followed up Thompson's concepts, requiring respondents to assess their knowledge of future environmental developments, consequences to the organization of incorrect decisions, and the likelihoods of future events that influence decisional success or failure. The work most frequently referred to as the seminal paper on perceived uncertainty, however, is by Lawrence and Lorsch (1967). They devised a scale for quantifying perceived uncertainty, relying on indicators such as time necessary for performance feedback, clarity of job-role requirements, and task difficulty. Other attempts to measure subjective uncertainty in this field have focused on perceived rate of change (Tung 1979), inability to assign probabilities to the likelihood of future events (Pfeffer and Salancik 1978, Pennings 1981), and inability to predict the outcomes of decisions (Downey and Slocum 1975, Schmidt and Cummings 1976).

As with the classical attempts, these subjective measures of uncertainty are vulnerable to several criticisms. First, they are nonsociological and in some cases are derived from the psychological and behavioral decision literatures. As a result, they begin by assuming that the constructs and environmental features underlying perceived uncertainty are identical for everyone and stable over time.

Secondly, they fail to address important definitional problems by ignoring crucial distinctions among different kinds of uncertainty and ignorance. Several researchers have investigated the correlation between the two most popular scales (Duncans' and Lawrence-Lorsch), only to find they are not strongly or reliably related (e.g., Downey and Slocum 1975). Given the construction of these two indices, however, this should come as no surprise. Clearly, they have tapped different aspects of a multi-dimensional construct. A related criticism is that the research results relating these measures of perceived uncertainty to other aspects of organizational performance or functioning have been unclear. Again, this outcome may well be due to the conceptual sloppiness of the measures themselves.

Another line of research has attempted to relate measures of perceived uncertainty to objective uncertainty. Tosi et al. found only low correlations between their index of volatility and the Lawrence-Lorsch scale (of magnitude less than 0.29). Duncan (1972) had suggested that turbulence is the best predictor of perceived uncertainty (see also Bourgeois 1980), but it is surprising that he did not utilize an objective version of Thompson's scheme. Duncan settled for a two-fold scheme of predictors in which high uncertainty results from a combination of system complexity and dynamicism.

Snyder and Glueck (1982) fared better by revising Tosi et al.'s indices and tailoring them to specific industries, and then using stock-brokers whose opinions were solicited in the form of subjective numerical estimates for each of the Tosi subscales. The crucial difference between this study and all the others was the more careful attention to compatibility between the subjective and objective operationalizations of uncertainty, with the expected consequence of a high correlation between the two (about 0.87).

Unfortunately, even a methodologically sound effort such as Snyder and Glueck's study cannot overcome the manifold problems with measures of uncertainty in the organizational literature. Although some recent reviews have attempted to surmount these difficulties (e.g., Jauch and Kraft 1986, Milliken 1987), they miss the main points raised here. Jauch and Kraft labor under the classical assumption that uncertainty is an objective feature of the external environment, while Milliken retreats to the position that all uncertainty amounts to the perceived inability to predict accurately. A reasonable question to ask, of course, is whether perceptions of uncertainty predict organizational responses better than so-called objective measures. Unfortunately, this question is almost impossible to answer, given the incoherent state of uncertainty measurement in this field. Thus, indirect answers such as are provided by Downey et al. (1977) to the effect that uncertainty perceptions are better explained by cognitive processes than by environmental conditions must be treated with considerable caution.

It is easy to criticize the organizational literature for its poor measurement practices. In all fairness, however, it should be borne in mind that the normative literatures from probability and statistics, decision theory, and even cognitive psychology offer little genuine assistance to the organizations theorist. A great deal of the uncertainty and ignorance encountered in real organizational life simply does not fit neatly into a probabilistic mold. Also, as March and Olsen (1976) amply demonstrate, many organizational decisions do not have the molar quality to them that individual decisions sometimes do. Finally, even those psychological instruments that purport to measure uncertainty perception, 'risk aversion' or other orientations towards uncertainty are neither sufficiently general nor context-sensitive enough to be imported into organizational studies.

6.4.2 Uncertainty, Organizational Structure, and Process

Despite its definitional intractability for the organizations theorists, 'uncertainty' is widely held to be an important concern for both researchers and practitioners. In an early programmatic statement Thompson (1967: 159) asserted that uncertainty is the most fundamental concern of top administrators. However, the literature reveals an asymmetry that is already heralded by its tendency to view uncertainty in asocial and even objective terms: There are far more studies of how organizations should or do adapt to uncertainty than there are of how uncertainty is generated in the first instance. I shall review these adaptational studies first before turning to the 'minority reports' on the genesis of ignorance in organizations.

As mentioned above, the dominant normative message from the classical works on organizational uncertainty is that effective managers either eliminate or absorb uncertainty. Prescriptive writers often begin by stipulating that in the face of uncertainty the re-establishment of equilibrium through structural changes is a primary goal (Keller, et al. 1974, Lorenzi, et al. 1981). Again, Thompson (1967:190) summarizes the main thrust of the classical view during the 1950's and 60's with his admonitions to eliminate uncertainty and buffer the technical core of the organization. Indeed, his list of strategies for adaptation to uncertainty consist of buffering, smoothing, forecasting, and rationing (pp.20-22). Forecasting and rationing are strategies for outright elimination of uncertainty. However, buffering and smoothing are attempts to absorb uncertainty. Buffering involves maintaining resources whose supply is uncertain (cf. Kopp and Litschert 1980), while smoothing is an attempt to reduce fluctuations or cyclical peaks and declines (Lev 1975).

Most of the normative pronouncements about organizational decision making during this period echo similar sentiments. Problems should be carefully and explicitly defined (e.g., Pounds 1969), goals should be specific (Locke et al. 1981), but alternatives should be generated in a spirit of competition (Mason and Mitroff 1981) and creativity (Nadler 1981). Much of the decisional literature in this field appears to have been informed by various versions of the SEU model.

Most of the early descriptive studies of organizational decision making appeared to support the uncertainty-eliminative view as well. Cyert and March (1963) summarized their dominant findings with the phrase "uncertainty avoidance." They claimed that organizations accomplish this in two ways. First, they evade the requirement of correctly anticipating events in the distant future by using decision rules that emphasize short-term responses to immediate feedback and solving 'pressing problems' instead of planning long-range policies. Secondly, they avoid the problem of anticipating future contingencies or outcomes by arranging an internal organizational environment that is self-confirming through standardization, fixed plans, and appeals to tradition.

To this, commentators such as Hofstede (1980) adds a list of 'uncertainty avoiding' rituals which delay decisions or delegate them to others. Other uncertainty-eliminating strategies include interlocking directorates to control environmental fluctuations (Schoorman et. al. 1981). Some studies have found that uncertainty is apparently ignored altogether (March 1981, March and Feldman 1981, Nutt 1984). Several reviewers of this literature have also remarked that the criterion that has loomed large as the measure of organizational adjustment has been the establishment of equilibrium rather than performance as such (Jauch and Kraft 1986), which would indicate a bias against dynamicism.

There is some evidence that organizations also reduce uncertainty, and ignorance as well (e.g., in the form of secrecy), by reducing internal competition (Merton 1968, see also Ouchi 1980). Likewise, inducing conformity to an authority-driven set of rules or ideological tenets may also serve as a buffer against uncertainty (an insight that goes back to Max Weber). In this kind of organizational climate, only those decisions that carry the weight of authority and ideological legitimacy are considered binding (Rothschild-Whitt 1979: 512) and this criterion alone may determine which choices get made in the face of uncertainty or ignorance.

In the economically oriented literature on this matter, diversification has long been a highly regarded strategy for militating against investment risk. Indeed, a finding from the famous Lawrence and Lorsch (1967) study is that organizations operating in uncertain environments tend to be more highly differentiated (diversified) than those operating in relatively routine environments. However, subsequent studies failed to replicate their findings.

More dramatically, conventional economic wisdom on these matters has recently been challenged. Not only has diversification been undermined as a credible hedge against risk, but the maxim that risk and return are positively correlated has been empirically disconfirmed (Bowman 1980). A recent review of empirical studies by Fiegenbaum and Thomas (1988) indicates that the overall record to date is mixed. These authors attempt to explain these findings by invoking prospect theory, with the finding that firms having returns below target experience a negative risk-return correlation while the reverse is true for those having returns above target. This is an ironic use of prospect theory, whose own empirical validity still is doubtful.

Fiegenbaum and Thomas' results point to a recent realization in research on organizational adaptation to uncertainty which parallels similar developments in the psychological and decisional literature: Not all adaptive strategies are risk-averse. Miles, Snow, and Pfeffer (1974) Miles et al. (1978) suggest that some high-performing managers actively search for change and uncertainty. They call these managers 'prospectors', and their descriptions of them are not far afield from Mars' 'hawks' (1982, see next section). Presumably, these managers strategically select uncertain environments in which they will have a competitive advantage. In a similar vein, Jauch and Kraft (1986) suggest that organizations may choose strategies that create uncertainty but also result in a relative information advantage for themselves. Hickson et al. (1971) earlier provided an internal version of this thesis when they observed that people or departments that cope effectively with uncertainty tend to increase their own power base.

While the earlier open systems theories held that uncertainty itself is bad for organizations and that uncertainty reduction or absorption is adaptive, more recent writers have disagreed with the latter. Some of them argue that those organizations that reduce feedback, induce conformity, and reduce internal competition begin to behave like closed systems and thereby make themselves even more vulnerable (Forrester 1968, Weick 1979). Those which attempt to control the sources of uncertainty in the environment via boundary-spanning activities, smoothing, and buffering also are trying to turn the environment into a deterministic, closed system. Finally, overconformity in perceptions and goals has been roundly criticized for its maladaptive potential, the most well-known critique being Janis' (1972) account of groupthink.

In this connection, March and Olsen (1976) criticize uncertainty-avoiding practices for their inhibiting effects on organizational learning. They point out that overcontrol of behavior and standardization of procedures may prevent individual members from responding creatively to novel situations or knowledge. Secondly, since learning occurs most rapidly when the coupling between the organization and individual is loose, increasing regulation can retard the acquisition of knowledge. On the other hand, in the event of extremely loose couplings between organizational actions and environmental responses, 'superstitious learning' may result from members' overinterpretation of the causal relevance of those actions. Finally, in a theme which echoes the psychological literature on 'learned helplessness', March and Olsen point out that 'ambiguous learning' may occur under conditions of extreme uncertainty when organizations become not only risk-averse but outright decision-averse. In that event, decisions tend to be delayed, to "accrete" in Weiss' sense, and both their origins and rationales become diffuse and unclear.

A somewhat more balanced and empirically informed viewpoint is offered by Bourgeois (1985), who addresses several strategic issues concerning the management of uncertainty and its relationship with economic performance. Bourgeois' conclusions, as he admits, must be considered tentative, given the small sample size and questionable operationalizations of crucial variables. Nevertheless, he provides an interesting contingency hypothesis concerning the effects of perceived uncertainty and goal consensus on economic performance. His findings are that performance is enhanced when managers correctly perceive environmental uncertainty, and that the effects of consensus in perceptions of uncertainty and goals depend on the congruence between perceived and actual uncertainty. Specifically, if this congruence is high then diverse perceptions and goals enhance economic performance.

Even the normative perspectives in the organizations literature have softened somewhat on whether uncertainty is to be treated adversatively. Most of the impetus for this shift seems to have come from writings on technological risk and hazard management. Collingridge (1980:19) claims that in the face of technological innovation many organizations and regulatory agencies find themselves caught in a 'dilemma of control'. On the one hand, in the early stages when the technology is easily controllable, too little is known about its potential effects to warrant imposing controls. On the other, by the time those effects have become apparent control has become costly and difficult.

Wilson (1980) adds an historical dimension to this dilemma by observing that while in the past regulation consisted primarily of limiting monopolies and, later, limiting competitive strategies, the most recent stage has consisted of the direct regulation of industrial products and processes. The result has been to place environmental and human risks in direct opposition to the ideals of economic progress and technological innovation, rendering innovation more costly. Against this background, Collingridge's prescription is not for choices that maximize SEU, but decisions which permit monitoring of effects, allow time for correction, and involve manageable costs for corrigibility. These prescriptions imply that we should favor decisions that lead to flexible outcomes (system freedom, which in fact is positively labeled uncertainty), decisions that are insensitive to error, and that are hedged (in systemic terms, vague).

As mentioned before, the mainstream organizational literature, like its psychological counterpart, has treated uncertainty and ignorance as if they are external 'enemies', to be vanquished, controlled, or at least managed by the besieged organization. As in other fields, there is a noticeable shift during the last 15 years away from an eliminative view to a managerial view about uncertainty. If the source of uncertainty is within the organization, then the normative and descriptive perspectives are both taken from the viewpoint of the manager who must cope with it. However, there is a 'gadfly' literature which presents another side to this picture by suggesting that some organizational members (or indeed even some organizations) find ignorance useful and not only seek it out, but set about creating it.

6.4.3 Creating and Using Ignorance in Organizations

No mainstream theorist in the sociology of organizations appears to distinguish between using (or coping with) existing ignorance and actually creating it. Perhaps the latter is too much at odds with the zeitgeist of the mainstream literature. Indeed, the classical open systems perspective deals only with attempts to eliminate or manage uncertainty, or with the kinds of organizational structures associated with routinization and predictability (e.g., standardization, formalized control, centralized authority, high degree of specialization, and intensive supervision or regulation of workers: Burns and Stalker 1961, Woodward 1965, and Perrow 1970). Yet to any intelligent observer who has worked in organizations, the idea that organizational members and units strategically use and create ignorance for a variety of purposes should come as no surprise at all.

This section, then, investigates the social scientific research which has any bearing on the issue of how ignorance is created and used in organizational life. There is little research of this kind, and most of it only indirectly addresses this issue. We begin with the symbolic interactionists and some related findings from role-theoretic studies in organizational contexts. The most informative research of this kind, however, has come from studies of the hidden economy, and the bulk of this section is devoted to them.

The symbolic interactionists seem to have brought the notion of strategically created ignorance to other social scientists' attention, in connection with their portrayals of impression management and public social interaction. Goffman's dramaturgical perspective leads on to the concept of organized deception quite naturally. He remarks (1959: 104-105) that every team needs its secrets maintained in order to function effectively, and in this connection we may observe that every organizational role brings with it at least two ignorance-bearing requirements: (1) injunctions to selectively attend only to those parts of the environment that pertain to the job at hand (the 'need to know' basis for distributing knowledge), and (2) adherence to privacy and secrecy arrangements which limit the power of the role-taker. Workers often are informed about what consitutes relevant or irrelevant knowledge for their jobs; they are instructed not to be distracted nor to inquire into matters that are not their concern.

Furthermore, organizations are structured so as to enforce mandatory ignorance by the efforts of special personnel whose roles involve controlling information flow. There is a small literature on this kind of role, which is termed 'boundary-spanning' (e.g., Aldrich and Herker 1977, Tushman and Scanlan 1981), and descriptions of this role also dovetail with the traditional concept of 'informal structure' in the sociology of organizations.

The other side of role-requirements with respect to ignorance concerns the extent to which roles permit room for strategically using and creating ignorance. Bowyer (1982) makes an anecdotal case for the claim that in most games and certain other organized activities people distinguish between permissible deception and outright cheating (e.g., bluffing in cards or body-fakes in ballgames). It might seem unproblematic to extend this to organizational roles. However, as developments in the past two decades have amply demonstrated, there is a fuzzy area between the two that is sufficiently broad to encompass entire 'shadow' economies.

The literature on the hidden economy and related activities is relevant here because those activities are based on the creation and maintenance of systematic ignorance. Theories and interpretive accounts of the hidden economy, fiddling, and semi-organized crime are therefore rich sources of insights into the organizational, social, and economic concomitants of intentionally created ignorance.

The economic literature is mainly concerned with attempting to estimate the size and variety of the hidden economy and the structural forces that shape it. There are, however, some useful explanations of how and why the hidden economy has flourished in both the East and West during the past 25 years. Although shadow economies are nothing new (see the historical material in Henry 1978), all evidence points to their rapid growth during a period of marked economic downturn. Most explanations for why the hidden economy has expanded during this time invoke some combination of the rise in unemployment, the increased tax burden in many countries, increasing regulation of industrial practices and processes by governments, increased bureaucratic red-tape, and an upsurge in both legal and illegal immigration (Mattera 1985).

The rise of unemployment actually is symptomatic of a much broader trend towards the destabilization of work throughout the modern world. Employment is considerably less secure and certain than it was during the 1950's and 60's, not only in terms of whether a person will find or keep work, but also whether the job requirements will remain stable over time. Along with increases in unemployment and retrenchment, we also have seen increases in the proportion of jobs being redefined and of the workforce being retrained.

At the same time, increasingly complex and restrictive governmental regulations have made legitimate innovation and change more expensive and difficult. Given the pressures generated by insecurity and change in the marketplace, there are both economic and social incentives for innovation. But would-be innovators are faced with the classic Mertonian (1968) recipe for criminality: Legitimate avenues are either blocked or made difficult by over-regulation. Consequently, they turn to illicit and therefore secretive innovation.

Put in organizational terms, the above account suggests that when a high rate of change and great insecurity are coupled with increasing regulation of behavior (one of the most frequently observed responses of organizations to environmental volatility), organization members will innovate in secret, thereby increasing the extent to which members of the organization are ignorant of one another's activities and goals. If the source of regulation is outside the organization, then the entire organization may revert to a 'secret society' (Simmel 1950). In turn, this hypothesis seems evocative of March and Olsen's (1976) 'garbage-can' model of decision making under uncertainty. That is, it suggests that environmental uncertainty and hyper-regulation may lead to increased intra-organizational ignorance, looser couplings between members' actions and outcomes, greater chances for systematic misperceptions of cause and effect, and finally a diminished capacity for the organization to learn.

An obvious question to ask at this point is whether there are organizational factors or responses that encourage or prevent these developments. The best accounts of how organizational structure and culture influence the kinds of illicit activities that members can engage in have been provided by the social anthropologist Mars (1982). Interestingly, he categorizes "fiddles" according to a typology based on Douglas' grid/group constructs which she developed out of her work on taboos. This is a reasonable connection to make, since Douglas' main theme was that taboo-ridden societies also tend to be very hierarchical and defensively cohesive. Mars' work, as well as the explanations given above for the upsurge in the hidden economy, indicate that such societies may be the most secretive as well.

According to Mars (1982: 29), jobs that are 'high-grid' tend to be low on autonomy, high on supervision and regulation, insulated, high on reciprocal obligations, and uncompetitive. They tend to induce greater propensities for individual secrecy, while low-grid jobs accord greater rights to individual privacy. High-grid jobs are more likely to incur surveillance; low-grid job roles may even be protected from surveillance by status-related norms. The 'group' dimension, on the other hand, refers mainly to the cohesion of the immediate work-group and the maintenance of boundaries between it and other groups. High-group jobs limit the opportunities for individual deception but at the same time provide greater scope for group collusion against the rest of the organization or the environment.

Figure 6.1. shows Mars' taxonomy and his zoological terminology. The Donkeys are isolated, highly regulated, and often their duties are routine. They can least afford detection and must regulate their deception more carefully than the other kinds of job-holders. Wolves, on the other hand, must operate in packs or not at all. Like Donkeys, their deceptions must be carefully regulated to avoid detection and as a consequence 'wolfpacks' often have an informal hierarchy dominated by an autocratic leader who determines the norms for deception and the distribution of its rewards. Vultures are less regulated by their own group when engaging in deception, but instead operate according to shared group norms and goals. For both vulture and wolf groups, retaining one's job may require participation in group fiddles. Hawks usually enjoy the highest status in their legitimate roles, which partly explains the fact that their scope for individual deception is greatest. They tend to engage in 'entrepreneurial' deception.

GRID	Donkeys High Grid Low Group	Wolves High Grid High Group
	Hawks Low Grid Low Group	Vultures Low Grid High Group

GROUP

Figure 6.1. Mars' (1982: 29) Taxonomy of Fiddles

Mars (1982: 138-158) also specifies some structural weak points and conditions in organizations that are conducive to deception and fiddling: (1) passing trade, in which people transact only once; (2) holding special expertise or hidden knowledge; (3) gatekeeping or boundary-spanning roles and positions; (4) triadic operations involving an employer, employee, and client or customer; (5) conditions where surveillance, regulation, or control is expensive and/or difficult (including large scale or size); (6) "ambiguity" in job requirements or performance criteria (for Mars, ambiguity means any kind of imprecision or nonspecificity); (7) goods and services which can be converted, disguised, or hidden; and (8) anonymity or confidentiality.

As with the mainstream sociological literature on organizations and uncertainty, Mars and his colleagues have provided a number of provocative insights and themes concerning ignorance rather than a systematic theory. Their importance for a wider picture, however, is that they raise issues not considered in the normative or psychological accounts of ignorance. Even descriptive accounts of the intentional creation or use of ignorance in organizations---or even everyday life---are rare (e.g., Bok 1978, Bowyer 1982), and they reveal a fundamental dilemma concerning ignorance and control.

While Collingridge characterizes this dilemma in terms of tradeoffs between ease of regulation and ignorance of effects, he ignores a deeper problem facing both organizations and regulatory agencies: Illegitimate innovations and even fiddling are often profitable and beneficial in other ways as well. In the case of the hidden economy, a number of national governments are caught between the desire to eradicate or milk the hidden economy and the realization that its unregulated activities contribute substantially to the kind of economic performance for the nation that wins votes while preventing the poor and unemployed from becoming rebellious.

In the U.S., Reagan publicly winked at tax-evaders, despite a recent step-up in the I.R.S. campaign to identify and prosecute them. The Tories in the U.K. have seriously considered declaring 'freeports' and 'enterprise zones' where a number of labor, tax, and other laws are suspended. In countries such as Italy where the hidden economy accounts for an estimated 30% of the GNP, the government tends to publicly wash its hands of the issue of mere control, let alone eradication.

Critics of current control practices (e.g., Mattera 1985, Mars 1982) make a parallel argument to those given in many other modern writings about ignorance, only theirs contains a more normatively radical message. They say that ignorance (and in particular, the intentional creation and use of it) is here to stay; it can be turned into a collective benefit rather than a threat. Smart managers should recognize the benefits of unregulated, secret innovation, and set up systems of 'soft control' that are receptive to fiddling but also monitor and establish limits for it. Likewise, the widespread use of on-the-job secrecy in the service of getting things done should signal managers that something is wrong with how those jobs are structured. Proprietary rights over innovations are recognized at the organizational level; perhaps more such rights should be accorded to individuals within the organization so that innovations are rewarded without requiring secrecy that cuts the organization out of any benefit.

Whether these and other similar recommendations are feasible cannot be debated here. The main point is that an emerging normative perspective holds that even deliberately created and used ignorance is not necessarily evil or detrimental. Indeed, it is an understandable response to the naive and clumsy regulatory impulse of large-scale organizations in the face of managerial ignorance and social change.

The flip-side of the response to governmental regulation is the debates and concerns over the privacy of citizens from governments. Michael (1982) remarks of the U.S. and U.K. that they are both nearly as secretive as it is possible to be and still qualify as democracies. The past two decades have seen increasing worries expressed about the extent to which governments and employers routinely invade personal privacy (cf. Woodman et al.'s 1982 study of employee attitudes and beliefs about personal information stored and used by their employers' companies). And of course the microcomputer and telecommunications revolutions have alarmed many people because of their implications for the easy storage and transfer of large amounts of information (e.g., Moshowitz 1980, Laver 1980).

Finally, it is noteworthy that governments and regulatory agencies themselves appear caught in a bind when it comes to educating or informing the public about important uncertainties, risks, or hazards. In a recent letter to the editor of the journal *Risk Analysis*, Stallen and Coppock (1987) underscore the conflicting goals behind communications policies and practices concerning public risks. On the one hand, recent policies in the West that favor open and full communication about risks seem to have been impelled by normative concerns (i.e., people ought to know and they have a right to know about risks that could affect them), motivational arguments (the belief that people want to know), and expectation (the public expects the government to perform this communicative role).

Unfortunately, other concerns operate at cross-purposes to these motives for full communication. First, regulatory agencies are worried about engendering an 'information overload'. Secondly, they do not wish to convey conflicting or disagreeing scientific judgments and assessments; they would rather that the public receive a consensual and unambiguous message about the risk concerned. Thirdly, they would like the public to respond quickly and unhesitatingly to local authorities who are placed in charge of the situation.

Reser and Smithson (1988) point to a more sinister aspect of this dilemma. When a risk involves organizational or governmental security, the 'education' of the public about it conflicts directly with organizational or governmental interests in secrecy. This is why so many government publications and pronouncements on nuclear weapons are at best partially informative and in some cases disinformative. Clearly for some sectors of society the manufacture of ignorance is not only considered beneficial; it is believed essential.

Although these descriptive studies of how people in organizations create and utilize ignorance are instructive, they suffer from the same lack of sophistication about ignorance as their predecessors in the mainstream literature. Ignorance seldom is explicitly dealt with, and when it is, the concept is treated as a unitary category (usually labeled 'uncertainty', in fact). I will briefly indicate some potential benefits that could accrue from applying the taxonomy and concepts developed in Chapter 1 to the study of ignorance in organizations.

First, no one has systematically pursued the observation that the kind of ignorance that is created in organizations may be a matter of strategic choice. Should Mars' entrepreneurial workers, for instance, require absolute secrecy for their activities, then they might need to confer *meta-ignorance* on their superiors and co-workers. High-status hawks, on the other hand, may be able to afford merely keeping others consciously ignorant because they have been accorded some rights to privacy. Meta-ignorance and conscious ignorance are likely to differ considerably in their organizational consequences, so the distinction is important for any workable theory of how organizations operate under ignorance.

Likewise, the choice between generating *irrelevance* and *error* also has strategic import. Consider, for example, a manager who wishes certain matters to be kept from subordinates. Should the subordinates be told to ignore the matters concerned, or should they be provided with erroneous information about those matters? If the manager elects to provide erroneous information, should it be outright *distortive* or merely *incomplete*? Not only are these questions prescriptively relevant, but they quite likely are similar to the strategic decisions managers actually have to make. There is scope for studying how members of organizations decide what kind(s) of ignorance they will create.

Those tactical decisions, in turn, are likely to depend on the purposes for which ignorance is to be used and the norms within the organization. Illustrations of this claim may be found in deliberations over the formulation of a new policy or rule. An almost universal point of debate is how specific and explicit the rule should be. The purpose of the rule is regulation, but the purpose of building some ignorance into the rule is to permit some degree of freedom (e.g., for the exercise of discretion in applying the rule). What kind of ignorance is likely to be built into the rule? *Irrelevance* may be used only to a limited extent, since the rule must invoke its subject to at least some extent. Typical organizational norms (and perhaps rules) do not permit the rule to be stated in a misleading way, so *distortion* is out. *Incompleteness* would seem to be the obvious choice, with *vagueness, ambiguity*, and even outright *absence* comprising the strategic possibilities here. Incomplete (e.g., vague or ambiguous) rules confer freedom of interpretation and, hence, of action, without contravening organizational norms that pertain to the construction of rules.

These remarks should indicate by now that unpacking ignorance immediately suggests several productive avenues of research and theory construction that are otherwise inaccessible. It also implies that organizations operate under ignorance that is in good part socially constructed by its own members. To return to an earlier point, however, a fully-fledged account of ignorance requires a revision of the traditional view of knowledge as synonymous with absolute truth. The sociology of organizations literature, for the most part, has not moved far from this view. Ironically, although perhaps predictably, the sociological field of inquiry which has most effectively applied the fundamental insight from the sociology of knowledge is the study of science itself.

6.5 Coda: Ignorance and the Sociology of Science

The sociology of science began as an offshoot from the sociology of knowledge. It might seem strange at this point in this chapter to return to the sociology of knowledge. Without the material from the organizations literature, however, the recent currents in the sociology of science that are relevant to ignorance are difficult to explain. Since the 1960's, this subfield has drawn heavily on both its older roots in the sociology of knowledge and from the sociology of organizations for its central concepts, which is understandable given that much scientific activity occurs in large-scale complex organizations.

The sociology of science prior to the early 1970's was influenced heavily by what Scheffler (1967) calls the 'standard view' of science (see also Mulkay's 1979 review of the sociology of science), and because this view was widely shared by classical sociologists of knowledge it effectively exempted scientific knowledge from sociological investigation. Instead, as De Gre (1955: 37) made quite explicit, the sociology of science was to be concerned with the social conditions that facilitate or hinder the discovery of objective knowledge.

Like the early sociologists of knowledge, sociologists of science treated ignorance simply as the absence of objective knowledge. They considered that truly scientific activity makes inroads into ignorance, and cumulative scientific progress is measured by the extent to which ignorance has been reduced. Consequently, their accounts of ignorance were a simple byproduct of their explanations of how objective knowledge is (dis)favored by various social norms, structures, or rules. Merton's (1973: 270) maxims are the best known. He claims that four institutional imperatives comprise the ethos of science: Universalism, communism, disinterestedness, and organized skepticism. Universalism refers to the 'impersonality' of criteria for judging knowledge-claims, while communism means an absence of ownership claims on scientific knowledge or information.

One might also add that the belief that the scientific method converges on the truth constitutes a pro-consensus view of science as well. In this tradition, the production of ignorance in the form of secrecy, fraud, or prejudice is considered 'deviant' and unusual among scientists. Likewise, serious dissent among high-performing scientists about truth in their fields is viewed as at most a transitional phenomenon; a properly constituted scientific field at any point in time enjoys a consensus on what the best available theory and research program are.

It is well known that this view has little currency now, either among philosophers or sociologists of science. It gave way under a succession of philosophical attacks on its central tenets (cf. Hanson 1965, Duhem 1962, Kuhn 1962, Lakatos 1970, Feyerabend 1975). Without detouring into these reformulations, let us consider the effects they have had on the implicit views sociologists have about the nature and role of ignorance in scientific activity.

One spinoff has been the destruction of the consensualism myth. Scientific disagreements and controversies have been shown to be much more common and long-standing than was previously thought. This contentiousness extends even to the level of epistemology. The standards and criteria for judging knowledge claims (and therefore ignorance as well) may not be agreed upon by competing scientists in the same field. Instead, scientific knowledge, like other kinds of knowledge, is now considered to be a social product and subject to nonrational negotiation.

An important early empirical study of technical dissent among applied scientists is Collins (1974), who found that in the discovery phase of developing a TEA laser, scientists who had developed a working prototype simply were unable to convey their methods to colleagues through formal publications. Not only was tacit (inexplicable) knowledge required to get a laser to work, but the scientists present at such events were unable to agree on which pieces of information were crucial. Even more importantly, Collins was able to show that information and results regarded by one scientist as crucial might be rejected by another as unimportant, and by a third as fraudulent. This network of scientists demonstrably lacked a common assessment of the validity or even relevance of experimental work in their field.

As a consequence of the demise of the myth of consensus, some analysts have asked whether scientists may differ in their readiness to reject conventional scientific wisdom or to perceive 'anomalies' as counter-examples to established theories. Both are not only precursors to new knowledge claims, but they constitute redefinitions of scientific ignorance. Clearly the scientist who is able to establish an 'ignorance claim' that convinces colleagues that there scope and need for research and theory-building is one who can create the possibility for novel knowledge, ideas, or even theories.

What social or psychological factors predispose a scientist to challenge orthodox scientific viewpoints? Frankel (1976) produces evidence that scientists who are in marginal positions within the scientific community (low status, insecure jobs, or on the 'fringe') are more likely to be receptive to alternative interpretations of empirical results, and Chubin (1976) likewise finds that many important scientific advances have originated with scientists in such marginal positions. They argue that these researchers have more to gain and less to lose than their more established counterparts by opposing established views.

Another aspect of the 'modern look' at scientific ignorance is the realization that communism and organized skepticism are by no means widespread norms. Aside from accounts by scientists themselves which challenged these views (e.g., Watson 1968), Mitroff (1974) claims his empirical findings supported the existence of 'counter-norms' in the scientific community that favor secrecy and commitment to ideas. By keeping their research secret, especially in its early stages, scientists are able to maintain their advantage and claims to priority over competitors. They also purchase the time in which to verify their results and establish their validity. Commitment to ideas (as against skepticism), although it can result in meta-ignorance, is valuable because it may see the researcher through considerable efforts that do not immediately confirm those ideas.

Only recently, however, have sociologists in this area paid much attention to ignorance itself (and even so, Merton (1987) recently castigated the field for neglecting this topic). Two approaches have been initiated. One of them inquires into how scientists negotiate strategically useful forms of ignorance as a precursor to research. While conscious ignorance is a required precursor for deliberate learning, Merton's major claim is that this ignorance must be 'specified' in such a way that scientists believe further knowledge is attainable, at least in principle. He argues that scientific communities have institutionalized norms for specified ignorance. Indeed, he observes that the production of specified ignorance claims is an expected outcome of knowledge production. However, although he acknowledges that scientific fields differ considerably on the criteria for what constitutes specified ignorance, Merton has not yet pursued this important claim systematically. Clearly those criteria are at stake in debates over decisional 'rationality' in the face of ignorance.

A similar thesis has been advanced at the same time by the philosopher Ravetz (1987). Not only does he coin similar phrases ('relevant' and 'usable' ignorance) to describe the same phenomenon, but claims that the art of selecting research problems may hinge on the scientist's ability to sense where the border separating knowledge from ignorance may be penetrated and to what depth (pg. 110). Furthermore, he argues that this kind of ignorance is generated by scientific activity *more rapidly* than is knowledge. While basic researchers and even applied scientists may have the luxury of selecting only those problems they believe they can solve, Ravetz points out that increasingly scientists are called upon by government and business to take on problems that the scientists themselves may not believe to be solvable.

His programmatic solution involves an abandonment of the ideal of scientific consensus on knowledge in favor of a 'clinical' model of knowledge acquisition and use. What he seems to intend by this term is the admission of "nonquantifiable and even nonspecifiable expert judgments" into risk assessments (pg. 104). Ravetz appears to be saying that in the arena of policy-directed science, 'usable' ignorance need not be entirely 'specifiable', and, like Collingridge (1980), he argues that decision makers are better off with an accurate picture of how extensive their ignorance is than a falsely precise assessment.

The other recent approach to ignorance in the sociology of science is Whitley's (1984) attempt to import concepts from organizational sociology into the study of how uncertainty and the organization of scientific fields affect one another. He begins by defining two dimensions of "scientific uncertainty". "Technical uncertainty" concerns the extent to which scientists fail to agree that research techniques are reliable and well understood (see also Rip 1982). "Strategic uncertainty" refers to uncertainty or disagreement about intellectual priorities, the significance of research problems, and ways of tackling them. Whitley claims that high technical uncertainty usually entails high strategic uncertainty but the reverse does not generally hold.

Like the organizations theorists, Whitley is concerned to follow out the organizational consequences of both kinds of 'uncertainty'. Thus, he argues that high technical uncertainty limits the size of reputational networks, places more reliance on personal contacts and controls (particularism, in Merton's terms), inhibits standardization of methods and criteria, and encourages diffuse and creative intellectual contributions. High strategic uncertainty, on the other hand, decentralizes control over research goals or priorities, and yields greater theoretical diversity. Most of the remainder of his book is devoted to elaborating an a priori typology of organizational schemas in scientific fields as a function of uncertainty and interdependence.

More interestingly, Whitley claims that political or economic pluralism engenders high strategic uncertainty, while high technical uncertainty tends to result when scientific standards are influenced by lay audiences or nonscientific interest groups. There is scope here for investigating the connections between his thesis and Ravetz's normative arguments about policy-directed research.

The distinctive contribution that the sociology of science offers to any emerging perspectives on ignorance, in fact, centers on just this issue. Most of the modern research and discussions about ignorance involve debates about what kinds of ignorance or uncertainty permit 'rational' choices to be made. Scientifically 'usable' ignorance is a special case in these debates. If scientific ignorance is not only socially constructed but also influences the conduct of scientific research, then we require an understanding of how this construction occurs. Moreover, such an account must be at least in part sociocultural. The next chapter considers some possible theoretical avenues to this understanding, before returning to a consideration of the relationship between normative and descriptive perspectives on ignorance.

Chapter 7:
A Dialog with Ignorance

"The sage is full of anxiety and indecision in undertaking anything, and so he is always successful." Chuang-tzu.
"In the case of uncertainty, the definition of rationality becomes problematic." J.G. March and H.A. Simon

7.1 The New Preoccupation with Ignorance

What are we to make of the recent intellectual ferment over ignorance? Clearly its explanation presents a fascinating problem in social psychological history, and the social psychological study of science in particular. While we are far from being able to obtain a complete account, I should like to suggest what such an account might look like. After all, this problem should hold more than the promise of intellectual interest; the more we know about why we are collectively and individually preoccupied with ignorance, the more able we will be to adopt mindful strategies and choices regarding ignorance.

7.1.1 Techno-Rational Explanations

So far, I have provided two fragmentary and rather commonsensical explanations for the recent upsurge in 'ignorance studies', both of which were introduced in Chapter 1. The first could be called the 'techno-rational' argument, which simply says that increasingly complex technological systems combined with advances in our scientific understanding of the environment have made us more aware of how uncertain those systems and that environment are, how little we know, and how drastic the consequences of making wrong decisions could be. In lieu of attaining total knowledge of these matters, becoming knowledgeable about our own ignorance makes strategic and economic sense.

I have already pointed out one shortcoming of this explanation, namely that preoccupations with ignorance have also emerged in fields that seem devoid of these commercial and social imperatives. Another, and more important, counter-argument has been put by Douglas and Wildavsky (1982: 9-15). To them, the above explanation begs motivational questions. They ask whether life has become demonstrably less safe or more unhealthy than it was in the past, and of course find that the evidence does not yield an affirmative answer. If anything, in many ways life now is less risky and more healthy than it has ever been. At the same time they do find considerable evidence to suggest that Western societies have become rather more risk-averse since World War II. The amount of risk that the public is willing to accept has substantially decreased in some areas, while in others the question of risk acceptability is being debated where it was never raised before.

If 'real' risks cannot be said to have markedly increased, could the rise in public concern, outcries, and political pressure explain the corresponding shifts in ignorance paradigms within certain applied sciences and various professions? After all, such outcries and pressures are widely cited as having influenced many political and economic decisions about environmental management, nuclear power, and the food and chemical industries. Perhaps similar but less obvious dynamics underlie the debates in engineering, artificial intelligence, medicine, the human sciences, and so forth. If so, then it seems important to account for the increased public concern. That is one of the main tasks set for risk perception researchers by Douglas and Wildavsky and re-emphasized in Douglas' (1985) review of risk acceptability literature from the social sciences.

Whether one is convinced by their particular explanations or not, Douglas and Wildavsky persuasively argue against several obvious candidates for explaining the rise of public risk-aversion. First, they point out that it does not necessarily correspond to increased overall awareness of risks or even of ignorance itself. Instead, only certain risks have become the selective focus of public concern and debate, and this selectivity also is buffeted by fads and fashion. These risks are not the most consequential. Nor are the selected risks always involuntary or irreversibly harmful, even though it is well-established that those features do make people more averse to risk. Furthermore, some of the greatest public concern arises in cases where actual knowledge about the extent and kind of ignorance or risk involved is very scanty. Conversely, attempts at public education about various kinds of risk have met with little success in areas where public concern is nevertheless very high. The connection, then, between concern and knowledge or awareness is either nonobvious or nonexistent.

Secondly, Douglas and Wildavsky dismiss the 'technological change' argument, noting in passing that this is one of our favorite explanations for attitude change. If low technology sets low expectations in such matters as health and longevity, then why have so many premodern cultures utterly rejected the concept of 'natural' or 'normal' death when postmodern society is just beginning to challenge that? Those changes in risk-related attitudes that are linked to technological developments seem more plausibly to have arisen from a simultaneous erosion in public trust of large institutions. While Douglas and Wildavsky do not venture an explanation of that development, there is widespread evidence of it.

The account they give involves a growing fear on the part of sections within the public of the technical and political capabilities of modern large-scale institutions, feelings that the economic and political agendas of those institutions have resulted in mistaken or even willful destruction, and the need to place blame and seek redress. Their argument concludes that the recent upsurge in calls for risk assessment, impact studies, accountability, and system monitoring has arisen out of those fears and the conflicts of interest engendered by them.

7.1.2 The Stage Model and Motivational Accounts

The second account of the paradigm shift offered earlier in this book is a somewhat crude but suggestive stage-model based on an historical interpretation of the crisis in modern mathematics. This model proposes that strategies for coping with ignorance are developed and selected in response to a sequential unfolding of crises and difficulties. Another way of using the model, however, is as a contingency theory of strategic responses. Thus, the so-called traditional ways of coping with ignorance, namely banishment, reductionism, and normative pragmatism, hinge on the assumption that ignorance can ultimately be vanquished altogether. Banishment and pragmatism are usually the first strategies tried, presumably because they require little effort or cost. Where their objectives clash, i.e., where there are pragmatic justifications for not ignoring ignorance, then reductionist efforts are mounted to rehabilitate problematic ideas or anomalies in terms of established frameworks.

Defeat at this stage then motivates attempts to understand the nature of ignorance itself, the goal being to rehabilitate the entire domain. These attempts involve what I have called 'second-order' strategies and descriptions, and the impulse is 'managerial' rather than eliminative. Ignorance is presumed to be manageable in the end, even though irreducible. A failure of second-order strategies leads to an ongoing process of living with the crisis, entailing relativistic coexistence, epistemic pluralism, and a type of specialization I have termed 'suburbanization'.

In Chapter 4 this stage-model was unpacked somewhat further, the third (or second-order, managerial) stage especially. In particular, developments in alternatives to probability have been portrayed as pragmatic expansions in attempts to 'tame' ignorance through rigorous description and quantification. Modern probabilists have at times pumped for outright banishment or at least a reductionist counter-movement. Other researchers have literally engaged discussion at the second-order level of ignorance. In the wake of these developments, spokespeople from a variety of camps have proposed schemes for pluralism (anticipating suburbanization) or synthesis (a second-order managerial approach).

The stage-model addresses a different question than the techno-rational account, in that its primary goal is to explain why researchers have selected certain strategies for dealing with ignorance. The techno-rational argument, along with Douglas and Wildavsky's socio-cultural accounts, focuses on reasons for concern with ignorance in the first place. It is initially tempting to combine the two perspectives and hope that a cogent framework emerges, but such an attempt is doomed. Both explanations are ideational and cognitive; they lack motivational, cultural, and social underpinnings. The source material for those underpinnings lies in the human sciences. Of course this is not to say that such material provides 'ready-made' explanations, but it comprises the stuff of which explanations must be made.

Perhaps the most obvious port of call is the sociology of science. My own judgment is, however, that much of the most famous work there cannot aid the construction of an explanation for the paradigmatic shifts in the conceptualization of ignorance. The pre-Kuhnian views are no help at all, for reasons outlined in Chapter 6. The Kuhnian framework itself is so tied to 'empirical' science that it is nearly impossible to apply it here. The debates among the competing schools of probability (or for that matter, statistical inference, as pointed out by Oakes 1986) did not arise over anomalies in the Kuhnian sense of "the recognition that nature has somehow violated the paradigm-induced expectations that govern normal science" (Kuhn 1970: 52-53). They were essentially normative debates concerning the range of situations to which the probability calculus could legitimately be applied. Even the most empirically oriented among the relative frequentists (von Mises) never faced criticisms to the effect that the laws of probability were contradicted by nature.

And so it has gone for all modern frameworks for representing and managing ignorance. The only place where empirical observations are brought to bear on the debates is in studies of the correspondence between human intuition and various normative standards. Although there is some feeling that at least some correspondence ought to be maintained between legitimate techniques and ordinary intuition, those impulses are drastically tempered both by normative restrictions and the realization that human intuition is quite variable. The discovery that probability is counter-intuitive has nothing like the force of a difference between controlled observation and theoretical expectation in empirical science.

Post-Kuhnian perspectives are somewhat closer to the mark, insofar as they emphasize the negotiated nature of scientific reality, the scope for disagreements among scientists on both the meaning and value of observations or ideas, and the need for an account of scientific research that is grounded in motivational, social, and political influences. To date, however, most such accounts are specific to particular scientific domains and periods in history. Indeed, some social scientists claim that this must always be so, in which case students of ignorance studies are thrown onto their own resources.

A motivational account of the recent paradigmatic upheavals regarding ignorance must recognize that researchers' motives may well conflict with one another in at least one respect. On the one hand, the professional ethos of scientific research stipulates the goal of finding answers to important and/or difficult questions. The same ethos, however, also exhorts researchers to create and pose important and difficult questions. The traditional motivational accounts in psychology (section 5.1) and early work in the sociology of science (section 6.5) tended to emphasize the first while largely ignoring the latter. Such accounts might be used to argue that as scientists and professionals discover that ignorance is irreducible they settle for knowing as much as they can about ignorance as a fall-back strategy.

Let us pursue this line of thought. Suppose the primary motivation for researchers in setting questions and creating ignorance is utterly performative, i.e., success-oriented. Then we should expect researchers to generate only questions that they believe they can answer, subject to the constraint that those questions be perceived by colleagues and superiors to be nontrivial. Accordingly, all that Merton's 'specifiable' and Ravetz's 'usable' ignorance really amount to is improved problem formulation. The recent efforts to find new ways of classifying, defining, representing, and estimating ignorance would be accounted for as part of a strategic quest for specifiable, usable ignorance in the service of successful problem-solving.

Unfortunately, there are at least two difficulties with this picture. One is that scientists can and do take on problems that they are not convinced they can solve. This is the case even in 'pure' research, although one could rather cynically argue that such instances may be explained away as miscalculations by researchers of their own abilities or the difficulty of the problems. More to the point, however, Ravetz (1987: 99) claims that in policy-related research, problems are "thrust upon" scientists without heed to the feasibility of attacking them. This is the same dilemma faced by applied mathematicians on more than one occasion in recent history, as we saw in Chapter 1.

Insofar as researchers are motivated to take on policy research (for reasons of monetary gains, social conscience, or whatever), their treatment of ignorance cannot be understood simply as an attempt to solve problems. When philosophers such as Ravetz and Collingridge, or applied scientists and engineers such as Zadeh, plead for the use of 'soft' methods in representing ignorance, they frequently justify their positions by pointing out that otherwise the scientist has nothing authoritative to say to policy-makers.

And it is the scientists' authority that is indeed at stake. There are cases in point illustrating the fact that conventional quantitative risk assessment methods, for instance, are widely regarded as suspect and manipulable by policy-makers. They do not carry the weight of authority and apparent objectivity that they once did (see, e,g., Carter's 1979 account of the disputes over the quantification of risks from carcinogens in the workplace, or Self's 1975 searching review of the politics of cost-benefit analysis in Great Britain). Likewise, recall Whitley's (1984) thesis that political and economic pluralism generate 'strategic uncertainty' while control of research standards by nonscientists results in 'technical uncertainty'. If there is truth in this, then quite possibly the venture by applied scientists into policy-related research has forced them to confront increased uncertainty or other kinds of ignorance within their own fields.

In the light of these developments, at least some of the search for new ways of representing and thinking about ignorance may be motivated by applied researchers' concerns over the criticisms raised against older methods and their attempts to negotiate a workable consensus with policy-makers. It may also reflect the shift in the scientists' role in such research, away from answer-producing and towards problem- and risk-assessment (and, therefore, the description and representation of ignorance). Scientists, of course, also disagree with one another on how best to represent and account for ignorance; such disagreements have provided most of the material for this book. Those disagreements in themselves are fuel for motivations to recast methods for assessing or managing ignorance.

How do we explain the specific nature of the methods that have been tried? Can the 'stage-model' be persuasively elaborated to provide a socio-cognitive account? I have already hinted several times that at least one analogy may be drawn between this problem and a field that has been beset as no other by recalcitrant issues of prediction and control. That field is the study of criminal deviancy.

7.1.3 The Deviancy Analogy

At first glance, this might seem an odd choice indeed for source material. More obvious candidates might include environmental psychology or perhaps the psychology of control itself. But criminal deviance and its cognate subdisciplines provide a rich set of ideas and accounts concerning how societies have attempted to eliminate, subdue, predict, control, or at least manage those elements deemed culpably deviant. There are several intriguing parallels between those attempts and the strategies that have been selected for coping with ignorance.

First, the dominant cultural orientation towards 'wrongful' behavior and 'wrong' ideas has been unremittingly negative: both criminal deviance and ignorance are widely regarded as problematic and dysfunctional. Tradition-bound strategies for dealing with criminal deviance and ignorance also resemble one another. As Cohen (1985) puts it, the dominant social impulses regarding miscreants are either to expel them from society or to absorb and rehabilitate them. He calls these impulses "exclusionary" and "inclusionary" respectively. We have seen nearly identical descriptions in the sociology of organizations literature of the traditional methods for dealing with uncertainty; the organization either eliminates uncertainty or absorbs it (e.g., by buffering or smoothing). I have drawn similar distinctions between the 'banishment' and 'reductionist' strategies for dealing with problematic mathematical constructions, and very similar descriptions (but different terms for them) have been applied to the same phenomena by Lakatos (1976).

The positivist impulse regarding both wrong behavior and ignorance also is the same, namely to harness scientific inquiry in the service of ultimately eliminating or controlling these 'problems'. This agenda gives rise to a striking resemblance between the two in the asymmetric applications of explanatory frameworks. In studies of social pathology it is widely assumed that 'wrong' behavior requires a special kind of explanation apart from those explanations used to account for unproblematic behavior (see Walker 1977 for a philosophical critique of this assumption). Thus, accounts of deviant acts and identities often bear scant resemblance to accounts of 'normal' acts and identities. There are special subdisciplines devoted solely to the study of deviants and deviancy rather than incorporating those studies into mainstream social or psychological theories.

As we have seen, much the same may be said of the classical approaches to ignorance. The sociology of knowledge either ignored ignorance or treated it marginally, with special explanations for 'false consciousness', 'ideology', and the like (the major modern exception is Bloor's 1976, although his is a programmatic statement). In psychology the study of illusions and delusions likewise has occupied a sideshow status on its own. Moreover, proponents of normative perspectives on ignorance (in particular, on uncertainty) largely have relegated the study of non- or anti-normative perspectives to the human sciences, reflecting an established presupposition that accounts of rationality are fundamentally different, and set apart, from accounts of nonrationality.

In the face of the dominant bias against ignorance, the view that it may be inevitably built-in to our knowledge frameworks and that this may have beneficial consequences has been entertained only recently and reluctantly. Correspondingly, the notion that a certain amount of deviance may be inevitable and even functional in some ways for the social order is distinctly modern. Within criminology, we see the first serious proponent of such an outlook in Durkheim near the turn of the century and later elaborated by Merton (1957). Durkheim's posited functions of deviancy are twofold. First, the criminal act and its consequences demonstrate to others in society precisely where the boundaries of acceptable behavior are; they make them apparent and real. Secondly, insofar as criminal acts reflect changing social mores and rebellions against outmoded laws, a certain level of criminal deviancy may indicate a healthy non-stagnating society.

These claims echo recently articulated sentiments about the benefits of consciously acknowledged and articulated ignorance, especially by sociologists and philosophers of science. Articulated ignorance, so the argument runs, points out where the real questions and problems are; it makes the boundaries of knowledge apparent and real. Likewise, healthy scientific disciplines should expect to be surrounded with a penumbra of ignorance. A field that has no further questions or problems is no longer an active field, but in Lakatos' terms, a "decaying paradigm".

If we are willing to entertain, for the moment, an analogy between the socio-cultural responses to wrong acts and wrong ideas, then we may mine the historical accounts and explanations that have been developed in the field of deviancy for possible accounts concerning the strategic responses to ignorance. Rather than attempt a review here, I shall rely on the review by S. Cohen (1985).

Cohen claims that two major 'transformations' in social control have occurred during the past two centuries. The first, which he calls the Great Incarceration, involved a shift in emphasis from exclusionary to inclusionary modes of control over deviants. Centralized prison and asylum systems made their appearances during this transformation, and the object of incarceration became rehabilitation rather than simply punishment. Although imprisonment itself remained exclusionary in the sense of removing the deviant from society, the mode of control inside the prison itself was made more inclusionary than before, with the ultimate aim of returning the criminal to society as a socially productive and rehabilitated individual.

The second transformation, Cohen argues, began during the 1960's and was hailed as a 'decarceration' movement. Inclusionary control measures were to be taken out of the prisons and into the community in the form of 'soft' or 'humanitarian' alternatives to imprisonment. The community itself would rehabilitate the criminal (with some help, of course, from professionals). The old monolithic prison and justice systems would ultimately be decentralized or even dismantled altogether. Interestingly, Cohen claims both transformations failed in their 'official' objectives. The first transformation neither effectively rehabilitated criminals nor even deterred criminal activities. The second transformation not only failed to destroy the old system, but it merely grafted a new 'soft' control system onto the old and resulted in a much more encompassing set of controls which have been applied to a greater range of citizens than ever before.

Cohen's discussion of the strategic merits of exclusionary and inclusionary control methods may have some pertinence for ignorance management. He points out that exclusion is attractive because it is relatively inexpensive and easy to implement, requiring only a crude technology. Inclusionary control, on the other hand, requires surveillance and information-gathering, vigilance, and takes both time and effort. The price paid by exclusion, of course, is in the loss of potentially productive people and an increasing threat of rebellion, drawbacks which successful inclusionary control avoids.

Somewhat similar observations have come from many quarters concerning the older exclusionary strategies for dealing with ignorance. They are inexpensive and easy to execute but the long-term price is increasingly regarded as unendurable. The traditional inclusionary strategy of reductionism has not worked either. In a sense, the developments of the past 25 years or so could be viewed as a kind of 'decarceratory movement' with respect to ignorance, whereby some of the safer and more manageable forms of ignorance have been allowed back into the respectable community of ideas and knowledge, provided that they are carefully monitored, classified, and managed.

There are two possible reasons why this decarceration movement might have greater likelihood of succeeding than its criminal justice counterpart. The most obvious factor that has made a decarceratory form of inclusionary control more feasible in the management of ignorance is the information technology revolution. Within a very short time, it has become much easier to represent and manipulate both quantitative and qualitative information about anything, including states or degrees of ignorance. The second, of course, has been the proliferation of new ideas and frameworks for doing exactly that.

The analogy between social control and ignorance management extends the stage-model articulated in Chapter 2 somewhat by suggesting that modern managerial strategies may have the same kind of strategic appeal that the decarceratory impulse did in the 1960's and 70's. The Hilbert program in mathematics could be viewed as an ultimate inclusionary control strategy. Likewise, the subsequent 'suburbanization' stage may be linked with the present situation as Cohen sees it for criminal justice, in which social control operates on a much grander scale than before with inclusionary controls applied to 'manageable' cases and exclusion reserved for 'hard' cases or failures.

This analogy is valuable in that it complements Douglas' 'purity and danger' thesis by exploring the strategic options which have evolved over two centuries for dealing with the threats and polluting influences embodied in criminal and pathological deviants. It also imports an evocative image, taken from Levi-Strauss' (1977) metaphor of the cannibals who devour their enemies and thereby take on their powers (inclusionary control) and the people who vomit up their enemies and are cured or purified as a result (exclusionary control). Cohen's analysis also reminds us that the paradigmatic shifts reviewed thus far apply only to particular kinds of ignorance (mainly varieties of uncertainty), so that by and large the 'hard' cases are still being handled by banishment or reduction. In that sense the new frameworks have been grafted onto the old; banishment and reductionism have by no means vanished as strategic options.

The utility of this analogy declines, however, once we move past the general stage model and attempt to apply these insights to specific fields or disciplines, as does the value of much of the explanatory material considered thus far. The problem is that we lack accounts of what is considered manageable ignorance in particular fields, and we need only recall the comparison of the engineering and legal professions in Chapter 2 to remind ourselves that professions can differ widely on this issue. Clearly there is much work to be done here in a social psychological vein. In the meantime, it may be possible to draw some pragmatic lessons from the overview provided by this book about how best to pursue constructive dialogs between normative and descriptive frameworks for dealing with ignorance. The remainder of this chapter returns once again to the problematic connection between the normative and the descriptive.

7.2 An Example of A Normative-Descriptive Dialog: The Study of Second-Order Uncertainty
7.2.1 The Normative versus the Descriptive

I will begin with an extended example of a dialog between the normative and descriptive perspectives on second-order uncertainty, partly because this is relatively unexplored territory. It is also an opportunity to demonstrate how the material (and disciplines) surveyed in Chapters 4 and 5 may be brought together. For our present purposes, I will constrain this dialog, for the most part, to the intra-psychic level.

Second-order uncertainty is sufficiently diverse that no single normative perspective dominates the field. Instead, there are multiple competing and complementary perspectives. Moreover, virtually none of these bring with them a persuasive psychological theory of how people think of second-order uncertainty. One need only recall that these consist of generalizations of Shannonian entropy, relative variation coefficients, and the Shaferian measures of conflict. All of these are more difficult to comprehend, calculate, and communicate to nonspecialists than probability theory. The sole possible exception is the humble intervalic expression (i.e., "from x to y" or [x,y]). Unfortunately, as can be appreciated by dipping into a text on intervalic mathematics (e.g., R.E. Moore 1966) or considering generalizations of [x,y] as an expression of relative freedom (Smithson 1988), the hope that simultaneous intervalic uncertainty will be easy to understand and communicate quickly evaporates. In applications we may expect to run afoul of the Rationalist Quandary, Zadeh's Thesis, and Zeleny's Tradeoff (see 7.4). Indeed, insofar as normative standards insist on quantitative representations of second-order uncertainty, these problems loom quite large.

All of this implies that we should expect an even sharper tension between intuitive judgments and normative standards than exists regarding ordinary first-order probability. There is no hope at all that people are going to be intuitively computing relative variances or generalized fuzzy entropies, let alone a formula like

$$FG = (1-ne_1-ne_2-ne_3)^2 - \max(0,1-po_1-ne_2-ne_3)^2$$
$$- \max(0,1-po_2-ne_1-ne_3)^2 - \max(0,1-po_3-ne_1-ne_2)^2$$
$$+ \max(0,1-po_1-po_2-ne_3)^2 + \max(0,1-po_1-po_3-ne_2)^2$$
$$+ \max(0,1-po_2-po_3-ne_1)^2 \quad (7.1)$$

which is the formula for relative freedom in a 3-parameter system of probabilities, where the po_i and ne_i represent the upper and lower bounds of the ith probability p_i and FG is the relative freedom of the p_i to vary given the restriction that the p_i must sum to 1 (Smithson 1988). Educating, retraining, or calibrating human judges is out of the question here. If normative standards are to be considered authoritative, then they must be implemented in a purely technical fashion without the aid of human judges; the role of humans becomes more passively interpretive than diagnostic or judgmental.

A further implication is that if we do find that human judgment or preference covaries with one of these normative measures of second-order uncertainty, we should be wary of attributing causal status to that correlation. Edwards identified some of these concerns in his 1962 paper on variance preferences. He considered people's choices between pairs of bets that are identical in SEU terms (and incidentally, reported preference patterns that anticipated Kahneman and Tversky's Prospect Theory). He introduced Atkinson's (1957) model of decision making when a person perceives outcomes of a risky act as contingent on his skill, and derived from that the proposition that people may prefer bets with less variance. Nor was Edwards the first to suggest this; he acknowledged that critics from Allais onward had raised this point against standard SEU theory.

Edwards himself (1954) as well as others (e.g., Coombs and Pruitt 1960) conducted experiments to test for variance preferences. Edwards (1962b), however, criticized these investigations for failing to take into account that for first-order probabilities and two-outcome gambles, variance is confounded with both probability and utility. This is easily seen by considering, for instance, the offer of $100 for sure with the gamble for $200 with probability 0.5 ($0 otherwise). Edwards was skeptical that variance accounts for the preference for the $100 sure-thing offer, but he does not provide an air-tight reason. Nevertheless, he concluded that available evidence and plausibility arguments favor the hypothesis that people with high achievement motivation and low need to avoid failure will prefer high-variance bets while the reverse will be true for people with low achievement motivation and high need to avoid failure.

If we consider Edwards' concerns alongside Ellsberg's paradox and other psychological commentaries on second-order uncertainty, even the restricted normative framework of probability theory yields an embarrassment of riches. For example, should 'rational' decision makers and gamblers be averse to variance in the probabilities (as Gardenfors and Sahlins 1982 would have it), or should they be averse to variance in utilities? It is easy to see that these are by no means the same thing. In fact, Ellsberg's paradox is 'paradoxical' not only in SEU terms but also in higher-order moments such as the variance of the utilities.

Consider first the ordinary two-outcome bet with a probability p of getting amount A and otherwise nothing. Denoting the utility of this gamble by X, the expected utility $E(X) = pA$, and the variance $E[(X-E(X))^2] = A^2 p(1-p)$. Now consider Gardenfors and Sahlins' 'ambiguous probability' scheme, which we will express in a general form by saying that the probability of gaining A itself has two possible values, u and s, which it takes with probability of 0.5. Denoting this variable probability by t, this means our gamble is $P(X=A) = t$ and $P(X=0) = 1-t$, where $p(t=s) = p(t=u) = 0.5$. If the average of u and s is p, then $E(X) = pA$ as before. The variance $E[(X-E(X))^2] = A^2[(s+u)/2][1 - (s+u)/2] = A^2 p(1-p)$, which is just the same as the variance for the first gamble. So, variation in the probabilities does not increase the variance of the utilities.

Finally, consider the 'vague probability' scheme of Ellsberg fame, which we may express in a general form by having the probability t of getting A take a uniform probability distribution over the interval $p-h \leq t \leq p+h$. The density for t over that interval is $f(t) = 1/2h$, and a simple derivation shows that the variance of X is again $A^2 p(1-p)$. Neither vague nor ambiguous probabilities influence the expected variation of the utilities. Which, then, is the preferred 'rationality': To be worried about variation in probabilities or utilities?

If we compare probabilistic with nonprobabilistic representations of ignorance, the difficulties involved in specifying a unique normative guideline increase. Which guideline should be used for purposes of declaring two representations 'equal' in uncertainty, for instance? Consider vague probabilities as expressed by a simple interval [p-h, p+h] and vague utilities as expressed by the interval [A-k, A+k]. How should k and h be chosen so as to make the resulting gambles normatively equivalent? If we choose to make the SEU of the endpoints the same, then we should equate $A(p\pm h)$ with $p(A\pm k)$, so that Ah = pk. But the variance of the utilities will then be confounded with p.

On the other hand, if we require SEU equivalence as well as equal variances in the utilities, then we encounter additional problems. Specifically, the variance for the uniform distribution over [A-k, A+k] is $k^2/3$, and as I have argued previously the variance for the intervalic probability scheme is $A^2p(1-p)$. Now if we wish A-k > 0 (that is, positive intervals only), we have two requirements: k < Ap (to make the gambles equivalent in SEU terms as well as ensuring A-k > 0) and $k^2/3 = A^2p(1-p)$ (to ensure equal variances in the utilities). Simple algebra suffices to demonstrate that these requirements entail p > 3/4. Therefore these two normative requirements restrict our comparisons to the upper fourth of the range of probability. Another way of putting it is that if we wish to re-express intervalic (or even first-order!) probability schemes for gambles in terms of simple utility intervals, once the p falls below 3/4 we cannot obtain equality in both SEU and utility variance without using negative lower-bounds for the utility intervals.

7.2.2 A Study of Preference Patterns

Despite the plurality of normative guidelines and the inevitable confounds among them, we may still usefully ask whether intuitive judgments correlate with any such guides (or a combination of them). I conducted an experiment involving 153 volunteer subjects from a first-year psychology class, none of whom had a background in probability or university-level mathematics. These students were randomly assigned either to a 'gain' condition or a 'loss' condition. The 'gain' group was asked to choose between pairs of gambles involving prospects of gain while the 'loss' group was given identical pairs couched in terms of prospective loss.

The bets themselves were arranged so as to replicate the Edwards-type comparisons and the Ellsberg paradox, but also included two kinds of second-order uncertainty representations: simple intervals and vague probabilities. All bets were, of course, equivalent in SEU (which equals $100 for all of them). The second-order probability and utility interval schemes were made SEU-equivalent in their endpoints, and one of the interval schemes was made approximately equal in utility variance to one of the first-order probability schemes. The experimental groups clearly were intended to test the 'reflection hypothesis' of Prospect Theory (the cross-subjects version) on various kinds of second-order uncertainty. Subjects were presented with all possible pairs of gambles and asked to choose which one they would prefer in each pair.

Table 7.2 Preference Patterns for Pairs of Bets

Gamble	Gain Condition						s.d.
1. $100 for sure							0
2. .2 prob. $500	.88						200
3. .5 prob. $200	.87	.36					100
4. .8 prob. $125	.73	.25	.32				50
5. $80 to $120	.49	.18	.18	.27			11.5
6. $20 to $180	.83	.20	.44	.61	.87		46.2
7. .4-.6 pr. $200	.83	.39	.56	.85	.87	.67	100
8. .1-.9 pr. $200	.83	.49	.67	.80	.88	.67	.54	100
	1.	2.	3.	4.	5.	5.	7.

Gamble	Loss Condition						s.d.
1. $100 for sure							0
2. .2 prob. $500	.70						200
3. .5 prob. $200	.80	.54					100
4. .8 prob. $125	.71	.47	.24				50
5. $80-$120	.64	.39	.31	.38			11.5
6. $20-$180	.83	.43	.45	.59	.68		46.2
7. .4-.6 pr. $200	.76	.64	.51	.70	.64	.57	100
8. .1-.9 pr. $200	.71	.39	.47	.62	.64	.54	.53	100
	1.	2.	3.	4.	5.	6.	7.

Table 7.2 shows the subjects' preference patterns for all pairs of bets, as well as the specific nature of the bets themselves. The proportions in the 'gain' matrix refer to

preference for the *column* option, while the proportions in the 'loss' matrix refer to preference for the *row* option. If the reflection hypothesis is true, then these two matrices should be identical.

The right-hand margins of the matrices in Table 7.2 show the standard deviation in expected utility for each of the gambles. We could easily be lulled into thinking that the difference between the standard deviations of the two paired bets explains these results. After all, 22 out of 28 comparisons in the gain matrix and 17 out of 28 comparisons in the loss matrix are significant in the expected direction at the 0.05 level, while only 2 comparisons run significantly in the contrary direction (taking into account the reflection hypothesis). Better still, a regression model using the standard deviations of each bet as predictors of the proportion preferring a given bet yields a multiple R = .805 and unstandardized coefficients of 0.002 and -0.002 for the two standard deviations, indicating that a simple difference model is indeed the best of the linear models. For the gain matrix R = 0.90 and for the loss matrix R = 0.72. These correlations are of a magnitude that frequently attracts claims of 'strong support' for an hypothesis in the psychological judgment literature.

However, there are good reasons for not accepting this result at face value. First, it includes some apples-and-oranges comparisons. Second, there are some important findings that, while not directly contradicting the variance hypothesis, point quite strongly to other explanatory factors.

We have on the one hand a corroboration of Edwards' and Coombs and Pruitt's preference findings; the comparisons among gambles 1, 2, 3, and 4 in the gain matrix reveal reasonably strong preferences in the expected direction. The reflection effect, however, does not emerge clearly in the loss matrix except in comparisons involving gamble 1 (the $100 sure-thing option).

But the waters begin to muddy somewhat when we consider some comparisons across different types of uncertainty. First, we may observe that there is only a weak correspondence between variance and the strength of preference for the sure-thing option itself. Consider also the gambles involving first-order and intervalic probabilities. Gambles 3, 7, and 8 have identical standard deviations in utility as well as utilities, as argued previously. Preferences are approximately evenly split between them with the exception of the comparison between 3 and 8 in the gain matrix. Comparisons with gamble 2 on the other hand, which has twice the standard deviation of 3, 7, and 8, do not support the variance hypothesis. Gamble 7 is preferred to gamble 2 in both the gain and loss matrices, and there is no clear preference for gamble 8 over 2 in the gain matrix, despite gamble 2 having much the greater standard deviation.

If we turn now to the comparison of simple intervalic uncertainties with the other kinds, there is only one pair that has approximately equal standard deviations: gambles 4 and 6. This is an artifact of the previously mentioned restriction such that utility interval gambles may have the same variance as first-order probability gambles only if the probabilities are higher than 3/4. There is a moderate tendency for subjects to prefer the first-order probability version in the gain condition and to reverse that preference on the loss side. It seems doubtful that variance is the explanatory factor here, especially given that the preferences actually run in the opposite direction of that expected on the basis of the minor difference between the standard deviations of these gambles.

Although we cannot squeeze a completely unconfounded test for whether subjects prefer one kind of ignorance representation over another out of this study, the results given above suggest further inquiry into this possibility. Table 7.3 shows the relevant comparisons between types of representation, and close inspection reveals some intriguing trends. The proportions in the Gain column refer to the fraction of subjects preferring the second gamble, and the proportions in the Loss column refer to those preferring the first gamble.

We may first observe that we obtain only modest support for the Ellsberg-type effect. The four relevant comparisons are the ones with standard deviation differences of 0 and 3.8, since while they do not differ in the variance of utilities, they do in terms of probability variation. The only really consistently supportive pair is the comparison between .8 prob. of $125 and the interval $20 to $180. Likewise, in the two comparisons between utility intervals and second-order probability gambles that are SEU-equivalent in their endpoints, the preference patterns indicate that endpoint-equivalence does not correspond to subjects' choices.

Table 7.3 Preferences for Representational Style

Comparison	Gain	Loss	Std. Dev. Diff.
.2 vs. 80-120	.82	.61	188.5
.5 vs. 80-120	.82	.69	88.5
.8 vs. 80-120	.73	.62	38.5
.2 vs. 20-180	.80	.57	153.8
.5 vs. 20-180	.56	.55	53.8
.8 vs. 20-180	.39	.41	3.8
.2 vs. .4-.6	.61	.36	100
.5 vs. .4-.6	.44	.49	0
.8 vs. .4-.6	.15	.30	-50
.2 vs. .1-.9	.51	.61	100
.5 vs. .1-.9	.33	.49	0
.8 vs. .1-.9	.20	.38	-50
.4-.6 vs. 80-120	.87	.64	88.5
.4-.6 vs. 20-180	.67	.57	53.8
.1-.9 vs. 80-120	.88	.64	88.5
.1-.9 vs. 20-180	.67	.54	53.8
.4-.6 vs. .1-.9	.46	.47	0

I will argue that the trends in Table 7.3 suggest the following hypothesis: Second-order uncertainty is somewhat dispreferred to first-order uncertainty when variances are comparable. Evidence for this claim stems from the three comparisons involving choices between first-order and second-order gambles for which the variances are equal or nearly equal, and we have seen that there is a modest supportive trend there.

A second more extensive set of comparisons involves variance differences of approximately 50. These are interesting because the variance difference is held more or less constant, so that we may compare the strength of preference when the variance difference favors a second-order gamble with that favoring a first-order gamble. First, we note that in the two cases involving comparisons with the sure-thing option, the proportion favoring the sure-thing over the 0.8 probability of

$125 was .73 in the gain condition, which was reflected to a 0.71 preference in the opposite direction in the loss condition (standard deviation difference = 50). But the proportions involved in the comparison between the sure-thing and the $20-$180 gamble were 0.83 and 0.83 (std. dev. diff. = 46.2).

Proportions emerging from choices between the 0.8 probability of $125 versus 0.4-0.6 probability of $200 and 0.1-0.9 probability of $200 (std. dev. diff. = 50) were 0.85 and 0.70, and 0.80 and 0.62 respectively for the gain and loss conditions. These comparisons both favored the first-order gamble on the gain side and the second-order gamble on the loss side. The effects that emerged for a choice favoring a second-order gamble on the gain side and first-order on the loss side, however (.5 probability of $200 versus $20-$180, std. dev. diff. = 53.8) yielded proportions of only 0.56 and 0.55. Moreover, two other choice pairs of this kind (.2 prob. of $500 versus .4-.6 prob. of $200, and versus .1-.9 prob. of $200, both having a std. dev. diff. = 100) yielded inconsistent preference proportions (0.61 for gain and 0.36 for loss, and 0.51 for gain and 0.61 for loss respectively), despite the fact that the standard deviation difference was twice the size of the other comparisons.

The evidence concerning whether subjects favour vague probabilities or vague utilities is, unfortunately, scanty. The entire matter rests entirely on two comparisons near the bottom of Table 7.3. For a standard deviation difference of 53.8, a comparison favoring the $80-$210 gamble over the .4-.6 prob. and .1-.9 prob. of $200 gambles on the gain side yielded preference proportions of 0.67 (gain) and 0.57 (loss), and 0.67 (gain) and 0.54 (loss), respectively. Unfortunately this study has no comparisons favoring second-order probabilities over intervals on grounds of variance.

A second pilot experiment was therefore conducted, asking 236 subjects (from the same pool as the previous study) to choose from pairs of the following three gambles:
(1) .8 prob. of gaining (losing) $200
(2) .65-.95 prob. of gaining (losing) $200
(3) gaining (losing) $20-$300
These gambles are SEU-equivalent, and also equivalent in the variance of their utilities. Subjects were randomly assigned to the gain and loss conditions (124 in the former and 112 in the latter).

The results, surprisingly, seem to indicate a preference for gamble (3) over the other two under the gain condition and a corresponding reflection effect for the loss condition. The effect is less marked for the comparison with gamble (1) than with gamble (2) (64.2% favoring (3) over (1) versus 73.5% favoring (3) over (2) in the gain condition, and 61.7% favoring (1) over (3) versus 66.2% favoring (2) over (3) in the loss condition). The comparison involving gambles (1) and (2) are in the expected direction; 59.7% favor (1) over (2) in the gain condition and 73.2% favor (2) over (1) in the loss condition. Finally, the cross-subject patterns described above in the gain and loss conditions also described the modal subjects in both conditions (39% and 41%, respectively). These findings suggest that vague utilities are preferred and vague disutilities dispreferred in comparison with both first-order and second-order probabilities.

Despite the presence of confounds, these studies indicate that when variance is taken into account, subjects may differentially prefer gambles on the basis of how uncertainty is represented. Should these hypotheses hold up under more rigorous tests, their implications for the uses of quantitative second-order representations of ignorance are intriguing.

As indicated in the beginning of this section, they reinforce the intuitive impression that such representations are even more remote from human intuition and ordinary discourse than is probability theory. If, say, expert risk assessors or medical professionals were to adopt the language of quantitative second-order uncertainty in public discussions of ignorance, we should expect that the kinds of problems with comprehension and communication engendered by using probabilities would be exacerbated.

Secondly, these findings present us with a 'truth-in-advertising' quandary. Given at least three SEU- and variance-equivalent ways of expressing second-order uncertainty, which should be adopted for purposes of communicating this uncertainty to various audiences? In knowledge engineering, which representation should be used for eliciting second-order uncertainty judgments from experts? Clearly, the subjective impact of second-order uncertainty may be manipulated by selecting one or another frame.

7.2.3 Framing Effects in Nonprobabilistic Uncertainty

The results of the previous study suggest that we inquire into presentation and framing effects along the lines of the studies reviewed in Chapter 5, taking into account some of the

criticisms raised against such studies. One of the most obvious kinds of effects to study is embodied in the proverbial allusion to calling the glass half-empty or half-full. At least three such framing effects could be usefully examined:
(1) When uncertainty is expressed in connection with a desired outcome is it more positively evaluated than when it is expressed in terms of an undesired outcome?
(2) If relative freedom is expressed in terms of possibility is it perceived as greater than if it is expressed in terms of restriction?
(3) Is relative freedom perceived differently when it is regarded as a utility than when it is considered a disutility?
These hypotheses could have implications for the psychology of control and decision theory as well as obvious practical spinoffs.

    The study reviewed above also contained a within- and between-subjects investigation into the first two questions. Hypothesis (1) was tested by presenting subjects with one or the other uncertainty-equivalent version of the following statements:
Version A: (1) The medical treatment for a fatal disease is known to fail at least 1/3 of the time.
(2) The medical treatment for another fatal disease is known to succeed at least 2/3 of the time.
Version B: (1) The medical treatment for a fatal disease is known to succeed at most 2/3 of the time.
(2) The medical treatment for another fatal disease is known to fail at most 1/3 of the time.
In each of these situations, the subject was asked to rate on a scale from 0 to 100 the effectiveness of the treatment. The normatively correct responses are that (1) is less effective than (2), and that versions A and B of (1) and (2) are equivalent, respectively. The normative within-subjects hypothesis (our null hypothesis) is that subjects will consistently rate (1) more highly than (2), and the normative between-subjects hypothesis is that the mean ratings for (1) should be equal for Versions A and B, and likewise the means for (2). The alternative hypotheses, of course, are that subjects will rate the versions of (1) and (2) as more effective when success is referred to and less effective when failure is mentioned.

    Readers familiar with the psychological literature on judgment reviewed in Chapter 5 will recognize the similarity between this 'disease' problem and the famous study by

Kahneman and Tversky reviewed in Chapter 5. The crucial difference, of course, is that their problems dealt solely with pointwise probabilities and sure-thing outcomes, while this study utilizes second-order uncertainty (which, for some people, might be represented as intervalic probability). It is worth noting that the second-order descriptions involved in this study are more plausible, even more 'grammatically correct' than those in the Kahneman-Tversky problems. As Macdonald (1986) observed, estimates of treatment effectiveness do not take the form "400 people will be saved with probability 2/3, otherwise no one will be saved." If they include probabilities at all, they refer to the individual rather than to some collection of people. Simple intervalic descriptions of the kind given in this study, however, are intuitively plausible.

The framing effects clearly emerge for both the between- and within-subjects setups. For the between-subjects effect-size, the most conservative indication is that the means for (1) and (2) differ between the two versions by 6.64 and 4.56 standard errors of the mean with the largest s.e., respectively (and by 13.56 and 6.07 standard errors if we use the smaller s.e.'s). The within-subjects effect is probably most easily grasped by examining Table 7.4, which shows the extent to which subjects rated (1) as more, less, or equal in effectiveness to (2). The percentages are almost entirely explained by the framing effect, with no clear 'main effect' from a tendency to favor (2) over (1). It is also noteworthy (a la Scholz 1987) that the numerical information seems to have distracted a large portion of the subjects. In all four rating distributions the modal response was approximately 67 (49%, 46%, 33%, and 61% gave a rating of 66-67 in the four conditions). This turns out to account for the majority of those subjects who gave the weakly counternormative response of rating (1) adn (2) identically.

Table 7.4 Within-Subjects Comparisons of (1) and (2)

	Version A		Version B	
(1) < (2)	38	45%	5	7%
(1) = (2)	44	52%	37	54%
(1) > (2)	2	2%	27	39%

The second hypothesis also was tested using a within- and between-subjects design. A set of intervalic constraints on percentages was presented in four different versions, wherein the manipulation was to express those constraints possibilistically in one instance and in terms of necessity in the other. The manipulation was achieved by using two scenarios:
(1) You are in charge of a field-trip which takes 100 children to a museum. The museum has three (3) exhibits, but there is time to take the children to only one of them. If the exhibits could hold more than 100 children at a time then they could all choose the exhibit they wanted. However,
Version A: exhibit 1 can hold up to 70 children, exhibit 2 can hold up to 50, and exhibit 3 can hold up to 90.
Version B: 30 children cannot have access to exhibit 1, 50 cannot have access to exhibit 2, and 10 cannot have access to exhibit 3.
On a scale from 0 (no freedom) to 100 (total freedom), how much freedom do you think there is for the kids as a group to see whichever exhibits they want?

(2) In a certain community there is a subpopulation of elderly citizens who can no longer drive their own automobiles. There are three possible means of transport for these people: Bus, taxi, and being driven by a friend/relation. However, not all of them have access to all three of these options:
Version A: 10% cannot take the bus, 30% cannot use the taxi, and 50% have no one to drive them around.
Version B: 90% may take the bus, 70% may use the taxi, and 50% have someone to drive them around.
On a scale from 0 (no freedom) to 100 (total freedom) how free are these elderly as a group to choose among these three transportation options?

The normative (null) hypothesis is that subjects should perceive these scenarios as equivalent and give identical ratings for all of them (using the formula for FG, the normative rating would be 65). Instead, both framing effects came through quite strongly. The between-subjects effect-sizes are at least 3.65 and 4.35 standard errors' difference between the means in the expected directions. However, it is again noteworthy that some of the subjects appear to have been using the numerical information as a basis for their ratings. In all four distributions 50 and 70 were modal responses, accounting for 33%, 33%, 37%, and 48% of the respondents. These results suggest that some people may be using rules such as "pick the biggest restriction" or "pick the middle one".

The within-subjects effect was not as strongly supportive of the hypothesis as the between-subjects results. In Version A, 52% of the subjects rated the possibilistic scenario more highly than the necessitarian one, while 58% did so in Version B. However, 18% and 23% (in Versions A and B respectively) gave the normatively correct response of rating them equally. The remaining percentages whose ratings differed in the opposite of the expected direction are 30% and 26% respectively. The trend is in the right direction, but only moderately so, and a substantial plurality of subjects' responses are not anticipated by either the normative or framing effect hypotheses.

These tentative steps towards a dialog between normative and descriptive accounts of second-order uncertainty reinforce the importance of examining how the representation of ignorance influences perceptions of and discourse about it. I have suggested here that individuals may have patterned preferences for one representation over another, that their perceptions of the extent of ignorance can be influenced by its expression, and likewise by whether ignorance is regarded as a utility or disutility. Only a limited variety of second-order uncertainties have been investigated here; obviously we would benefit from normative-descriptive inquiries into relative entropy, confusion, ambiguity, and conflict as elucidated in the fuzzy set, possibility, and Shaferian belief function frameworks.

7.2.4 Expanding the Dialog

Despite the space given here to a pilot study that in some ways apes the worst aspects of laboratory-bound studies on subjects who are utterly unrepresentative of the general population, I am convinced that the best direction for future dialogs on second-order uncertainty (and ignorance generally) points *away* from this kind of approach in two respects. First, we must shed the older preoccupation with measuring humans against some normative benchmark. The insights gained by such studies are, for one thing, too impoverished. For another, in second-order uncertainty particularly, there are a number of alternative normative frameworks that could be used. We have found, for instance, that variation in probabilities does not seem to explain subjects' preference patterns, while variance in utilities does. Which uncertainty is the more normatively important?

Worse still, we have found that subjects may prefer one kind of representation of second-order uncertainty over another, even when they are equivalent in both SEU and utility variance (e.g., intervalic utilities versus intervalic probabilities). And the second study indicates that subjects' perceptions of second-order uncertainty depends on whether it is linked with desirable outcomes, or even whether the uncertainty itself is perceived as desirable. No matter what normative framework for representing uncertainty we adopt, it is vulnerable to 'framing effects'.

There is a truth-in-representation problem, but its solution seems to require a meta-rationalism along the lines suggested by Good (cf. sections 4.5.2 and 4.5.3). The difficulty with Good's quasi-utilities and 'type II' meta-rationality, of course, is that it leads to the same infinite regress problems and paradoxes that Russell's theory of types was supposed to solve. It would seem appropriate, then, to study the 'grammar' of ignorance representations in the discourse of both lay and expert populations. Why are some kinds of uncertainty, for example, represented using subjective probabilities while others are not? Why are engineers willing to quantify their probabilities while lawyers and judges refuse? A systematic inquiry along these lines might well reveal motivational and other reasons for selecting one seemingly equivalent representation over another, as well as enhancing our ability to communicate effectively about ignorance.

The second direction away from the older experimental paradigm leads into the real world and also, more importantly, to the inclusion of situational and social factors. At the social psychological level of analysis, there is ample scope for exploring situational influences on the perception of and response to various kinds of ignorance. Attribution theory, for instance, has produced several hypotheses during the last 20 years concerning the different accounts of human behavior that actors and observers provide. The most well-known of these is that actors tend to attribute their own behavior to situational contingencies and influences, while observers attribute those same actions to the actor's internal dispositions.

Are there actor-observer differences in the types of ignorance that are salient where human behavior is concerned? One plausible hypothesis is that for observers, probabilistic uncertainty is salient because they are attempting to predict what the actor will do next. Actors themselves, however, may not be asking themselves what they are *probably* going to do next. Instead, the kind of uncertainty salient for them most likely concerns the range of possible options and restrictions on their choice of behaviors. They are concerned with what is *possible* to do. For the observer, then, the salient uncertainty is probabilistic; for the actor, it is possibilistic (I am indebted to a conversation with R.M. Farr for this idea). Not only is this an interesting research question, but it pertains to some fundamental methodological questions in social psychology itself. After all, if this hypothesis is valid, then the use of probability to model uncertainty in human behavior presumes an observer's point of view.

Finally, there is a rich source of applied sociological research material to be found in cross-pollinating the sociology of ignorance framework developed throughout this book with any of the mainstream research questions in the sociology of organizations and professions. Very little systematic investigation has been done into how and why various kinds of ignorance are constructed and utilized in organizations and professions, or what the consequences and functions of specific kinds of ignorance are. Likewise, aside from a few psychologists' attempts to map ignorance-bearing phrases such as 'very likely' onto numerical scales, almost no research has been done on how ignorance is represented collectively (in the sense of the European 'social representations' literature) or how it is presented to the public by powerful institutions and the media. Sociologists and social psychologists would be well-placed to conduct such studies, and would be more likely than other kinds of researchers to faithfully capture the viewpoints of their respondents.

7.3 What Goals Are Served by Which Normative Frameworks?

The risk perception literature in recent years has begun to pay attention to social, political, and cultural agendas, although its main focus still is at the intra-psychic level. Nevertheless, this change is salutary and may be reasonably generalized by observing that the strategic value of debates among competing perspectives on ignorance cannot be understood without comprehending the social or political goals and cultural assumptions which underly those debates. In other words, we must know something about the kinds of plans, hopes, and fears held by relevant sectors of society for the future.

Wildavsky (1985) borrows a useful dichotomy from ecology to distinguish between two ways of coping with an uncertain future: anticipation and resilience. Anticipation requires that we ascertain future dangers in order to ward them off before they arise. It is most compatible with both an exclusionary mode of control and inclusion geared towards total rehabilitation and, therefore, with the traditional ignorance-reducing strategies of banishment and reduction. Organizational theorists who write about organizational change in the traditional mold echo the agendas of anticipation when they cite buffering, smoothing, forecasting, and rationing as the primary adaptive strategies for managing the unknown. Anticipation also is the agenda of at least some of the second-order strategies for dealing with ignorance, most notably the Hilbert-like attempts to guarantee that the unknown will be well-behaved by subsuming it under formal axiomatic systems that are both complete and consistent.

Resilience, on the other hand, refers to learning how to "bounce back" by utilizing change to cope with the unknown, and it entails dealing with dangers as they arise. In the managerial and helping professions this is often referred to as 'crisis management'. Far from relying on uniformity and regulation as does anticipation, resilience becomes most feasible under conditions of variability and spontaneity, or at least freedom of action. Miles et al.'s (1978) "prospectors" and Mars' (1982) "hawks" require such environments in order to be innovative and flexible, as do the "adhocracies" of Silicon Valley fame. Resilience, therefore, is compatible with those frameworks for ignorance that stress the decarceratory kind of inclusionary control: the eclectic second-order management strategies that characterize what I have termed 'suburbanization'.

As Wildavsky points out, the debates over risk assessment and regulation involve a strategic conflict between visions of anticipation and resilience as preferred modes for dealing with the future. This conflict also is evident in battles between adjudicatory and regulatory law (J.C. Smith 1983). In most Western legal systems, the criminal law model is anticipatory in that it establishes a priori standards for legitimate acts and penalties for particular kinds of contraventions of those standards. While criminology itself has undergone major transformations of the kind outlined by S. Cohen, the criminal justice system remains largely influenced by the Enlightenment version of classical criminology, which embodies regulation and either exclusionary or carceratory inclusionary control. The tort law model, by contrast, attempts to settle conflicting claims over rights and damages as those conflicts arise and on a case-by-case basis. This model operates as a resilience strategy, and is therefore compatible with decarceratory inclusionary control. As Cohen would argue, the more 'criminal' the act, the more likely the anticipation model will be invoked, while the resilience model tends to be applied to 'civil' cases.

The point of all this is that the relevance (and perhaps even the character) of normative debates about ignorance representation and management hinges on the comparative value of control strategies (i.e., exclusionary, carceratory inclusionary, and decarceratory inclusionary), which in turn depends on whether the dominant planning style is anticipation or resilience. Anticipation demands initially that we reduce ignorance as much as possible. It therefore entails a realist epistemology, and favors a narrow conceptualization of ignorance as objectively measurable. It also seems to favor an exclusive focus on uncertainty (and usually probabilistic uncertainty at that) rather than ignorance *sui generis*, in a manner which locates uncertainty in the environment rather than in people's cognitive schema. Many of the normative debates over how to assess probabilities of harmful events, how to correctly calibrate subjective judgments of probability, and which belief-updating logic to use serve the goals of anticipation and ultimately regulation. Likewise, when the normative and descriptive are compared, the chief purpose is to expose counter-normative errors and deficiencies in how people deal with the unknown, and to prescribe methods for retraining them.

Resilience, on the other hand, requires that we be concerned with faithfully representing the values, perceptions, and coping strategies of the people affected by change or danger. The epistemological stance is nominalist, or at least partly relativist. Ignorance is not considered objectively measurable but understandable mainly in subjective terms. The unknown may well be 'out there' in the environment, but its representation as a form of ignorance resides in the mind. Resilience demands that we establish a common linguistic and conceptual framework for discussing and comparing people's perceptions of ignorance. Normative debates that contribute to resilience focus on standards for ignorance representation in which the main criteria are the adequacy of those representations for capturing the variety of people's subjective assessments and beliefs. The goals of descriptive studies are primarily to find out how people think about and deal with the unknown rather than exposing cognitive 'errors'. Insofar as logical consistency or other formal normative criteria are brought in, their merits tend to be debated on a case-by-case basis rather than being decided in general.

Like any dichotomy on this scale, the anticipation-regulation vs. resilience-adjudication distinction has been necessarily oversimplified here for purposes of argumentation. Most complex socio-technological systems involve a mixture of the two, and the nature of the mixture may well be crucial. Nevertheless, before moving on to consider specific kinds of mixes and their implications for ignorance debates, it is worthwhile to consider briefly some of the arguments that have been raised for and against these two modes.

The most obvious arguments favoring anticipation appeal to desires for security, safety, and stable controlled systems that are error-free. In fact, anticipation is the preferred strategy of classical utopian planning which involves a vision of an ultimate social order which includes all good and banishes all evil. Utopias are secure and safe, free of error, risk, or pain, and, of course, stable. Wildavsky further argues that people tend to favor anticipation when they believe risks are high or when they wish to defend something valuable.

Wildavsky goes on to argue, however, that both ecological evidence and logic indicate that anticipation may actually increase risk rather than guaranteeing safety. First, a preoccupation with avoiding errors inhibits learning and therefore decreases innovative capacity. Regulation and anticipatory control require elaborate rules, large-scale bureaucratic structures and centralized authority, all of which make innovation costly and difficult. One consequence of this is decreased variability or diversity.

A standard ecological argument has it that interspecies diversity and nonspecialization within species are pro-survival while their opposites are not adaptive. This argument has been extended to the human management of natural environments by ecologists such as Clark (1980), who pointed out the disadvantages that accrued from the classical managerial assumption that the removal of "uncertainty or variability" from the environment is an "unmitigated good" (pg. 298). These over-managed environments prove more vulnerable to unforseen change or catastrophe than those which have experienced and adapted to high variability before. Wildavsky caps these observations with yet another metaphor, this time from the insurance industry. While parts of society may be insured, one cannot insure the whole. Anticipation might be workable for parts of society but only if some other parts are resilient to unanticipated problems.

Other arguments against anticipation have been provided by critics of classical utopian planning. Sennett (1970), for instance, claims that utopian plans fail to allow for unanticipated change or growth, and as a consequence they embody political and social values that their proponents do not intend. Like Wildavsky, they point out that anticipation requires a 'big brother' approach to government (it was precisely this thesis that was the point of Orwell's *1984*). Its ethos is intolerant of change, conflict, deviance, or even civil liberties. Its implicit model of human action is a stochastic determinism which ignores individual self-determination and requires people to act as extensions of advanced technology.

Wildavsky's arguments favoring resilience echo the ecological models of adaptation borrowed for his purposes. He claims that resilience outperforms anticipation in long-term insurance of safety (especially in turbulent times), although it may not do so well during periods of short-term stability. Resilience virtually demands innovation, improvisation, and creativity. It also permits an adjudicatory model of justice to operate in a case-wise manner, fitting the recompense to the crime as it were. Critics of classical centralized utopian planning adopt a 'small is beautiful' position, and argue that laissez-faire, ad hoc approaches to management and control are conducive to decentralization of power, small-scale organizations that are close to the people they serve, and maximization of individual freedoms. S. Cohen (1985) points out that the ideological arguments for the decarceration movement and modern inclusionary control of deviants appealed to many of these values: small-scale community-based self-help rather than large-scale state-controlled regulation and incarceration.

It is Cohen, however, who provides the most subtle arguments against inclusionary control. They make an interesting contrast with Wildavsky's ecological vindication of the resilience strategy. First, he points out that the humane, democratic values associated with inclusionary control are not carried through if inclusionary control is performed by the state itself. It is liberal only if it is anti-state. The more widely it is implemented, the more people, time, and effort are taken up in surveillance, testing, intervention, therapy, social constraint, and evaluation. Secondly, Cohen warns that several of the psychological and social functions of exclusionary control (e.g., scapegoating, rule clarification, boundary maintenance, and social solidarity) cannot be fulfilled by inclusionary control on its own. Thirdly, the 'laissez-faire' ethic can be twisted into an excuse to ignore real maldistribution of harm and other related injustices. His final point is a telling observation: As in Orwell's dystopia, Western society tends to reserve inclusionary control for the middle class and the 'soft-core' offenders while subjugating the lower classes and 'hard-core' cases to exclusionary measures.

Is there, then, an 'optimal mixture'? Wildavsky's prescription is for government to intervene only when there is evidence that intervention will yield actual benefits as well as reducing or avoiding harm. But this seems to require exactly what he claims he wants society to avoid: the need for guarantees before committing acts. To be fair, Wildavsky recognizes the choice as a dilemma and his discussion echoes some of the themes raised by control theorists ranging from Collingridge to Cohen. Also, Wildavsky points out that the social version of a resilience strategy requires public trust of the institutions involved, which is in short supply. Cohen himself does not provide a clear answer, but confesses that he is committed to furthering the search for a humane alternative to either exclusionary or state-run inclusionary control. He would like to find a workable version of decarceratory inclusion.

An additional dimension to the dilemma of control emerges if we combine the observations by Cohen, Collingridge, Wildavsky, and Douglas with the trends discussed in the organizational sociology literature and the recent work by Henry, Mars, and Mattera on the hidden economy. I can only provide a thumbnail sketch of this dimension here. During the most recent few decades, public trust in major institutions has eroded while awareness of and aversion to risks have increased. Moreover, the traditional styles of social control (exclusion and state-run incarceratory inclusion) have been supplemented by enlarged systems of surveillance (I am including diagnostic testing, impact assessment, evaluation and assessment research, and various forms of data-gathering for purposes of social control as special kinds of surveillance) and regulation, both at the individual and organizational level. Information has begun to resemble hard currency, and the post-modern 'information society' is also an increasingly regulated, litigious society.

At the same time, individuals and organizations have resisted external regulation, bureaucratic red tape, and the threat of liability in at least three ways. First, they have lobbied for stronger rights and guarantees of privacy, anonymity, and confidentiality. When deprived of privacy, they have become increasingly secretive, increasing their 'irregular' activities. Both responses entail an increase in intentionally created ignorance, specifically in the forms of distortion, incompleteness, and taboo. Finally, individuals and lower-level organizational units have agitated for greater freedom of access to information held by their higher-level counterparts. Likewise, the force of Linnerooth's and Downs' observations about political agendas that tempt bureaucrats and organizations to be less than honest about the true extent of ignorance comes from the fact that all of those agendas are responses to external pressures for legitimation, accountability, apparent certitude, and apparent control.

The new dilemma, then, is that the control strategies that are designed to eliminate, absorb, or reduce *unintentional* ignorance may actually increase *intentional* ignorance. Insofar as those control strategies are not effective in reducing unintended ignorance, the net amount of ignorance in the system may well be increasing.

The absence of a generalized, simple solution for the whole system suggests that effective mixtures of these strategies must be tailored to subsystems and even specific problems. In short, these dilemmas and problems make suburbanization an attractive way out, despite the fact that it begs the question of what to do about the big picture. In the next section, we reconsider some of the debates described in previous chapters in view of the control issues discussed here.

7.4 The Social Nature of Rationality

We lack an agreed-upon technology for making judgments and decisions under ignorance (cf. the final paragraph in Chapter 5) and there are no helpful organic models for cognition that would provide us with competency-performance distinctions and benchmarks. Moreover, the behaviorist and evolutionist attempts to ground probability in alternative 'rationalities' by reducing it to maximization of expected rewards or long-run survival chances are fraught with the same difficulties as the older attempts to reduce mathematics to logic. Reductionism may serve particular goals or values, but those goals and values do not have objective status. Rationality is an inescapably social creation.

Thus, we are thrown back into normative considerations about 'rationality' and 'meta-rationality'. Mathematicians, psychologists, and decision theorists have offered only a narrowly cognitive, individualistic conception of meta-rationality (i.e., computational effort, ease of understanding, communicability, amount of information, explanatory power, and minimization of apparent error). I have argued in Chapter 6 and in the preceding section that the meta-rationality debate must be expanded to include social, political, and cultural factors. These need not (or perhaps cannot) be resolved in general, but at least they may be explicitly considered in discussions of specific problems, professions, or even entire disciplines.

For want of a better way to summarize the tradeoffs, polarizations, and issues, I have elected to present them as dichotomies in the table below with more or less compatible values, assumptions, and viewpoints placed in the same column. The elements in the same column should not be taken to be causally linked, but moderately correlated and/or ideologically connected. This is not to say that cross-column combinations are unlikely, let alone impossble. I would not wish to defend these dichotomies and their groupings too vigorously, but instead offer them as an aid to thought and discussion. The most important links are between specific entries in the "Control Orientations" and "Control Style" portions of the table, on the one hand, and those in the "Ignorance Orientations" portion, on the other.

Table 7.1 Issues in Control and Ignorance
CONTROL ORIENTATIONS

Philosophical orientation:

Idealism, realism	Nominalism, pragmatism

Values orientation:

Security, safety	Liberty, innovation

Justice model:

Criminal law requiring establishment of guilt 'beyond reasonable doubt'	Adjudication or tort law, establishing guilt 'on a balance of probabilities'

Model of Society:

Equilibrium, homeostasis Societal or organizational well-being evaluated by stability or rate of return to equilibrium	Conflict or dynamicism Societal or organizational well-being evaluated by performance or growth-rate

Dominant Organizational Structures:

Hierarchical, formalized bureaucratic, high grid-group	Heterarchical, informal adhocratic, low grid-group

Model of Human Behavior:

Stochastically determined or rule-obeying; People are ignorance-averse	Voluntaristic, reflexive, and/or rule-using; People are not always ignorance-averse

---

CONTROL STYLE

Planning Style:

Anticipation, defense	Resilience, crisis management

Control Model:

Exclusion or state-run incarceratory inclusion	Decarceratory inclusion

Governing Style:

Regulatory, benevolent dictatorship	Laissez-faire, participatory democracy

Decision-Making Style:

Molar, go/no-go, executive; Decisions by individual expert or elite on behalf of public	Accretive, garbage-can, group-made; Decisions via public negotiation and debate

---

IGNORANCE ORIENTATIONS

Normative Positions:

Empiricist, realist	Personalist, subjective
Insists on standard logic	Permits alternative logics or systems
Emphasizes initial precision	Emphasizes final precision
Restricts representation of ignorance to quantifiable and probabilistic uncertainty	Emphasizes representation of nonprobabilistic and/or unquantified ignorance
Emphasizes first-order certainty	Emphasizes second-order certainty
Banishes natural language from ignorance representation and discourse	Admits natural language representions of and discourse on ignorance

Normative-Descriptive Relations:

Normative framework provides absolute guides to correct judgment under ignorance	Normative framework forms basis for social consensus on fair or adequate judgment under ignorance
Normative is used to expose errors in intuitive judgment and to set standards for training or education	Normative is used to adjudicate intuitive judgment or as possible explanations for judgment processes

Normative provides the basis for education, training, correction, and calibration of judges	Normative provides the common language for dialog and negotiation among judges
Descriptive and explanatory perspectives used to anticipate judgmental errors and reveal strategies for correcting or preventing them	Descriptive and explanatory perspectives used in dialog with normative frameworks to account for particular judgments and the normative frameworks that evolve

I shall also attempt to summarize the various control and representation dilemmas in terms of their immediate implications for the debates about ignorance. Most of these take the form of goals or values that cannot be simultaneously maximised. While most of them have been pointed out by at least several authors, I have given them names for purposes of easy reference.

Collingridge's Dilemma: The less well-entrenched a system is and the shorter the time it has been operating, the more easily and inexpensively it can be changed; but the greater is our ignorance of its likely effects or problems. By the time ignorance of those effects has been reduced, it is too expensive and difficult to change the system. A corollary would be that normatively adequate first-order descriptions of ignorance itself may take too long or be too costly.

Wildavsky's Maxim: System resilience requires toleration of ignorance. Attempts to reduce ignorance by anticipation weaken the system's resilience to the unanticipated. Again a simple corollary requires members of the resilient system to tolerate the prospect of some meta-ignorance.

Mattera's Dilemma: The greater the attempts to regulate behavior and thereby increase predictability or control, the more reactive people become and the more they attempt to generate ignorance by way of preserving their freedom. A climate favoring creativity, initiative, and entrepreneurship requires the toleration of ignorance (including deception). Attempts to gain information about people may be proactive, motivating people to withold information or to give false information.

Ravetz' Law: The more socially or more politically important the research questions, the less amenable they are to unique answers and the greater the magnitude and variety of ignorance to cope with. The more diverse the public interest, the less likely that a normative consensus will be achieved.

Rationalist Quandary: The more specialized and normatively restricted the language for representing ignorance, the more people are excluded from discussion or debate, and the less effective the language becomes for communicating about ignorance to nonspecialists.

Zadeh's Thesis: Beyond a certain system size, precision and relevance in the description of the system become incompatible. The more normatively adequate the description, the narrower its focus, the less relevant it is, and the less ignorance is represented or recognized.

Zeleny's Tradeoff: The more context-free (hence unambiguous) the representation of ignorance, the lesser are its representational capabilities. The more context-sensitive, and therefore descriptively rich, the more difficult it is to attain clarity of expression. As more than one wag has put it, we have a choice between being vague and wrong.

While these dilemmas and other tradeoffs realizable from Table 7.1 may not have simple solutions, they nevertheless give some guidance for future debates and dialogs between normative and descriptive perspectives on ignorance. Four of these will be discussed below.

First, the kind of debate over a formalism such as Bayesian probability that we find in psychology can never be resolved at the general level. There will always be applications or problems for which first-order quantified probability judgments are not deemed communicatively appropriate or descriptively adequate. The conditions most favorable to the Bayesian line are those in which the main control agendas are conservative, anticipatory and/or regulatory (as per the left-hand column in Table 7.1); people agree that the relevant kinds of ignorance can all be considered as probabilistic uncertainty, which in turn is considered to be quantifiable; (ironically) the ultimate basis for probabilistic judgments is thought to lie in physical reality; and the people making the judgments are adequately trained in probability theory. Thus, weather-forecasting is an example fulfilling all these conditions. Nuclear power risk analysis, on the other hand, encounters Collingridge's Dilemma along with all the representational dilemmas from Ravetz' Law to Zeleny's Tradeoff, so that experts may equally validly argue that probabilistic risk assessment must be the normative standard, or that such an approach is inadequate to represent the full extent and variety of relevant ignorance (as in Turner 1978).

Second, we need a more critical and socially informed inquiry into the nature of those normative agreements and conflicts that do exist in particular professions or disciplines. To return to an earlier question, why do engineers almost always quantify their probabilities while lawyers and judges generally resist quantification? To simply say "training" and leave it at that begs too many questions. To what extent are the current arguments about risk assessment techniques explicable in terms of how control, regulation, justice, and the like are viewed by the disputants? Many risk assessment problems are beset by Collingridge's Dilemma and Wildavsky's Maxim; where they concern tradeoffs between benefits and risks then Mattera's Dilemma may enter in as well. The recent admission of socio-cultural considerations into this domain should be encouraged and broadened. After all, the day may not be far away when representations of ignorance are widely regulated by law in various technological, economic, and professional spheres.

Third, the dialogs between the normative and descriptive have been far too dominated by an exclusively psychological (individual-cognitive) model of judgment and decision making. There is a great need for psychologists and organizational sociologists to communicate with each other. Likewise, applied mathematicians, computer scientists, engineers, and other creators of innovative normative frameworks need to divest themselves of the assumption that 'behavioral science' equals cognitive psychology. Their frameworks also have been dominated by a preoccupation with an individualistic image of decision making.

Fourth, these same disciplines and professions need to become more acquainted with what social scientists have to say about intentionally created ignorance, and they should expand their conceptual armamentarium to encompass irrelevance as well. The social scientists, on the other hand, need to become better-informed about the competing normative frameworks, particularly those schools of probability that do not underly the Neyman-Pearson school of statistical inference, and the nonprobabilistic uncertainty formalisms also. They must also shift their emphasis from the comparatively easily observed phenomenon of intentional ignorance to studying unintended ignorance, and they should begin evolving workable distinctions among different kinds of ignorance rather than treating it as merely the absence or distortion of 'truth'. The organizations literature especially requires an investigation into the influence of problems in ignorance representation and discourse as exemplified by Zadeh's Thesis, the Rationalist Quandary, and Zeleny's Tradeoff.

Let us consider an example of how some of the material in this chapter may be applied to a normative framework. In Chapter 5, I claimed that fuzzy set research has seen a relatively interactive and mutually respectful dialog between its normative and descriptive wings. Nonetheless, fuzzy set and possibility theory both exemplify the confusions and dilemmas that beset such dialogs, and it is worth examining some of them here.

We have already seen that fuzzy set theorists have made efforts to ground their framework in mathematically respectable foundations. But much of that effort sits uneasily alongside the rhetoric behind the introduction of fuzzy sets in the first place. Zadeh's Thesis is taken from his "incompatibility principle" and a programmatic statement justifying his apparent departure from conventional methods of representing ignorance (1973: 687): "... we need approaches which do not make a fetish of precision, rigor, and mathematical formalism, and which employ instead a methodological framework which is tolerant of imprecision and partial truth." Zeleny (1984) and Kochen (1984) claim that fuzzy set theorists have made a fetish of precision, rigor, and formalism; and they argue that it has done so at the expense of its relevance to the very type of applications for which it was intended. They ask how a theory of imprecision can be so precise, quantitative, and dependent on the tools of ordinary mathematics.

Their conclusion is that fuzzy set theory should become essentially a behavioral science, its problems and results arising from empirical questions about human systems. Zeleny maintains that it would be more fruitful to rebuild fuzzy set theory by reformulating it in the context of specific problems where it is required than by trying to apply a context-free version of fuzzy set theory to empirical problems. These sentiments are echoed in the same issue of *Human Systems Management* by Amarel.

Kochen and Zeleny also castigate the language of fuzzy set writings for its needlessly forbidding complexity and formality. Kochen echoes Zeleny's Tradeoff and the Rationalist Quandary when he says he no longer has the time or energy to follow the mathematics in *Fuzzy Sets and Systems* articles (a problem I share with him). I also have first-hand experiences of similar criticisms from even mathematically capable colleagues in the human sciences. One of my strongest motivations to write a book on fuzzy sets for the behavioral and social sciences was to breach this barrier by re-expressing the basic ideas in terms and through examples that human scientists would appreciate.

The more moderate positions taken by Majumder and Bookstein in the *Human Systems Management* debate underscore certain dilemmas faced by would-be users of fuzzy set theory. Majumder and Bookstein both agree with Zeleny's point about making a fetish out of rigor and formalism, but they do not believe that the mathematization of imprecision is incompatible with an accurate representation of it. Majumder points out that making fuzzy set theory context-dependent might be attained by using appropriate mathematical representations, and Bookstein observes that probability is after all a precise formalization of uncertainty. What is missing, both separately argue, is a sufficiently rich set of alternative ways for representing fuzziness that can be successfully selected or modified to fit into real-world problems in particular disciplines.

Consider the use of the unit interval [0,1] for representing membership. This seems overly foreign and restrictive to many behavioral and social scientists in my experience, and so they often stop right there. In Smithson (1987: 86-89), I have attempted to coax them by showing how one could have membership values that are not restricted to [0,1] and arguing that one can apply fuzzy set operators to conventional bounded scales, ordinal scales, and even to unbounded interval scales (e.g., z-score transformations). To be sure, some of the rigor and mathematical richness of fuzzy set theory is lost in making these transitions. But this is simply the expected tradeoff for enhanced comprehensibility and applicability.

I should like to add that the dilemmas and tradeoffs reviewed here are not always unresolvable in specific contexts. As a contrast with the usual kind of ignorance representation problem, consider the difficulty faced by social survey takers when their promises of confidentiality and anonymity are threatened. In a number of Western nations, survey data can be subpoenaed if deemed admissible as court evidence. In surveys where respondents may have admitted to illegal acts or intentions, there is at least an implicit credibility problem with the interviewer who assures the respondent of confidentiality or anonymity. Ironically survey evidence itself (cf. Turner 1982) indicates a general increase in public mistrust of such assurances, not only from surveys taken by private organizations but also from social scientists, census-takers, and other government officials. This evidence fuels long-standing doubts on the part of survey analysts about the truthfulness of answers to intimate or awkward questions. Moreover, some lobbyists have mounted an in-principle argument that even with informed consent, certain kinds of questions amount to an invasion of privacy.

The standard ploy for surveys that are accessed by others (e.g., the census) is simply not to permit access to individual cases, but instead to limit this to a certain aggregate size. Of course, this is an obvious extension of the custom among social scientists of never reporting individual cases in publications based on such surveys, and both are unsatisfactory in the event of, for instance, a subpoena of the data and master-list of respondents. The problem, then, is to find a way of concealing the respondent's information in such a way as to render it unidentifiable but still usable for analysis.

One solution package includes using intervals instead of point-valued response formats (e.g., $20,000-25,000 for recording an income of, say, $21,000), and thereby replaces precision with vagueness. Another family of solutions, known as 'randomized response methods', makes it impossible to identify an individual's response for certain. For instance, a 'yes-no' question involves having the respondent take a bead from an urn (without the interviewer knowing what color the chosen bead is). A red bead, say, means the respondent must tell the truth, while a white bead means the respondent must answer 'yes' no matter what the truth is and a black bead means she or he must answer 'no'. The survey-taker knows what the portions of red, white, and black beads are but of course no one knows which respondents drew which ones. The result is that the analyst can subtract out the corresponding proportions of false 'yeses' and 'no's', knowing that in crosstabulations they will be independently distributed. There are several variations on this approach (see, e.g., Warner 1965,

1971, Greenberg et al. 1969, Poole 1974, and Boruch and Cecil 1982).

The remarkable aspect of these solutions, and the literature in which they are discussed, is that we see researchers colluding with their respondents and clients in the strategic manufacture of ignorance. Moreover, a social consensus on how this is to be done is achieved in the face of seemingly incompatible needs. In this particular instance, a normative agreement on the frameworks for ignorance representation is reached by compromising on the goals of analyzability and privacy protection, in such a way that the amounts of uncertainty or imprecision are estimatable to the satisfaction of the researchers and their clients.

Both the investigation of specific problems and examinations of normative and descriptive debates over a particular ignorance representation framework are rather limited in what they can tell us about the possibilities for dialog between the normative and the descriptive. Elucidations of particular applications provide only anecdotes and case-studies, while explorations of fuzzy set theory, probability, or belief functions on their own do not always generalize to other frameworks. Far better would be a dialog that encompasses a particular kind of ignorance (taken from a workable taxonomy along the lines suggested in Chapter 1), incorporating more than one framework for its representation, and including sociocultural as well as psychological factors.

In view of the social nature of rationality and the apparent links between various normative perspectives and socio-political orientations, the explanation and interpretation of our recent cultural and intellectual upheaval over ignorance deserves careful attention and creative research. The stage model proposed in this book may serve as a starting point, and the material in Table 7.1 suggests part of a research agenda. I would like to be able to present examples of investigations into ignorance that deal with social or political factors and crucial populations (e.g., do certain professions exhibit preferences for representing ignorance in specific ways? How do those preferences articulate with a given profession's codes of ethics and standards? How are its members taught to think about and cope with ignorance?), but I know of almost none as yet. Clearly, such investigations should be conducted, and they should be well-informed by current normative debates on the representation of ignorance without becoming enslaved by any particular normative outlook.

These research possibilities point to the need for *interdisciplinary* (as distinct from multidisciplinary) efforts involving mathematicians, cognitive scientists, engineers, psychologists, sociologists, and practitioners in relevant professions. The dialogs between normative and descriptive perspectives on ignorance cannot succeed without such boundary-spanning work. There is no other feasible way of coping with dilemmas and quandaries of the kind reviewed in this section. Insofar as there are any reasonable, humane answers to the questions posed by our post-modern ignorance explosion, it is unrealistic to expect them to arrive miraculously intact in any one specialized field. Specialization, after all, is a form of systematic ignorance.

Bibliography

Abramson, L.Y. and Alloy, L.B. 1980 "Judgment of contingency: Errors and their implications." In A. Baum and J.E. Singer (eds.) *Advances in Environmental Psychology, Vol.2: Applications of Personal Control.* Hillsdale, N.J.: Erlbaum.

Abramson, L.Y., Seligman, M.E.P., and Teasdale, J.D. 1978 "Learned helplessness in humans: Critique and reformulation." *Journal of Abnormal Psychology*, 87: 49-74.

Adams, J.B. 1976 "A probability model of medical reasoning and the MYCIN model." *Mathematical Biosciences*, 32: 177-186.

Adorno, T.W., Frenkel-Brunswik, E., Levinson, D.J., and Sanford, R.N. 1950 *The Authoritarian Personality.* New York: Harper.

Ajzen, I. 1977 "Intuitive theories of events and the effect of base-rate information on prediction." *Journal of Personality and Social Psychology,* 35: 303-314.

Aldrich, H. and Herker, D. 1977 "Boundary spanning roles and organization structure." *Academy of Management Review*, 2: 217-239.

Allais, M. 1953 "Le comportement de l'homme rationnel devant le risque; critique des postulats et axiomes de l'Ecole Americaine." *Econometrica*, 21: 503-546.

Alloy, L.B. and Tabachnik, N. 1984 "Assessment of covariation by humans and animals: The joint influence of prior expectations and current situational information." *Psychological Review*, 91: 112-149.

Allport, G.W. and Postman, L. 1947 *The Psychology of Rumor.* New York: Henry Holt.

Alston, W.P. 1964 *Philosophy of Language.* Englewood Cliffs, N.J.: Prentice-Hall.

Althusser, L. 1976 *Essays in Self-Criticism.* Trans. G. Locke. London: Humanities Press.

Amarel, S. 1984 "On the goals and methodologies of work in fuzzy sets theories." *Human Systems Management*, 4: 309.

Amsel, A. and Stanton, M. 1980 "The ontogeny and phylogeny of the paradoxical reward effects." In J.S. Rosenblatt et al. (eds.) *Advances in the Study of Behavior*. New York: Academic Press.

Anderson, N.H. 1981 *Foundations of Information Integration Theory*. New York: Academic Press.

Anderson, N.H. 1982 *Methods of Information Integration Theory*. New York: Academic Press.

Anderson, N.H. and Shanteau, J.C. 1970 "Information integration in risky decision making." *Journal of Experimental Psychology*, 84: 441-451.

Arkes, H.R. and Harkness, A.R. 1983 "Estimates of contingency between two dichotomous variables." *Journal of Experimental Psychology: General*, 112: 117-135.

Ashby, E. 1981 "Introductory remarks." *Proceedings of the Royal Society of London.*, v.376, no.1764.

Ashby, E. 1982 "Opening address." *Seminar on Risk, Cost and Pollution at Oxford*. London: Harwell.

Atkinson, J.M. 1968 "On the sociology of suicide." *Sociological Review*, 16: 83-92.

Atkinson, J.W. 1957 "Motivational determinants of risk-taking behavior." *Psychological Review*, 64: 359-372.

Bacharach, M. 1972 "Scientific disagreement." Unpublished manuscript, Oxford.

Backer, E. 1978 *Cluster analysis by optimal decomposition of induced fuzzy sets*. Delft: Delft University Press.

Badia, P., Harsh, J., and Abbott, B. 1979 "Choosing between predictable and unpredictable shock conditions: Data and theory." *Psychological Bulletin*, 86: 1107-1131.

Baird, D. 1985 "Lehrer-Wagner consensual probabilities do not adequately summarize the available information." *Synthese*, 62: 47-62.

Baldwin, J.F. 1986 "Support logic programming." In A.I. Jones, et al. (eds.) *Fuzzy Sets: Theory and Applications.* Dordrecht: Reidel.

Bar-Hillel, M. 1973 "On the subjective probability of compound events." *Organizational Behavior and Human Performance*, 9: 396-406.

Bar-Hillel, M. 1980 "The base-rate fallacy in probability judgments." *Acta Psychologica*, 44: 211-233.

Bar-Hillel, M. 1983 "The base-rate fallacy controversy." In R.W. Scholz (ed.) *Decision Making Under Uncertainty.* Amsterdam: North-Holland.

Bar-Hillel, M. 1984 "Representativeness and fallacies of probability judgment." *Acta Psychologica*, 55: 91-107.

Barnes, B. 1977 *Interests and the Growth of Knowledge.* London: Routledge and Kegan Paul.

Barnett, V. 1973 *Comparative Statistical Inference.* New York: Wiley.

Barnlund, D. 1975 "Communication styles in two cultures: Japan and the United States." In A. Kendon, R.M. Harris, and M.R. Key (eds.) *Organization of Behavior in Face-to-Face Interaction.* The Hague: Mouton.

Barr, M. 1986 "Fuzzy set theory and topos theory." *Canadian Mathematical Bulletin*, 29: 501-508.

Barron, F. 1953 "Complexity-simplicity as a personality dimension." *Journal of Abnormal and Social Psychology*, 68: 163-172.

Barron, F.H. 1987 "Influence of mising attributes on selecting a best multiattributed alternative." *Decision Sciences*, : 194-214.

Bartlett, M.S. 1962 *Essays on Probability and Statistics*. London: Methuen.

Bass, B.M., Cascio, W.F., and O'Connor, E.J. 1974 "Magnitude estimation of expression of frequency and amount." *Journal of Applied Psychology*, 59: 313-320.

Baum, W.M. 1974 "On two types of deviation from the matching law: Bias and undermatching." *Journal of the Experimental Analysis of Behavior*, 22: 231-244.

Beach, B.H. 1975 "Expert judgment about uncertainty: Bayesian decision making in realistic settings." *Organizational Behavior and Human Performance*, 14: 10-59.

Beach, L.R. and Mitchell, T.R. 1978 "A contingency model for the selection of decision strategies." *Academy of Management Review*, 3: 439-449.

Behn, R.D. and Vaupel, J.W. 1982 *Quick Analysis for Busy Decision Makers*. New York: Basic Books.

Bellman, R. and Giertz, M. 1973 "On the analytic formalism of fuzzy sets." *Information Sciences*, 5: 149-156.

Benenson, F.C. 1984 *Probability, Objectivity, and Evidence*. London: Routledge and Kegan Paul.

Berenstein, C., Kanal, L.N., and Lavine, D. 1986 "Consensus rules." In L.N. Kanal and J.F. Lemmer (eds.) *Uncertainty in Artificial Intelligence*. Amsterdam: North Holland.

Berger, P.L. and Luckmann, T. 1967 *The Social Construction of Reality*. New York: Doubleday.

Berkeley, D. and Humphreys, P. 1982 "Structuring decision problems and the 'bias heuristic'." *Acta Psychologica*, 50: 201-252.

Beyth-Marom, R. 1981 *The Subjective Probability of Conjunctions* (rep. no. 81-12). Eugene, OR: Decision Research.

Beyth-Marom, R. 1982 "How probable is probable? A numerical translation of verbal probability expressions." *Journal of Forecasting*, 1: 257-269.

Beyth-Marom, R., Dekel, S., Gombo, R., and Shaked, M. 1985 *An Elementary Approach to Thinking Under Uncertainty*. Hillsdale, N.J.: Erlbaum.

Bezdek, J.C. 1974 "Numerical taxonomy with fuzzy sets." *Journal of Mathematical Biology*, 1: 57-71.

Bieri, J. and Blackner, E. 1967 "The generality of cognitive complexity in the perception of people and inkblots." In D.N. Jackson and S. Messick (eds.) *Problems in Human Assessment*. New York: McGraw-Hill.

Billig, M. 1982 *Ideology and Social Psychology*. Oxford: Basil Blackwell.

Birkhofer, A. 1980 "The German risk study for nuclear power plants." *International Atomic Energy Agency Bulletin*, 22.

Birnbaum, M.H. 1983 "Base-rates in Bayesian inference: Signal-detection analysis of the cab problem." *American Journal of Psychology*, 96: 85-94.

Birnbaum, M.H. and Mellers, B.A. 1983 "Bayesian inference: Combining base rates with opinions of sources who vary in credibility." *Journal of Personality and Social Psychology*, 45: 792-804.

Black, M. 1937 "Vagueness: An exercise in logical analysis." *Philosophy of Science*, 4: 427-455.

Black, M. 1967 "Probability." *Encyclopedia of Philosophy*. New York: MacMillan and Free Press.

Blockley, D. 1980 *The Nature of Structural Design and Safety*. Chichester: Ellis Horwood.

Blockley, D., Pilsworth, B.W., and Baldwin, J.F. 1983 "Measures of uncertainty." *Civil Engineering Systems*, 1: 3-9.

Bloor, D. 1976 *Knowledge and Social Imagery.* London: Routledge and Kegan Paul.

Blumberg, H.H. 1972 "Communication of interpersonal evaluations." *Journal of Personality and Social Psychology*, 23: 157-162.

Blumer, H. 1969 *Symbolic Interactionism: Perspective and Method.* Englewood Cliffs, N.J.: Prentice-Hall.

Bok, S. 1978 *Lying: Moral Choice in Public and Private Life.* New York: Pantheon.

Bonissone, P.P. and Decker, K.S. 1986 "Selecting uncertainty calculi and granularity: An experiment in trading off precision and complexity." In L.N. Kanal and J.F. Lemmer (eds.) *Uncertainty in Artificial Intelligence.* Amsterdam: North-Holland.

Bookstein, A. 1984 "Relevance of fuzzy sets theory in information retrieval." *Human Systems Management*, 4: 323-324.

Bordley, R.F. and Wolff, R.W. 1981 "On the aggregation of individual probability estimates." *Management Science*, 27: 959-964.

Borgida, E. and Brekke, N. 1981 "The base-rate fallacy in attribution and prediction." In J.H. Harvey, W.J. Ickes, and R.F. Kidd (eds.) *New Directions in Attribution Research, Vol.3.* Hillsdale, N.J.: Erlbaum.

Boruch, RF. and Cecil, J.S. 1982 "Statistical strategies for preserving privacy in direct inquiry." In J.E. Sieber (ed.) *The Ethics of Social Research: Surveys and Experiments.* New York: Springer Verlag.

Bourgeois, L.J. III 1978 "The environmental perceptions of strategy makers and their economic correlates." *Working*

Paper No. 273. Graduate School of Business, University of Pittsburgh.

Bourgeois, L.J. III (1980) "Strategy and environment: A conceptual integration." *Academy of Management Review*, 5: 25-39.

Bourgeois, L.J. III 1985 "Strategic goals, perceived uncertainty, and economic performance in volatile environments." *Academy of Management Journal*, 28: 548-573.

Bowman, E.H. 1980 "A risk/return paradox for strategic management." *Sloan Management Review*, 23: 17-31.

Bowyer, J.B. 1982 *Cheating*. New York: St. Martin's Press.

Braine, M.D.S. 1978 "On the relation between the natural logic of reasoning and standard logic." *Psychological Review*, 85: 1-21.

Braroe, N. 1970 "Reciprocal exploitation in an Indian-White community." In G.P. Stone and H.A. Farberman (eds.) *Social Psychology through Symbolic Interaction*. Waltham, Mass.: Xerox College Press.

Brehm, S.S. and Brehm, J.W. 1981 *Psychological Reactance: A Theory of Freedom and Control*. New York: Academic Press.

British Psychological Society 1986 "Report of the Working Group on the use of the polygraph in criminal investigation and personnel screening." *Bulletin of the British Psychological Society*, 39: 81-94.

Brown, P. and Levinson, S. 1978 "Universals in language use: Politeness phenomena." In E.N. Goody (ed.) *Questions and Politeness*. London: Cambridge University Press.

Bryant, G.D. and Norman, G.R. 1980 "Expressions of probability: Words and numbers." (Letter) *New England Journal of Medicine*, 302: 411.

Brzustowski, T.A. 1982 "Risk assessment in largbe chemical energy projects." In N.C. Lind (ed.) *Technological Risk*. Waterloo, Ontario: University of Waterloo Press.

Buchanan, B.G. and Shortliffe, E.H. (eds.) 1984 *Rule-Based Expert Systems: The MYCIN Experiments of the Stanford Heuristic Programming Project*. Reading, Mass.: Addison-Wesley.

Budescu, D.V. and Wallsten, T.S. 1985 "Consistency in interpretation of probabilistic phrases." *Organizational Behavior and Human Decision Processes*, 36: 391-405.

Budner, S. 1962 "Intolerance of ambiguity as a personality variable." *Journal of Personality*, 30: 29-50.

Burgess, D., Kempton, W., and MacLaury, R.E. 1983 "Tarahumara color modifiers: Category structure presaging evolutionary change." *American Ethnologist*, 10: 133-149.

Burns, T. and Stalker, G.M. 1961 *The Management of Innovation*. London: Tavistock.

Cacioppo, J.T. and Petty, R.E. 1982 "The need for cognition." *Journal of Personality and Social Psychology*, 42: 116-131.

Caltabiano, M. and Smithson, M. 1983 "Variables affecting the perception of self-disclosure appropriateness." *Journal of Social Psychology*, 120: 119-128.

Cameron, K.S., Kim, M.U., and Whetton, D.A. 1987 "Organizational effects of decline and turbulence." *Administrative Science Quarterly*, 32: 222-240.

Campbell, B. 1983 "Uncertainty in science and policy: The case of biologists as experts in the Mackenzie Valley pipeline inquiry." *Project on Expertise, Legal Processes, and Public Policy*. Department of History, University of Lancaster.

Campbell, D.T. 1975 "On the conflicts between biological and social evolution and between psychology and moral tradition." *American Psychologist*, 30: 1103-1126.

Carnap, R. 1950 *Logical Foundations of Probability*. Chicago: University of Chicago Press.

Carrega, J.C. 1983 "The categories set-H and Fuz-H." *Fuzzy Sets and Systems*, 9: 327-332.

Carroll, J.S. and Siegler, R.S. 1977 "Strategies for the use of base-rate information." *Organizational Behavior and Human Performance*, 19: 392-402.

Carter, L.J. 1979 "Dispute over cancer risk quantification." *Science*, 203: 1324-1325.

Cascells, W., Schoenberger, A., and Grayboys, T. 1978 "Interpretation by physicians of clinical laboratory results." *New England Journal of Medicine*, 299: 999-1000.

Chaikin, A.L. and Derlega, V.J. 1974 "Variables affecting the appropriateness of self-disclosure." *Journal of Consulting and Clinical Psychology*, 42: 588-593.

Chandler, A. 1962 *Strategy and Structure*. Cambridge, Mass.: MIT Press.

Cheeseman, P. 1985 "In defense of probability." *Proceedings of the Ninth International Joint Conference on Artificial Intelligence*. Los Angeles: 1002-1009.

Cheeseman, P. 1986 "Probabilistic versus fuzzy reasoning." In L.N. Kanal and J.F. Lemmer (eds.) *Uncertainty in Artificial Intelligence*. Amsterdam: North-Holland.

Chew, S. and Waller, W. 1986 *Journal of Mathematical Psychology*, 30: 55.

Chicken, J.C. 1975 *Hazard and Control Policy in Britain*. Oxford: Pergamon Press.

Chicken, J.C. 1978 "Safety requirements for remote and continuous inspection." Paper presented to the BNES Conference on Radiation Protection in Nuclear Power Plants and the Fuel Cycle, London.

Christensen-Szalanski, J.J.J. and Beach, L.R. 1984 "The citation bias: Fad and fashion in the judgment and decision literature." *American Psychologist*, 39: 75-78.

Christensen-Szalanski, J.J.J. and Bushyhead, J.B. 1981 "Physicians' use of probabilistic information in a real clinical setting." *Journal of Experimental Psychology: Human Perception and Performance*, 7: 928-935.

Chubin, D.E. 1976 "The conceptualization of scientific specialties." *The Sociological Quarterly*, 17: 448-476.

Cialdini, R.B. 1984 *Influence: How and Why People Agree to Things*. New York: Quill.

Clark, W.C. 1980 "Witches, floods, and wonder drugs: Historical perspectives on risk management." In R.C. Schwing and W.A. Albers (eds.) *Societal Risk Assessment: How Safe Is Safe Enough?* New York: Plenum.

Cohen, J. 1964 *Behaviour in Uncertainty*. London: Allen and Unwin.

Cohen, J., Chesnick, E.I., and Haran, D. 1972 "A confirmation of the inertial-ψ effect in sequential choice and decision." *British Journal of Psychology*, 63: 41-46.

Cohen, J., Dearnaley, E.J., and Hansel, C.E.M. 1956 "The addition of subjective probabilities: The summation of estimates of success and failure." *Acta Psychologica*, 12: 371-380.

Cohen, J., Dearnaley, E.J., and Hansel, C.E.M. 1958a "Skill and chance: Variations in estimates of skill with an increasing element of chance." *British Journal of Psychology*, 49: 319-323.

Cohen, J., Dearnaley, E.J., and Hansel, C.E.M. 1958b "The mutual effect of two uncertainties." *Durham Research Review*, 2: 215-222.

Cohen, J., Dearnaley, E.J., and Hansel, C.E.M. 1958c "A quantitative study of meaning." *British Journal of Educational Psychology*, 28: 141-148.

Cohen, J. and Hansel, C.E.M. 1959 "Preferences for different combinations of chance and skill in gambling." *Nature (London)*, 183: 841-842.

Cohen, J. and Hickey, T. 1979 "Two algorithms for determining volumes of convex polyhedra." *Journal of the Association for Computing Machinery*, 26: 401-414.

Cohen, L.J. 1977 *The Probable and the Provable*. Oxford: Clarendon Press.

Cohen, L.J. 1979 "On the psychology of prediction: Whose is the fallacy?" *Cognition*, 7: 385-407.

Cohen, L.J. 1981 "Can human irrationality be experimentally demonstrated?" *The Behavioral and Brain Sciences*, 4: 317-331.

Cohen, L.J. and Christie, I. 1970 *Information and Choice*. Edinburgh:

Cohen, M. and Jaffray, J.-Y. 1988 "Is Savage's independence axiom a universal rationality principle?" *Behavioral Science*, 33: 38-47.

Cohen, M., Jaffray, J.-Y., and Said, T. 1987 "Experimental comparison of individual behavior under risk and under uncertainty for gains and for losses." *Organizational Behavior and Human Decision Processes*, 39: 1-22.

Cohen, M.D., March, J.G., and Olsen, J.P. 1972 "A garbage-can model of organizational choice." *Administrative Science Quarterly*, 17: 1-25.

Cohen, M.S. 1986 "An expert system framework for non-monotonic reasoning about probabilistic assumptions." In L.N. Kanal and J.F. Lemmer (eds.) *Uncertainty in Artificial Intelligence*. Amsterdam: North-Holland.

Cohen, P.R. 1985 *Heuristic Reasoning about Uncertainty: An Artificial Intelligence Approach*. London: Pitman.

Cohen, P.R. and Grinberg, M. 1983 "A theory of heuristic reasoning about uncertainty." *AI Magazine*, 4: 17-24.

Cohen, S. 1985 *Visions of Social Control*. Cambridge: Polity Press.

Collingridge, D. 1980 *The Social Control of Technology*. Milton Keynes: Open University Press.

Collins, H.M. 1974 "The TEA set: Tacit knowledge and scientific networks." *Science Studies*, 4: 165-185.

Consultancy, J.C. 1986 *Risk Assessment for Hazardous Installations*. Oxford: Pergamon Press.

Cooley, C.H. 1922 *Human Nature and the Social Order*. New York: Scribner's.

Coombs, C.H. and Pruitt, D.G. 1960 "Components of risk in decision-making: Probability and variance preferences." *Journal of Experimental Psychology*, 60: 265-277.

Cooper, W.S. 1987 "Decision theory as a branch of evolutionary theory: A biological derivation of the Savage axioms." *Psychological Review*, 94: 395-411.

Cornfield, J. 1951 "A method of estimating comparative rates from clinical data." *Journal of the National Cancer Institute*, 11: 1269-1275.

Covello, V.T. 1984 "Actual and perceived risks: A review of the literature." In P.F. Ricci, et al. (eds.) *Technological Risk Assessment*. The Hague: Martinus Nijhoff.

Cox, R.T. 1946 "Probability, frequency, and reasonable expectation." *American Journal of Physics*, 14: 1-13.

Crocker, J. 1981 "Judgments of covariation by social perceivers." *Psychological Bulletin*, 90: 272-292.

Curley, S.P., Yates, J.F., and Abrams, R.A. 1986 "Psychological sources of ambiguity avoidance." *Organizational Behavior and Human Decision Processes,* 38: 230-256.

Cyert, R.M. and March, J.G. 1963 *A Behavioral Theory of the Firm*. Englewood Cliffs, N.J.: Prentice-Hall.

Dalkey, N.C. 1972 "An impossibility theorem for group probability functions." P-4862, Santa Monica, California: Rand Corporation.

D'Amato, M.R. and Safarjan, W.R. 1979 "Preference for information about shock duration." *Animal Learning Behavior*, 7: 89-94.

Davis, J.D. 1977 "Effects of communication about interpersonal process on the evolution of self-disclosure in dyads." *Journal of Personality and Social Psychology*, 35: 31-37.

Davis, L.N. 1979 *Frozen Fire*. San Francisco: Friends of the Earth.

Dawid, A.P. 1982a "Intersubjective statistical models." In G. Koch and F. Spizzichino (eds.) *Exchangeability in Probability and Statistics*. Amsterdam: North Holland.

Dawid, A.P. 1982b "The well-calibrated Bayesian (with discussion)." *Journal of the American Statistical Association*, 77: 605-613.

De Gre, G. 1955 *Science as a Social Institution*. New York: Random House.

DeGroot, M.H. 1974 "Reaching a consensus." *Journal of the American Statistical Association*, 69: 118-121.

De Luca, A. and Termini, S. 1972 "A definition of nonprobabilistic entropy in the setting of fuzzy sets theory." *Information and Control*, 20: 301-312.

Dempster, A.P. 1967 "Upper and lower probabilities induced by a multivalued mapping." *Annals of Mathematical Statistics*, 38: 325-339.

Derber, C. 1979 *The Pursuit of Attention*. Oxford: Oxford University Press.

Derlega, V.J. and Chaikin, A.L. 1975 *Sharing Intimacy: What We Reveal to Othes and Why*. Englewood Cliffs, N.J.: Prentice-Hall.

Derlega, V.J., Wilson, M., and Chaikin, A.L. 1976 "Friendship and disclosure reciprocity." *Journal of Personality and Social Psychology*, 34: 578-582.

Dixon, K. 1980 *The Sociology of Belief: Fallacy and Foundation*. London: Routledge and Kegan Paul.

Douglas, J.D. 1967 *The Social Meanings of Suicide*. Princeton: Princeton University Press.

Douglas, M. 1966 *Purity and Danger*. London: Routledge and Kegan Paul.

Douglas, M. 1973 *Natural Symbols*. London: Barrie and Jenkins.

Douglas, M. 1985 *Risk Acceptability According to the Social Sciences*. London: Routledge and Kegan Paul.

Downey, H.K., Hellreigel, D., and Slocum, J.W. Jr. 1977 "Individual characteristics as sources of perceived uncertainty variation." *Human Relations*, 30: 161-174.

Downey, H.K. and Slocum, J.W. Jr. 1975 "Uncertainty: Measures, research and sources of variation." *Academy of Management Journal*, 18: 562-578.

Downs, A. 1966 *Inside Bureaucracy*. Boston: Little, Brown and Co.

Dubois, D. and Prade, H. 1980 *Fuzzy Sets and Systems: Theory and Applications*. New York: Academic Press.

Dubois, D. and Prade, H. 1983 "Twofold fuzzy sets: An approach to the representation of sets with fuzzy boundaries based on possibility and necessity measures." *Journal of Fuzzy Mathematics (China)*, 3: 53-76.

Dubois, D. and Prade, H. 1985a "A review of fuzzy set aggregation connectives." *Information Sciences*, 36: 85-121.

Dubois, D. and Prade, H. 1985b "A note on measures of specificity for fuzzy sets." *International Journal of General Systems*, 10: 279-283.

Dubois, D. and Prade, H. 1986 "Recent models of uncertainty and imprecision as a basis for decision theory: Towards less normative frameworks." In G. Mancini and D. Woods (eds.) *Intelligent Decision-Aids in Process Environments*. Berlin: Springer-Verlag.

Dubois, D. and Prade, H. 1987a "Twofold fuzzy sets and rough sets: Some issues in knowledge representation." *Fuzzy Sets and Systems*, 23: 3-18.

Dubois, D. and Prade, H. 1987b "Properties of measures of information in evidence and possibility theories." *Fuzzy Sets and Systems*, 24: 161-182.

Duda, R.O., Hart, P.E., and Nilsson, N.J. "Subjective Bayesian methods for rule-based inference systems." *Proceedings of the 1976 National Computer Conference, AFIPS*, 45: 1075-1082.

Duhem, P. 1962 *The Aim and Structure of Physical Theory*. New York: Atheneum.

Duncan, R.B. 1972 "Characteristics of organizational environments and perceived environmental uncertainty." *Administrative Science Quarterly*, 17: 313-327.

Dunn, J.C. 1976 "Indices of partition fuzziness and the detection of clusters in large data sets." In M.M. Gupta (ed.) *Fuzzy Automata and Decision Processes*. New York: American Elsevier.

Durbin, J. 1985 "Evolutionary origins of statisticians and statistics." In A.C. Atkinson and S.E. Fienberg (eds.) *A Celebration of Statistics: The ISI Centenary Volume*. New York: Springer-Verlag.

Durbin, J. 1988 "Is a philosophical consensus for statistics attainable?" *Journal of Econometrics*, 37: 51-61.

Dutta, A. 1985 "Reasoning with imprecise knowledge in expert systems." *Information Sciences*, 37: 3-24.

Dweck, C.S. and Elliott, E.S. 1983 "Achivement motivation." In P. Mussen and E.M. Hetherington (eds.) *Carmichael's Manual of Child Psychology: Social and Personality Development*. New York: Wiley.

Edwards, W. 1954 "Variance preferences in gambling." *American Journal of Psychology*, 67: 441-452.

Edwards, W. 1961 "Behavioral decision theory." *Annual Review of Psychology*. Palo Alto: Annual Reviews, Inc.

Edwards, W. 1962a "Subjective probabilities inferred from decisions." *Psychological Review*, 69: 109-135.

Edwards, W. 1962b "Utility, subjective probability, their interaction, and variance preferences." *Conflict Resolution*, VI: 42-51.

Edwards, W. 1968 "Conservatism in human information processing." In B. Kleinmutz (ed.) *Formal Representation of Human Judgment*. New York: Wiley.

Edwards, W. 1972 "N=1: Diagnosis in unique cases." In J.A. Jacquez (ed.) *Computer Diagnosis and Diagnostic Methods*. Springfield, Ill: Thomas.

Edwards, W. 1983 "Human cognitive capabilities, representativeness, and ground rules for research." In P. Humphreys, O. Svenson, and A. Vari (eds.) *Analysing and Aiding Decision Processes*. Amsterdam: North-Holland.

Edwards, W. 1984 "How to make good decisions." *Acta Psychologica*, 56: 7-10.

Edwards, W. and von Winterfeldt, D. 1986 "On cognitive illusions and their implications." In H.R. Arkes and K.R. Hammond (eds.) *Judgment and Decision Making: An Interdisciplinary Reader*. New York: Cambridge University Press.

Eggleston, R. 1978 *Evidence, Proof, and Probability*. London: Weidenfeld and Nicolson.

Einhorn, H.J. and Hogarth, R.M. 1981 "Behavioral decision theory: Processes of judgment and choice." *Annual Review of Psychology*, 32: 53-88.

Einhorn, H.J. and Hogarth, R.M. 1985 "Ambiguity and uncertainty in probabilistic inference." *Psychological Review*, 92: 433-461.

Einhorn, H.J. and Hogarth, R.M. 1986 "Judging probable cause." *Psychological Bulletin*, 99: 3-19.

Elig, T. and Frieze, I. 1979 "Measuring causal attributions for success and failure." *Journal of Personality and Social Psychology*, 37: 621-634.

Ellsberg, D. 1961 "Risk, ambiguity and the Savage axioms." *Quarterly Journal of Economics*, 75: 643-669.

Emery, F.E. and Trist, E.L. 1965 "The causal texture of organizational environments." *Human Relations*, 18: 21-32.

Eytan, M. 1981 "Fuzzy sets: A topos-logical point of view." *Fuzzy Sets and Systems*, 5: 47-67.

Feller, W. 1968 *An Introduction to Probability Theory and its Applications* (3rd Edition, Vol. 1). New York: Wiley.

Fellner, W. 1961 "Distortion of subjective probabilities as a reaction to uncertainty." *Quarterly Journal of Economics*, 75: 670-692.

Feyerabend, P. 1975 *Against Method*. London: NLB.

Fiegenbaum, A. and Thomas, H. 1988 "Attitudes toward risk and the risk-return paradox: Prospect theory explanations." *Academy of Management Journal*, 31: 85-106.

Filmer, P. 1972 "On Harold Garfinkel's ethnomethodology." In P. Filmer, et al. (eds.) *New Directions in Sociological Theory*. Cambridge, Mass.: MIT Press.

Fine, T.L. 1973 *Theories of Probability*. New York: Academic Press.

Finetti, B. de 1974 *Theory of Probability, V.1*. Trans. A. Machi and A. Smith. New York: Wiley.

Fischhoff, B. 1980 "For those condemned to study the past: Reflections on historical judgment." In R.A. Shweder and D.W. Fiske (eds.) *New Directions for the Methodology of Social and Behavioral Science: No.4, Fallible Judgment in Behavioral Research*. San Francisco: Jossey-Bass.

Fischhoff, B. and Bar-Hillel, M. 1984 "Diagnosticity and the base-rate effect." *Memory and Cognition*, 12: 402-410.

Fischhoff, B. and Beyth-Marom, R. 1983 "Hypothesis evaluation from a Bayesian perspective." *Psychological Review*, 90: 239-260.

Fischhoff, B. and MacGregor, D. 1980 *Judged Lethality* (Report no. 80-4). Eugene, OR: Decision Research.

Fischhoff, B. and Slovic, P. 1980 "A little learning...: Confidence in multi-cue judgment." In R. Nickerson (ed.) *Attention and Performance VIII*. Hillsdale, N.J.: Erlbaum.

Fischhoff, B. Slovic, P., and Lichtenstein, S. 1977 "Knowing with certainty: The appropriateness of extreme confidence." *Journal of Experimental Psychology: Human Perception and Performance*, 3: 552-564.

Fishburn, P.C. 1984 "SSB utility theory and decision-making under uncertainty." *Mathematical Social Sciences*, 8: 253-285.

Fisz, M. 1963 *Probability Theory and Mathematical Statistics* (Third Edition). New York: Wiley.

Foddy, W.H. and Finighan, W.R. 1980 "The concept of privacy from a symbolic interactionist perspective." *Journal for the Theory of Social Behavior*, 10: 1-10.

Forrester, J. W. 1968 *Principles of Systems*. Cambridge, Mass.: Wright-Allen.

Fox, J., Barber, D.C., and Bardhan, K.D. 1980 "Alternative to Bayes? A quantitative comparison with rule-based diagnostic inference." *Methods of Information in Medicine*, 19: 210-215.

Fox, R. 1980 "The evolution of medical uncertainty." *Health and Society*, 58: 49.

Frankel, E. 1976 "Corpuscular optics and the wave theory of light: The science and politics of a revolution in physics." *Social Studies of Science*, 6: 141-184.

French, S. 1985 "Group consensus probability distributions: A critical survey." In J.M. Bernardo, et al. (eds.) *Bayesian Statistics 2*. Amsterdam: North Holland.

Fromm, E. 1947 *Man for Himself*. New York: Rhinehart.

Fung, R.M. and Chong, C.Y. 1986 "Metaprobability and Dempster-Shafer in evidential reasoning." In L.N. Kanal and J.F. Lemmer (eds.) *Uncertainty in Artificial Intelligence*. Amsterdam: North-Holland.

Furby, L. 1973 "Interpreting regression toward the mean in developmental research." *Developmental Psychology*, 8: 172-179.

Gaines, B.R. 1975 "Stochastic and fuzzy logics." *Electronic Letters*, 11: 188-189.

Gaines, B.R. 1976 "Foundations of fuzzy reasoning." *International Journal of Man-Machine Studies*, 8: 623-688.

Gardenfors, P. and Sahlins, N.-E. 1982 "Unreliable probabilities, risk taking, and decision making." *Synthese*, 53: 361-386.

Garfinkel, H. 1967 *Studies in Ethnomethodology*. Englewood Cliffs, N.J.: Prentice-Hall.

Genest, C. and Schervish, M.J. 1985 "Modeling expert judgments for Bayesian updating." *Annals of Statistics*, 13: 1198-1212.

Genest, C. and Zidek, J.V. 1986 "Combining probability distributions: A critique and an annotated bibliography." *Statistical Science*, 1: 114-148.

Giles, R. 1976 "Lukasiewicz logic and fuzzy set theory." *International Journal of Man-Machine Studies*, 8: 313-327.

Giles, R. 1979 "A formal system for fuzzy reasoning." *Fuzzy Sets and Systems*, 2: 233-257.

Giles, R. 1982 "Semantics for fuzzy reasoning." *International Journal of Man-Machine Studies*, 17: 401-415.

Ginosar, Z. and Trope, Y. 1980 "The effects of base rates and individuating information on judgments about another person." *Journal of Experimental Social Psychology*, 16: 228-242.

Glasstone, S. and Jordan, W.H. 1980 *Nuclear Power and Its Environmental Effects*. Washington, D.C.: American Nuclear Society.

Goffman, E. 1959 *Presentation of Self in Everyday Life*. Garden City, N.Y.: Doubleday.

Goffman, E. 1963 Stigma: *Notes on the Management of Spoiled Identity*. Englewood Cliffs, N.J.: Prentice-Hall.

Goffman, E. 1974 *Frame Analysis*. Cambridge, Mass.: Harvard University Press.

Goguen, J.A. 1969 "The logic of inexact concepts." *Synthese*, 19: 325-373.

Goldsmith, R.W. 1978 "Assessing probabilities of compound events in a judicial context." *Scandinavian Journal of Psychology*, 19: 103-110.

Goldstein, W.M., and Einhorn, H.J. 1987 "Expression theory and the preference reversal phenomenon." *Psychological Review*, 94: 236-254.

Good, I.J. 1950 *Probability and the Weighing of Evidence*. New York: Hafner's.

Good, I.J. 1952 "Rational decisions." *Journal of the Royal Statistical Society, Series B*, 14: 107-114.

Good, I.J. 1960 "Weight of evidence, corroboration, explanatory power, information and the utility of experiments." *Journal of the Royal Statistical Society, Series B*, 22: 319-331.

Good, I.J. 1962 "Subjective probability as the measure of a nonmeasurable set." In E. Nagel, et al. (eds.) *Logic, Methodology, and Philosophy of Science*. Stanford: Stanford University Press.

Good, I.J. 1971a "46656 varieties of Bayesians." *American Statistician*, 25: 62-63.

Good, I.J. 1971b "The probabilistic explication of information, evidence, surprise, causality, explanation, and utility." In V.P. Godambe and D.A. Sprott (eds.) *Foundations of Statistical Inference*. Toronto: Holt, Rinehart, and Winston of Canada.

Good, I.J. 1976 "The Bayesian influence, or how to sweep subjectivism under the carpet." In C.A. Hooker and W. Harper (eds.) *Foundations of Probability Theory, Statistical Inference, and Statistical Theories of Science, V.2*. Dordrecht: Reidel.

Good, I.J. 1983 *Good Thinking: The Foundations of Probability and Its Applications*. Minneapolis: University of Minnesota Press.

Goodman, I.R. 1982 "Fuzzy sets as equivalence classes of random sets." In R.R. Yager (ed.) *Fuzzy Set and Possibility Theory: Recent Developments*. New York: Pergamon.

Goodman, I.R. and Nguyen, H.T. 1985 *Uncertainty Models for Knowledge-Based Systems*. Amsterdam: North-Holland.

Goody, E.N. (ed.) 1978 *Questions and Politeness*. London: Cambridge University Press.

Graham, I. 1987 "Fuzzy sets and toposes--- Towards higher order logic." *Fuzzy Sets and Systems*, 23: 19-32.

Graham, I. 1988 "A reply to Peter Johnstone's 'Open Letter'." *Fuzzy Sets and Systems*, 27: 250-251.

Greenberg, B.G., Abul-Ela, A.A., Simmons, W.R., and Horvitz, D.G. 1969 "The unrelated question randomized response model: Theoretical framework." *Journal of the American Statistical Association*, 64: 520-539.

Grether, D.M. and Plott, C.R. 1979 "Economic theory of choice and the preference reversal phenomenon. " *American Economic Review*, 69: 623-638.

Groner, R., Groner, U., and Bischof, W.F. 1983 "The role of heuristics in models of decision. In R.W. Scholz (ed.) *Decision Making Under Uncertainty*. Amsterdam: North-Holland.

Grosof, B.N. 1986 "Evidential confirmation as transformed probability." In L.N. Kanal and J.F. Lemmer (eds.) *Uncertainty in Artificial Intelligence*. Amsterdam: North-Holland.

Guiasu, S. 1977 *Information Theory and Applications*. New York: McGraw-Hill.

Gunnar, M. 1980 "Contingent stimulation: A review of its role in early development." In S. Levine and H. Ursin (eds.) *Coping and Health*. New York: Plenum.

Hacking, I. 1967 "Slightly more realistic personal probability." *Philosophy of Science*, 34: 311-325.

Hacking, I. 1975 *The Emergence of Probability*. Cambridge: Cambridge University Press.

Hahn, H. 1956 "The crisis in intuition." In J.R. Newman (ed.) *The World of Mathematics, V. 3*. New York: Simon and Schuster.

Hagen, O. 1979 "Towards a positive theory of preferences under risk." In M. Allais and O. Hagen (eds.) *Expected Utility Hypotheses and the Allais Paradox*. Holland: Reidel.

Hakel, M. 1968 "How often is often?" *American Psychologist*, 23: 533-534.

Hammerton, M. 1973 "A case of radical probability estimation." *Journal of Experimental Psychology*, 101: 252-254.

Hanson, N.R. 1965 *Patterns of Discovery*. Cambridge: Cambridge University Press.

Harre, R. 1970 "Probability and confirmation." In *Principles of Scientific Thinking*. Chicago: University of Chicago Press.

Hartley, R.V.L. 1928 "Transmission of information." *Bell System Technical Journal*, 7: 535-563.

Hebb, D.O. 1949 *The Organization of Behavior: A Neuro-Psychological Theory*. New York: Wiley.

Heckerman, D. 1986 "Probabilistic interpretation for MYCIN's certainty factors." In L.N. Kanal and J.F. Lemmer (eds.) *Uncertainty in Artificial Intelligence*. Amsterdam: North-Holland.

Heider, F. 1958 *The Psychology of Interpersonal Relations*. New York: Wiley.

Henslin, J.M. 1967 "Craps and magic." *American Journal of Sociology*, 73: 316-330.

Henry, S. 1978 *The Hidden Economy: The Context and Control of Borderline Crime*. Oxford: Martin Robertson.

Herrnstein, R.J. 1961 "Relative and absolute strength of response as a function of frequency of reinforcement." *Journal of the Experimental Analysis of Behavior*, 4: 267-272.

Hersh, H.M. and Caramazza, A.A. 1976 "A fuzzy set approach to modifiers and vagueness in natural language." *Journal of Experimental Psychology: General*, 105: 254-276.

Hershey, J. and Schoemaker, P. 1980 "Prospect theory's reflection hypothesis: A critical examination." *Organizational Behavior and Human Performance*, 25: 395-418.

Hickson, D.J., Hinings, C.R., Lee, C.A., Schneck, R.E., and Pennings, J.M. 1971 "A strategic contingencies theory of intraorganizational power." *Administrative Science Quarterly*, 16: 216-229.

Higashi, M. and Klir, G.J. 1982 "On measures of fuzziness and fuzzy complements." *International Journal of General Systems*, 8: 169-180.

Higashi, M. and Klir, G.J. 1983 "On the notion of distance representing information closeness: possibility and probability distributions." *International Journal of General Systems*, 9: 103-115.

Hisdal, E. 1986 "Infinite-valued logic based on two-valued logic and probability. Part 1.1. Difficulties with present-day fuzzy set theory and their resolution in the TEE model." *International Journal of Man-Machine Studies*, 25: 89-111.

Hoch, S.J. 1984 "Availability and interference in predictive judgment." *Journal of Experimental Psychology: Learning, Memory and Cognition*, 10: 649-662.

Hofstede, G. 1980 *Culture's Consequences: International Differences in Work-Related Values*. Beverly Hills, California: Sage Publications.

Hogarth, R.M. 1975 "Cognitive processes and the assessment of subjective probability distributions (with discussion)." *Journal of the American Statistical Association*, 70: 271-294.

Hogarth, R.M. 1980 *Judgment and Choice: The Psychology of Decision*. Chichester: Wiley.

Hogarth, R.M. 1981 "Beyond discrete biases: Functional and dysfunctional aspects of judgmental heuristics." *Psychological Bulletin*, 90: 197-217.

Hogarth, R.M. 1986 "Generalization in decision research: The role of formal models." *IEEE Transactions on Systems, Man, and Cybernetics*, 16: 439-448.

Hogarth, R.M. and Makridakis, S. 1981 "The value of decision making in a complex environment: An experimental approach." *Management Science*, 27: 92-107.

Hohle, U. 1982 "Entropy with respect to plausibility measures." *Proceedings of the 12th I.E.E.E. Symposium on Multiple Valued Logic*: 167-169.

Horvitz, E.J. and Heckerman, D.E. 1986 "The inconsistent use of measures of certainty in artificial intelligence research." In L.N. Kanal and J.F. Lemmer (eds.) *Uncertainty in Artificial Intelligence*. Amsterdam: North-Holland.

Horvitz, E.J., Heckerman, D.E., and Langlotz, C. 1986 "A framework for comparing alternative formalisms for plausible reasoning." Memo KSL-86-25, Stanford University School of Medicine.

Houston, D.B. 1964 "Risk, insurance, and sampling." *The Journal of Risk and Insurance*, 535-538.

Howell, W.C. 1972 "Compounding uncertainty from internal sources." *Journal of Experimental Psychology*, 95: 6-13.

Howell, W.C., and Burnett, S.A. 1978 "Uncertainty measurement: A cognitive taxonomy." *Organizational Behavior and Human Performance*, 22: 45-68.

Huber, O. 1983 "The information presented and actually processed in a decision task." In P. Humphreys, O. Svenson, and A. Vari (eds.) *Analyzing and Aiding Decision Processes*. Amsterdam: North-Holland.

Huff, D. 1959 *How to Take a Chance*. Harmondsworth, Middlesex: Penguin.

Imada, H. and Nageishi, Y. 1982 "The concept of uncertainty in animal experiments using aversive stimulation." *Psychological Bulletin*, 91: 573-588.

James, G. 1941 "Relevance, probability, and the law." *California Law Review* 29: 689-699.

Janis, I.L. 1972 *Victims of Groupthink*. Boston: Houghton Mifflin.

Janis, I.L. and Mann, L. 1977 *Decision Making: A Psychological Analysis of Conflict, Choice, and Commitment*. New York: Free Press.

Jauch, L.R. and Kraft, K.L. 1986 "Strategic management of uncertainty." *Academy of Management Journal*, 11: 777-970.

Jaynes, E. 1968 "Prior probability." *IEEE Transactions on Systems Science and Cybernetics*, SSC-4: 227-241.

Jaynes, E. 1979 "Where do we stand on maximum entropy?" In R.L. Levine and M. Tribus (eds.) *The Maximum Entropy Formalism*. Cambridge, Mass.: M.I.T. Press.

Jenkins, H.M. and Ward, W.C. 1965 "Judgment of contingency between responses and outcomes." *Psychological Monographs: General and Applied*, 79 (1, Whole no. 549).

Johnson, D.M. 1972 *A Systematic Introduction to the Psychology of Thinking*. New York: Harper and Row.

Johnson, R.W. 1986 "Independence and Bayesian updating methods." In L.N. Kanal and J.F. Lemmer (eds.) *Uncertainty in Artificial Intelligence*. Amsterdam: North-Holland.

Johnstone, P.T. 1988 "Open letter to Ian Graham." *Fuzzy Sets and Systems*, 27: 244-250.

Jourard, S. 1971 *The Transparent Self, Second Edition*. New York: Van Nostrand.

Jungermann, H. 1983 "The two camps on rationality." In R.W. Scholz, (ed.) *Decision Making Under Uncertainty*. Amsterdam: Elsevier.

Kadane, J.B., Dickey, J.M., Winkler, R.L., Smith, W.S., and Peters, S.C. 1980 "Interactive elicitation of opinion for a normal linear model." *Journal of the American Statistical Association*, 75: 845-854.

Kahneman, D. and Tversky, A. 1972 "Subjective probability: A judgment of representativeness." *Cognitive Psychology*, 3: 430-454.

Kahneman, D. and Tversky, A. 1973 "On the psychology of prediction." *Psychological Review*, 80: 237-251.

Kahneman, D. and Tversky, A. 1979 "Prospect theory: An analysis of decision under risk." *Econometrica*, 47: 263-291.

Kahneman, D. and Tversky, A. 1982a "On the study of statistical intuitions." *Cognition*, 11: 123-141.

Kahneman, D. and Tversky, A. 1982b "Variants of uncertainty." *Cognition*, 11: 143-157.

Kahneman, D. and Tversky, A. 1982c "The simulation heuristic." In D. Kahneman, P. Slovic, and A. Tversky (eds.) *Judgment Under Uncertainty: Heuristics and Biases*. New York: Cambridge University Press.

Kalbfleisch, J.D., Lawless, J.F., and MacKay, R.J. 1982 "The estimation of small probabilities and risk assessment." In N.C. Lind (ed.) *Technological Risk*. Waterloo, Ontario: University of Waterloo Press.

Kates, R.W. 1962 *Hazard and Choice Perception in Flood Plain Management* (Research Paper No. 78). Chicago: University of Chicago, Department of Geography.

Kaufmann, A. 1975 *Introduction to the Theory of Fuzzy Subsets, Vol. 1*. New York: Academic Press.

Kay, P. and McDaniel, C. 1975 *Color categories as fuzzy sets* (Working Paper No. 44). Berkeley: University of California, Language Behavior Research Laboratory.

Keeney, R.L., and Raiffa, H. 1976 *Decisions with Multiple Objectives: Preferences and Value Tradeoffs.* New York: John Wiley.

Keller, R.T., Slocum, J.W. Jr., and Susman, G.I. 1974 "Uncertainty and type of management system in continuous process organizations." *Academy of Management Journal,* 17: 56-68.

Kempton, W. 1978 "Category grading and taxonomic relations: A mug is a sort of a cup." *American Ethnologist,* 5: 44-65.

Kendall, D.G. 1974 "Foundations of a theory of random sets." In E.F. Harding and D.G. Kendall (eds.) *Stochastic Geometry.* New York: Wiley.

Keren, G. and Wagenaar, W.A. 1987 "Violation of utility theory in unique and repeated gambles." *Journal of Experimental Psychology: Learning, Memory and Cognition,* 13: 387-391.

Keynes, J. M. 1962 *A Treatise on Probability.* New York: Harper and Row.

Kickert, W.J.M. 1978 *Fuzzy Theories on Decision-Making.* Leiden: Martinus Nijhoff.

Kim, J. 1984 "PRU measures of association for contingency table analysis." *Sociological Methods and Research,* 13: 3-44.

Klaua, D. 1965 "Uber einen Ansatz zur mehrwertigen Mengenlehre." *Monatsber. Deut. Akad. Wiss. Berlin,* 7: 859-876.

Klayman, J. and Ha, Y.-W. 1987 "Confirmation, disconfirmation, and information in hypothesis testing." *Psychological Review,* 94: 211-228.

Kline, M. 1980 *Mathematics: The Loss of Certainty.* Oxford: Oxford University Press.

Klir, G.J. 1987 "Where do we stand on measures of uncertainty, ambiguity, fuzziness, and the like?" *Fuzzy Sets and Systems*, 24: 141-160.

Knorr-Cetina, K. 1981 *The Manufacture of Knowledge*. Oxford: Pergamon.

Kochen, M. 1979 "Enhancement of coping through blurring." *Fuzzy Sets and Systems*, 2: 37-52.

Kochen, M. 1984 "Possibility theory, management and artificial intelligence." *Human Systems Management*, 4: 306-308.

Koestler, A. 1964 *The Act of Creation*. New York: MacMillan.

Kohlberg, L. 1976 "Moral state and moralization." In T. Lickman (ed.) *Moral Development and Behavior*. New York: Holt, Rinehart, and Winston.

Koopman, B. 1940 "The bases of probability." *Bulletin of American Mathematical Society*, 46: 763-774.

Kopp, D.C. and Litschert, R.J. 1980 "A buffering response in light of variation in core technology, perceived environmental uncertainty, and size." *Academy of Management Journal*, 23: 252-260.

Kraft, C., Pratt, J., and Seidenberg, A. 1959 "Intuitive probability on finite sets." *Annals of Mathematical Statistics*, 30: 408-419.

Kramer, E.E. 1970 *The Nature and Growth of Modern Mathematics*. Princeton, N.J.: Princeton University Press.

Krantz, D., Luce, R.D., Suppes, P., and Tversky, A. 1971 *Foundations of Measurement*. New York: Academic Press.

Krantz, D.S. and Schulz, R. 1980 "A model of life crisis, control, and health outcomes: Cardiac rehabilitation and relocation of the elderly." In A. Baum and J.E. Singer (eds.) 1980 *Advances in Environmental Psychology, Vol. 2: Applications of Personal Control*. Hillsdale, N.J.: Erlbaum.

Kruse, R. and Meyer, K.D. 1987 *Statistics with Vague Data*. Dordrecht: Reidel.

Kuhn, T. 1962 *The Structure of Scientific Revolutions*. Chicago: University of Chicago Press.

Kwaakernak, H. 1978 "Fuzzy random variables 1: Definitions and theorems." *Information Sciences*, 15: 1-29.

Kyburg, H. 1961 *Probability and the Logic of Rational Belief*. Middletown: Wesleyan University Press.

Kyburg, H. 1985 "Bayesian and non-Bayesian evidential reasoning." *Computer Sciences Technical Report 139*. Rochester: University of Rochester.

Kyburg, H. 1987 "Representing knowledge and evidence for decision." In B. Bouchon and R.R. Yager (eds.) *Uncertainty in Knowledge-Based Systems: Lecture Notes in Computer Science No. 286*. New York: Springer Verlag.

Kyburg H., and Smokler, H.E. (eds.) 1964 *Studies in Subjective Probability*. New York: Wiley.

La Breque, M. 1980 "On making sounder judgments." *Psychology Today*, 6: 33-42.

Lakatos, I. 1970 "Falsification and the methodology of scientific research programmes." In I. Lakatos and A. Musgrave (eds.) *Criticism and the Growth of Knowledge*. Cambridge: Cambridge University Press.

Lakoff, G. 1973 "Hedges: A study in meaning criteria and the logic of fuzzy concepts." *Journal of Philosophic Logic*, 2: 458-508.

Langer, E.J. 1975 "The illusion of control." *Journal of Personality and Social Psychology*, 32: 311-328.

Langer, E.J. 1977 "The psychology of chance." *Journal for the Theory of Social Behaviour*, 7: 185-207.

Laver, M. 1980 *Computers and Social Change*. London: Cambridge University Press.

Lavine, T.Z. 1962 "Reflections on the genetic fallacy." *Social Research*, XXIX: 321-336.

Lawrence, P.R. and Lorsch, J.W. 1967 *Organization and Environment*. Homewood, Illinois: Irwin.

Lee, R.C.T. 1972 "Fuzzy logic and the resolution principle." *Journal of the Association of Computing Machines*, 19: 109-119.

Lehrer, K. and Wagner, C.G. 1981 *Rational Consensus in Science and Society*. Dordrecht: D. Reidel.

Lemmer, J.F. 1986 "Confidence factors, empiricism, and the Dempster-Shafer theory of evidence." In L.N. Kanal and J.F. Lemmer (eds.) *Uncertainty in Artificial Intelligence*. Amsterdam: North-Holland.

Lev, B. 1975 "Environmental uncertainty reduction by smoothing and buffering: An empirical verification." *Academy of Management Journal*, 18: 864-871.

Levi, I. 1977 "Four types of ignorance." *Social Research*, 44: 745-756.

Levi, I. 1980 *The Enterprise of Knowledge: An Essay on Knowledge, Credal Probability, and Chance*. Cambridge, Mass.: M.I.T. Press.

Levi, I. 1984 *Decisions and Revisions*. Cambridge: Cambridge University Press.

Levi-Strauss, C. 1977 *Tristes Tropiques*. Harmondsworth: Penguin.

Levin, I.P. Johnson, R.D., Deldin, P.J., Carstens, L.M., Cressey, L.J., and Davis, C.R. 1986 "Framing effects in decisions with completely and incompletely described alternatives." *Organizational Behavior and Human Decision Processes*, 38: 48-64.

Levin, I.P., Johnson, R.D., Russo, C.P., and Deldin, P.J. 1985 "Framing effects in judgment tasks with varying amounts of information." *Organizational Behavior and Human Decision Processes*, 36: 362-377.

Lichtenstein, S. and Fischhoff, B. 1980 "Training for calibration." *Organizational Behavior and Human Performance*, 26: 149-171.

Lichtenstein, S., Fischhoff, B., and Phillips, L.D. 1982 "Calibration of probabilities: The state of the art to 1980." In D. Kahneman, P. Slovic, and A. Tversky (eds.) *Judgment Under Uncertainty: Heuristics and Biases*. New York: Cambridge University Press.

Lichtenstein, S. and Newman, J.R. 1967 "Empirical scaling of common verbal phrases associated with numerical probabilities." *Psychonomic Sciences*, 9: 563-564.

Lichtenstein, S. and Slovic, P. 1971 "Reversal of preference between bids and choices in gambling decisions." *Journal of Experimental Psychology*, 89: 46-55.

Lichtenstein, S. and Slovic, P. 1973 "Response-induced reversal of preference in gambling: An extended replication in Las Vegas." *Journal of Experimental Psychology*, 101: 16-20.

Lindesmith, A.R., Strauss, A., and Denzin, N. 1975 *Social Psychology: Fourth Edition*. Illinois: Drysdale.

Lindley, D.V. 1965 *Introduction to Probability and Statistics from a Bayesian Viewpoint, Part I*. Cambridge: Cambridge University Press.

Lindley, D.V. 1971 "The estimation of many parameters." In V.P. Godambe and D.A. Sprott (eds.) *Foundations of Statistical Inference*. Toronto: Holt, Rinehart, and Winston.

Lindley, D.V. 1982 "Scoring rules and the inevitability of probability (with discussion)." *International Statistical Review*, 50: 1-26.

Lindley, D.V. 1985 "Reconciliation of discrete probability distributions." In J.M. Bernardo, et al. (eds.) *Bayesian Statistics 2*. Amsterdam: North Holland.

Linnerooth, J. 1984 "The political processing of uncertainty." *Acta Psychologica*, 56: 219-231.

Locke, E.A., Shaw, K.N., Saari, L.M., and Latham, G.P. 1981 "Goal setting and task performance, 1969-1980." *Psychological Bulletin*, 90: 125-152.

Loo, S.G. 1977 "Measures of fuzziness." *Cybernetica*, 20: 201-207.

Lopes, L.L. 1981 "Decision making in the short run." *Journal of Experimental Psychology: Human Learning and Memory*, 7: 377-385.

Lopes, L.L. 1982 "Doing the impossible: A note on induction and the experience of randomness." *Journal of Experimental Psychology*, 8: 626-637.

Lord, C., Ross, L., and Lepper, M. 1979 "Biased assimilation and attitude polarization: The effect of prior theories on subsequently considered evidence." *Journal of Personality and Social Psychology*, 37: 2098-2109.

Lorenzi, P., Sims, H.P. Jr., and Slocum, J.W. Jr. 1981 "Perceived environmental uncertainty: An individual or environmental attribute?" *Journal of Management*, 7: 27-41.

Loui, R.P. 1986 "Interval-based decisions for reasoning systems." In L.N. Kanal and J.F. Lemmer (eds.) *Uncertainty in Artificial Intelligence*. Amsterdam: North-Holland.

Luce, R.D. 1967 "Sufficient conditions for the existence of a finitely additive probability measure." *Annals of Mathematical Statistics*, 38: 780-786.

Lynn, S.J. 1978 "Three theories of self-disclosure." *Journal of Experimental Social Psychology*, 14: 466-479.

Lyon, D. and Slovic, P. 1976 "Dominance of accuracy information and neglect of base-rates in probability estimation." *Acta Psychologica*, 40: 287-298.

MacCrimmon, K.R. and Larsson, S. 1979 "Utility theory: Axioms versus 'paradoxes'." In M. Allais and O. Hagen (eds.) *Expected Utility Hypotheses and the Allais Paradox*. Holland: Reidel.

Macdonald, R.R. 1986 "Credible conceptions and implausible probabilities." *British Journal of Mathematical and Statistical Psychology*, 39: 15-27.

Machina, M.J. 1982 "'Expected utility' analysis without the independence axiom." *Econometrica*, 50: 277-323.

Machina, M.J. 1987 "Decision making in the presence of risk." *Science*, 236: 537-543.

MacKinnon, D.W. 1962 "The personality correlates of creativity: A study of American architects." In G.S. Nielson (ed.) *Proceedings of the 14th International Congress of Psychology*. Copenhagen: Munksgaard.

Majumder, D.D. 1984 "On the relevancy of fuzzy mathematical systems." *Human Systems Management*, 4: 315-323.

Mandelbrot, B.B. 1983 *The Fractal Geometry of Nature*. New York: W.H. Freeman.

Manes, E.G. 1982 "A class of fuzzy theories." *Journal of Mathematical Analysis and Applications*, 85: 409-451.

Manicas, P.T. and Rosenberg, A. 1985 "Naturalism, epistemological individualism, and 'the strong programme' in the sociology of knowledge." *Journal for the Theory of Social Behaviour*, 15: 76-101.

Mannheim, K. 1936 *Ideology and Utopia*. New York: Harcourt Brace and World.

March, J.G. 1981 "Footnotes to organizational change." *Administrative Science Quarterly*, 26: 563-577.

March, J.G. and Feldman, M.S. 1981 "Information in organizations as signals and symbols." *Administrative Science Quarterly*, 26: 171-186.

March, J.G. and Olsen, J.P. 1976 *Ambiguity and Choice in Organizations*. Bergen: Universitetsforlaget.

March, J.G. and Simon, H.A. 1958 *Organizations*. New York: Wiley.

Markowitz, H. 1952 "The utility of wealth." *Journal of Political Economy*, 60: 151-158.

Mars, G. 1982 *Cheats at Work*. London: George Allen and Unwin.

Martin, N. 1981 Personal communication regarding fieldwork at Aurukun, Queensland, Australia.

Mason, R.O. and Mitroff, I.L. 1981 *Challenging Strategic Planning Assumptions*. New York: Wiley Interscience.

Matheron, G. 1975 *Random Sets and Integral Geometry*. New York: Wiley.

Mattera, P. 1985 *Off the Books: The Rise of the Underground Economy*. London: Pluto Press.

Mazur, J.E. 1984 "Tests of an equivalence rule for fixed and variable reinforcer delays." *Journal of Experimental Psychology: Animal Behavior Processes*, 4: 426-436.

McArthur, D.S. 1980 "Decision scientists, decision makers, and the gap." *Interfaces*, 10: 110.

McCall, M. and Simmons, T. 1966 *Identities and Interactions*. Glencoe, Illinios: The Free Press.

McCarthy, J. 1980 "Circumscription: A form of nonmonotonic reasoning." *Artificial Intelligence*, 13: 27-39.

McCauley, C. and Stitt, C.L. 1978 "An individual and quantitative measure of stereotypes." *Journal of Personality and Social Psychology*, 36: 929-940.

McClelland, D.C., Atkinson, J.W., Clark, R.A., and Lowell, E. 1953 *The Achievement Motive*. New York: Appleton-Century-Crofts.

McConway, K.J. 1981 "Marginalization and linear opinion pools." *Journal of the American Statistical Association*, 76: 410-414.

McDermott, D. and Doyle, J. 1980 "Non-monotonic logic I." *Artificial Intelligence*, 13: 41-72.

Mead, G.H. 1934 *Mind, Self, and Society*. Chicago: University of Chicago Press.

Meehl, P. and Rosen, A. 1955 "Antecedent probability and the efficiency of psychometric signs, patterns, or cutting scores." *Psychological Bulletin*, 52: 194-216.

Mehan, H. and Wood, H. 1975 *The Reality of Ethnomethodology*. New York: Wiley.

Merton, R.K. 1957 *Social Theory and Social Structure* (revised and enlarged edition). New York: The Free Press.

Merton, R.K. 1973 *The Sociology of Science*. Chicago: University of Chicago Press.

Merton, R.K. 1987 "Three fragments from a sociologist's notebooks: establishing the phenomenon, specified ignorance, and strategic research materials." *Annual Review of Sociology*, 13: 1-28.

Michael, J. 1982 *The Politics of Secrecy*. Harmondsworth: Penguin.

Milburn, M.A. 1978 "Sources of bias in the prediction of future events." *Organizational Behavior and Human Performance*, 21: 17-26.

Miles, R.E., Snow, C.C., Meyer, A.D., and Coleman, H.J. Jr. 1978 "Organizational strategy, structure, and process." *Academy of Management Review*, 3: 546-562.

Miles, R.E., Snow, C.C., and Pfeffer, J. 1974 "Organizational environment: Concepts and issues." *Industrial Relations*, 13: 244-264.

Miller, D.W. and Starr, M.K. 1967 *The Structure of Human Decisions*. Englewood Cliffs, N.J.: Prentice-Hall.

Miller, R.R., Greco, C., Vigorito, M., and Marlin, N.A. 1983 "Signaled tailshock is perceived as similar to a stronger unsignaled tailshock: Implications for a functional analysis of classical conditioning." *Journal of Experimental Psychology: Animal Behavior*, 9:105-131.

Miller, S.M. and Mangan, C.E. 1983 "Interacting effects of information and coping style in adapting to gynecologic stress: Should the doctor tell all?" *Journal of Personality and Social Psychology*, 45: 223-236.

Milliken, F.J. 1987 "Three types of perceived uncertainty about the environment: State, effect, and response uncertainty." *Academy of Management Review*, 12: 133-143.

Mills, C.W. 1959 *The Sociological Imagination*. New York: Oxford University Press.

Mineka, S. and Henderson, R.W. 1985 "Controllability and predictability in acquired motivation." *Annual Review of Psychology*, 36: 495-529.

Minsky, M. 1975 "A framework for representing knowledge." In P. Winston (ed.) *The Psychology of Computer Vision*. New York: McGraw-Hill.

Mises, R. von 1957 *Probability, Statistics, and Truth* (Second Edition). London: Allen and Unwin.

Mitroff, I. 1974 *The Subjective Side of Science*. Amsterdam: Elsevier.

Montgomery, H. and Svenson, O. 1976 "On decision rules and information processing strategies for choices among multiattribute alternatives." *Scandinavian Journal of Psychology*, 17: 283-291.

Moore, R. 1985 "Semantical considerations on non-monotonic logic." *Artificial Intelligence*, 25: 75-94.

Moore, R.E. 1966 *Interval Analysis*. Englewood Cliffs, N.J.: Prentice-Hall.

Moore, W.E. and Tumin, M.M. 1949 "Some social functions of ignorance." *American Sociological Review*, 14: 787-796.

Morgan, M.G. 1981 "Risk assessment." *IEEE Spectrum* (in two parts), Nov.: 58-64, and Dec.: 53-60.

Morris, P.A. 1977 "Combining expert judgments: A Bayesian approach." *Management Science*, 23: 679-693.

Morris, P.A. 1986 "Comment on Genest and Zidek." *Statistical Science*, 1: 141-144.

Morrison, D.G. 1967 "On the consistency of preferences in Allais' paradox." *Behavioral Science*, 12: 373-383.

Moshowitz, A. (ed.) 1980 *Human Choice and Computers 2*. Amsterdam: North-Holland.

Moskowitz, H. 1974 "Effects of problem representation and feedback on rational behavior in Allais and Morlat type problems." *Decision Sciences*, 5: 225-242.

Mowen, J.C. and Gentry, J.W. 1980 "Investigations of the preference-reversal phenomenon in a new product introduction task." *Journal of Applied Psychology*, 65: 715-722.

Mulkay, M. 1979 *Science and the Sociology of Knowledge*. London: Allen and Unwin.

Murphy, A.H. and Winkler, R.L. 1977 "Can weather forecasters formulate reliable probability forecasts of precipitation and temperature?" *National Weather Digest*, 2: 2-9.

Murthy, C.A., Pal, S.K., and Majumder, D.D. 1985 "Correlation between two fuzzy membership functions." *Fuzzy Sets and Systems*, 17: 23-38.

Nadler, G. 1981 *The Planning and Design Approach*. New York: Wiley.

Nagel, E. 1939 *Principles of the Theory of Probability, V.1. no.6 of International Encyclopedia of Unified Science*. Chicago: University of Chicago Press.

Nagel, E. 1961 *The Structure of Science*. London: Routledge and Kegan Paul.

Nagy, T.J. and Hoffman, L.J. 1981 "Exploratory evaluation of the accuracy of linguistic versus numeric risk assessment of computer security." *Technical Report GWU-IIST-81-07*. St. Louis: George Washington University, Computer Security Research Group.

Nakao, M.A. and Axelrod, S. 1983 "Numbers are better than words." *American Journal of Medicine*, 74: 1061-1065.

Natvig, B. 1983 "Possibility versus probability." *Fuzzy Sets and Systems* 10: 31-36.

Navarick, D.J. 1987 "Reinforcement probability and delay as determinants of human impulsiveness." *Psychological Record*, 37: 219-225.

Navon, D. 1978 "The importance of being conservative: Some reflections on human Bayesian behavior." *British Journal of Mathematical and Statistical Psychology*, 31: 33-48.

Newman, J.R. (ed.) 1956 *The World of Mathematics*, V.1. New York: Simon and Schuster.

Nguyen, H.T. 1977 "On fuzziness and linguistic probabilities." *Journal of Mathematical Analysis and Applications*, 61: 658-671.

Nguyen, H.T. 1978 "On random sets and belief functions." *Journal of Mathematical Analysis and Applications*, 65: 531-542.

Nguyen, H.T. 1984 "On modeling of linguistic information using random sets." *Information Sciences*, 34: 265-274.

Nickerson, R.S. and McGoldrick, C.C., Jr. 1965 "Confidence ratings and level of performance on a judgmental task. *Perceptual Motor Skills*, 20: 311-316.

Nisbett, R.E. and Ross, L. 1980 *Human Inference: Strategies and Shortcomings of Social Judgment.* Englewood Cliffs, N.J.: Prentice-Hall.

Nutt, P.C. 1984 "Types of organizational decision processes." *Administrative Science Quarterly*, 29: 414-450.

Oakes, M. 1986 *Statistical Inference: A Commentary for the Social and Behavioural Sciences.* London: Wiley.

Olson, C.L. 1976 "Some apparent violations of the representativeness heuristic in human judgment." *Journal of Experimental Psychology: Human Perception and Performance*, 2: 599-608.

Osborn, R.N., Hunt, J.G., and Jauch, L.R. 1980 *Organization Theory: An Integrated Approach.* New York: Wiley.

Osherson, D.N. and Smith, E.E. 1981 "On the adequacy of prototype theory as a theory of concepts." *Cognition*, 9: 35-58.

Oskamp, S. 1965 "The relationship of clinical experience and training methods to several criteria of clinical prediction." *Psychological Monographs*, 76.

Ouchi, W.G. 1980 "Markets, bureaucracies, and clans." *Administrative Science Quarterly*, 25: 129-141.

Overmier, J.B., Patterson, J., and Wielkiewicz, R.M. 1980 "Environmental contingencies as sources of stress in animals" (cited in Gunnar 1980).

Overmier, J.B. and Seligman, M.E.P. 1967 "Effects of inescapable shock upon subsequent escape and avoidance responding." *Journal of Comparative and Physiological Psychology*, 63: 28-33.

Parsons, T. 1959 "An approach to the sociology of knowledge." Reprinted in J. Curtis and J.W. Petras (eds.) *The Sociology of Knowledge: A Reader*. New York: Praeger, 1970.

Pawlak, Z. 1982 "Rough Sets." *International Journal of Information and Computer Science*, 11: 341-356.

Pawlak, Z. 1985 "Rough sets and fuzzy sets." *Fuzzy Sets and Systems*, 17: 99-102.

Payne, J.W. 1982 "Contingent decision behavior." *Psychological Bulletin*, 92: 382-402.

Payne, J.W., Braunstein, M.L., and Carroll, J.S. 1978 "Exploring predecisional behavior: An alternative approach to decision research." *Organizational Behavior and Human Performance*, 22: 17-44.

Pennings, J.M. 1981 "Strategically interdependent organizations." In P.C. Nystrom and W.H. Starbuck (eds.) *Handbook of Organizational Design, Vol. I*. New York: Oxford University Press.

Pennington, N. and Hastie, R. 1986 "Evidence evaluation in complex decision making." *Journal of Personality and Social Psychology*, 51: 242-258.

Pennington, N. and Hastie, R. 1988 "Explanation-based decision making: Effects of memory structure on judgment." *Journal of Experimental Psychology: Learning, Memory, and Cognition*, 14: 521-533.

Pepper, S. 1981 "Problems in the quantification of frequency expressions." In D. Fiske (ed.) *New Directions for Methodology of Social and Behavioral Sciences: Problems with Language Imprecision.* San Francisco: Jossey-Bass.

Pepper, S. and Prytulak, L.S. 1974 "Sometimes frequently means seldom: Context effects in the interpretations of quantitative expressions." *Journal of Research in Personality,* 8: 95-101.

Perrow, C. 1970 *Organizational Analysis: A Sociological View.* Belmont, Calif.: Brooks-Cole.

Peterson, C.R., and Beach, L.T. 1967 "Man as an intuitive statistician." *Psychological Bulletin,* 68: 29-46.

Pettigrew, T. 1958 "The measurement and correlates of category width as a cognitive variable." *Journal of Personality,* 26: 531-544.

Pettigrew, T. 1982 "Cognitive style and social behavior: A review of category width." In L. Wheeler (ed.) *Review of Personality and Social Psychology, Vol.3.* Beverly Hills, California: Sage.

Pfeffer, J. and Salancik, G. 1978 *The External Control of Organizations: A Resource Dependence Approach.* New York: Harper and Row.

Phillips, L.D. 1973 *Bayesian Statistics for Social Scientists.* London: Nelson.

Phillips, L.D. 1983 "A theoretical perspective on heuristics and biases in probabilistic thinking." In P.C. Humphreys, O. Svenson, and A. Vari (eds.) *Analysing and Aiding Decision Processes.* Amsterdam: North Holland.

Phillips, L.D. 1984 "A theory of requisite decision models." *Acta Psychologica,* 56: 29-48.

Pickart, R.C. and Wallace, J.B. 1974 "A study of the performance of subjective probability assessors." *Decision Sciences,* 5: 347-363.

Pill, J. 1971 "The Delphi method: Substance, context, a critique and an annotative bibliography." *Socio-economics and Planning Sciences*, 5: 57-71.

Pirsig, R. 1974 *Zen and the Art of Motorcycle Maintenance*. New York: William Morrow.

Pitts, A. 1982 "Fuzzy sets do not form a topos." *Fuzzy Sets and Systems*, 8: 101-104.

Pitz, G.F. 1974 "Subjective probability distributions for imperfectly known quantities." In L.W. Gregg (ed.) *Knowledge and Cognition*. New York: Wiley.

Poincare, H. 1913 *The Foundations of Science*. New York: The Science Press.

Pollatsek, A., Konold, C.E., Well, A.D., and Lima, S.D. 1984 "Beliefs underlying random sampling." *Memory and Cognition*, 12: 395-401.

Ponasse, D. 1983 "Some remarks on the category Fuz(H) of M. Eytan." *Fuzzy Sets and Systems*, 9: 199-204.

Poole, W.K. 1974 "Estimation of the distribution function of a continuous type random variable through randomized response." *Journal of the American Statistical Association*, 69: 1002-1005.

Popper, K.R. 1959 *The Logic of Scientific Discovery*. New York: Basic Books.

Poulton, E.C. 1968 "The new psychophysics: Six models for magnitude estimation." *Psychological Bulletin*, 69: 1-19.

Pounds, W. 1969 "The process of problem finding." *Industrial Management Review*, 1-19.

Rachlin, H., Castrogiovanni, A., and Cross, D. 1987 "Probability and delay in commitment." *Journal of the Experimental Analysis of Behavior*, 48: 347-353.

Rachlin, H., Logue, A.W., Gibbon, J. and Frankel, M. 1986 "Cognition and behavior in studies of choice." *Psychological Review*, 93: 33-45.

Raiffa, H. 1968 *Decision Analysis: Introductory Lectures on Choices Under Uncertainty*. Reading, Mass.: Addison-Wesley.

Ralescu, A.L. and Ralescu, D.A. 1984 "Probability and fuzziness." *Information Sciences*, 34: 85-92.

Rasmussen, N.C. 1975 *Reactor Safety Study*. Rept. WASH-1400 (NUREG-75/014). Washington D.C.: National Research Council.

Ravetz, J.R. 1987 "Usable knowledge, usable ignorance." *Knowledge: Creation, Diffusion, Utilization*, 9: 87-116.

Real, L. and Caraco, T. 1986 "Risk and foraging in stochastic environments." *Annual Review of Ecological Systems*, 17: 371-390.

Reichenbach, H. 1949 *The Theory of Probability*. Berkeley: University of California Press.

Reiter, R. 1980 "A logic for default reasoning." *Artificial Intelligence*, 13: 81-131.

Renyi, A. 1970 *Probability Theory*. Amsterdam: North-Holland.

Reser, J.P. and Smithson, M. 1988 "When ignorance is adaptive: not knowing about the nuclear threat." *Knowledge in Society* (accepted for publication).

Rip, A. 1982 "The development of restrictedness in the sciences." In N. Elias, et al. (eds.) *Scientific Establishments and Hierarchies*, Sociology of the Sciences Yearbook 6. Dordrecht: Reidel.

Rodin, J. and Langer, E. 1977 "Long-term effects of a control-relevant intervention with the institutionalized aged." *Journal of Personality and Social Psychology*, 35: 897-902.

Rokeach, M. 1960 *The Open and Closed Mind*. New York: Basic Books.

Rosen, S. and Tesser, A. 1970 "On reluctance to communicate undesirable information: The MUM effect." *Sociometry*, 33: 253-263.

Rosch, E. 1978 "Principles of categorization." In E. Rosch and B.B. Lloyd (eds.) *Cognition and Categorization*. Hillsdale, N.J.: Erlbaum.

Rosch, E. and Mervis, C.B. 1975 "Family resemblance: Studies in the internal structure of categories." *Cognitive Psychology*, 7: 573-605.

Ross, M. and Fletcher, G.J.O. 1985 "Attribution and social perception." In G. Lindzey and E. Aronson (eds.) *Handbook of Social Psychology, Third Edition, Vol. II*. New York: Random House.

Ross, M. and Sicoly, F. 1979 "Egocentric biases in availability and attribution." *Journal of Personality and Social Psychology*, 37: 322-336.

Rothbaum, F., Weisz, J.R., Snyder, S. 1982 "Changing the world and changing the self: A two-process model of perceived control." *Journal of Personality and Social Psychology*, 42: 5-37.

Rothschild-Whitt, J. 1979 "The collectivist organization: An alternative to rational bureaucratic models." *American Sociological Review*, 44: 509-527.

Rotter, J.B. 1954 *Social Learning and Clinical Psychology*. Englewood Cliffs, N.J.: Prentice-Hall.

Roubens, M. 1978 "Pattern classification problems and fuzzy sets." *Fuzzy Sets and Systems*, 1: 239-253.

Russell, B. 1919 *Introduction to Mathematical Philosophy*. London: Allen and Unwin.

Russell, B. 1958 *Portraits from Memory and Other Essays*. London: Allen and Unwin.

Savage, L.J. 1954 *The Foundations of Statistics.* New York: Wiley.

Savage, L.J., et al. 1962 *The Foundations of Statistical Inference.* London: Methuen.

Schank, R. and Abelson, R. 1977 *Scripts, Plans, Goals, and Understanding.* Hillsdale, N.J.: Erlbaum.

Scheffler, I. 1967 *Science and Subjectivity.* New York: Bobbs-Merrill.

Schmidt, S.M. and Cummings, L.L. 1976 "Organizational environment, differentiation, and perceived environmental uncertainty." *Decision Sciences,* 7: 447-467.

Schneider, K. and Posse, N. 1982 "Risk-taking in achievement-oriented situations: Do people really maximize affect or competence information?" *Motivation and Emotion,* 6: 259-271.

Scholz, R.W. 1987 *Cognitive Strategies in Stochastic Thinking.* Dordrecht: Reidel.

Schoorman, D.F., Bazerman, M.H., and Atkin, R.S. 1981 "Interlocking directorates: A strategy for reducing environmental uncertainty." *Academy of Management Review,* 6: 243-251.

Schulz, R. and Hanusa, B.H. 1980 "Experimental social gerentology" A social psychological perspective." *Journal of Social Issues,* 36: 30-46.

Schustack, M.W., and Sternberg, R.J. 1981 "Evaluation of evidence in causal inference." *Journal of Experimental Psychology: General,* 110: 101-120.

Schwartz, B. 1968 "The social psychology of privacy." *American Journal of Sociology,* 6: 741-762.

Seaver, D.A., von Winterfeldt, D., and Edwards, W. 1978 "Eliciting subjective probability distributions on continuous variables." *Organizational Behavior and Human Performance*, 21: 379-391.

Self, P. 1975 *Econocrats and the Policy Process: The Politics and Philosophy of Cost Benefit Analysis*. New York: MacMillan.

Seligman, M.E.P. 1975 *Helplessness: On Depression*. San Francisco: Freeman.

Seligman, M.E.P. and Maier, S.F. 1967 "Failure to escape traumatic shock." *Journal of Experimental Psychology*, 74: 1-9.

Sennett, R. 1970 *The Uses of Disorder*. New York: Vintage Books.

Shackle, G.L.S. 1952 *Expectation in Economics*. Cambridge: Cambridge University Press.

Shafer, G. 1976 *A Mathematical Theory of Evidence*. Princeton: Princeton University Press.

Shafer, G. 1984 *The Combination of Evidence*. Lawrence, Kansas: University of Kansas, Department of Mathematics.

Shafer, G. 1985 "Belief functions and possibility measures." In J.C. Bezdek (ed.) *The Analysis of Fuzzy Information, Vol. 1*. New York: CRC Press.

Shafer, G. 1986 "Comment on Genest and Zidek." *Statistical Science*, 1: 135-137.

Shaklee, H. and Mims, M. 1982 "Sources of error in judging event covariations: Effects of memory demands." *Journal of Experimental Psychology: Learning, Memory, and Cognition*, 8: 208-224.

Shannon, C.E. 1948 "The mathematical theory of communication." *Bell System Technical Journal*, 27: 379-423.

Shortliffe, E.H. 1976 *Computer-Based Medical Consultations: MYCIN*. New York: North-Holland.

Shortliffe, E.H. and Buchanan, B.G. 1975 "A model of inexact reasoning in medicine." *Mathematical Biosciences*, 23: 351-379.

Shweder, R.A. 1977 "Likeness and likelihood in everyday thought: Magical thinking in judgments about personality." *Current Anthropology*, 18: 637-648.

Simmel, G. 1950 (tr. K. Wolff) *The Sociology of Georg Simmel*. New York: The Free Press.

Simon, H. 1978 "Rationality as a process and as a product of thought." *American Economic Review*, 68: 1-16.

Simon, R.J. and Mahan, L. 1971 "Quantifying burdens of proof." *Law and Society Review*, 5: 319-322.

Simpson, R.H. 1944 "The specific meanings of certain terms indicating differing degrees of frequency." *Quarterly Journal of Speech*, 30: 328-330.

Simpson, R.H. 1963 "Stability in meanings for quantitative terms: A comparison over 20 years." *Quarterly Journal of Speech*, 49: 146-151.

Skala, H., Termini, S., and Trillas, E. (eds.) 1984 *Aspects of Vagueness*. Dordrecht: Reidel.

Slovic, P. 1969 "Manipulating the attractiveness of a gamble without changing its expected value." *Journal of Experimental Psychology*, 79: 139-145.

Slovic, P. 1972 "From Shakespeare to Simon: Speculations--- and some evidence--- about man's ability to process information." *Oregon Research Institute Monograph*, 12.

Slovic, P. and Lichtenstein, S. 1971 "Comparison of Bayesian and regression approaches to the study of information processing in judgment." *Organizational Behavior and Human Performance*, 6: 649-744.

Slovic, P. and Tversky, A. 1974 "Who accepts Savage's axiom?" *Behavioral Science*, 19: 368-373.

Smedslund, J. 1963 "The concept of correlation in adults." *Scandinavian Journal of Psychology*, 4: 165-173.

Smedslund, J. 1966 "Note on learning, contingency, and clinical experience." *Scandinavian Journal of Psychology*, 7: 265-266.

Smets, P. 1982 "Probability of a fuzzy event: An axiomatic approach." *Fuzzy Sets and Systems*, 7: 153-164.

Smith, C.A.B. 1961 "Consistency in statistical inference and decision." *Journal of Royal Statistical Society, Series B*, 23:1-37.

Smith, J.C. 1983 "The process of adjudication and regulation: A comparison." In T. Machlan and M.B. Johnson (eds.) *Rights and Regulations*. San Francisco: Pacific Institute.

Smithson, M. 1980 "Interests and the growth of uncertainty." *Journal for the Theory of Social Behaviour*, 10: 157-168.

Smithson, M. 1982 "Applications of fuzzy set concepts to behavioral sciences." *Mathematical Social Sciences*, 2: 257-274.

Smithson, M. 1984 "Multivariate analysis using 'and' and 'or'". *Mathematical Social Sciences*, 7: 231-251.

Smithson, M. 1985 "Toward a social theory of ignorance." *Journal for the Theory of Social Behaviour*, 15: 151-172.

Smithson, M. 1987 *Fuzzy Set Analysis for Behavioral and Social Sciences*. New York: Springer Verlag.

Smithson, M. 1988 "Possibility theory, fuzzy logic, and psychological explanation." In T. Zetenyi (ed.) *Fuzzy Sets in Psychology*. Amsterdam: North-Holland.

Snyder, M. 1981 "Seek and ye shall find: Testing hypotheses about other people." In E.T. Higgins, C.P. Heiman, and M.P. Zanna (eds.) *Social Cognition: The Ontario Symposium on Personality and Social Psychology*. Hillsdale, N.J.: Erlbaum.

Snyder, M. and Swann, W.B., Jr. 1978 "Hypothesis testing in social interaction." *Journal of Personality and Social Psychology,* 36: 1202-1212.

Snyder, M., Tanke, E.D., and Berscheid, E. 1977 "Social perceptions and interpersonal behavior: On the self-fulfilling nature of social stereotypes." *Journal of Personality and Social Psychology,* 35: 656-666.

Snyder, N.H. and Glueck, W.F. 1982 "Can environmental volatility be measured objectively?" *Academy of Management Journal,* 25: 185-192.

Sochor, A. 1984 "The alternative set theory and its approach to Cantor's set theory." In H. Skala, et al. (eds.) *Aspects of Vagueness.* Dordrecht: Reidel.

Sorrentino, R.M. and Hewitt, E. 1984 "Uncertainty-related properties of achievement tasks as a function of uncertainty orientation and achievement-related motives." *Journal of Personality and Social Psychology,* 47: 884-899.

Sorrentino, R.M., Short, J.C., and Raynor, O. 1984 "Uncertainty orientation: Implications for affective and cognitive views of achievement behavior." *Journal of Personality and Social Psychology,* 46: 189-206.

Sorrentino, R.M. and Short, J.C. 1986 "Uncertainty orientation, motivation, and cognition." In R.M. Sorrentino and E.T. Higgins (eds.) *Handbook of Motivation and Cognition: Foundations of Social Behavior.* New York: Guilford Press.

Spencer-Brown, G. 1957 *Probability and Scientific Inference.* London: Longmans, Green.

Spencer-Brown, G. 1974 *Laws of Form.* New York: E.P. Dutton.

Stael von Holstein, C.-A.S. "An experiment in probabilistic weather forecasting." *Journal of Applied Meteorology,* 10: 635-645.

Stallen, P.J. and Coppock, R. 1987 "About risk communication and risky communication." *Risk Analysis,* 7: 413-414.

Stallings, W. 1977 "Fuzzy set theory versus Bayesian statistics." *I.E.E.E. Transactions on Systems, Man, and Cybernetics*, 7: 216-219.

Stark, W. 1958 *The Sociology of Knowledge*. London: Routledge and Kegan Paul.

Stephens, D.W. and Krebs, J.R. 1986 *Foraging Theory*. Princeton, N.J.: Princeton University Press.

Stone, D.R. and Johnson, R.J. 1959 "A study of words indicating frequency." *Journal of Educational Psychology*, 50: 224-227.

Stout, L.N. 1984 "Topoi and categories of fuzzy sets." *Fuzzy Sets and Systems*, 12: 169-184.

Straub, H. 1949 *A History of Civil Engineering*. Trans. E. Rockwell. London: Leonard Hill.

Strickland, L.H., Lewicki, R.J., and Katz, A.M. 1966 "Temporal orientation and perceived control as determinants of risk-taking." *Journal of Experimental Social Psychology*, 2: 143-151.

Sugeno, M. 1977 "Fuzzy measures and fuzzy integrals: A survey." In M.M. Gupta, G.N. Sardis, and B.R. Gaines (eds.) *Fuzzy Automata and Decision Processes*. Amsterdam: North-Holland.

Taylor, D.A. 1979 "Motivational bases." In G.J. Chelune (ed.) *Self Disclosure*. San Francisco: Jossey-Bass.

Taylor, S.E. 1983 "Adjustment to threatening events: A theory of cognitive adaptation." *American Psychologist*, 38: 1161-1173.

Terreberry, S. 1968 "The evolution of organizational environments." *Administrative Science Quarterly*, 12: 590-613.

Tesser, A. and Rosen, S. 1975 "The reluctance to transmit bad news." In L. Berkowitz (ed.) *Advances In Experimental Social Psychology, Vol. 8.* New York: Academic Press.

Tesser, A., Rosen, S., and Conlee, M.C. 1972 "News valence and available recipient as determinants of news transmission. *Sociometry*, 35: 619-628.

Theil, H. 1967 *Economics and Information Theory.* Chicago: Rand McNally.

Thompson, J.D. 1962 "Decision-making, the firm, and the market." In W.W. Cooper, H.J. Leavitt, and M.W. Shelly II (eds.) *New Perspectives on Organization Research.* New York: Wiley.

Thompson, J.D. 1967 *Organizations in Action.* New York: McGraw-Hill.

Thorngate, W. "Efficient decision heuristics." *Behavioral Science*, 25: 219-225.

Tierney, M. 1972 *Toposes, Algebraic Geometry, and Logic: Lecture Notes in Mathematics No. 274.* New York: Springer Verlag.

Tosi, H., Aldag, R., and Storey, R. 1973 "On the measurement of the environment: An assessment of the Lawrence and Lorsch environmental uncertainty subscale." *Administrative Science Quarterly*, 18: 27-36.

Trillas, E., Alsina, C., and Valverde, L. 1982 "Do we need max, min, and 1-J in fuzzy set theory?" In R.R. Yager (ed.) *Fuzzy Set and Possibility Theory: Recent Developments.* New York: Pergamon.

Trope, Y. 1975 "Seeking information about one's own ability as a determinant of choice among tasks." *Journal of Personality and Social Psychology*, 32: 1004-1013.

Trope, Y. 1979 "Uncertainty-reducing properties of achievement tasks." *Journal of Personality and Social Psychology*, 37: 1505-1518.

Tune, G.S. 1964 "Response preferences: A review of some relevant literature." *Psychological Bulletin*, 61: 286-302.

Tung, R. 1979 "Dimensions of organizational environments: An exploratory study of their impact on organizational structure." *Academy of Management Journal*, 22: 672-693.

Turner, A.G. 1982 "What subjects of survey research believe about confidentiality." In J.E. Sieber (ed.) *The Ethics of Social Research: Surveys and Experiments*. New York: Springer Verlag.

Turner, B.A. 1978 *Man-made Disasters*. London: Wykenham.

Tushman, M.L. and Scanlan, T.J. 1981 "Characteristics and external orientations of boundary-spanning individuals." *Academy of Management Journal*, 24: 83-98.

Tushman, M.L. and Scanlan, T.J. 1981 "Boundary-spanning individuals: Their role in information transfer and their antecedents." *Academy of Management Journal*, 24: 289-305.

Tversky, A. and Bar-Hillel, M. 1983 "Risk: The long and the short." *Journal of Experimental Psychology: Learning, Memory, and Cognition*, 9: 713-717.

Tversky, A. and Kahneman, D. 1971 "The belief in the 'law of small numbers'." *Psychological Bulletin*, 76: 105-110.

Tversky, A. and Kahneman, D. 1973 "Availability: A heuristic for judging frequency and probability." *Cognitive Psychology*, 5: 207-232.

Tversky, A. and Kahneman, D. 1974 "Judgment under uncertainty: Heuristics and biases." *Science*, 185: 1124-1131.

Tversky, A. and Kahneman, D. 1980 "Causal schemas in judgments under uncertainty." In M. Fishbein (ed.) *Progress in Social Psychology, Vol.1*. Hillsdale, N.J.: Erlbaum.

Tversky, A. and Kahneman, D. 1981 "The framing of decisions and the rationality of choice." *Science*, 221: 453-458.

Tversky, A. and Kahneman, D. 1982 "Judgments of and by representativeness." In D. Kahneman, P. Slovic, and A. Tversky (eds.) *Judgment Under Uncertainty: Heuristics and Biases*. New York: Cambridge University Press.

Tversky, A. and Kahneman, D. 1983 "Extensional versus intuitive reasoning: The conjunction fallacy in probability judgment." *Psychological Review*, 90: 293-315.

Unger, P. 1975 *Ignorance: A Case for Scepticism*. Oxford: Clarendon Press.

Vari, A. and Vecsenyi, J. 1984 "Pitfalls of decision analysis: Examples of R and D planning." In P.C. Humphreys, O. Svenson, and A. Vari (eds.) *Analysing and Aiding Decision Processes*. Amsterdam: North-Holland.

Vopenka, P. 1979 *Mathematics in the Alternative Set Theory*. Lepizig: Tubner Texte.

Wagner, C.G. 1982 "Allocation, Lehrer models, and the consensus of probabilities." *Theory and Decision*, 14: 207-220.

Walker, N. 1977 *Behaviour and Misbehaviour*. Oxford: Basil Blackwell.

Walley, P. 1981 "Coherent lower (and upper) probabilities." *Statistics Research Report*. University of Warwick.

Walley, P. and Fine, T.L. 1982 "Towards a frequentist theory of upper and lower probability." *Annals of Statistics*, 10: 741-761.

Wallsten, T.S. 1983 "The theoretical status of judgmental heuristics." In R.W. Scholz (ed.) *Decision Making Under Uncertainty*. Amsterdam: North-Holland.

Wallsten, T.S., Budescu, D., Rappoport, A., Zwick, R., and Forsyth, B. 1986 "Measuring the vague meanings of probability terms." *Journal of Experimental Psychology: General*, 115: 348-365.

Wallsten, T.S., Fillenbaum, S. and Cox, J.A. 1986 "Base-rate effects on the interpretation of probability and frequency expressions." *Journal of Memory and Language*, 25: 571-581.

Warner, S.L. 1965 "Randomized response: A survey technique for eliminating evasive answer bias." *Journal of the American Statistical Association*, 60: 63-69.

Warner, S.L. 1971 "The linear randomized response model." *Journal of the American Statistical Association*, 66: 884-888.

Warren, C. and Laslett, B. 1977 "Privacy and secrecy: A conceptual comparison." *Journal of Social Issues*, 33: 43-51.

Wason, P.C. 1960 "On the failure to eliminate hypotheses in a conceptual task." *Quarterly Journal of Experimental Psychology* 12: 129-140.

Wason, P.C. 1968 "On the failure to eliminate hypotheses: A second look." In P.C. Wason and P.N. Johnson-Laird (eds.) *Thinking and Reasoning*. Harmondsworth, Middlesex: Penguin.

Wason, P.C. and Johnson-Laird, P.N. 1972 *Psychology of Reasoning: Structure and Content*. London: Batsford.

Watson, J.D. 1968 *The Double Helix*. N.Y.: Atheneum.

Weatherford, R. 1982 *Philosophical Foundations of Probability Theory*. London: Routledge and Kegan Paul.

Weaver, W. 1963 *Lady Luck*. New York: Anchor Books.

Weerahandi, S. and Zidek, J.V. 1981 "Multi-Baysian statistical decision theory. *Journal of the Royal Statistical Society, Series A*, 144: 85-93.

Weerahandi, S. and Zidek, J.V. 1983 "Elements of multi-Bayesian decision theory." *Annals of Statistics*, 11: 1032-1046.

Weick, K.E. 1979 *The Social Psychology of Organizing, Second Edition*. Reading, Mass.: Addison Wesley.

Weiner, B. 1972 *Theories of Motivation: From Mechanism to Cognition*. Chicago: Markham.

Weinstein, D. and Weinstein, M.A. 1978 "The sociology of nonknowledge: A paradigm." In R.A. Jones (ed.) *Research in the Sociology of Knowledge, Sciences, and Art, V.1*. New York: JAI Press.

Weinstein, N.D. 1980 "Unrealistic optimism about future life events." *Journal of Personality and Social Psychology*, 39: 806-820.

Weiss, C. 1980 "Knowledge creep and decision accretion." *Knowledge: Creation, Diffusion, Utilization*, 1: 381-401.

West, M. 1984 "Bayesian aggregation. *Journal of the Royal Statistical Society, Series A*, 147: 600-607.

Whitley, R. 1984 *The Intellectual and Social Organization of the Sciences*. Oxford: Clarendon Press.

Wildavsky, A. 1985 "Trial without error: Anticipation vs. resilience as strategies for risk reduction." In M. Maxey and R. Kuhn (eds.) *Regulatory Reform: New Vision or Old Curse*. New York: Praeger.

Wilder, R.L. 1981 *Mathematics as a Cultural System*. New York: Pergamon.

Wilson, J.Q. (ed.) 1980 *The Politics of Regulation*. New York: Basic Books.

Winkler, R.L. 1968 "The consensus of subjective proability distributions." *Management Science*, 15: 61-75.

Winkler, R.L. 1981 "Combining probability distributions from dependent information sources." *Management Science*, 27: 479-488.

Winkler, R.L. 1982 "Information and modeling in risk assessment." In H. Kunreuther (ed.) *Risk: A Seminar Series*. Laxenburg, Austria: International Institute for Applied Systems Analysis.

Winkler, R.L. and Murphy, A.H. 1968 "Evaluation of subjective pecipitation probability forecasts." In *Proceedings of the First National Conference on Statistical Meteorology*. Boston: American Meteorological Society.

Wise, B.P. and Henrion, M. 1986 "A framework for comparing uncertain inference systems to probability." In L.N. Kanal and J.F. Lemmer (eds.) *Uncertainty in Artificial Intelligence*. Amsterdam: North-Holland.

Wittgenstein, L. 1958 *Philosophical Investigations* (Third edition). Oxford: Basil Blackwell.

Wittgenstein, L. 1983 *Remarks on the Foundations of Mathematics* (Revised edition). Cambridge, Mass.: MIT Press.

Woodman, R.W., Ganster, D., Adams, J., McCuddy, M.K., Tolchinsky, P.D., and Fromkin, H. 1982 "A survey of employee perceptions of information privacy in organizations." *Academy of Management Journal*, 25: 647-663.

Woodward, J. 1965 *Industrial Organization: Theory and Practice*. London: Oxford University Press.

Wright, G. and Ayton, P. 1987 "Task influences on judgemental forecasting." *Scandinavian Journal of Psychology*, 28: 115-127.

Wright, G. and Phillips, L.D. "Cultural variation in probabilistic thinking." *International Journal of Psychology*, 15: 239-257.

Wright, G.H. von 1941 *The Logical Problem of Induction* (Second Edition, 1957). Oxford: Basil Blackwell.

Wright, G.H., von 1962 "Remarks on the epistemology of subjective probability." In E. Nagel, P. Suppes, and A. Tarski (eds.) *Logic, Methodology, and Philosophy of Science.* Stanford: Stanford University Press.

Wyer, R.S., Jr. 1970 "Quantitative prediction of belief and opinion change: A further test of a subjective probability model." *Journal of Personality and Social Psychology*, 16: 559-570.

Yaari, M. 1987 "The dual theory of choice under risk." *Econometrica*, 55: 95-115.

Yager, R.R. 1979a "On the measure of fuzziness and negation. Part I: Membership in the unit interval." *International Journal of General Systems*, 5: 221-229.

Yager, R.R. 1979b "A note on probabilities of fuzzy events." *Information Sciences*, 18: 113-122.

Yager, R.R. 1982 "Measuring tranquility and anxiety in decision making: An application of fuzzy sets." *International Journal of General Systems*, 8: 139-146.

Yager, R.R. 1983 "Entropy and specificity in a mathematical theory of evidence." *International Journal of General Systems*, 9: 249-260.

Yager, R.R. 1984 "A representation of the probability of a fuzzy subset." *Fuzzy Sets and Systems*, 13: 273-283.

Yates, J.F., Jagaginski, C.M., and Faber, M.D. 1978 "Evaluation of partially described multiattribute options." *Organizational Behavior and Human Performance*, 21: 240-251.

Zadeh, L.A. 1965 "Fuzzy Sets." *Information and Control*, 8: 338-353.

Zadeh, L.A. 1968 "Probability measures of fuzzy events." *Journal of Mathematical Analysis and Applications*, 23: 421-428.

Zadeh, L.A. 1972 "A fuzzy set theoretical interpretation of hedges." *Journal of Cybernetics*, 2: 4-34.

Zadeh, L.A. 1973 "Outline of a new approach to the analysis of complex systems and decision processes." *I.E.E.E. Transactions on Systems, Man, and Cybernetics*, 3: 28-44.

Zadeh, L.A. 1975a "Fuzzy logic and approximate reasoning." *Synthese*, 30: 407-428.

Zadeh, L.A. 1975b "Calculus of fuzzy restrictions." In L.A. Zadeh, K.S. Fu, K. Tanaka, and M. Shimura (eds.) *Fuzzy Sets and their Application to Cognitive and Decision Processes*. New York: Academic Press.

Zadeh, L.A. 1975c "Foreword" In A. Kaufmann, *Introduction to the Theory of Fuzzy Subsets*. New York: Academic Press.

Zadeh, L.A. 1976 "A fuzzy algorithmic approach to the definition of complex or imprecise concepts." *International Journal of Man-Machine Studies*, 8: 249-291.

Zadeh, L.A. 1978 "Fuzzy sets as a basis for a theory of possibility." *Fuzzy Sets and Systems*, 1: 3-28.

Zadeh, L.A. 1979 "Precisiation of human communication via translation into PRUF." *Memo. No. UBB/ERL M79/73*. Berkeley: University of California, Electronics Research Laboratory.

Zadeh, L.A. 1980 "Fuzzy sets versus probability." *Proceedings of I.E.E.E.*, 68: 421.

Zadeh, L.A. 1982 "A note on prototype theory and fuzzy sets." *Cognition*, 12: 291-297.

Zadeh, L.A. 1983 "A computational approach to fuzzy quantifiers in natural languages." *Computers and Mathematics with Applications*, 9: 149-184.

Zadeh, L.A. 1984 "Review of Shafer's 'A Mathematical Theory of Evidence'." *AI Magazine*, 5: 81-83.

Zadeh, L.A. 1986 "Is probability sufficient for dealing with uncertainty in AI: A negative view." In L.N. Kanal and J.F.

Lemmer (eds.) *Uncertainty in Artificial Intelligence.* Amsterdam: North-Holland.

Zakay, D. 1983 "The relationship between the probability assessor and the outcomes of an event as a determiner of subjective probability." *Acta Psychologica*, 53: 271-280.

Zeleny, M. 1984 "On the (ir)relevancy of fuzzy sets theories." *Human Systems Management*, 4: 301-306.

Zimmer, A.C. 1983 "Verbal vs. numerical processing of subjective probabilities." In R.W. Scholz (ed.) *Decision Making Under Uncertainty.* Amsterdam: North-Holland.

Zimmer, A.C. 1984 "A model for the interpretation of verbal predictions." *International Journal of Man-Machine Studies,* 20: 121-134.

Zimmermann, H.-J. and Zysno, P. 1980 "Latent connectives in human decision making." *Fuzzy Sets and Systems*, 4: 37-51.

Zwick, R., Budescu, D.V., and Wallsten, T.S. 1988 "An empirical study of the integration of linguistic probabilities." In T. Zetenyi (ed.) *Fuzzy Sets in Psychology.* Amsterdam: North-Holland.

Zwick, R. and Wallsten, T.S. 1988 "Combining stochastic uncertainty and linguistic inexactness: Theory and experimental evaluation of four fuzzy probability models." *International Journal of Man-Machine Studies* (in press).

Name Index

Abelson, R. 150
Abramson, L.Y. 160
Adams, J.B. 106
Adorno, T.W. 154
Ajzen, I. 170
Aldrich, H. 251
Allais, M. 177-179, 276
Alloy, L.B. 160
Allport, G.W. 157
Alston, W.P. 113
Althusser, L. 5
Amarel, S. 303
Amsel, A. 161
Anderson, N.H. 190, 199
Arkes, H.R. 169
Ashby, E. 84, 90
Atkinson, J.M. 99
Atkinson, J.W. 276
Axelrod, S. 149
Ayton, P. 180

Bacharach, M. 81
Backer, E. 133
Badia, P. 160
Baird, D. 83
Baire, R. 32
Baldwin, J.F. 120, 129-130, 134, 144
Bar-Hillel, M. 171, 173, 182, 186-187, 194, 212
Barber, D.C. 149
Bardhan, K.D. 149
Barnes, B. 216
Barnett, V. 42, 58
Barnlund, D. 230
Barr, M. 139
Barron, F. 155
Barron, F.H. 201
Bartlett, M.S. 52
Bass, B.M. 164
Baum, W.M. 205
Bayes, T. 47-48, 63
Beach, B.H. 78

Beach, L.R. 98, 208
Beach, L.T. 169, 193
Behn, R.D. 149
Bellman, R. 114
Benenson, F.C. 69
Berenstein, C. 78
Berger, P.L. 5, 216, 220, 223, 235
Berkeley, D. 186, 211
Bernoulli, D. 176, 178
Bernoulli, J. 47-48, 119
Berscheid, E. 169
Beyth-Marom, R. 149, 162, 164-165, 168, 173, 181
Bezdek, J.C. 133
Bieri, J. 157
Billig, M. 155
Birkhofer, A. 88
Birnbaum, M.H. 183, 186
Black, M. 42, 94, 113-114
Blackner, E. 157
Blockley, D. 21-23, 137
Bloor, D. 5, 216, 220, 271
Blumberg, H.H. 231
Blumer, H. 223
Bok, S. 255
Bolyai, W. 30
Bonissone, P.P. 117
Bookstein, A. 304
Boole, G. 24, 32, 77
Bordley, R.F. 81
Borel, E. 32, 49, 119
Borgida, E. 170
Boruch, R.F. 306
Boulding, E. 152
Bourgeois, L.J. III 242-243, 249
Bowman, E.H. 248
Bowyer, J.B. 252, 255
Braine, M.D.S. 214
Braroe, N. 224
Braunstein, M.L. 180
Brehm, J.W. 233
Brehm, S.S. 233
Brekke, N. 170
Brouwer, L.E.J. 33

Brown, P. 228-229
Bryant, G.D. 149
Brzustowski, T.A. 87
Buchanan, B.G. 98-102, 104, 106-107
Budescu, D.V. 164-165
Budner, S. 157
Burgess, D. 115, 199
Burnett, S.A. 10
Burns, T. 241, 251
Burt, C. 184
Bushyhead, J.B. 167
Butler, J. 42

Cacioppo, J.T. 156
Caltabiano, M. 231
Cameron, K.S. 243
Campbell, B. 240
Campbell, D.T. 217
Cantor, G. 33-35
Caraco, T. 204
Caramazza, A.A. 115, 117
Carnap, R. 42, 62, 69-70, 100, 106
Carrega, J.C. 138
Carroll, J.S. 187
Carter, L.J. 269
Cascells, W. 171
Cecil, J.S. 306
Chaikin, A.L. 230
Chandler, A. 241
Cheeseman, P. 69, 136-137
Chesnick, E.I. 172
Chew, S. 178
Chicken, J.C. 86, 88
Chong, C.Y. 136-137, 140
Christensen-Szalanski, J.J.J. 167, 208
Christie, I. 24
Chuang-tzu 264
Chubin, D.E. 261
Cialdini, R.B. 233
Clark, W.C. 294
Cohen, J. 134
Cohen, J. 164, 167, 169, 172-173
Cohen, L.J. 24, 73-77, 96-97, 119, 137, 143, 163, 197, 202, 210

Cohen, M. 192, 210
Cohen, M.D. 240
Cohen, M.S. 123-124, 130, 143-145
Cohen, P.J. 35
Cohen, P.R. 149
Cohen, S. 270, 272-274, 292, 294-295
Collingridge, D. 249-250, 255, 262, 269, 295, 300-302
Collins, H.M. 260
Condorcet, M. de 77
Consultancy, J.C. 84-85, 87-88
Cooley, C.H. 223
Coombs, C.H. 276, 280
Cooper, W.S. 207-208
Coppock, R. 257
Cornfield, J. 89
Cournot, A.A. 24, 77
Covello, V.T. 91
Cox, J.A. 164
Cox, R.T. 61-64, 106, 113, 136
Crocker, J. 169
Cummings, L.L. 244
Curley, S.P. 202
Cyert, R.M. 236, 241, 246

Dalkey, N.C. 81
D'Amato, M.R. 160
Davis, J.D. 230
Davis, L.N. 87
Dawid, A.P. 79
Dearnaley, E.J. 164, 167, 172
Decker, K.S. 117
De Gre, G. 259
DeGroot, M.H. 80
De Luca, A. 132
De Moivre, A. 47
De Morgan, A. 32, 96, 110
Dempster, A.P. 120, 123-124, 128-130, 133-134, 136-137, 140, 143-145, 197
Denzin, N. 224
Derber, C. 233
Derlega, V.J. 230
Descartes, R. 18, 31
Diophantus 32

Dixon, K. 5, 221
Douglas, J.D. 99
Douglas, M. 8, 217, 222, 232, 234-235, 237, 242, 253, 264-266, 267, 274, 295
Downey, H.K. 242, 244-245
Downs, A. 239, 296
Doyle, J. 150
Dubois, D. 110-111, 114, 117-118, 128-129, 132-134, 200, 214
Duda, R.O. 100, 103
Duhem, P. 260
Duncan, R.B. 243-244
Dunn, J.C. 133
Durbin, J. 207-208
Durkheim, E. 272
Dutta, A. 142
Dweck, C.S. 160

Edwards, W. 99, 161-162, 167, 169, 171, 173, 178, 190, 196, 209, 276-277, 280
Eggleston, R. 23-24, 74
Einhorn, H.J. 97, 163, 188, 190, 192, 195, 201, 211
Elig, T. 175
Elliott, E.S. 160
Ellis, L. 41
Ellsberg, D. 95-96, 127, 197, 201, 277
Emery, F.E. 241
Euler, L. 29, 47
Eysenck, H. 157
Eytan, M. 139, 147

Farr, R.M. 290
Feldman, M.S. 247
Feller, W. 174
Fellner, W. 190
Fermat, P. de 47
Feyerabend, P. 260
Fiegenbaum, A. 248
Fillenbaum, S. 164
Filmer, P. 225
Fine, T.L. 42, 46, 49, 52, 54, 57, 67, 71-72, 125
Finetti, B. de 59, 61, 63-65
Finighan, W.R. 231
Fischhoff, B. 167-168, 175, 181, 190, 194

Fishburn, P.C. 190
Fisher, R.A. 65
Fisz, M. 54
Fletcher, G.J.O. 175
Foddy, M. xii
Foddy, W.H. 231
Forrester, J.W. 248
Fox, J. 149
Fox, R. 92
Fraenkel, A.A. 34-35
Frankel, E. 261
Frege, G. 31
French, S. 78
Freud, S. 154
Frieze, I. 175
Fromm, E. 17
Fung, R.M. 136-137, 140
Furby, L. 175

Gaines, B.R. 112, 141
Galileo 29
Gardenfors, P. 126-127
Garfinkel, H. 44, 218, 225
Gauss, K.F. 29-30
Genest, C. 78, 80-82
Gentry, J.W. 180
Giertz, M. 114
Giles, R. 113-114, 139
Ginosar, Z. 170
Glasstone, S. 86
Glueck, W.F. 244
Godel, K. x, 35-36
Goffman, E. 224, 251
Goguen, J.A. 149
Goldsmith, O. 216
Goldsmith, R.W. 173
Goldstein, W.M. 192, 195
Good, I.J. 42, 45, 57, 59, 63, 65-66, 106, 122, 125, 136-137, 140, 148, 289
Goodman, I.R. 142, 148
Goody, E.N. 234-235
Graham, I. 139-140
Grassmann, H.G. 30

Graunt, J. 43, 47
Grayboys, T. 171
Greenberg, B.G. 306
Grether, D.M. 180, 192
Grinberg, M. 149
Groner, R. 193
Grosof, B.N. 100, 103, 106, 136, 140
Guerin, B. xii
Guiasu, S. 131
Gunnar, M. 160

Ha, Y.-W. 169
Hacking, I. 43-44, 50, 73, 118-119
Hahn, H. 34
Hagen, O. 178
Hakel, M. 164
Hammerton, M. 187
Hamilton, W.R. 29-30
Hansel, C.E.M. 164, 167, 172
Hanson, N.R. 57, 260
Hanusa, B.H. 160
Haran, D. 173
Harkness, A.R. 169
Harre, R. 96, 100
Hart, P.E. 100, 103
Hartley, R.V.L. 131, 133
Hastie, R. 188
Heaviside, O. 38
Hebb, D.O. 232
Heckerman, D. 105-107, 136
Heider, F. 175
Helmholtz, H. von 30
Henderson, R.W. 160
Henrion, M. 137
Henry, S. 252, 295
Henslin, J.M. 175
Herker, D. 251
Herrnstein, R.J. 205
Hersh, H.M. 115, 117
Hershey, J. 191
Hewitt, E. 156
Hickey, T. 134
Hickson, D.J. 248

Higashi, M. 132-133
Hilbert, D. 28, 34-36, 39, 45, 274
Hisdal, E. 137, 139-140, 147
Hoch, S.J. 188
Hoffman, L.J. 149
Hofstede, G. 156-158, 236, 240, 247
Hogarth, R.M. 78, 97, 163, 188, 190, 195, 201, 211-212
Hohle, U. 134
Horvitz, E.J. 61, 105-107, 136
Houston, D.B. 91
Howell, W.C. 10, 167
Huber, O. 185
Hudde, J. 43, 47
Huff, D. 170
Hume, D. 16
Humphreys, P. 185, 211
Huygens, C. 43, 47

Imada, H. 160

Jackson, J.J. xii
Jaensch, E. 155
Jaffray, J.-Y. 192, 210
James, G. 75
James, W. 17
Janis, I.L. 98, 236, 248
Jauch, L.R. 242, 245, 247-248
Jaynes, E. 65, 131
Jeffreys, H. 69
Jenkins, H.M. 169
Johnson, B. xii
Johnson, D.M. 168
Johnson, R.J. 164
Johnson, R.W. 107
Johnson, S. 2
Johnson-Laird, P.N. 168
Johnstone, P.T. 139, 148
Jones, R.A. xii
Jordan, W.H. 86
Jourard, S. 231
Jungermann, H. 97, 163, 176, 213

Kadane, J.B. 79

Kahneman, D. 10, 97, 163, 170-175, 178, 180, 184, 187-192, 196, 198, 210, 214, 276
Kalbfleisch, J.D. 88
Kant, I. 29
Kastner, A.G. 29
Kates, R.W. 174
Katz, A.M. 175
Kaufmann, A. 111, 132, 147
Kay, P. 116, 199
Keeney, R.L., 82
Keller, R.T. 246
Kempton, W. 115, 117
Kendall, D.G. 142
Keren, G. 192, 212
Keynes, J. M. 23, 48, 69-70, 77
Kickert, W.J.M. 142
Kim, J. 131
Klaua, D. 108
Klayman, J. 169
Kline, M. 4, 29, 31, 36
Klir, G.J. 114, 132-133, 141
Klugel, G.S. 29
Knorr-Cetina, K. 5, 217
Kochen, M. 200, 303-304
Koestler, A. 155
Kohlberg, L. 157
Koopman, B. 63, 147
Kopp, D.C. 246
Kraft, C. 71-72
Kraft, K.L. 242, 245, 247-248
Kramer, E.E. 35
Krantz, D.S. 62, 160
Krebs, J.R. 204
Kronecker, L. 32
Kruse, R. 142
Kuhn, T. 260, 267-268
Kwaakernak, H. 142
Kyburg, H. 42, 140, 147

La Breque, M. 193-194
Lagrange, J.-L. 29, 47
Lakatos, I. 260, 271-272
Lakoff, G. 117
Lambert, J.H. 29
Langer, E.J. 160, 175
Laplace, P.-S. 24, 29, 47-48, 50-51, 77, 119
Larsson, S. 178
Laslett, B. 231-232
Laver, M. 256
Lavine, T.Z. 221
Lawrence, P.R. 243-244, 247
Lebesgue, H. 32
Lee, R.C.T. 112
Lehrer, K. 81
Leibniz, G.W. 24, 44, 49, 73, 119
Lemmer, J.F. 140
Lepper, M. 169
Lev, B. 246
Levi, I. 96, 124, 140, 143, 152
Levi-Strauss, C. 274
Levin, I.P. 201
Levinson, S. 228-229
Lewicki, R.J. 175
Lewis, C.I. 41
Lichtenstein, S. 164-165, 167, 171, 180, 190, 192
Lindesmith, A.R. 224
Lindley, D.V. 63, 66, 79, 82, 136
Linnerooth, J. 239, 296
Litschert, R.J. 246
Lobachevsky, N. 30
Locke, E.A. 246
Locke, J. 14
Loo, S.G. 132
Lopes, L.L. 184, 195, 209, 211-212
Lord, C. 169
Lorenzi, P. 246
Lorsch, J.W. 242-243
Loui, R.P. 140, 144
Lowenheim, L. 36-37
Luce, R.D. 71
Luckmann, T. 5, 216, 220, 223, 235
Lynn, S.J. 230

Lyon, D. 187

MacCrimmon, K.R. 178
Macdonald, R.R. 163, 185, 195, 210
MacGregor, D. 190
Machina, M.J. 176, 178-179, 191, 214
MacKinnon, D.W. 155
Mahan, L. 24
Maier, S.F. 159
Majumder, D.D. 304
Makridakis, S. 163, 195
Mandelbrot, B.B. 34
Manes, E.G. 142
Mangan, C.E. 161
Manicas, P.T. 217
Mann, L. 98
Mannheim, K. 216, 221
March, J.G. 236, 240-241, 245-247, 249, 264
Markowitz, H. 191
Mars, G. 248, 253-257, 291, 295
Martin, N. 234
Mason, R.O. 246
Matheron, G. 142
Mattera, P. 252, 256, 295, 300-302
Mazur, J.E. 206
McArthur, D.S. 209
McCall, M. 224
McCarthy, J. 150
McCauley, C. 170
McClelland, D.C. 155
McConway, K.J. 81
McDaniel, C. 116, 199
McDermott, D. 150
McGoldrick, C.C., Jr. 167
Mead, G.H. 223
Meehl, P. 170
Mehan, H. 225
Mellers, B.A. 186
Merton, R.K. 217, 220, 247, 253, 259, 261-262, 269, 272
Mervis, C.B. 115, 189
Meyer, K.D. 142
Michael, J. 256
Milburn, M.A. 180

Miles, R.E. 248, 291
Miller, D.W. 98
Miller, R.R. 160
Miller, S.M. 161
Milliken, F.J. 245
Mills, C.W. 235
Mims, M. 169
Mineka, S. 160
Minsky, M. 150
Mises, R. von 42, 50-57, 119, 184, 268
Mitchell, T.R. 98
Mitroff, I. 246, 261
Montaigne 16
Montgomery, H. 185
Montmort, P.R. de 47
Moore, R. 150
Moore, R.E. 275
Moore, W.E. 5, 220-221
Morgan, M.G. 83
Morris, P.A. 80, 82
Morrison, D.G. 178
Moshowitz, A. 256
Moskowitz, H. 178
Mowen, J.C. 180
Mulkay, M. 5, 217, 259
Murphy, A.H. 167
Murthy, C.A. 111

Nadler, G. 246
Nageishi, Y. 160
Nagel, E. 23, 42
Nagy, T.J. 149
Nakao, M.A. 149
Natvig, B. 137
Navarick, D.J. 205
Navon, D. 181
Newman, J.R. 32, 164-165
Newton, I. 29, 31
Nguyen, H.T. 142, 148
Nickerson, R.S. 167
Nilsson, N.J. 100
Nisbett, R. 169
Norman, G.R. 149

Nutt, P.C. 247

Oakes, M. 21, 58, 66, 267
Olsen, J.P. 240, 245, 249
Olson, C.L. 189, 193
Orwell, G. 294-295
Osborn, R.N. 242
Osherson, D.N. 115-117, 199
Oskamp, S. 167
Ouchi, W.G. 247
Overmier, J.B. 159-160

Parsons, T. 320-321
Pascal, B. 31, 44, 47
Pawlak, Z. 128
Payne, J.W. 98, 180, 196
Peano, G. 34
Peirce, C.S. 32, 218
Pennings, J.M. 244
Pennington, N. 188
Penner, J. 216
Pepper, S. 164
Perrow, C. 251
Peterson, C.R. 169, 193
Pettigrew, T. 157
Petty, R.E. 156
Pfeffer, J. 243-244, 248
Phillips, L.D. 66, 163, 196, 209, 214
Pickart, R.C. 167
Pidgeon, N. xii
Pill, J. 80
Pirsig, R. 233
Pitts, A. 138
Pitz, G.F. 167
Plato 1, 14, 18
Plott, C.R. 188, 192
Poincare, H. 31-32, 53
Poisson, S.D. 24, 77
Pollatsek, A. 196
Ponasse, D. 139
Poole, W.K. 306
Popper, K.R. 16, 62, 106, 136, 184
Posse, N. 156

Postman, L. 157
Poulton, E.C. 190
Pounds, W. 246
Prade, H. 110-111, 114, 117-118, 128-129, 132-134, 200, 214
Pruitt, D.G. 276, 280
Prytulak, L.S. 164
Pynchon, T. 1

Rachlin, H. 205-206
Raiffa, H. 77, 82
Ralescu, A.L. 141
Ralescu, D.A. 141
Ramsey, F.P. 59, 65
Rasmussen, N.C. 86
Ravetz, J.R. 233, 262, 269, 300-302
Raynor, O. 156
Real, L. 204
Reichenbach, H. 42, 52, 54, 56
Reiter, R. 150
Renyi, A. 131
Reser, J.P. xii, 211, 257
Riemann, G.B. 30
Rip, A. 263
Rodin, J. 160
Rogers, C. 154
Rogers, W. 92
Rokeach, M. 154-156
Rosen, A. 170
Rosen, S. 229-230
Rosenberg, A. 217
Rosch, E. 115, 189
Ross, L. 169
Ross, M. 175, 187
Rothbaum, F. 160
Rothschild-Whitt, J. 247
Rotter, J.B. 204
Roubens, M. 133
Rousseau, J. 17
Russell, B. 31-32, 289

Saccheri, G. 29
Safarjan, W.R. 160
Sahlins, N.-E. 126-127

Said, T. 192
Salancik, G. 243-244
Savage, L.J. 59-60, 65-66, 70, 77, 178, 207-208, 210
Scanlan, T.J. 251
Schank, R. 150
Scheffler, I. 135, 259
Schervish, M.J. 82
Schmidt, S.M. 244
Schneider, K. 156
Schoemaker, P. 191
Schoenberger, A. 171
Scholz, R.W. 181, 183-186, 189, 194-196, 209, 286
Schoorman, D.F. 247
Schulz, R. 160
Schustack, M.W., 169
Schwartz, B. 231
Seaver, D.A. 167
Self, P. 269
Seligman, M.E.P. 159
Sennett, R. 294
Shackle, G.L.S. 119-120, 143, 147
Shafer, G. 47, 59, 67-68, 83, 96, 98, 120-124, 128-130, 132-134, 136-137, 140, 143, 146, 197
Shaklee, H. 167
Shannon, C.E. 65, 131, 134
Shanteau, J.C. 190
Short, J.C. 156-157
Shortliffe, E.H. 98-102, 104, 106-107
Siegler, R.S. 187
Shweder, R.A. 169
Sicoly, F. 187
Simmel, G. 253
Simmons, T. 224
Simon, H. 233, 236, 241, 264
Simon, R.J. 24
Simpson, R.H. 164
Skala, H. 38
Skinner, B.F. 159
Skolem, T. 36-37
Slocum, J.W., Jr. 242, 244
Slovic, P. 97, 167, 171, 178-180, 187, 190, 192
Smedslund, J. 169
Smets, P. 142

Smith, C.A.B. 63, 125, 147
Smith, E.E. 115-117, 199
Smith, J.C. 291
Smithson, M. 3, 5, 98, 111, 115-117, 132-134, 200, 211, 231, 240, 257, 275-276, 304
Smokler, H.E. 42
Snow, C.C. 248
Snyder, M. 169
Snyder, N.H. 244
Sochor, A. 38
Socrates 6, 16
Sorrentino, R.M. 156-157
Spencer-Brown, G. 184, 236
Stael von Holstein, C.-A.S. 167
Stalker, G.M. 241, 251
Stallen, P.J. 257
Stallings, W. 117, 137
Stanton, M. 161
Stark, W. 216, 220-221
Starr, M.K. 98
Stephens, D.W. 204
Sternberg, R.J. 169
Stitt, C.L. 170
Stone, D.R. 164
Stout, L.N. 139
Straub, H. 21
Strauss, A. 224
Strickland, L.H. 175
Sugeno, M. 113-114
Svenson, O. 185
Swann, W.B., Jr. 169

Tabachnik, N. 160
Tanke, E.D. 169
Taylor, D.A. 230
Taylor, S.E. 160, 211
Termini, S. 132
Terreberry, S. 241
Tesser, A. 229-230
Theil, H. 133
Thomas, H. 248
Thomas, L. 92
Thompson, J.D. 243-244, 246

Thorngate, W. 211
Tierney, M. 138
Tosi, H. 242, 244
Trillas, E. 114
Trist, E.L. 241
Trope, Y. 156, 170
Tumin, M.M. 5, 220-221
Tune, G.S. 175
Tung, R. 244
Turing, A. 65, 122
Turner, A.G. 305
Turner, B.A. 87, 236, 302
Tushman, M.L. 251
Tversky, A. 10, 97, 163, 170-175, 178, 180, 184, 187-192, 196, 198, 210, 212, 214, 276

Unger, P. 2

Van Melle 104
Vari, A. 209
Vaupel, J.W. 149
Vecsenyi, J. 209
Venn, J. 52
Voltaire, A.M. 14
Vopenka, P. 38

Wagenaar, W.A. 192, 212
Wagner, C.G. 81
Walker, N. 271
Wallace, J.B. 167
Waller, W. 178
Walley, P. 83, 125
Wallsten, T.S. 164-165, 193, 200
Ward, W.C. 169
Warner, S.L. 306
Warren, C. 231-232
Wason, P.C. 168
Watson, J.D. 261
Weatherford, R. 42, 51, 54, 70
Weaver, W. 212
Weber, M. 247
Weerahandi, S. 79-80
Weick, K.E. 248

Weiner, B. 156
Weinstein, D. 3, 220
Weinstein, M.A. 3, 220
Weinstein, N.D. 180
Weiss, C. 239, 249
West, M. 82
Weyl, H. 33
Whitehead, A.N. 31
Whitley, R. 5, 262-263, 269
Wildavsky, A. 235-236, 264-267, 291-295, 300, 302
Wilder, R.L. 37-38
Wilkins, J. 42
Wilson, J.Q. 250
Winkler, R.L. 77, 80, 82, 84, 87, 167
Winterfeldt, D. von 162, 167, 173
Wise, B.P. 137
Witt, J. de 43, 47
Wittgenstein, L. 19, 28
Wolff, R.W. 81
Wood, H. 225
Woodman, R.W. 256
Woodward, J. 251
Wright, G. 180, 209
Wright, G.H. von 42, 66
Wyer, R.S., Jr. 173

Yaari, M. 190
Yager, R.R. 132-134, 142
Yates, J.F. 201

Zadeh, L.A. 94, 96, 98, 108-119, 123, 137, 140-143, 146, 165, 199, 269, 275, 301, 303
Zakay, D. 180
Zelazny, R. 14
Zeleny, M. 301-304
Zermelo, E. 33-34
Zidek, J.V. 78-82
Zimmer, A.C. 149, 165
Zimmermann, H.-J. 116-117, 200
Zwick, R. 116-117, 200
Zysno, P. 116-117, 200

Subject Index

Absence (of information) 1, 5, 7, 9, 93, 99, 221, 228, 258
Achievement 155-156
Admissibility of evidence 25-26
Absolutism 5, 29, 56-57, 217, 220-222, 224, 226, 237, 258
Aggregation of:
  Belief 122-124, 129-130
  Evidence 100-101, 105
  Fuzzy sets 110, 114-117, 198-200
  Individuals' judgments 77-82
  Probabilities 62, 71-72, 76, 125, 172-173
Ambiguity:
  General 7, 9, 93-94, 113-114, 133, 141, 146, 197, 229, 237, 255, 258, 288
  in Base-rate problems 181-183
  in Language 227-228
  in Probabilities 95, 277
  Intolerance of 154-155, 157-158, 201-203
  Measurement of 131, 133
Anchoring 190
Anonymity 255, 296, 304-305
Anthropology 12, 216, 218, 234-235
Anticipation strategy 291-295, 298, 300-301
Artificial intelligence xi, 3, 93, 108, 137, 149, 197
Attention 168, 226, 232-234
Attribution 175, 289
Authoritarianism 154-155
Availability heuristic 187

Banishment 30-35, 43, 45-46, 135-136, 266, 271, 274, 291
Bargaining 80
Base-rates 168, 170-172, 181-183, 194, 196
Behaviorism 197-198, 203-207
Belief:
  Degree of 59-64, 96, 98, 105-106
  Dempster-Shaferian 96, 98, 120-124, 128-130, 133-134, 197
  Erroneous 14, 18-19, 220-222
  Full 28
  Group-level 83
  Updating 98-100, 103, 105-107, 136-137
Betting 61, 95

Boundary-spanning 248, 251-252, 255

Calibration 79, 166-167, 276, 292, 300
Certainty factors 98-108, 136
Cheating 252
Circumscription 150
Cognitive psychology xi, 4, 96-98, 107, 216
Coherency 60-62, 64, 78
Completeness 36, 39
Complexity 4, 57, 98
Confidence 166-167
Confirmation 62, 96, 100, 106, 136
Confirmation bias 168-169
Conflict 122-124, 130, 143-144, 147, 154
Confusion 7, 9, 134
Consensus 78, 213-215, 223-227, 236-237, 260, 299, 306
Conservatism 168-170
Consistency 25, 36, 39, 64
Control 153, 158-161, 174-175, 218, 230-232, 234-235, 240, 247, 249-250, 255-256, 292-296
Control strategies (exclusionary and inclusionary) 30, 39, 272-274, 292-295, 298, 300
Creativity 155, 249, 294, 300, 306
Credibility 25
Criminal deviancy 270-274, 292

Deception 224, 251-255, 300
Decisions:
   Corrigible 250, 262
   Organizational 219, 239-241, 243-244, 245-250, 253, 258
   under Ambiguity 201-202
   under Ignorance 144, 262
Decision theories:
   Fuzzy 142
   Nonstandard 190-193, 276
   SEU 4, 84, 190-191, 207-212, 250, 276
Default reasoning 150
Delphi technique 80
Determinism 29, 51, 75, 294, 298
Deterministic chaos 174
Differentiation 247
Dishonesty 150, 238
Disinformation 229, 232, 257

Dissonance 134, 147
Distortion 1, 5, 7, 9, 93, 135, 150, 220-221, 227-228, 238, 242, 258, 296, 303
Dogmatism 14-19
Doubt 2, 14-16, 23, 29, 121
Dual residentialism 5, 221-222

Ecology 204, 208
Economics xi, 4
Endorsements 149, 228
Engineering 10, 20-23, 26-27, 97, 265, 269, 274
Entropy 65, 131-134, 275
Epidemiology 89-90
Equipossibility 45, 48-50, 119
Equiprobability 45, 48-50, 67
Error 2, 6-7, 9, 14, 16, 18, 26, 92, 140, 144, 163, 227, 235, 238, 258, 293, 297
Ethnomethodology 44, 225-227
Evidence 24-26, 59
Evolution 198, 204, 207-208
Expert systems xi, 3, 97-100, 107, 120, 149, 197
Expression theory 192-193

Fault tree 86
Fiddles 253-255
Formalism 35-36
Framing effects 175, 179-181, 284-288
Freedom 17, 119, 134, 234, 250, 258, 275-276, 284-288
FRIL 129
Functionalism 5, 220-222, 225, 234
Fuzziness 8-9, 112-114, 132-133
Fuzzy sets:
   Criticisms of 115-118
   Empirical study of 98, 115-117, 165-166, 199-200
   and Logic 111-112, 147-148
   versus Probability 94, 112-114, 137, 139-143
   Second order 109-110, 128
   Theory of x, 108-118, 303-304, 307
   and Topoi 138-139

Gambler's fallacy 174, 189
Game theory 83
Generality 94

Genetic fallacy 221-222
Geology 99
Groupthink 236, 248

Hedges 111, 117, 229
Helplessness 159-161
Heresy 14-15
Heuristics 97-98, 107, 163, 187-193, 211
Hidden economy 252-258

Ideology 221, 247, 271
Ignorance:
  Definition of 6
  Epistemological 6-7
  Informational 6-7
  Social construction of 6, 216-219, 250-255, 257-262
  Specified 262, 269
Incompleteness 1, 7-9, 51, 74, 152, 227-228, 238, 258, 296
Independence axiom 176-180, 190-191, 202
Indexicality 44-45, 218, 227
Induction 70, 100
Information:
  processing 107
  seeking 232, 234-235
  theory 65, 131-134
Innocence 15, 24, 85
Insurance 90-91
INTERNIST 107
Intersubjectivity 78
Intuition 34, 268
Intuitionism 32-33, 138
Irrelevance 7-9, 26-27, 87, 93, 150, 238, 242, 251, 258, 302
Irrationality 18

Judges 20, 23-27

Language 7, 226-229
Law 10, 23-27, 73-77, 279, 292, 298, 302
Law of Excluded Middle 33, 111-112, 115, 138, 147, 214
Law of large numbers 51
Likelihood ratio 89, 103, 106
Load factors 22
Logic:

Dualistic 135, 138-139, 147-148
　　　General 16, 31-35, 108, 110, 168, 198-199, 213-215
　　　Fuzzy 111-112
　　　Modal 75-76
Logicism 31-33, 35

Management xi, 3-4, 266-267, 274
Marxism 5, 19, 221-223
Matching law 205
Mathematics 4, 11, 28-40, 108
Medicine 98-100, 107
Meta-ignorance xii, 6, 15, 146, 258, 261
Modularity 102-103, 105, 107
MYCIN 99-104, 107

NASA 85
Necessity 17, 128-130, 287-288
Nonsense 18
Nonspecificity 8-9, 94, 113, 135, 146, 227-228, 255
Nuclear power 86-88, 257

Operationalism 53, 78, 161
Opinion 16, 78-83

PASCAL 227
Perception 213
Philosophy:
　　　General 1-2, 14-20, 62, 100, 108, 113, 218
　　　Nominalist 298
　　　Realist 217, 298
　　　of Science xi, 260
Plausibility 128, 134
Politeness 226-229
Positivism 18, 53, 57-58
Possibility:
　　　Concept of 45, 48-50, 114, 140, 285, 288
　　　Theory of 93, 96, 118-120, 128-129, 133-135, 142, 148-149, 165
Pragmatism 17, 19-20, 30, 45, 115, 266, 298
Prediction 158-161
Preference 77, 163, 190-191
Preference reversal 192-193
Principle of Indifference 48-50, 65, 67, 69

Prior distribution 65-67
Privacy 219, 226, 231-234, 251, 254, 256, 258, 296, 305-306
Probability:
  Aleatory 44, 47, 119, 210
  Bayesian (subjective) x, 22-23, 41-43, 58-69, 77, 82-83, 91, 112-114, 120, 136-137, 161-163, 209-210
  Classical 43-44, 47-52
  Epistemic 44, 210
  Inverse 51
  Intervalic 63, 67, 70, 83, 90, 96, 125-127, 136, 143-144
  Judicial 23-24, 73-77, 137
  Logical (a priori) 43, 69-70, 84-86, 91
  Nonquantitative 70-77
  Relative frequentist 42, 49, 52-58, 60, 84-86, 91, 112, 268
Probativity 25, 43, 75
PROLOG 129
PROSPECTOR 99, 103, 106-107
Prospect theory 191-192, 248, 276, 279
Psychoanalysis 152, 154-155

Quantification 22-24, 27, 60, 66, 76-77, 84-85, 149-150, 162-163, 164-168, 210, 267, 269, 289

Randomicity 54, 56-57, 174-175, 184, 189, 209
Random sets 142
Rationalism 18, 66, 210, 217, 289
Rationality 12, 18, 44, 64, 67, 76-77, 97, 163, 197, 208-213, 271, 277, 297-307
Reactance 233
Reality testing 235-236
Reductionism 30-35, 43, 45-46, 135, 138-140, 147, 203-208, 266, 271, 274, 291, 296-297
Reflection effect 191-192, 279
Reinforcement 159-161
Representation of ignorance 49, 68, 83, 88, 90-91, 94-96, 121, 136-137, 210, 268-269, 275, 278-279, 284, 288-289, 293, 299, 301-306
Representativeness heuristic 188-189
Resilience strategy 291-295, 298, 300-301
Reward delay 204-206
Risk xi, 3, 83-91, 176-179, 236, 243, 249, 257

Safety factors 22, 90

Sanity 18-19
Secrecy 15, 219, 224, 226, 231-234, 247, 251, 253-254, 256, 260-261, 296
Second order:
  Fuzzy sets 109-110
  Ignorance 93, 125-135, 201, 203
  Logic 35-36
  Strategy 35, 39, 266-267, 291
  Uncertainty 125-135, 140, 147-148, 201, 203, 275-290, 299
Self-disclosure 226, 230-231
Sensitivity analysis 22
Simulation heuristic 187
Skepticism 2, 14, 16-18, 259
Specialization 4, 37, 251, 307
Social structure 218, 226, 234
Sociology of ignorance x, xii, 5-6
Sociology of knowledge x, 5-6, 39, 216-226, 259, 271
Suburbanization in:
  Ignorance 267, 274, 291
  Mathematics 37-39, 148
  Probability 46
  Uncertainty 93, 138
Support logic programming 129-130
Surprise theory 93, 120
Symbolic interactionism 223-225, 227
Synthesis 135, 141-145, 267

Taboo 8-9, 15, 25, 234-235, 238, 242, 253, 296
Theory of types 32
Topoi 138
Turbulence 241, 243

Uncertainty:
  Accounts of 4-5, 245-250, 262-263
  Definitions of 7, 9-10, 112-114, 131-133, 141-142, 185, 241-245
  General x, 3-4, 15-16, 42-43, 45, 51, 78, 84, 92-93, 107-108, 153, 197
  Group level 83
  Measures of 131-135
Uncertainty avoidance 156, 237, 247-248
Undecidability 8-9, 135
Untopicality 8-9

Utility 64-65, 78, 84, 140, 161-162, 176-181, 190-193, 208, 211-212, 276-285, 288-289

Vagueness:
 Definitions of 94, 113-114, 141-142
 General: 2, 8-9, 24, 93-94, 135, 146, 197, 258, 305
 in Language 227-229
 Measure of 132-134
 in Probabilities 95, 202, 277-278, 283-284

Welfare functions 84

INVENTING WOMEN'S WORK: THE LEGACY OF THE CHARABANC THEATRE COMPANY

The 1980s is now seen as a golden decade for Northern Irish drama. Companies, playwrights, and theatres made the North, and Belfast in particular, a centre for dramatic arts similar to early twentieth century Dublin. And no group did more to foster that healthy theatrical climate than Charabanc Theatre Company which was formed in 1983 by five actors – Marie Jones, Maureen Macauley, Eleanor Methven, Carol Moore, Brenda Winter.[1] Company members first collected oral history, next developed their own plays and performance methods, and then toured those productions more extensively than any other Irish company in history. In the twelve years of the company's existence, 1983 to 1995, Charabanc produced and performed eighteen new works and four extant works.[2] They took these twenty-two productions throughout Ireland, to small community centres as well as to the main professional theatres and festivals. They also toured to the USSR, Germany, and Canada; four times to the USA; ten times to London; five times to Glasgow's Mayfest; and once each to the Edinburgh Festival, to the Brighton Festival, and to Cardiff. And at the Betteshangar Miner's Welfare in Kent during a strike, Charabanc performed *Lay Up Your Ends* (1983), their play about a Belfast mill strike. Despite their international profile, however, the fact that they presented plays to Irish people in their own neighbourhoods continues to be a source of pride to the Charabanc women.

This book contains four original Charabanc playscripts: *Now You're Talkin'* (1985), *Gold in the Streets* (1986), *The Girls in the Big Picture* (1986), and *Somewhere Over the Balcony* (1987). (A Glossary follows the play texts; the Appendix lists production and tour information.) The original plan for this book included the company's first play *Lay Up Your Ends*; however, one of the six authors of that play would not agree to the royalties offered him. Proposed percentages recognized the collaborative nature of the work and credited research as well as writing. Desiring to credit appropriately the contributions of all the company members, the publisher and the editor jointly decided to remove this early play from the collection. In the section of this introduction chronicling

INTRODUCTION

the beginnings of the company, I discuss *Lay Up Your Ends* at length, so readers of this volume will, in the end, learn sufficient detail about this first, somewhat flawed play.

The scripts in this collection represent the best of Charabanc from their third to their sixth play *Balcony*, which marked the company high point.[3] Never sombre no matter what the subject, these comedies demonstrate the broad range of the company's creative work: portraits of urban and rural women; early, mid-, and late twentieth century settings; and various social, religious, historical, political, or personal situations. The four plays raise cultural and gender issues with an unequaled wit and a distinctive style and deal fearlessly with a variety of recurring Northern Irish themes – emigration, labour disputes, political divisiveness, intransigence, and reconciliation. Women's power and resilience, regardless of the circumstances, are consistent themes throughout Charabanc's work. Plays written by women, especially collaborative works, are seldom published; however, these unique scripts from an unusual company with a creative method which bridges many disciplines have wide-spread interest.

The story of an organization can merit telling for any number of reasons – the fascinating people involved, the overall impact of the organization, or even the intriguing tale itself. The story of Charabanc Theatre Company needs to be told for all three of these reasons. This singular Belfast enterprise included during its existence many of the major theatre artists on the island (see Appendix). Considering how unstable the business of theatre is, how seldom companies last more than seven years with the best resources, the company's twelve-year survival is, in itself, noteworthy. Charabanc's history is a triumphant story of women creating their own work and in the process changing the shape of Irish drama. Katy Hayes, founder member of Glasshouse, and Lynne Parker, founder member and Artistic Director of Rough Magic, made the following extemporaneous comments 28 June 1995 during a panel discussion of Irish women theatre artists at the American Conference for Irish Studies (ACIS) meetings in Belfast, Northern Ireland.[4]

HAYES: When I was a student ten years ago working in the drama societies in UCD, certain elements in Irish theatre were very influential. One of them was Charabanc, which at that time was touring some of their early works, and I went to see a number of their plays. And I felt that these were the kind of stories that I was interested in hearing and would perhaps be interested in telling or helping tell.

INTRODUCTION

PARKER: I was born in Belfast but left, ironically, just at the time when something very interesting was about to happen in theatre for women in the North of Ireland and for all of Ireland. This was Charabanc. It was very interesting for me to see their work. And I'd have to admit that I was not interested in them so much as a women's theatre company or a company which worked for the community; quite simply, I hadn't seen quality like that before in Ulster Theatre. This was a real shot in the arm for me.

Hayes and Parker express the two aspects for which Charabanc Theatre Company is most known: the uniqueness of their stories and the excellence of their productions. That each of these directors, before commenting on her own work, took time to mention the work of another theatre company is some indication of the influence Charabanc has had on theatre throughout the island.

Hayes and Parker echo my own intrigue when in 1983 after seeing *Lay Up Your Ends* I conducted my first interviews with company principals. Trying to capture the Charabanc magic became my quest over the years. Marie Jones emerged as a writer during the development of that first play and was credited with the scripts which grew from the group's largely self-discovered method of collaboration. First, company members would identify a time period or an issue or a dramatic event; then they undertook extensive research – recording interviews, finding documents, viewing records of all sorts; finally, they crafted the play itself during the writing, rewriting, and rehearsal stages.[5] After seeing many of their productions and after numerous conversations with members of the group, I followed closely the development of *Somewhere Over the Balcony* which is now generally recognized as their finest play.[6] At that point, the company members were Marie Jones, Eleanor Methven, and Carol Moore. Then as a Charabanc board member for the company's final four years, I learned firsthand what led to their phenomenal success. Ironically, unknown to most of the ACIS panel participants, the planned agenda for the Charabanc board meeting the following night (29 June 1995) was to discuss how the company would cease operations. At that final board meeting, I was assigned the sad task of writing an upbeat press release (10 July 1995) announcing the company's decision to disband or to 'get off the bus' which I characterized as their doing 'the unthinkable'.[7]

What was remarkable about this particular mix of individuals; what prompted their entrepreneurial spirit; and what does Charabanc's highly unlikely success reveal about the potential for other women who want to create their own work? And what about the quality of

INTRODUCTION

the work itself? Do the plays produced in this creative but chaotic way hold up critically? What are the merits of these plays as theatre and as women's history? And then what of the company's unique methods – how were they developed and refined over time; how did they vary with different plays and collaborators; and how were researching, writing, rehearsing, performing, and touring integrated into the total process? Finally, what ingredients in Belfast made the work of Charabanc possible?

In this introduction, I explore these issues, as well as others, and discuss the workings of the company under four headings: **The beginning** section is the story of the development of the company and their first play. **The value** section is an analysis of the group's innovative methods and an assessment of the quality of the work. **The journey** section is a detailed discussion of the four plays in this collection, placing them within their Northern Irish contexts and rendering any additional preface to the plays themselves unnecessary. This section also contains a brief discussion of some of the company's work not included in this volume. **The legacy** section is an examination of the overall significance of the enterprise for theatre and for women. The magic of Charabanc lies in the stories and in the way they are told. The mythic stories a people tell themselves about themselves reveal the deep issues that concern any given society. In Ireland, theatrical narratives are frequently the means used to present the dilemmas confronting the culture, and Charabanc Theatre Company worked hard first to recapture and then to present theatrically these cultural narratives from the point of view of women. To tell the history of a company which was dedicated to the telling of history is an exercise which has a pleasurable circularity.

The beginning...

The lobby was buzzing before the premiere performance of *Lay Up Your Ends* at the Belfast Civic Arts Theatre on Botanic Avenue, 15 May 1983. But the theatregoers gathering there in friendly clusters that Sunday night were an unusual mix even for the Arts. Working-class accents vied with Malone Road ones, so voices commonly heard at Belfast mills were competing in the hubbub with those more typical at the theatre. With a mixture of wonder and shy curiosity, individuals eyed each other across the obvious class divide as they filed in and found their seats. Those who had told their stories of working in the linen mills, who had by their histories made this night possible, many who had never gone to a play at a downtown theatre before, were now rubbing shoulders with the theatre faithful.

INTRODUCTION

The one thing nearly everyone there that night had in common was that they had met the Charabanc five – Marie Jones, Maureen Macauley, Eleanor Methven, Carol Moore, Brenda Winter – the effervescent actors who had created this play about a 1911 Belfast linen mill strike as their own unique industrial protest. *Lay Up Your Ends* was the newly formed Charabanc Theatre Company's response to underemployment; in fact, starting the company was their concerted effort to overcome the lack of strong women's roles. Only three of them had suffered from unemployment in Belfast theatre, but they were all discouraged about the roles on offer. Rather than merely complain about the situation, however, they decided to do something positive and create their own work. So with a borrowed stake of £1000 from Ian McElhinney, they moved with dispatch to realize both a play and a company; from January 1983 when they first conceived the idea to May when they opened, the group researched, wrote, rehearsed, and then performed this play that changed Irish theatre history. They had initially thought of *Lay Up Your Ends* as a one-off and had no idea then of what lay ahead for them or for their fledgling theatre company.

These five women, while great for the *craic*,[8] are not what you would expect actors to be; they are sensible, dependable, personable. They are also, nonetheless, strong individuals with differing opinions, despite the widespread belief that Charabanc was an organization with five heads speaking in unison. Their first director Pam Brighton's rather odd recollection of them then emphasizes an ordinariness: 'all looked unlikely candidates for the stage – too big, too small, too scrawny, too plain'.[9] But these remarkable women had marshalled tremendous interest in their project during a very short time. The excitement they had generated was rare even for an opening night; unheard of on a Sunday, people queued down Botanic Avenue that May night and 560 squeezed into the theatre. From the moment the stage lights came up and these five exceptional looking women strode onto the stage singing – 'Mind your frame, look to your yarn, lay up your ends' – there was instant recognition in the audience. Women who had worked the looms were back in the York Street Mill again as the process was mimed on stage and the ear-splitting screech of one hundred machines filled the auditorium.

That shrill sound brought the reality of women's experience in the linen mills into full consciousness better than nearly anything else dramatized in the production. Even though the play represents many aspects of the oppressive working conditions – the ankle-deep water on the floor, the twelve-hour days, the overbearing foreman, the poor

INTRODUCTION

pay and concomitant money lender – nothing works quite as well as that unremitting shriek. Once the sound started, when I first saw the play during the Dublin Theatre Festival in October 1983, the only thing that made it possible for me to remain in the John Player auditorium was the knowledge that the women were going to go on strike so the unbearable noise would, no doubt, cease. And the relief when they turned off the machines after only a few minutes must have been somewhat like walking out of the mill at the end of a long shift, but the ringing in my ears seemed to continue throughout the play.

Lay Up Your Ends may have begun typically enough with the five women striding onto the stage singing on their way to work. Soon, however, the performance displayed unmistakable originality. The intensity of the script and the verve of the acting style were breathtaking – literally. Scenes flowed one into another; the wives of the mill owners singing Gilbert and Sullivan suddenly jolted forward *en masse* and became instantly the mill workers whose charabanc had just stopped. Sardonic Belfast wit crackled. Here were real, ribald women on stage – talking about sex, teasing each other unmercifully, pulling down the pants of the hapless young man sweeping the floor. When my landlord saw the play at the Arts Theatre in Belfast near the end of the run, Andy smiled when the young man came on stage: 'That was me. I had the job of cleaning up and bringing the women flax, and they were always at me', he said, chuckling. During *Lay Up Your Ends*, the five women performed many roles, including crowd scenes, and the raw energy of their performances was exhilarating.

Woven into the play are the stories of five central women characters, ranging in age and circumstance, Catholic and Protestant, but all struggling as mill workers to provide both necessities and the odd, small treat for their families. Although barely eking out an existence, they courageously walk out of the mill because of new, unfair rules and a proposed cut in hours and, therefore, in pay. But they are immediately overwhelmed by criticism and the increased privation the strike brings to their respective families. The play shows their growing awareness of the significance of their desperate act. They form a strike committee, choose spokeswomen, collect a strike fund, and effectively answer opposition to their actions. By agreeing to follow the advice of James Connolly, a Republican Trade Unionist based in Dublin, and to reject the pleas of their own Mary Galway to return to work, the women place themselves at odds with their Ulster culture, their husbands, and many of their coworkers. And when finally even James Connolly tells them to return to work, they walk in together singing in the face of the intransigent mill owner. They may appear to have made no headway, but they have gained a sense of their own power as workers. At the end of the play, they defy the new rules against talking or singing or combing their hair; they resolve to walk out again with anyone who is suspended unfairly; they profess their solidarity as women workers; and in Brechtian fashion, one of the actors steps out of the scene and reports directly to the audience that 'two weeks after those York Street Mill girls went back to work, the first branch of the Irish Transport and General Workers Union was formed, especially for the Belfast Mill girls'.

These five Charabanc women reinvented vital women's work. They found rich stuff for drama while working together in their embattled city. Maureen Macauley's note in the *Lay Up Your Ends* programme explains the political climate in Belfast during the early decades of the century:

Belfast was divided not only by economics, but by religion and politics. Unionists were suspicious that a growing Irish Nationalism would undermine Belfast's prosperity, and the whole country was split by the Home Rule question. Sectarian animosity flourished and rioting became endemic. This was perhaps, not the best atmosphere to foster Trade Unionism. Suspicion and fear divided the workforce in their fight for rights, better pay and conditions. Yet Belfast was not all drab poverty. Naturally not everyone could afford a weekly trip to the Alhambra or the Opera House, but most people could look forward to the occasional charabanc trip to Glengormley or Bangor.

The name for the theatre company occurred to them during the writing of this first play; the characters go on a charabanc to lift their spirits during the strike. Ian McElhinney, identified as the play's producer, gives their reasons for choosing such an unusual name for the company: 'Charabanc is a day out, a tour to remember, an occasion of song, merriment, a pick-me-up, a nostalgic reminder of the brighter side of Old Belfast' (programme note). *Lay Up Your Ends* and the company's subsequent work certainly fulfilled these merry expectations.

Very few theatrical productions enter the mythology of a people – a scene here maybe or a character there might – but in this case an entire play, its tour, and the birth of the company that wrote and performed it has become the stuff of myth. *Lay Up Your Ends* was a watershed experience in Irish theatre. Those who were in the audience on opening night can easily recapture the glow they felt. And all you need do is mention the evening to any of the company principals and faces radiate with joy. Nothing has diminished that unforgettable Sunday night at the Arts. And what led up to that triumphant night is a story in itself.

On 4 November 1983 in a ninety-minute interview at Brenda Winter's house in East Belfast, the five founding members, giddy with their unexpected success and finishing each other's sentences, described the company's beginnings. This interview, punctuated with their laughter, records the verbal style of the group – interruptions, banter, simultaneous talking, each member adding to the ideas of the others. (What follows, unless otherwise indicated, comes from that tape.) Most of the company's founders thrive on this easy give and take. Brenda and Maureen, though, admitted years later that they had sometimes felt overwhelmed and silenced, one reason, perhaps, why neither was associated with the company past their second play *Oul Delf and False Teeth*. Nonetheless, to attempt as they did a fully shared collaboration on ideas, research, writing, and management was brave and was also very nearly realized during the early years of the company.

Marie Jones, speaking above that exuberant chorus, explains the happenstance involved in the company's beginnings in January 1983: 'The five of us came together by chance rather than by design. Some of us knew each other well and others had worked together but all of us were in the same situation. We had all been out of work too much.' Sounding a unanimous note, they insist that if they had been asked to play one more cipher or long-suffering victim they would have screamed. Eleanor Methven colourfully describes her early desperation as an actor during the ACIS panel: 'In the first five years of my professional life, I played nothing but Noras and Kathleens. I did a fine line of them. I was told in no uncertain terms never to cut my hair or the work would dry up', clearly a reference to Eleanor's thick reddish mane. But then getting any acting role at all was difficult for Irish women. Speaking on the panel about Charabanc's beginnings, Eleanor emphasizes the necessity for the company: 'We were five very demoralized actresses who watched actresses being flown over from England to take three-line parts. And when that happens, you are no longer angry; you are just defeated. You think, "I am so bad that they cannot give me three lines. I'll just go put my head in the oven".'

During Sunday evening meetings, these five Charabanc founders talked first about the possibility of creating a small lunchtime show since they had no Arts Council grant. They discussed producing a few funny sketches by Martin Lynch, a working-class Belfast playwright

INTRODUCTION

who was just finishing a term as writer-in-residence at the Lyric Theatre in Belfast. Since Lynch had written mostly about men, however, they decided to create a play about their own Belfast background and ask Lynch to write it for them. Then all they would have to do, as Marie put it, 'was act and be quite cozy'. Lynch surprised the Charabanc women by saying he would help them write the play themselves. In the *Lay Up Your Ends* programme, Lynch describes the first time (February 1983) he met with the group: 'I listened as they told me how they wanted ME to write a play about the experience of Belfast women. The absurdity of this struck me immediately and I asked them why THEY couldn't sit down and write about THEIR experiences as Belfast women. This produced an instant silence, followed by laughter.' Even though apprehensive because they had not written anything before, the women agreed to collaborate on writing a play.

Because of the widespread belief that the city of Belfast had reached its peak at the beginning of the twentieth century, the group decided to set the play sometime during the first few decades. All five women began researching Belfast history, each ambitiously taking a ten-year chunk, accessing the newspaper archives at the Central and Linen Hall Libraries in Belfast. The mills naturally became a focus of their research because the linen industry was so prominent in Belfast during the early part of the twentieth century. A programme note explains that 'in 1912 there were 37,292 power looms in Ireland weaving yarn from hundreds of thousands of spindles'. Older Belfast women talked frequently about their mill experiences, and then some Charabanc members also had direct family connections with the mills. But Carol acknowledges that the wealth of information about the mills was actually a problem in early drafts of the play: 'there was too much colour'. They had to develop the discipline to leave out details, no matter how interesting, if the information was not necessary to advance the dramatic action.

The mills' prominence, however, was not the main reason the company used the linen industry as their starting point. In the industrial and politically divided city of Belfast, the mills were unique because they crossed all sectarian barriers. As Eleanor clarifies, 'Everybody worked in the mills; it was not a Catholic job or a Protestant job. Although workers in the shipyards tended to be Protestant because the shipyards were in East Belfast, many of the linen mills were in West and North Belfast.' So mill workers came from the most thoroughly mixed areas of Belfast. 'If you're a community theatre company, that's ideal', Eleanor explains, 'because you can take a production into nearly any area and it is relevant. In any audience, people will have worked in the mills or come from mill houses.' During those weekly meetings, the struggle of women in the Belfast mills captured their imaginations and emerged, without question, as the story the Charabanc women wanted to dramatize.

Although nearly everyone the Charabanc women interviewed seemed to have some connection to the linen mills and were excited that the mill story might finally be told, the company still needed a dramatic hook on which to hang the stories they were busily collecting. All the ingredients were there – they had their subject, the mix of individuals and the level of commitment was right – but the play needed a central dramatic conflict. Then while visiting Belfast City Councillor Paddy Devlin, they found their focus; Devlin told them about the linen mill strike of 1911 when 2000 women walked out of mills all over Belfast. They went back to the papers once they had something specific to look for. But trying to find any mention of a massive, two-week strike of women mill workers was very difficult. Marie explains that the strike was recorded in only one Belfast paper on 'one column of about two inches. And nothing in the *News Letter*. *The Irish Times* was a wee bit better; it told how many days the strike had been going on and how many mills were involved and how many thousands of people were still out of work. And then it gave a few of the confrontations between Mary Galway and James Connolly – her saying you should go back and him saying no stay out.'

Because the strike was not well documented, the women talked to more people about the strike itself and 'invented' some things that probably would have happened. However, Eleanor emphasizes the overall authenticity of the play: 'The confrontations were all factual; Sadie Patterson [an elderly former mill worker and labour organizer] was a huge help in that sort of thing'. Marie adds, 'The 1911 article in *The Irish Times* reported that Miss Mary Galway said James Connolly should get back south where he belonged and asked what had he to do with Belfast anyway since he was an extremist'. Extremism of a republican stripe was especially frightening in Ulster. But the women were incredibly strong, as Marie says: 'They wanted their rights. The most important thing to them was their rights as workers, and they held out for two weeks.' Besides the newspaper accounts, other important written sources the group used to construct the play were *Picking Up the Linen Threads* by Betty Messenger and *The Illustrated History of Belfast* by Johnathan Bardon.

INTRODUCTION

So the project evolved in those weekly planning meetings held by this time at the Ravenhill Park home shared by three of the principals – Eleanor Methven, Marie Jones, and Marie's husband Ian McElhinney. Speaking for many in the company, Brenda describes the energy those meetings required: 'Charabanc meetings take more out of me than performances'. But Marie also speaks for the group by emphasizing their total commitment: 'They're tough but you have to be there'. Although at the beginning they thought in terms of doing one play for the fun of it, the idea of a trilogy of Belfast plays and of actually forming a theatre company to produce them gradually came into focus. One front room of the Ravenhill Park house became the company office. At those Sunday evening meetings, the organization solidified, and Martin Lynch, Cherry McIlwaine, and Ian McElhinney were named company directors.

And then finally they found the money. They needed to fund salaries which the £1000 McElhinney had loaned them for production costs and the small grant from the Belfast City Council Paddy Devlin had helped them secure would not cover. After devoting months, entirely unpaid, to researching and writing the play, they needed to hire a stage crew and an administrator to handle the tour. Lynch received a commission worked out by his agent for whatever writing he did for the group. These five women founders were professionals; in spite of their desperate lack of funds, they wanted to start out right and not exploit anyone, not even themselves. At the suggestion of Alex Clarke, their Equity Secretary, they approached Action for Community Employment (ACE). The possibility of an ACE grant was clearly one reason they organized into an official theatre company, but ACE was an organization that was more accustomed to financing small businesses which produce a tangible product rather than something as ephemeral as a play. Charabanc's challenge was to convince ACE that their project was also worthwhile to the community. Janet McIvar quickly caught the vision, sold the idea to her bosses at ACE, and helped the newly formed company qualify for a scheme which finances small businesses hiring the unemployed. At this early stage of the company, being paid to do the work they wanted to do was almost a surprise. And despite their move over the years toward increasing professionalism and financial success, doing good work remained the group's primary motivation; getting paid was simply an added bonus.

The phrase *lay up your ends* means to join the yarn. A new strand is joined to the strands where the spinning left off the night before, or if the yarn breaks while the machine is running, then the strands are joined and the machines are started again. The phrase designates what a worker does when spinning new yarn. The appropriateness of *Lay Up Your Ends* being the name for the initial play of a new endeavour like Charabanc dedicated to spinning the yarns of the culture is clear. The group's purpose was to describe the joined strands of women's lives. What kind of prescience was the company demonstrating then by picking this particular weaving term to name a play about a strike when 2000 women walked out of mills throughout Belfast, showing a solidarity unknown before in the industry. A desire for a similar connection with each other and with the women of Belfast prompted these five founding members to produce this play. These Charabanc women from all sides of the cultural divide were tireless in fulfilling their policy priority to 'create and perform strong female roles in Irish theatre'. The pictures of them standing casually in front of the York Street Mill published in the souvenir programme for that year could not have been further from the truth. Company members were always on the move – interviewing sources, collecting information, calling venues, distributing leaflets. In fact, their indefatigable energy is the reason Martin Lynch gives for not delivering what he had promised during the writing process for the second play, requiring Marie Jones to step in as principal writer; Lynch says he was literally unable to keep up with the women and dropped out of the company.

What foreknowledge, as well, caused them to give the company the obscure often mispronounced name Charabanc (pronounced 'Sharabang'). No other designation could be more apt to name a company which took the 'brighter side of Belfast' throughout Ireland and the world. In his programme note, Ian McElhinney emphasizes their awareness of the resonance of both the name of the company and the title of the first play. Explaining that the titles were among their final decisions, he writes, 'Charabanc seemed to capture precisely the nature of the group's intentions. We are of this community, we create from this community experience, we write for this community, and we hope to play to your local centre and throughout the province, to go further afield and present this community's history to others.' Going on a charabanc for a much needed break from work is still a tradition in Northern Ireland, but the company gave the word new meaning. During those twelve years and twenty-two productions, the history of Charabanc Theatre Company became inescapably linked to the women's history they were dramatizing; their very existence joined the strands of women's lives.

INTRODUCTION

The value . . .

Charabanc's plays focus on women committed to improving their own lives and strengthening their communities. The company's method of first collecting oral history and then writing a play and performing it for their sources creates situations and dialogue that ring true, even to the women whose history they were retelling. Charabanc was, without doubt, the most successful touring company throughout Irish theatre history, but they hated that they were frequently referred to as the 'best women's touring' company – there were no others. Lynda Henderson reported in *Theatre Ireland* in 1983 that, after the 'more than capacity audience' for the first performance at the Arts Theatre, *Lay Up Your Ends* 'went on to tour community and leisure centres as part of the Belfast City Festival. On several nights it was attracting larger audiences in those non-theatre spaces than were the two producing theatres in the city combined – proof that this particular "alternative" has both caught the public imagination and is touching new audiences.'[10] If all those who claim to have seen this first play had actually been present at a performance, then *Lay Up Your Ends* would have had to run much longer than the five months from 15 May to 22 October 1983. When they finally laid it to rest that year after a week back in the Arts, 13,515 people had seen one of those 96 performances in the 59 different venues.

To see a Charabanc production was such a pleasure, such a wonderful mix of story and song – confrontive, dramatic, entertaining. I had first seen *Lay Up Your Ends* on the last night of the Dublin Festival run, and Marie commiserated: ' It was cold and wet and miserable, and we'd a job trying to warm the audience up. It was a real challenge to say, right – they're not going home in the miserable way they came in.' That effort was certainly not lost on me; the play seemed to leap off the stage into the auditorium. Charabanc had promptly developed a clear theatrical vision. Company members knew how theatre could simultaneously be daring and fun; their plays manifest a highly evolved Belfast sense of the ludicrous. Belfast playwright Stewart Parker claims that educating in an entertaining way is the primary goal of drama. Parker's work, although very different from Charabanc's, nonetheless also uses a mixture of story and song and displays a similar sardonic humour. Parker believes that for the word *drama* should be substituted the word *play*, saying 'that we learn nothing of consequence other than through play'. By being such a strong advocate of the ludicrous (derived from the Latin *ludere* – to play), Parker shares an affinity with Charabanc. He argues that 'play is how we test the world and register its realities. Play is how we experiment, imagine, invent, and move forward. Play is above all how we enjoy the earth and celebrate our life upon it.'[11] Parker's belief in the power of play is echoed in Charabanc's theatrical vision.

Lay Up Your Ends was reviewed favourably but also patronizingly. The script does present problems. Similar to many message plays, the emphasis is on telling rather than showing. Despite the freshness of the company's performance style, too much of the play's action takes place off stage; for example, during one long scene, the five actors must react to a sound tape of imagined political speeches by James Connolly and Mary Galway, presumably debating the merits of the strike at a mass meeting. Some reviewers were quite critical of the company once Charabanc continued to produce plays; they unfavourably compared later works to this first play, which became enshrined in mythic proportions. As an example of the too ready critical dismissal, David Nowlan in his review in *The Irish Times* (29 December 1987) refers to *Somewhere Over the Balcony*, the company's sixth play, as 'cobbled together' rather than written. Although *cobbled together* might aptly describe the construction of *Lay Up Your Ends*, the expression seems unfairly dismissive of *Balcony*, a script which displays a finely honed dramatic sophistication. But then women's plays seldom receive their critical due. Eleanor Methven describes the critical response to Charabanc in her 1995 ACIS panel remarks:

The work was dismissed as written by committee. It did not seem to matter whether it was good, bad, or indifferent; it was just, well, 'It is not really a play, is it', because it had not been written by a person sitting at a typewriter. I mean obviously somebody wrote down the dialogue [usually Marie Jones], but the early drafts of our scripts have at least four or five people's handwriting in them because we did whatever we had to do to get them finished. . . . But we were continuously dismissed in that vein. And it was also the accent, and I mean that not just literally but as the cultural base from which we were coming. We did very much concentrate in the early days on an urban Belfast working-class approach and style of language for the productions. Again, that was almost denigrated.

What Eleanor says about the accent or cultural base causing the work to be belittled was borne out later during the discussion period following the ACIS panel. When asked how Charabanc was received by working-class women in Belfast since there was no theatre tradition in Protestant culture, Eleanor addressed the subtext underlying that false perception – Catholics have culture but Protestants do not. Charabanc members were largely from a

INTRODUCTION

Protestant ethos, especially after Brenda Winter and Martin Lynch left the company, so, in essence, this widely-held misperception disenfranchises the company culturally. But then many theatre artists, including Stewart Parker who was Protestant, suffered from this prevalent cultural misunderstanding.

Choosing a play, buying a ticket, and visibly going to a particular venue implies a commitment to listen and that choice is, at times, a political statement in itself. Attitudes can change under the influence of entertaining, confrontive work like Charabanc's. A performance can cause audience members to question their behaviour, including political intransigence. During the panel discussion, Eleanor acknowledges the company's vital role in a polarized community: 'For the last twelve years, we have been examining where we come from and why and trying to reflect the community through our work back at the community. Theatre is about the life of the imagination, certainly; the political happenings here over the last twenty-five years cannot allow for imagination.' Charabanc used the Belfast 'accent' to pose uncomfortable dilemmas for audiences. Charabanc plays explode stereotypical thinking, drawing audiences into a story about themselves by presenting troubling aspects of the culture humorously. Eleanor identifies the question fundamental to their work: 'Must we as a culture continue to behave this way?' Their plays, however, also teach truths about Northern Irish culture to outsiders who are hungry for understanding, so the group was successful in part because of their Belfast roots. Charabanc was presenting vitally important stories to receptive, enthusiastic audiences.

By fall 1983 the mythmaking process was already firmly in place. Several different individuals told me they had formed this unusual women's theatre company. Everyone seemed to own it. Although in charge from the beginning, the 'Charabanc girls', as they were often called, had sparked widespread, proprietary interest in their project. Many were probably charmed by the graciousness of these women, by their unaffected, self-deprecating manner. Others may have been attracted to the rousing good spirits of any Charabanc gathering. Ownership, however, implies more than a fun time was had by all. Even though many of the principals severed their official association with Charabanc years ago, the *idea* of the company continues to be an issue for most of those who were involved in the beginning. For some, the idea has to do with which type of plays are Charabanc plays; for others, the idea concerns the style of the productions themselves; for some, the idea is more what venues Charabanc should have played or how extensive the tour should have been; for still others, the idea centres on the political stance of the work itself or on the working-class status of various company members.

Martin Lynch, for example, was somewhat critical of Charabanc's later work, believing they had abandoned their original brief as a community theatre company. In that 1983 interview, however, Marie says, 'So we got Martin over and he said, "Oh yeah, that's exactly what I've been wanting to do for a long time. I've wanted to do community theatre." We weren't actually thinking of community theatre at that time. We just wanted work.' Each of these strong individuals brought their own *idea* to the company. Lynch came to the project with his idea of a politically-based community theatre and was disappointed when economic conditions caused them eventually to tour fewer community centres. But politics had intruded. After the Anglo-Irish Agreement outlining a framework for cooperation between Dublin and London on Northern issues was signed on 15 November 1985, certain Unionist political groups within Northern Ireland protested the Agreement by refusing to cooperate; the protest strangled City Councils and no business was conducted, including a total curtailment of local arts funding. Charabanc was then left in the bizarre position of being able to obtain grants to tour internationally but not locally. Lynch argued that Charabanc should continue to tour community centres without financial support, even though the company struggled merely to break even financially. His suggestion clashed with the founding members' *idea* of the company's professionalism, committed as they were to the quality of their work and to the welfare of their employees. Nonetheless, because of Charabanc, an important phenomenon in Northern Ireland is taking theatre directly to the people, getting them involved in writing the plays, interviewing them, using their pictures or their lives as models to tell history in a new way – a project Martin Lynch has continued with amateur groups.

Director Pam Brighton, who left the company before their tour to Toronto with *Now You're Talkin'* in 1986,[12] wrote a controversial 1990 update on the company for *Theatre Ireland,* just before she formed DubbleJoint Theatre Company. Brighton condemns Charabanc's artistic direction, the choice of scripts, the acting, a 'jobs for the girls' attitude, and even the political stance of the participants. She complains in particular that Charabanc performed *Cauterised*, a new play by Neill Spears, for the 1989 Belfast Festival. Perhaps because this play is about sexual not national politics, Brighton found it 'disappointing – a thin derivative piece' and 'the antithesis of everything that Charabanc had set out to do'.[13] Yet in a previous

INTRODUCTION

issue, *Theatre Ireland* published two other reviews of *Cauterised*, ones vastly different from Brighton's: Brian Baird did not care for the play while Jane Coyle did, but both praise the 'deft' pace, 'impressionistic subtlety', precise 'timing', and 'uncannily lifelike performances'.[14] There is no hint in their reviews of the 'lack-luster' or 'caricatured performances' Brighton describes.

In response to Brighton's article, Maureen Jordan, then Chairperson of the Charabanc Board of Directors, wrote the editor: 'Ms Brighton's last professional contact with the company was directing *Now You're Talkin* (the company's third production) in 1985. From that time until 1989, to our knowledge, Ms Brighton saw none of the intervening six productions, despite invitations to every subsequent London opening. It is therefore unsurprising that the bulk of the company's work is omitted from the "profile".'[15] Without identifying the source, Brighton suggests in her article that Charabanc is criticized for 'pursuing a doggedly non-sectarian line', and for avoiding controversy. Yet nothing could have been further from the truth; the company was fearless in their search for issues and individuals needing a voice. Curiously, Brighton does not mention *Somewhere Over the Balcony* (1987), the only Charabanc play to focus on only one side of the cultural divide. How could this politically charged play about West Belfast Catholic women be characterized as an effort to avoid controversy? But Brighton does assert that 'Charabanc itself has set the criteria for committed, robust, imaginative theatre'. She also states that *Lay Up Your Ends* 'was one of those pieces which, as a director, you can plan, plot and calculate for but never achieve unless the particular mix of personalities, time, place and material are absolutely right. The whole theatrical experience of this piece amounted to more than the sum of its constituent parts. It was a rare evening in the theatre where the story and story tellers had you by the throat: by the need to tell the story and by the urgency and humor with which it was told.' Despite Brighton's argument that Charabanc later departed from the socialism which she believed to be the *idea* of the work, Brighton, nonetheless, recognizes here what a rare experience it was to be involved in the production of the company's first play.

Charabanc's collaborative writing process had broken down by the time Brighton wrote her 1990 article for *Theatre Ireland*. Artistic differences developed during the writing of *The Hamster Wheel* (1990), and Marie Jones was in the process of leaving the company to form DubbleJoint with Brighton. Brighton's comments do point out other even more subtle difficulties artists face on a small island like

INTRODUCTION

Ireland. Jealousies, ideosyncracies, disaffections, politics – all force companies to compete rather than be mutually supportive. And a company like Charabanc which appeared to get its share of both funds and fame ends up being fair game. Like other criticisms of Charabanc, Brighton's article raises the issue of ownership, of legitimacy, of appropriate 'accent'. Charabanc's aims were always much more in flux than criticisms of the organization imply. The company did perform committed theatre, but as the principals have repeatedly insisted, they never intended for Charabanc's work to advocate either a Nationalist or Unionist ethos. Most of their plays have both Catholic and Protestant characters, and company members came from both cultures as well. Charabanc wanted their work to build bridges between communities, so labelling the group *antisectarian* rather than 'non-sectarian', as Brighton does, would be more apt. They also had no intention of forming a socialist theatre company, even if socialism, through the influence of Lynch and Brighton, was unquestionably an aspect of their first play. Although Charabanc members are politically aware, putting politics on stage was never their primary goal. The five founders espouse a politics with, as they express it, 'a lower-case *p*'. Unlike many theatre artists in Britain during the 1970s and early 1980s who organized collectives for socialist/feminist reasons, the Charabanc women organized as artists to create good theatre work for themselves and for other women. The company was never a collective even though they worked together collaboratively.

The company's early often chaotic way of working was frequently misunderstood as well. During the six-week writing period for the second play *Oul Delf and False Teeth*, Brighton complains about the 'predictable Charabanc need for everybody to overextend themselves'; she describes Charabanc's working process 'like someone newly delivered of quads'.[16] But the excitement generated by the company's creative process was without doubt infectious. Their method of working imitated the language of chaos theory: nonlinear, irregular, boundaryless, turbulent, unpredictable, complex. Because they were women juggling many life roles as they worked closely together, they developed a nonlinear, unpredictable creative style that served them and their work well; their chaotic system became a pathway to heightened creativity. The chaos was born out of necessity or perhaps was dictated by certain personalities within the group. Mounting a play is a frenetic process under any circumstances, and Marie Jones, Eleanor Methven, and Carol Moore, at least, thrived on the thrill of each new project and the tour that accompanied it. Much like the intricate, computer-generated pictures of chaotic systems, Charabanc plays display a unique fluidity as did their working life.

Charabanc's way of working, however, would seem foreign to those who privilege the creative model of a single playwright whose vision is then produced by a director and design team. Even Martin Lynch, despite his first-hand knowledge of how the company created its work, was insisting on the model of the solitary writer during royalty negotiations over this book, saying, 'I went away and wrote the play'. Carol Lloyd, in her study of creativity, labels that model 'The Monocled Monk' but says in the current job market fewer artists can maintain a single-minded style and must instead play many roles. Some of Lloyd's other creative models – 'The Interdisciplinarian', 'The Tightrope Walker', 'The Whirling Dervish' – aptly describe Charabanc's ways of working. Lloyd's outline of ten creative types under two headings, Collaborative and Individual Creativity, reads like a list of the various skills Charabanc members brought to their projects.[17] But in response to criticism, some members of the group became persuaded that their methods were somehow not legitimate. Over the years, the company gradually adopted a more conventional, linear approach and abandoned the chaotic style that generated the fine early work collected in this volume. Scientists define chaos as order without predictability; the essence of chaos, much like the theatrical experience, is the delicate balance between forces of stability and forces of instability. Charabanc's work itself became the ordering principle for the group; the conscious play on stage and off with the expected and the unexpected created order out of the seeming disorder.

Victoria White offers yet another reason why a company like Charabanc was critically dismissed: the woman-focussed content of their stories. White believes that the only real opinion-makers for the Irish theatre are *The Irish Times* and *The Sunday Tribune* reviewers, writing for their white-collar readerships; 'the critics have a fairly reliable tendency', White says, 'to make short work of feminist or woman-oriented plays'.[18] Lack of critical enthusiasm for Charabanc's work was also, in part, due to their boldness; dare they write and perform their own plays; dare they court a new audience and still compete successfully for the established one; dare they enjoy that level of success and adulation as women. The lack of appreciation for women's work and the subtle anti-feminist attitude White identifies among many reviewers is true even for a company like Charabanc whose feminism was unconscious, or at least largely unspoken. In

1983, all of them expressed concern about the company being labelled feminist; they did not want to alienate their broadly based audience, although certainly as individuals they embrace an aware, non-radical feminism.

A hallmark of the company – the women playing the male characters as they viewed them – became a distinctive stylistic technique over the first several years. But this unique example of Charabanc's *female gaze* came about not from any overt feminist attitude but out of simple necessity when they were unable to find unemployed male actors who fit the ACE scheme and who also wanted to be involved in the company. So the women put on cloth caps and jackets and played the male roles themselves, and both male and female audience members loved the audacity of it. This approach allowed them to suggest many characters and attitudes in a fluid, well-crafted way, showing everything from the female point of view – even the men. After centuries of the reverse, the turnabout was quite refreshing. Their performances became psychodramas of a sort. In later productions, unfortunately, they discontinued this inventive practice; they became convinced by criticism that to be taken seriously as actors they needed to play single, sustained characters.

The too-ready identification of Charabanc as a feminist company created its own often odd expectations. For example, the group never claimed to be separatist, and yet, because as women they employed male actors, writers, and directors, they were criticized, as well. What the word *feminist* meant to the various groups using it helped or hurt Charabanc's image and labelled members of the company either bad or good *girls*. Gaining deserved recognition as responsible women was tough on nearly all fronts. During the ACIS panel, Eleanor Methven describes the company's approach to gender and theoretical constructs:

When Charabanc started, it was from a very pragmatic economic base and was completely actor led.... It was not a theoretical base. We really didn't come at it from an academic point of view.... We didn't think of it in any feminist terms – it was an unconscious feminism, if you like. It was really strange where the coincidence was our gender. The five of us were women actors. The other coincidence was that we were all from this place, not necessarily specifically Belfast but from the six counties. And it was in the eighties, and the six counties were undergoing what this [entity] had been undergoing for the past, well-nigh, twenty-five years.... It was interesting to watch the other companies, that were coming up, Field Day, of course, [was] formed before Charabanc, but only slightly before, and on a very different basis, on an academic basis, on an inspiration of making a statement. Very, very interesting work. We came along, almost contemporaneously, from completely the other end of the spectrum. They had academic and literary heavyweights on their board, and we had local trade union leaders and anybody who had been nice to us along the way. And it worked. It certainly did work. But we were always praised for the rawness and the energy. There was just a slight edge of patronization there.

The different reception Field Day, Brian Friel's company, received was never far from their awareness; a report in *Theatre Ireland* about Charabanc's April 1987 tour to Boston with *Gold in the Streets* outlines workshops, presentations, and even mentions a poetry reading they gave one evening, ending with the question, 'Is Charabanc becoming intellectual? – We'll be publishing pamphlets next'.[19]

The seriousness Charabanc developed about themselves and their work increasingly became part of the dialogue. In 1983, they discussed the miracle of the company almost apologetically, as if they could not understand their good luck. But they now have a clearer sense of their creative strengths. Charabanc did not come into being because they were lucky; the company existed because they had a vision of their dramatic future and were willing to work hard to realize it. Rejecting the idea that women's stories are secondary or simply supportive of men's, they show complex women in their plays. Women are allowed to be weak, strong, angry, nurturing, superstitious, knowledgeable, naive, serious, funny, loving, and earthy. Charabanc plays place no limit on the role. In each, the female characters seem fully aware of their power and are willing to use it. But women are also shown to be soft and caring – recast into characters who, however tough, still yearn for love and children and home. That a women's company presented a fuller view of Irish womanhood is not surprising. What is most telling is how Charabanc rewrote the stereotypes; gone is the depiction of women as either unattainable saints or bewitching sinners; gone is the mystery surrounding their behaviour.

Fintan O'Toole raises and partially answers objections this discussion of gender may provoke: 'Whenever anyone talks about issues of gender in relation to the theatre there is an air of barely suppressed exasperation in the response. Sure enough, the standard response goes, it would be nice to have more women playwrights, but does it make any real difference to the plays? Haven't male playwrights created superb women characters? What does it matter who the writer is, so long as women are represented on stage?'[20]

INTRODUCTION

Although admitting that male playwrights have written splendid roles for women, O'Toole argues that the perspective on those lives 'remains a male one'; he then lists the types of female relationships missing from the Irish stage. The variety of female relationships in Charabanc's plays demonstrates what the work of one women's theatre company can do to overcome the lack of a full representation of women's lives on stage. By recognizing this primarily male perspective on women, O'Toole is describing a phenomenon referred to as the *male gaze* in film criticism. The argument is that 'films construct male subjects gazing at female objects on three levels: characters within the film, the camera's "eye", and spectators looking at the film'.[21] In theatre, the camera's eye is less controlled and fine-tuned than in film; the 'eye' would be analogous to how the play is presented.

Ordinarily, these three categories of male *looking* – characters, eye, and audience – are fairly easy to transfer to the stage, especially where male writers and directors are involved. The eye of the camera in a theatre production would normally be controlled by the director because he represents the views not only of the writer but also of the production team and of the actors. The argument could then be made that plays written, directed, and designed by women would employ an 'eye' or a perspective which viewed its subject differently. But whether or not women can escape the male filter is an ongoing debate since women supposedly view themselves through an imagined male lens, an idea which, if true, would cast doubt on the possibility of a *female gaze* and render the maleness or femaleness of the production team inconsequential. Nonetheless, Charabanc's work, with its pervasive focus on the female perspective from research through writing to production, came close to a female gaze, especially when the female actors played male characters as they saw them. In fact, this concept of *eye* could explain why some plays written and/or directed by men were less satisfying to Charabanc and their audiences. Peter Sheridan was one of their more successful male directors; his ability to enter into both their unique female perspective and their collaborative process was the reason, perhaps, why they worked with him several times and expressed pleasure in the experience.

Charabanc's work was revolutionary. Driven by their desire to play strong female roles on stage, they created proactive female protagonists. Laura Mulvey, in what has become a definitive early work on the male gaze, explains that a male protagonist is nearly always 'forwarding the story, making things happen'. Women then become objectified, according to Mulvey, on two levels: 'as erotic objects for the characters within the screen story, and as erotic objects for the spectator within the auditorium'.[22] The power to decide what is desirable is how the male gaze robs women of their experience. But Charabanc's work, by overturning the expectation of active males responding to passive females, provides a remedy; their portrayals which reclaim women as dynamic, passionate beings in their own right help overcome the misrepresentation of women. In *The Feminist Spectator as Critic*, Jill Dolan explains the need to rediscover women's desire: 'left passive in a narrative articulated by men who control its linguistic, social, political, and psychological power, women become objects pursued for the fulfilment of male desire. If male desire is the underlying principle driving narrative, then to disrupt the cinematic and narrative patterns that rob women of their subjectivity, women's desire must somehow find its place in representation'.[23] What women want is repressed or misrepresented in nearly every culture, but especially so in Ireland. Charabanc, with their fully articulated portrayals, brought an understanding of women's desires to the stage, to both the women and the men in their audiences. And they performed these characters they understood and appreciated with a compelling verve; their *jouissance* infused the work and the lives of those they touched.

A persuasive performance can convince women not only of what they are but also of what they are not. If gender is in itself a social construct, then Simone de Beauvoir's 1949 statement in *The Second Sex* – 'One is not born, but, rather, becomes a woman'[24] – is all the more true within that particular context. The female child is an apprentice, gradually learning both the boundaries of her gender role and the adequate performance of it. As Judith Butler demonstrates, the way one becomes a woman is through observing and performing the gender role *woman*:

The act that one does, the act that one performs, is, in a sense, an act that has been going on before one arrived on the scene. Hence, gender is an act which has been rehearsed, much as a script survives the particular actors who make use of it, but which requires individual actors in order to be actualized and reproduced as reality once again. The complex components that go into an act must be distinguished in order to understand the kind of acting in concert and acting in accord which acting one's gender invariably is.[25]

Acting one's gender then, off stage or on, is a largely unconscious, socialized act. The artistic representations of women, the behavior of known women, and the expectations of a given culture are all aspects of the social construct; what is deemed appropriate gender behavior is learned and then repeated, much like acting out a particular role every night on the stage. And theatre performances, like life performances, teach the gender role both to women who consciously or unconsciously contain their own behavior within the implied appropriate limits of the expressed role and to men who reinforce what they believe is acceptable behavior. Whether women can have perceptions of themselves divorced from ideas about what they should be in any given society is doubtful. Charabanc's fulfillment of their priority to 'create and perform strong female roles in Irish theatre' may therefore be the company's most significant contribution. Through their portrayals of unique but believable women, they demonstrate a variety and range of behaviours which not only broadens the possibilities for women but also reinterprets the gender role *woman*.

Nevertheless, fostering a female gaze or treating women as subject rather than object and focussing on issues vital to women – all seem an uphill battle. The limited number of women writing for the stage seems to render unattainable, at present, the goal of producing enough new plays by women to act as a necessary corrective to the overwhelmingly male interpretation of the female, yet another reason to decry the demise of Charabanc Theatre Company. Lost is not simply Charabanc's perspective on women or their professionalism or even their entertaining verve; also lost is the desire of the company to collect and tell women's stories that would otherwise be left unrealized and thus forgotten. Unless the mythic stories of a society capture or at least make room for all voices, they can never represent a culture accurately. Telling the story of Charabanc's dynamic theatrical enterprise through this collection of four plays is an attempt to redress the inherent imbalance in theatre history. The plays included will show how the stories Charabanc created place women central in Northern Irish culture, thus revealing not only the company's restoration of women to a fuller representation but also how their work produces an encompassing cultural critique of the Northern ethos.

The journey ...

The remedy Charabanc Theatre Company offers goes beyond redefining gender issues. The company was devoted to exploring the various issues confronting their polarized Northern Irish community; in fact, many of their original plays and productions of extant works address the political divisions directly. Charabanc was not alone in the hope that theatre might help reconcile disparate individuals or groups. I suspect that the recurring violence was one reason the 1980s produced an upsurge in theatre in Northern Ireland. Like Charabanc, most artists focussed their efforts on creating a climate for peace; some, unfortunately, capitalized on the difficulties and used violent situations as a means to sell work to eager producers. Since the various groups in Northern Ireland view even talking together a dangerous activity, however, the possibility that Protestants and Catholics might somehow be reconciled through a theatrical encounter seems preposterous. And yet, theatre is a dynamic process through which participants experience themselves and others; it discloses the inner self which otherwise would remain obscured and is the center, or womb as the work of women implies, where change gestates. Good theatre can tap deeply into the issues that matter most to a culture, but the subtle effect theatre has on its participants is both illusive and cumulative. Richard Schechner argues that 'theatre is a model of, or an experimentally controlled example of, human interactions. It is something else too: a reflection of, or mediation among, those interactions . . . where through exaggeration, repetition, and metaphorization they can be displayed and handled'. The instinct to use theatre as a means to 'handle' explosive situations is especially pronounced in Ireland with its vibrant theatre tradition. Schechner, in describing the human interactions which are the focus of theatre, could be describing Northern Ireland: 'the interactions played out in the theatre are those which are problematical in society, interactions of a sexual, violent, or taboo kind concerning hierarchy, territory, or mating'. According to Schechner, therefore, theatre 'is not a model of all human action, but of the most problematical, taboo, difficult, liminal and dangerous . . . where clarity of signal is needed most'.[26] Theatre in Northern Ireland is talk of the most dangerous kind – entertaining, seductive, insidious, repetitive, compelling talk. Using what is an inherently appealing medium, a company like Charabanc might actually be able to effect change, regardless of how difficult quantifying that change proves to be.

Many Irish plays deal implicitly with the idea of reconciliation, but Charabanc developed ***Now You're Talkin'*** (1985) to explore this issue explicitly. At this point, the company consisted of three remaining founding members – Marie Jones, Eleanor Methven, and Carol Moore. The play, set in a Northern Ireland reconciliation centre run by a well-meaning but misinformed Irish American Carter O'Donaghue, asks if five women from various socio-economic and cultural backgrounds can come together in Peace and Understanding.

INTRODUCTION

Charabanc's work consistently undercuts superficial solutions with a biting irony. The North American Tour programme notes identify the theme of this play as 'the perceptions which "outsiders" bring to Northern Irish people and politics. Many are well-intentioned if ill-informed. Carter's mid-Atlantic brand of love and harmony is an unlikely antidote to the women's brand of problems.' Carter, a trendy, pop psychologist, elicits from the women first lust and then total disrespect. The send-up is uncomfortable for any American in the audience or for anyone, for that matter, who has thought communication games might be a helpful way to get people talking. And Carter introduces them to the sappiest games; all the hand-holding, shell-gathering, May-Poling, singing, and dancing only bring these women's differences more clearly into focus. At one point, Carter divides the women into farmers and cowmen and has them sing 'The Farmer and the Cowmen Should be Friends'. He insists, incidentally, that the Protestants play farmers and the Catholics cowmen, exposing his silly, stereotypical thinking even further.

The exasperated women take over and lock Carter out of the room, replacing his childish songs with political ones. When they then begin talking in earnest, they discover no basis for agreement and put aside anything of substance to talk about later. Although the three politically moderate women try to talk, the two who represent the political extremes become silent. Thelma, the Protestant wife of a window washer whose main concern has been whether anyone is stealing her new £60 coat, silences herself by crawling onto her bunk and covering her head when the discussion becomes heated. Veronica, the Catholic republican who came to the centre this weekend to take a 'wee vacation' from her five children, is silenced by the group; they bind and gag her when she sarcastically sings 'We Shall Overcome'. Regardless of the respite silencing her stridency provides, the play demonstrates that eliminating contentious individuals or groups from the discussions is not the answer. As *The Guardian* review (15 September 1985) states, 'Tying up and gagging the hard-line and very vocal Sinn Fein supporter amongst them in a desperate attempt to get some sort of consensus is a potent symbol'. The play highlights the worst aspects of the cultural divide; every viewpoint is deflated, even the moderates' well-meaning earnestness. And the press is lampooned as well. One of the more telling details of the play is that Thelma, who has largely been uninvolved, becomes their spokesperson because of the media's confusion.

Charabanc performed *Now You're Talkin'* in Ireland, England, and Scotland. But when their planned tour to Canada, spring 1986,

INTRODUCTION

required that the play be restaged because Mary Marcus was replacing Ann Forsythe in the role of Thelma, director Pam Brighton informed Charabanc she was unavailable, thus precipitating her break with the company. During the ensuing rehearsals, the group along with Ian McElhinney as director changed the ending significantly. Comparing the original script with this new ending[27] documents the changes which the group members believe make the play more 'honest'. (This alternative and preferred ending is included in this volume following the original playscript.) The fluid workings of the company made adjusting to this type of dynamic script change easy. The rewriting shows clearly that no reconciliation occurs, and the women go their separate ways at the end of the play. But the changes they made are not large ones. The original ending left Jackie, the moderate Unionist, and Veronica, the Sinn Fein activist, talking. They are holding hands as they look out of the window watching Thelma make a meaningless statement to the press. Thelma is attesting to a new level of understanding achieved by the group through talking out their differences, something that the press wants to hear but which did not occur. Charabanc members believe this original ending gave the false impression that Veronica could have brought Jackie over to the republican side of the divide. In fact, they believe Pam Brighton forced that resolution on the play. Although the new ending still shows Jackie and Veronica earnestly speaking from their own viewpoints and trying to understand one another, there is no hand holding now, just the grim realities of the battle. At the end, 'They look at each other with nothing left to say. Veronica leaves. Jackie leaves slowly.' And so the final irony is stronger as the music swells, playing, 'The Farmer and the Cowman Should be Friends'.

Now You're Talkin' was successful both before and after the script changes. All elements of the conflict converge in this play which gives its audiences a safe way to experience the many diverse opinions in Northern Ireland and imagine what it might be like to be a member of an opposing group. Nonetheless, no one in the group talks enthusiastically about the play or its reconciliation theme; in fact, they describe it as their least satisfying production, even after rewriting the ending. Their lack of enthusiasm is surprising in light of the audience reaction and uniformly positive reviews: 'a funny, fast-moving, satirical look at politics and the problems of reconciliation in Northern Ireland' (*Belfast Telegraph*, 18 March 1985); 'within Charabanc's own terms of reference – to examine Belfast's cultural and social history – the play is a runaway success and certainly their most likely box office hit to date. In a wider dramatic context, it can

INTRODUCTION

hold its own against all local comers' (*The Irish News*, 20 March 1985); 'It may well become both one of the most entertaining, interesting, and dare I say without putting anyone off, socially important plays of our time' (*The News Letter*, 3 July 1985); '*Now You're Talkin*' is the most comically entertaining, politically relevant (the autumn talks) and socially compassionate play about Northern Ireland since Friel's *Freedom of the City*' (*Evening Press*, 1 October 1985). Although the subject of Northern Ireland is often treated as sacrosanct by outsiders, that fact alone cannot account for the overwhelmingly positive reaction from reviewers outside Ireland: '*Now You're Talkin*' is charged with . . . honesty, brave impartiality, and native wit' (*Glasgow Herald*, 20 August 1985); it 'moves with wit and intelligence from pessimism to optimism and back, and through its fresh, imaginative approach brings alive problems which for many people have been dulled by repetition' (*The Times*, 31 August 1985).

But despite the decidedly positive reaction, Charabanc members believe the play, even when revised, did not work. Performing *Now You're Talkin*' may have been a dissatisfying experience for several reasons. A contemporary play like this did not capitalize on their oral-history approach nor could they experience closure since the story was still being written. Even now the conflict depicted continues to be played out in the streets and in political gatherings. The nature of the subject made artistic distance nearly impossible to achieve; the various cultural backgrounds of the actors became intertwined with the characters and with the issues. As frequently happened, Protestants were cast playing Catholics; one of the Catholics in the group played a Catholic in the production. But because the women played only one character, there was less fluidity than in both earlier and later productions. Performing only one role may also have made them feel more constrained by the particular beliefs of their characters. Also, because the plot requires exploration of the diverse Belfast backgrounds, the characters seem less subtle and the situations more broadly drawn than those in their other plays, so the roles were probably less gratifying to perform. And then each of them believed that reconciliation in Northern Ireland could not be achieved merely through talking. The differences the play dramatizes can overwhelm any hope for a solution to the problem; reality is often sobering. Honest as Charabanc is in all of their work, if they believe the central dramatic focus of *Now You're Talkin*' is misleading, they will first want to change it and then they will want to abandon it, which they did. As soon as their contracts permitted, they moved on to their fourth production, *Gold in the Streets*.

INTRODUCTION

Now You're Talkin', even in its earlier form, was certainly not viewed as a superficial gloss over difficult problems. The *Irish Independent* (1 October 1985) observes that 'the realism of this play does not stop to allow a happy ending. Instead, the audience is left with a depressing conclusion that, inevitably, the fighting will go on.' And *The Irish News* (20 March 1985) states that 'if the play has a message, it is that, beneath the surface of the average Belfast person lies not a neighbour longing to reach out, but a bigot'. In the context of all the talking about talking that takes place in Northern Ireland, the play's satire seems even more pointed. Sam McAughtry in his review in *The Irish Times* (27 March 1985) sums up the prevailing reaction to the play when he claims that *'Now You're Talkin'* has more laughs to the minute than any home-produced work I've seen this year'. But apparently, only the audience was laughing. Regardless of the company's discomfort, however, audience members wanted to believe in the hopeful message of the play. McAughtry concludes his review with this assertion: 'There was no reason to expect anything from the show other than a pleasant two hours watching Charabanc mature as players and writers, but I came away with a most unexpected bonus: a blazing insight, a new possibility for Northern Ireland, when I'd imagined that nothing new under the sun could enter the mind again'. When asked what they thought McAughtry's 'blazing insight' was, company members' response was bewildered shrugs. But the play was popular because it hinted at a possible solution, a way to end the stalemate, and offered hope. In a divided society like Northern Ireland, the hunger for solutions is, in itself, hopeful. If the play helped renew a desire for change, then it performed a useful role.

I include *Now You're Talkin'* in this volume, despite the slight protests from company members, so that readers can experience McAughtry's 'blazing insight'. This entertaining play lays out the dispute with rare depth of understanding, a dispute, incidentally, which has not changed much in the intervening years. In addition, the play gives a clear sense of the larger context of Northern Ireland with its enigmatic and yet fascinating society. At the beginning of Act Two, Scene Three, the opposing characters address this seductive Belfast quality:

JACKIE: You know, given all this trouble, I really love Belfast. Me and my Bobby were thinkin' of going to Canada when things were getting too much here . . . but we just couldn't do it. . . .
VERONICA: Aye, it's the Belfast people, isn't it? Like, you see when they want to hate, they can really friggin' hate . . . but, you see when they wanna love, they can do it in an even bigger way. Belfast . . . it's friggin' magic!

INTRODUCTION

To avoid the violence in Northern Ireland, as the play points out, what is imperative is not more talk nor better talkers but more and better listening. How clearly and honestly the issues are presented either in life or on stage could mean the difference between developing understanding or rendering existence even more difficult. Given what is at stake, reliable, responsible theatre is doubly important and may very well be one of the last and best hopes for creating change. But neither this play nor any other could adequately answer the question of whether or not a lasting reconciliation in Northern Ireland is finally a viable possibility; it cannot be imposed by movements, agreements, or a play. Still in light of events, *Now You're Talkin'* seems almost prophetic.

Charabanc first conceived *Gold in the Streets* (1986) as three fifteen-minute plays which focussed on Irish emigration to England, a recurring, troublesome issue on both sides of the Irish sea. A programme note defines the play as being 'about Ireland's biggest and most constant export – her people'. Although the present population of the island is just over five million, more than seventy million throughout the world trace Irish ancestry. The short version of *Gold in the Streets* was commissioned by Camden Borough Council, and the company performed it during their *Now You're Talkin'* London tour September 1985 in venues as diverse as the Kilburn community centre, the London Irish Women's Conference, and the Rose Bruford Drama College. Overwhelmingly positive reactions encouraged the group to expand the plays. The final version, which their director Ian McElhinney helped shape, is included in this volume and consists of three one-act plays set in Belfast during 1912, 1950, and 1985, respectively. Each act has as its core a central woman character: 'Each woman has, for different reasons, found herself faced with the decision to become an exile. The time and characters change. Sadly, the motivations to leave are perennial' (programme note). The title of the play is based on a popular song 'The Mountains of Mourne', which also explores the pain of displacement, forced as Irish immigrants often are to dig for 'gold in the streets' of London. Ironically, this play was a golden time for the company, marking the period of their greatest prosperity.

In Act One, Agnes Mullen brings her family to Belfast in 1912 to live with her sister when her husband's work diminishes because the linen mills introduce new machinery. As Agnes says, 'My John Joe was the best handloom weaver this side of the Bann. In fact, he still is.' But no one in Belfast wants to hire either of them because they are Catholic. No longer able to endure the discrimination and brutality

INTRODUCTION

from 'neighbours', Agnes and her family emigrate to England because 'over there nobody will care what we are'. In Act Two, Mary Connor returns to Belfast in 1950 after nine years away. She had married a British soldier who was then killed in World War II. She brings her young daughter back home, as she explains to her mother, 'to be with you, to be in Belfast where I belong'. Her mother will not forgive her, however, for marrying a Protestant Englishman and for drawing away from Catholicism. In the end, Mary reluctantly returns to England to protect her daughter from the prejudice of the place. Act Three, set in 1985, focuses on Sharon McAllister whose husband was employed by the DeLorean Motor Company until the factory closed. Davy has been out of work for four years. But nonetheless, Sharon cannot cope with her fear when he becomes a policeman – 'Six months he's been with the police / And it's taken six years off my life'. Strung out on Valium, Sharon finally takes their children to England rather than live with the daily possibility that Davy will be killed by paramilitaries. 'None of which sounds very cheerful', *Time Out* (3-10 December 1986) comments, 'but the excellent cast perform with such skill and humour, squeezing laughs from despair, that one cannot help but be moved and entertained. Sharing a multitude of characters between them they display a remarkable versatility and skill.' *The Times* (27 May 1986) finds 'the contrast between the resilient, gritty humour running through the performance and the irreparable damage it portrays, is beautifully managed'. Despite the subject matter, Charabanc's sardonic humor and entertaining verve surfaces, making this perhaps their play most accessible to outsiders.

Reviewing the premiere at the Belfast Arts Theatre, the *Belfast Telegraph* (14 January 1986) describes *Gold in the Streets* as a 'simple uncluttered production' with 'a strong narrative line and plenty of dramatic tension'. The play was written and produced in Brechtian style, except for the acting which was naturalistic. These three plays employ many distancing aspects; besides singing songs, the actors comment on the action, tell Irish jokes, play children's games, and recite poetry. The multiple roles are played by four women actors who are always visible; when they are not occupying the playing space, they sit on chairs around the edges of the stage, speaking some lines while seated. The transparent style of the production was undoubtedly one reason for the play's success. With 'impressive' versatility, the four 'play a bewildering variety of roles of both sexes, from millworkers to shipyard workers, children to elegant ladies, taxi drivers to publicans, Protestant louts to Catholic louts. With a change of a cap or coat, an Orange Order sash or a soccer team's scarf, they

INTRODUCTION

slide into characters with definably different accents, religions and world views' (*The Boston Globe*, 14 February 1986). None of this multiplicity seemed at all confusing. Reviewing Charabanc's American premiere on Long Island, *The Smithtown News* (26 June 1986) declares the actors 'a treat to behold'; 'The fact that what could be fragmented and choppy is a finely woven tapestry, has a great deal to do with the four performers who play all the parts – women, men, children – and even do all the sound effects'.

Reviews like these might have gone to their heads but did not. Even though this was clearly a giddy period of unexpected success, Charabanc members continued to be generous and down to earth, handling receptions and press interviews with panache. And they were delighted to shed their simple costumes and dress in finery for the parties. They published a commentary on their dizzying North American tour (10 June to 19 July 1986) in *The Girls in the Big Picture* programme with the caveat: 'Don't imagine, though that we were enjoying ourselves. This relentless social round was all in the cause of promoting the show – they're very big on Public Relations in the States so we related to the public as much as we could in the interest of Theatre!' A good example of this heady time was the performance of *Gold in the Streets* sponsored by the Horizons Theatre Company in Georgetown 30 June 1986. The day before, Charabanc had been feted at a formal garden party. Then the night of the performance, Sir Oliver Wright, the British Ambassador, and Paddy McDernon, the Irish Ambassador, were present with their wives, sitting near each other in the small theatre and laughing and crying along with the audience. At the end of the play, the audience thundered an appreciation for the play – clapping, stomping, whistling – a common occurrence, and yet the actors seemed almost surprised by the response. And the party in the theatre following the performance was also one of many, but these women, dressed in glamorous formals, treated every occasion with spontaneity, as an opportunity to laugh, sing, tell stories. And that night was no different, except that the party was followed by a moonlight tour of the Washington D.C. monuments. They readily cast off their sophistication along with their uncomfortable shoes to walk barefoot to the Vietnam Memorial, running their hands slowly along the wall. Later they solemnly read all of the quotations in the Jefferson Memorial, marvelling at the rights carved on the walls.

Charabanc's fifth play *The Girls in the Big Picture* (1986) developed from character rather than from theme or event; the rural setting was also a departure. Marie Jones, Eleanor Methven, and Carol Moore

INTRODUCTION

had for some time been improvising country women to break the monotony while on tour, developing fully articulated lives for these characters. They would slip into their respective roles as they shopped, went to the cinema, or ate in diners in the small towns where they toured. The roles may have been a stretch for Marie and Carol who had grown up in or near Belfast, but for Eleanor who was from the small country town of Magherafelt, Big Jean came naturally. *The Girls in the Big Picture* evolved from this imaginary world they had created. Their long association with Jean Moore (Eleanor Methven), Mary Jo McIlfatrick (Marie Jones), and Margaret Anderson (Carol Moore) may account for why these three founders still unanimously declare this their favourite Charabanc work. Any discussion of a volume of plays hinged on my including *The Girls in the Big Picture*. Although the play is sad and lacking the resilient spirit of the company's other work, the quality writing, the realistic situations, and the honest portrayals are reasons enough to include it in this representative volume. Any discomfort with how circumscribed these women's lives are may, in the end, be the play's underlying purpose. Certainly, the play's value as a reliable record of women living in rural Northern Ireland is indisputable. It presents a year in the life of these three women, all unmarried and in their thirties: 'the pressures from friends and families that their "spinsterhood" attracts, creates a sense of failure, conflict and desperation' (programme note). They deliberately set the play during the 1960s, just before the present violent struggle; this small town in rural Ulster is influenced by the rhythms of the farming seasons – harvest, lambing, hay-making. The town's social life revolves around basket teas, Women's Institute gatherings, and the cinema. The lives of these three women friends are comfortably predictable until a strange man comes into town and throws everything out of kilter.

The Girls in the Big Picture displays a typical Charabanc style but is more complicated structurally and dramatically. Again, the actors play many roles, bridging various ages and backgrounds but not genders. Enough male actors were cast so the female actors played only women characters in this play. Voice-overs were used extensively for the various social activities, a difficult dramatic choice to make convincing. A highly romantic, hilarious screenplay was developed for the voice-over when the characters attend the cinema. (Although specific screenplay dialogue is included in the playscript where appropriate, the text of the entire screenplay follows the text of the play in this volume.) With at least sixteen characters, in addition to nine in the screenplay, five sets, and intricate lighting and sound, this might have been their most challenging play to produce. Owing perhaps to their director Andrew Hinds' preferences or to better funding after *Gold in the Streets*, this production, rather than displaying the usual Charabanc economy, seemed more caught up in the traditional, external trappings of theatre. Reviews were mixed: the *Evening Press* (30 September 1986) says the play is 'beautifully written', the acting 'marvellous', the production 'lovely' and 'ingeniously directed too – just two cupboards providing several different settings'; however, *Fortnight* (November 1986) declares the set 'dull' and calls for someone to 'cut the dead wood from this production', criticizing 'all that fiddling about with a table (to make it bare for the cafe and loaded for the farmhouse). It finds its apotheosis as the balcony on which the rude boys sit in the cinema.'

The *Fortnight* reviewer, nonetheless, does praise Charabanc, saying that 'boldly and with remarkable success, they have moved out of the city and into the country to explore the frustrations of three women, trapped in a small town, who meet once a week in the cinema to feed on the phoney glamour of Hollywood and gossip about their lives. . . . I used to think that one person sweating in private was the only way to make a good play, but the communal plays of Charabanc are far more interesting and even coherent than most of the plays written by individuals in Ulster recently'. In contrast, *The Irish Times* (24 September 1986), even though claiming that the play retains 'Charabanc's sharp and unpitying sense of humour', still argues that the play suffers from the company's 'earlier documentary style' and is at best 'a transitional work using old techniques to explore territory new to the company'. Other reviews were unequivocally positive: the 'splendid' production shows Charabanc has 'come a long way from the early socio-political themes which dominated their work' (*Irish Independent*, 30 September 1986); 'splendid natural acting, a kind of droll, sardonic humour and hints of unconsciously cruel tyranny over the lives of people in a small rural community come through in the Charabanc Company's Festival offering' (*Evening Herald*, 30 September 1986); 'the Charabanc magic formula motors on from success to success . . . woven through what is often hilarious comedy and instantly recognisable homespun and homely situations, there are moments of pathos and tragedy that tug at the heartstrings in the way that only the best theatre can' (*News Letter*, 23 September 1986); 'another glittering success for Charabanc . . . one of the highlights in Ulster theatre in 1986' (*Belfast Telegraph*, 24 October 1986). This 'departure' into new territory seems, apparently, to have garnered for the company widespread praise of its 'splendid' aspects.

INTRODUCTION

Somewhere Over the Balcony (1987), the fifth and final play in this volume, seems in hindsight like the creative culmination of the best qualities of Charabanc's early work. Marie Jones, Eleanor Methven, and Carol Moore shone in the portrayals of Ceely Cash, Rose Marie Noble, and Kate Tidy – West Belfast women in their thirties. *Balcony* sparkles with wit, song, and monologue, fitting the particular skills of Charabanc's three artistic directors perfectly, a synergy that may be achievable in the theatre only when you develop the roles yourself. This unique play expresses the farcical quality of the Northern Irish political struggle and sprang in part from a dream Eleanor had of slap-stick, pie-throwing clowns which exposed the senselessness of the ludicrous situations surrounding her. Then Marie tells of seeing children on their way home from chapel on a beautiful Sunday morning, eating ice cream cones while stepping around the bomb fragments and burning cars from the night before. From that all too familiar yet surreal craziness this irrepressible, irreverent, highly entertaining account of Belfast living was born. A programme note explains that *Balcony* 'examines the bizarre kind of existence that passes for normality when people are trying to live their lives in a crazy, incomprehensible, uncontrollable situation'. The setting is reminiscent of but not exclusive to the run-down Divis Flats in Catholic West Belfast; company members want the associations to be broader than the Divis tower blocks. This play was Charabanc's first to focus on only one side of the divide; as Protestants playing Catholic women, they were understandably concerned that they portray Catholic viewpoints appropriately. They spent considerable time at Divis and came to delight in the indomitable spirit of the friends they made there. Peter Sheridan, a Catholic director from Dublin, had a major influence on this play, helping them develop the script and avoid offending Catholic sensibilities. Because of his own experiences with group play development, Sheridan was also fully responsive to and respectful of Charabanc's collaborative methods.

In *Balcony*, events occur with break-neck speed: helicopters hover; joy riders steal an ambulance and a saracen; the elevators do not work; the water is off; the trash chutes smoke and blaze; a dole snooper disguised as a paratrooper is discovered; a car is blown up; the walls in Kate's flat crumble around her from the impact of the blast; a child holding a tricolor falls from an empty flat while a visiting German is taking his picture; a wedding, very nearly a birth, and then a funeral take place in the chapel, which is under Army siege; the best man is on the run; a bingo game continues – 'armalite gun, 21 . . . on

INTRODUCTION

the run, 31 . . . doin' time, 29 . . . Prods are dirty, number 30 . . . Pope's a Mick, 26 . . . Brits are thick, 36 . . . Valium Heaven, number 7 . . . blocked up loo, 42' (Act Two); and the British Army watch and listen to it all from their post on the top of the tower block. Sounding much like Divis residents, the three characters take it all in stride. Kate speaks for all of them: 'That's what I love about this place. On a day like to-day you could be anywhere' (Act Three). Like a surreal Greek chorus, Ceely, Rose, and Kate comment on the absurd action taking place just over their balcony. In fact, the similarity to Greek drama is remarkable since all the action takes place off stage and the unities of time and place and character are adhered to. But no matter how much Kate might seem like a modern-day Cassandra bemoaning the fall of her walls, the connection is not intentional; Charabanc does not work that way. They are certainly not self-consciously academic or theoretical in their approach, so the Greek associations are happy accidents.

Because *Balcony* takes place on internment anniversary, the play offers an insider's view of the activities surrounding the political parades in the North.[28] Ceely, broadcasting to the flats on her pirate radio station, tries to find beds for foreigners visiting West Belfast for the internment parade: 'There are sixteen visitors sittin' over in that chapel two arms the one length and they still haven't bin claimed yet . . . now I want them out and lodged before Charlene and Danny get married . . right? . . . Three "Troops Out" from Manchester, men, middle aged. Three Basque Separatists; they want to stay together. Six of the Communist Party of Great Britain; they don't mind what way they are split up' (Act One). The entire play teems with witty social commentary like this. Nearly all of the men in the play are called Tucker, which is Belfast slang for Tommy – Little Tommy Tucker. Ceely's Granda Tucker was fifty-six when interned, 'too old . . nearly killed him . . hasn't spoke from the day and hour they let him out (looks up at the army post) . . cos of themmins listening . . claims they even bug the confession boxes' (Act One). Consistent with the rest of the play, *Balcony* has a hilarious but, nonetheless, ominous scene where Rose is arrested and interrogated for the shooting of a policeman by a woman with a twin buggy: 'Arrested! Me arrested! And I never even done nothin'. I says, "Mister, I'm only polishing my man's working' clothes, his riot shield and his helmet". Next thing I'm led down to a saracen with my two wee twins dukin' over the balcony . . . my poor Tucker and the twins were wavin' like I was going on a bus trip to Portrush' (Act Two).

Interspersed throughout all this insanity are songs which put to rest

INTRODUCTION

for all time the question of whether or not the Charabanc women can sing. The *Cork Examiner* (24 November 1987) mentions the 'rich pricelessly funny stream' of song and then declares simply that 'they sing well'. The song 'Somewhere Out There' late in Act Two is especially moving. Again, a sampling of reviews spells out the play's unqualified success: the company offers 'welcome proof that plays about the Troubles do not need to be airless, weighty diatribes' (*The Independent*, 23 September 1987); 'the play, wildly funny and surrealistic, mercilessly lays bare the now accepted conditions' (*News Letter*, 16 October 1987); *Balcony* shows 'absurd humour, impossible dreams and an unbreakable community spirit' (*The Irish News*, 13 November 1987); of all their productions, this is 'one of the most cogent, yet one of the least sane', 'improbability piles upon unlikelihood. Yet a crazy authenticity is maintained throughout. The humour is pithy and dark and incisively unsentimental' (*The Irish Times*, 29 December 1987); 'three excellent actresses put enormous effort into a work that depends on layers of brilliant, vulgar humour to lift its dark tale' (*Evening Herald*, 29 December 1987); 'the acting and singing is slick and polished throughout' (*Belfast Telegraph*, 11 November 1987); 'Charabanc has again confirmed itself as a powerful theatrical force to be admired and reckoned with' (*Time Out*, 16-23 September 1987); the play is 'breathtaking' and the actors' 'talents seem to be limitless' (*The San Francisco Progress*, 26 February 1988); 'the dynamic Belfast trio . . . strikes the right balance of sorrow and joy' (*The Boston Globe*, 20 March 1988); 'while entertaining us with songs and jokes and making us laugh and cringe, the actresses make us see our own world in a new way' (*The Baltimore Chronicle*, 3 February 1988); these are 'remarkable performers . . . who jump agilely back and forth between broad slapstick and true human emotion'. Some of the play's best moments are the 'close-knit harmonies of their musical numbers. In the end, though, it's the dark comedy we remember. We laugh so much at these women and their plight that we realize there's nothing at all funny about life in Divis Flats' (*St. Paul Pioneer Press Dispatch*, 12 March 1988).

Many reviews extol the production style of *Somewhere Over the Balcony*, which after *The Girls in the Big Picture* marked a return to a sparse approach, relying on lighting to establish the scene rather than on set or props. The *San Francisco Examiner* (18 February 1988), for example, praises everything 'from the assurance of each of the performers to the wonderfully astringent pacing' and says the play 'moves with an ease most political theatres in this country would covet' and is 'as accessible as it is unpretentious'. No matter which

INTRODUCTION

cultural background the audience or what venue, response was overwhelmingly positive, with one exception. As so often happens in Northern Ireland, violence changed everything. After a successful London run, Charabanc opened *Balcony* in Belfast at the Arts Theatre on 10 November 1987, just two days after a thirty-pound bomb exploded in an unused school in the centre in Enniskillen, bringing the building down on those gathering at a war memorial to celebrate Remembrance Day; eleven were killed and sixty-three were injured. Company members could barely perform the play; some speeches which they found impossible to say they cut. They were playing to complete silence; it took days before audiences began to laugh. That first night at the Arts when they finally came down into the bar to greet friends, everyone was just sitting there; no one seemed able to talk or look each other in the eye. But despite this difficult period, *Somewhere Over the Balcony* was clearly their most successful production and the artistic high point of the company. Like Charabanc's other plays, *Balcony* celebrates the resilience of Belfast women; it also punctures the impotent posing and macho exploits of the men. The play title suggests Somewhere over the Rainbow', and the rainbow Charabanc offers in their work could be the best remedy possible – laughter.

During the summer of 1988 following the resounding success of *Balcony*, these three founders took a much needed hiatus from the company to 'pursue other interests'. That November, Charabanc performed *The Terrible Twins' Crazy Christmas*, a show written by Marie Jones but without her joining the cast. Then Eleanor and Carol performed in *The Stick Wife* (February 1989) and *Cauterized* (November 1989), again without Marie who was concentrating more on writing than acting. Two new Jones' plays, *Weddin's Wee'ins and Wakes* (September 1989, November 1990) and *The Blind Fiddler of Glenadauch* (August 1990), were the result. During the research and writing process for *The Hamster Wheel* (February 1990), Charabanc's collaborative method came apart. Marie's writing and personal commitments prevented her from joining Eleanor and Carol in researching this play about women who take care of disabled family members. The easy give and take broke down, and Marie's resulting script was not what Eleanor and Carol had envisioned. Nonetheless, they both still regret the strong critique they gave Marie just before rehearsals were to begin about what they believed to be an overly sentimental treatment of carers. The three would probably have been able to repair the breach had Marie joined the cast of *The Hamster Wheel* as planned. Eleanor and Carol believe Marie would then have

INTRODUCTION

quickly recognized, as she ordinarily did, changes she needed to make in the script. Instead, their fluid relationship changed into a more traditional separation between Marie as writer and Eleanor and Carol as producers, leaving little room for the company's usual method of collaborating fully on research, plot and character development, writing, rewriting, and rehearsal.

Into this difficult mix came American director Bob Scanlan, a friend of Marie and Ian, who had little understanding of the company's collaborative method or appreciation for what he called the 'episodic quality' of Charabanc productions. In a taped interview on 16 March 1993, Scanlan expresses disappointment with *Somewhere Over the Balcony*, calling it merely 'an entertainment'. He wanted *The Hamster Wheel* to have a more 'conventional' shape and a 'romantic consistency', goals which seem out of sync with Charabanc's dramatic vision. He also wanted the play to treat its subject more seriously, thus totally missing the point of Charabanc's ironic, irreverent cultural critique. During the rehearsal period, Scanlan stayed at Marie's Ravenhill Park home, where Pam Brighton was also staying. (Immediately following this period, Brighton published her *Theatre Ireland* article critical of the company.) Although he admits that his brainstorming sessions about the play with Marie and Pam may have been 'a violation of the old contract' since they excluded Eleanor and Carol, those late-night discussions were still 'the most memorable and enjoyable parts of the creation' for him. Valuing a more linear, hierarchical approach, Scanlan had little tolerance for what he called the 'creative flux' of the Charabanc process or for the ideas of Eleanor and Carol who felt responsibility as artistic directors of the company. They knew that their audiences would readily catch mistakes resulting from his unawareness of geographical and cultural details and that, although Marie was not coming to rehearsals, she would want to change small aspects of the play which they tried to bring to Scanlan's attention. But he would not allow the usual Charabanc give and take and says, in the end, he 'willfully interfered with the dynamics of the company', interpreting his role as protecting Marie as the 'sole' writer. For example, Scanlan reports walking out one day late in the rehearsal period: 'It was calculated at the time as a way of getting everybody's attention and taking charge. I did one of those grandiose stage walk outs to establish discipline. And that's a male thing; there's no question about it. I was actually trying to say I'm the director here and discussion stops or I'm out of here.' Although conceding later that this was probably a 'miscalculation' on his part, his actions astounded and demoralized Eleanor and Carol. Clearly, Bob Scanlan was the wrong director for this particular company. But Scanlan has no regrets, apparently, about the break up of the company, simply saying, 'I didn't think it was my responsibility to keep Charabanc together'.

The Hamster Wheel toured Ireland but had an abbreviated run elsewhere. Health care issues explored in the play are unique to Ireland, making the play difficult to sell elsewhere, especially in America. Charabanc brought *Somewhere Over the Balcony* to Houston, Atlanta, and Amherst late March and early April 1990, suspending *The Hamster Wheel* tour for a few weeks. As it turns out, those eight American performances of *Balcony* were the last times all three – Marie, Eleanor, and Carol – were on stage together 'as Charabanc actors'. Everything had changed utterly.[29] Soon after, Marie decided to leave the company and concentrate on her writing. Eleanor and Carol became Charabanc's joint artistic directors and maintained the company for five more years. Although Charabanc continued to be known for its excellent productions of both commissioned plays and extant work (see Appendix), losing Marie Jones as resident writer changed the company focus. And any deviation from such a winning proposition, especially performing scripts by outsiders, seemed to some like a betrayal of the company's aims. Trying to hold a developing company like Charabanc to one way of working might be tempting but will always prove to be an impossibility. Theatre is a dynamic process, demanding constant reinvention. Both Eleanor and Carol worked with fine writers on various projects for the company, but the collaborative process which was responsible for the brilliance of *Somewhere Over the Balcony* never again worked quite so well. These three friends, now busily engaged in their own theatre and film projects, share not only a valuable work experience as reflected in this volume of plays but also the understanding that Charabanc made a difference to their community.

The legacy ...

Irish theatre has subtly changed its course because of the impact of Charabanc Theatre Company. Many more women are actively involved in creating and producing their own work; plays by women are now seen on the main Irish stages; women directors and designers have increased in number. Charabanc members provide inspiring examples of what women might accomplish; their insatiable curiosity and hard work led to the remarkable success of their company. And that success motivated those who worked with them as well. Many of

INTRODUCTION

their directors, designers, and technicians continue to benefit from having associated with the company and remain involved in theatre. After leaving the company, Janet Mackle, Charabanc's first administrator, continued to be a successful theatre administrator, as did Patricia McBride, the company's second administrator. Of the five founding members, only Carol Moore and Eleanor Methven were still with the company at the end, serving as joint artistic directors and sharing one salary. Although Carol still acts on occasion, she has become a director for stage and screen and has also completed an MA in Irish Studies at Queen's University, Belfast. Eleanor enjoys working with writers on new work and has begun writing herself; she is now concentrating on a successful acting career on stage and in film. Brenda Winters left the company because of family responsibilities after the first tour of *Oul Delf and False Teeth* but rejoined briefly to participate in the Russian tour of *Lay Up Your Ends* in October 1984; Brenda became Artistic Director of Replay, a Theatre in Education company which she formed. Maureen Macauley left after the company's second play closed and after the Russian tour; she is a sought after stage manager. Marie Jones officially resigned from the company 11 August 1990. Over the nearly eight years of her involvement, Marie's talent as a writer emerged; she was recognized as the company playwright until she left after writing *The Hamster Wheel*. Marie formed DubbleJoint Theatre Company with Pam Brighton. Now no longer with DubbleJoint, Marie continues to write highly successful plays, winning the Olivier Award in 2001 for *Stones in His Pockets*. All five Charabanc founders are therefore still absorbed in a variety of theatre pursuits which they developed for themselves, and all profess to have gained immeasurably from their involvement with Charabanc Theatre Company.

Despite Charabanc's amazing success, however, the company closed its doors the summer of 1995 because of financial difficulties. Although then receiving support from the Northern Ireland Arts Council, mismanagement by a new manager once discovered became simply too extensive to recover from. Nonetheless, during its twelve-year existence, Charabanc provided a compelling example of women triumphing against the odds with their wonderful mix of creativity, skill, perseverance, and audacity. Ian McElhinney, in the *Lay Up Your Ends* programme, warned theatregoers not to be deceived: 'The charabanc may have offered a glimpse of pleasure but it was only a brief relief from the tedium and stress of life, for those who knew what poverty, hardship and gruelling working conditions were all

INTRODUCTION

about'. In a polarized society where the various groups do not have many chances to interact, theatre can provide a safe beginning. And for this reason, the community theatre movement which Charabanc addressed better than any other company is especially important. (See Appendix for tour information.) The busses stop running too early to take people back to their homes from the professional theatres, but in their community centres, they are still at home, relatively safe. In addition, theatre can give outsiders a chance to hear the stories important to a culture, providing information that could be gathered in no other way. These mythic stories tell what a people believe is the truth about themselves, and if listened to carefully, the accumulated stories can help both outsiders and insiders respond in fresh, more helpful ways.

If theatre is a dynamic discovery process for all participants like it was for the Charabanc women and their audiences, then message and effect and participation take precedence over aesthetics. These aims are not mutually exclusive, of course; the best theatre achieves them all. That Charabanc could reach such a high level excellence in all areas of theatre production bodes well for other women who want to create their own work. Following the success of their first play, these women continued to write plays, incorporating into the action the collected experiences of Northern Irish women, refining along the way their creative methods. *Lay Up Your Ends* is an honest effort. Anything that has entered the mythic domain, as this play has, is nearly impossible to evaluate.[30] Carol and Eleanor resisted suggestions to revive the play for the company's ten-year anniversary, fearing that in light of later, better written work, *Lay Up Your Ends* might disappoint. But nonetheless Carol endorses the creative process they first developed in 1983; she points out that *Lay Up Your Ends* is an example of what you can do when you have absolutely no fear. For any enterprise, unfortunately, that luxury of fearlessness lasts for only the first endeavour; from then on, expectations – one's own and others' – influence the results. But the four scripts in this collection are fine scripts, and their value as women's history reaches far beyond their potential for theatre performance. Readers of this volume can now join Charabanc in celebrating the lives of women through *Now You're Talkin'*, *Gold in the Streets*, *The Girls in the Big Picture*, and *Somewhere Over the Balcony*.

And regardless of anything that transpired afterward, the *Lay Up Your Ends* opening at the Arts Theatre was a night to remember. The experience was pure theatre; it was condensed action which required a stage, demanded a performance, stunned an audience,

INTRODUCTION

and realized fully the promise of the theatrical form. That initial performance did not receive an obligatory first-night ovation which flows gradually through the audience until everyone is standing to avoid embarrassment. Instead, the audience rose together simultaneously and cheered and then called the women back again and again. The mill workers they had interviewed were standing there next to business people and politicians and theatre folk and friends and family – united in their appreciation of a significant Belfast story well told. Only then did those five women on stage know they had achieved something. Before that stomping, raucous response, they had simply felt lucky to have gotten through the performance with a script that had kept changing right up until curtain. Having never before created their own work, they were unsure how it would be received. But with that ovation, they knew that they had accomplished what they had set out to do, and after all these years, that knowledge still empowers the Charabanc five.

CLAUDIA W. HARRIS
February 2006

INTRODUCTION
NOTES

1. During the first few years, Marie Jones used the acting name Sarah Jones because of another equity actor named Marie Jones. Carol Moore changed from her married name Scanlan to her maiden name Moore during the final years of the company.
2. Charabanc productions in order of their first performance were *Lay Up Your Ends* (May 1983), *Oul Delf and False Teeth* (February 1984), *Now You're Talkin'* (March 1985), *Gold in the Street* (January 1986), *The Girls in the Big Picture* (September 1986), *Somewhere Over the Balcony* (September 1987), *The Terrible Twins' Crazy Christmas* (November 1988), *The Stick Wife* (February 1989), *Weddins, Weeins, and Wakes* (September 1989), *Cauterised* (November 1989), *The Blind Fiddler of Glenadauch* (August 1990), *The Hamster Wheel* (February 1990), *Me and My Friend* (February 1991), *Frontline Cafe* (August 1991), *Bondagers* (October 1991), *Skirmishes* (October 1992), *October Song* (April 1992), *The House of Bernarda Alba* (February 1993), *The Illusion* (October 1993), *Vinegar Fly* (August 1994), *Iron May Sparkle* (November 1994), *A Wife, a Dog, and a Maple Tree* (February 1995). Whether *Vinegar Fly* (1994) can be considered a bona fide Charabanc production is an open question; the decision to tour a full production of the play following a staged reading was manager Maureen Jordan's – a decision, incidentally, which contributed to the company's financial straits and its eventual closure. (See the Appendix for fuller production and tour information arranged alphabetically by play.)
3. The section includes the best collaborative work of the company. Company members did not want me to include *Oul Delf and False Teeth* (1984); although expressing affection for the play, they believe it would need extensive rewriting.
4. All panel quotations are from a transcript of the panel – 'Is Ireland a Matriarchy or Not? The Experience of Irish Women as Theatre Artists,' Organizer and Moderator, Claudia W. Harris, ACIS conference dates, 25 June-1 July 1995.
5. Going with them and director Peter Sheridan to the Tyrone Guthrie Centre in Annaghmakerrig (an arts centre in Co. Monaghan supported by arts councils north and south) where they began to shape the material they had gathered gave me a glimpse of their working methods. I was at Carol Moore's house for the first reading of the *Balcony* script and was present on several occasions, both before and after that read through, to observe their innovative collaborative process.
6. I produced the Atlanta, Georgia, 31 March-1 April 1990 production of *Somewhere Over the Balcony* when I was on the board of Theatre Gael which sponsored the tour.
7. Sources for information, unless specified otherwise, are company papers, programmes, letters, and interviews.
8. *Craic*: Informal Irish word (pronounced crack) meaning the storytelling, wisecracking, verbal-rapping that lifts many a night in an Irish pub to the sublime.
9. *Theatre Ireland*, **23** (Autumn 1990), p. 41.
10. *Theatre Ireland*, **3** (June-September 1983), p. 132.
11. Stewart Parker, *Dramatis Personae*, John Malone Memorial Lecture, Queen's University, Belfast, 1986, p. 6.
12. In an interview on 30 April 1994, Brighton described herself as being 'very angry and very hurt' about this break with Charabanc in 1986.
13. *Theatre Ireland*, **23**, p. 42; unless noted otherwise, subsequent quotations from Brighton are from this article.
14. *Theatre Ireland*, **21**, p. 31.

INTRODUCTION

15. *Theatre Ireland*, **24** (Winter 1991), p. 5.
16. *Theatre Ireland*, **6** (April/June 1984), p. 144-147.
17. Carol Lloyd, *Creating a Life Worth Living: A practical course in career design for artists, innovators, and others aspiring to a creative life*, Harper, New York, 1997.
18. *Theatre Ireland*, **30** (Winter 1993), p. 28.
19. *Theatre Ireland*, **12** (Summer 1987), p. 72.
20. Fintan O'Toole, *The Irish Times*, 20 September 1994, p. 10.
21. Gayle Austin, *Feminist Theories for Dramatic Criticism*, University of Michigan Press, Ann Arbor, 1990, p. 81.
22. Laura Mulvey, 'Visual Pleasure and Narrative Cinema', in *Art after Modernism: Rethinking Representation* (ed. Brian Wallis), David R. Godine, Boston, 1984, pp. 361-373 [reprinted from *Screen* **16** (Autumn 1975), pp. 6-18], p. 367.
23. Jill Dolan, *The Feminist Spectator as Critic*, University of Michigan Press, Ann Arbor, 1988, p. 49.
24. Simone de Beauvoir, *The Second Sex* (trans. and ed. H. M. Parshey, 1949), Jonathan Cape, London, 1953; reprinted Penguin, Harmondsworth, 1974, p. 295.
25. Judith Butler, 'Performative Acts and Gender Constitution: An Essay in Phenomenology and Feminist Theory', in *Performing Feminisms: Feminist Critical Theory and Theatre* (ed. Sue-Ellen Case), Johns Hopkins University Press, Baltimore, 1990 [pp. 270-282], p. 277.
26. Richard Schechner, *Performance Theory*, Routledge, New York, 1988, p. 213-4.
27. Marie Jones wrote out this alternative ending for me backstage before a performance of *Gold in the Streets* in Washington D.C., 30 June 1986. Others in the company helped her reconstruct the new dialogue as Marie wrote in the back of my *Now You're Talkin'* playscript. This then may be the only existing copy the alternative ending. 'This then may be the only existing copy with the alternative ending.'
28. The often volatile marching season stretches from mid July to mid August and includes Protestant Orange marches celebrating the Battle of the Boyne (12 July) and the Apprentice Boys of Derry (12 August) as well as Catholic marches commemorating Internment (9 August). (See Glossary.)
29. As their Atlanta host for four days during the run of *Balcony*, I learned three versions of the story. Nonetheless, the company had a grand time as usual: a reception at the British Consulate, a late-night party in Underground Atlanta, a run on the shopping malls, and a Sunday morning visit to the Martin Luther King, Jr., Church and Memorial.
30. The Linen Hall Library held a retrospective on Charabanc Theatre Company, 17 September 2000, during their Millennium Festival 'Open Door' to publicize the library's Charabanc archive. Following a panel discussion about the company, Brenda Winter directed students in excerpts from *Lay Up Your Ends*. The performance was moved from the library to the larger Group Theatre because of audience interest.

NOW YOU'RE TALKIN'

PLAY IN TWO ACTS

BY
MARIE JONES
AND
THE COMPANY

CHARACTERS

VERONICA – republican Catholic
COLETTE – moderate Catholic
MADELINE – Catholic in heritage, doesn't really care about sectarian debate
JACKIE – moderate Protestant
THELMA – uncompromising Protestant
CARTER – American supervisor of reconciliation centre
ISAAC – punk waiter

ACT I

Scene One:	Portrock Reconciliation Centre, Northern Ireland
Scene Two:	Seaside near reconciliation centre
Scene Three:	A pub
Scene Four:	At the reconciliation centre
Scene Five:	Reconciliation centre, an hour later

ACT II

Scene One:	Reconciliation centre
Scene Two:	At the centre, later the same evening
Scene Three:	Reconciliation centre, late that night
Scene Four:	Reconciliation centre, next morning
Scene Five:	At the centre
Scene Six:	Reconciliation centre, later that morning
Scene Seven:	Reconciliation centre, moments later

NOW YOU'RE TALKIN'

ACT I – SCENE I

SCENE: *SFX bus pulling away, seagulls, sea sounds, etc. LX up.*

VERONICA (*looks around the room admiringly then looks out the window. To the audience*). This is really beautiful. It reminds me of when was a youngster goin' on an outin' – that first smell when you get out of the city – the hedges, the grass and the smell of the sea. Two days of peace and quiet, away from everything, no kids. . . . Oh I love them, but you know how it is, I'm startin' to say 'toat' instead of 'coat'. Jesus, this is really friggin' beautiful! (JACKIE *enters. She and* VERONICA *acknowledge each other.*)

JACKIE (*to audience*). I've always wondered what a Reconciliation Centre would look like. Sandra, my chum, has been here four times already and she thinks it's the best thing that's ever happened to Northern Ireland. Anyway, I'm just going to watch and listen, exchange ideas, get a few pointers....got my wee pocket dictionary with me . . . I hate that – people using words you don't understand – you look around and everybody is nodding in agreement, then you find out later that they didn't know the meanin' of the words either. (*Looks in her dictionary.*) I just finished the B's last week – this week I'm on the C's . . . 'Convolute', 'to complicate'. (*Thinking aloud.*) 'With my first child I had convolutions' – no, that doesn't right . . . 'Convoke', 'bring together' (*Thinking aloud.*) 'We have been convoked here for the weekend' . . . Jesus!

COLLETTE (*entering. Addresses audience*). The Group Leader's a man . . . that's just typical . . . we're a group of women – this is about bringing women together! There are some women who just cannot assert themselves – they won't even see the ridiculousness of the situation – they'll just think it's par for the course . . . (*looking over at* JACKIE.) Oh, God! Look at yer woman in the lemon tracksuit . . . well, if there is anything physical, I'll just say I left my leotard at home – I can always do it in my dress . . . would just look so stupid!

THELMA (*entering. Addresses audience*). I shouldn't be here – he won't manage on his own. It was wee Mrs. Watson and our Olive coaxed me intil it! He was all for it, too, but wait'll he see them dishes pilin' up and the dust startin' t'gather . . . he'll miss me, I know he will. I'm not used to strange beds either. . . . Oh God, I hope I don't have to take off me fornenced people! (*Sniffs.*) Smells like an old people's home! (*To the others.*) Excuse me (*indicating the name badges*), are yousins yella?

COLLETTE (*handing* THELMA *a programme*). Yes.

THELMA. What's that?

COLLETTE. Programme for the weekend.

THELMA (*now very worried*). I've come here with our wee Church group and we've all been split up. Wee Mrs Watson, our leader, will be very annoyed.

VERONICA (*still gazing at the view*). I wouldn't worry about it, love, everybody's been split up. This is really beautiful, isn't it?

JACKIE. I think it's a good idea – people bein' split up – otherwise you would just form wee 'caucuses' and never get to know anybody.

VERONICA. Wonder how many points you need to get one o' these?

COLLETTE. Points?

VERONICA. Aye, the Housin' Executive! 'Think I'll put in an official complaint. I always wanted to live in a big house overlookin' the sea.

COLLETTE. They're liable to tell you to go pitch a tent in the middle of the Irish Sea for being so ungrateful.

THELMA. Ah, holy flip! My good coat! I've left it out in that big hall and I noticed one of them other women had one just like it but mine is only brand new and I wouldn't like them to get mixed up.

COLLETTE. Sure, why don't you go and get it?

THELMA. But the man didn't say we could leave the room.

VERONICA. This is definitely for nothin', isn't it? Like, all this and the grub threw in? Frig, somebody nip me, make sure I'm here!

THELMA. Marks and Spencers – it was right and dear!

COLLETTE (*getting up*). Think I'll go out and get a breath of that fresh air.

THELMA. Are you allowed?

COLLETTE. Who's going to stop me?

JACKIE. You're not supposed to. Carter is going to be here in a minute and we have to do everything in groups.

COLLETTE. Carter who?

THELMA. The man.

JACKIE. He's going to be our Group Leader.

NOW YOU'RE TALKIN' – ACT I SCENE I

VERONICA. That the Yank?
JACKIE. Aye, he seems to be a very nice man.
COLLETTE. He's all right.
THELMA. It was the guts of sixty pound, too.
VERONICA. What?
THELMA. My coat.
VERONICA. Jesus! Sixty pound! I'll go and get if for you if you'll give me a reward.
JACKIE. Why did yous all come? I've been tryin' to get the women in my wee community centre involved in an information group, but you couldn't get them dug out of the house.
THELMA. Ach, wee Mrs Watson'll know it for she remarked on it and I told her it was brand new.
COLLETTE (*to* JACKIE, *ignoring* THELMA). This is hardly a wee community centre?
JACKIE. Aye, but they seem to be gettin' people together.
COLLETTE. No wonder! Look at it – big mansion overlooking the sea, free food, chance to get away from it all.
VERONICA. Tell you the truth, that's why I came. Like, I've got five kids, all under eight. Jesus, anything to get my head showered! Frig, I could be joinin' the 'Moonies', for all I know!
THELMA (*horrified*). The 'Moonies'? . . . Hope it's nothin' like that . . . we are supposed to be discussin' spiritual awareness among ourselves. . . .
VERONICA. Why did you come, Collette, is it?
COLLETTE. No, it's Joan Collins, I'm incognito for the weekend! . . . No, seriously, I'd heard a lot about this place and just thought I'd come and see it for myself and, I suppose, with all this peace and quietness, do a bit of thinking.
VERONICA. Aye, I'm the same, only I've decided to think about nobody but me for a change.
CARTER (*entering*). Hi everybody, I'm Carter. Welcome to Portrock Reconciliation Centre. (*Checking* THELMA'S *badge*.) Hallo, Thelma, nice to have you here.
THELMA (*quietly*). Excuse me, Mr Carter. . . .
CARTER. No, no, just 'Carter', Carter O'Donaghue, but you must call me 'Carter'.
THELMA. Yes . . . well . . . I think I'm in with the wrong group?
CARTER. You'll be fine just here, Thelma. Jackie? Hi!
JACKIE. Hello.
CARTER. Collette, good to have you. . . . Ah, Veronica. Hi!
VERONICA. Hi ya.

CARTER. Okay. . . . If you would like to take a chair and gather round, we can begin. (JACKIE grabs a chair and pushes her way in beside CARTER.) I want you to treat Atlantic View as your home.

VERONICA. I'm your woman, Carter!

CARTER. I want you to relax, forget about the outside world and really get to know each other. We are going to conduct some 'explorations' with the object of achieving communication. Now, let me explain. For instance, if I were to say the word, 'love', it would mean something different to each of us, but, by expressing why we feel the way we do, we can learn something about each other and ourselves.

COLLETTE. Is that not a wee bit personal?

CARTER. Well it's important to realise that everyone's feelings have value. It's only when we feel insecure we feel our thoughts are somehow shameful or silly.

VERONICA. What do you mean – like, I 'love' this place or I'm 'in love' with the milkman?

CARTER. I want just exactly what you feel. Okay, now, the word 'freedom'. Take your time. . . .

JACKIE (*jumping in immediately*). Well, 'freedom' to me would be getting away from the housework and the routine, coming here and meeting people, finding out new things in an atmosphere that's . . . conducive. . . .

CARTER. Very good. . . . Thelma?

THELMA. Well, now, I have a very good man and I have all the 'freedom' I want. Now, I know there's many another one that is chained to the house, but I was very lucky with my Ronnie. You see, you never know who you're gettin' til you get them, and once you've got, you've got, and there's no gettin' out of it.

CARTER. Collette?

COLLETTE. Well . . . 'freedom' to me is the opportunity to make choices . . . you see, I chose to come here this weekend as opposed to going somewhere else. I'm very lucky to have that freedom. A lot of people don't.

CARTER. Veronica?

VERONICA. Freedom for Ireland! (*There is an awkward silence.*)

CARTER. I want you to look out there and tell me what you see. (*There is a long silence as they all look out the window not knowing what to say.*)

COLLETTE. Are we supposed to be looking for anything in particular?

CARTER. No, just tell me what you see.

JACKIE. Azure sky . . . rain-filled clouds . . . undulating hills . . . ?

CARTER. Anything else?

THELMA. Ammm . . . sheep?

CARTER. Isn't it beautiful? Nature . . . the beauty of nature . . . and it's ours to enjoy. Through nature we can find our real selves. It is something that is in all of us and belongs to all of us.

VERONICA. You're dead lucky though, Carter. When I luk out my window y'see a dirty great big wall with 'Fuck the U.V.F.' written on it!

JACKIE. Carter, I think I know what you mean. Those birds flyin' through the air . . . that's 'freedom' in a way. . . .

CARTER (*starting to move expressively around the room*). I want you to look out and listen to the silence . . . it's the silence of the land . . . that no words can convey. The silence of land, the water, the rocks. Feel the passage of the lonely bird flying across the ravine . . . the wash of wind on a field of small purple flowers . . . the land . . . our land yours and mine. Now, can anybody tell me what the word 'unite' means?

THELMA. It means 'join together'.

CARTER. And we are going to join together. (CARTER *joins their hands together. Everyone looks awkward*). Join together in enjoying the silence. . . . Now, does anyone notice anything strange or unusual about each other?

JACKIE. No, not really.

CARTER. All the same species . . . we eat the same food, we live in the same land and, united as we are now, we can feel as one the joy and fulfilment of the beauty of nature. (*They all sit in silence, holding hands.* CARTER *is feeling every moment of it. The women are bemused.*)

LX FADE.

ACT I – SCENE II

SCENE: *SFX seagulls and sea. LX for the seaside.*

JACKIE (*addressing the audience*). Collecting shells to express yourself? I'm not sure exactly what it means but I'm sure I'll understand the philosophy eventually. . . . I hope these other ones do . . . it only takes one to spoil everything . . . I mean, anything can work if everybody believes in it enough?

THELMA (*carrying twigs, shells etc. Addressing the audience*). God, that woman sayin', 'Freedom for Ireland' – I near had rickets! You never know who you're keepin' company with. . . . I'm not used to all these new-fangled things, sittin' down and holdin' hands! Still, you don't like til object for I'm sure he knows what he's doin'. You're heart feared to say anything for fear they think you're daft.

COLLETTE (*addressing the audience*). They're all like sheep, really. He says, 'go out and express yourselves with nature and', just as a suggestion, 'collect some shells' and they're all doing it. I'm going to write a poem. He probably thinks we're all wee thick Belfast women.

VERONICA (*addressing the audience*). That oul' doll near had a fit when I said, 'Freedom for Ireland'! Frig her! Exploration exercises? Soul searchin'? What the hell for? Oh God! This place really is out of this world!

CARTER (*entering*). Well, everybody, how are we getting on?

JACKIE. Oh, Carter, this view really is magnificent! All this beauty would make you want to lose yourself. (*Holding up a shell.*) This is what you call cornucopic, isn't it?

VERONICA. Aye, if we had all this in Ballymurphy, there'd be changes all right.

COLLETTE. Yea, they'd be drownin' people instead of shootin' them! (VERONICA *does not appreciate the joke. LX fade. LX go up on the same room. SFX soothing music.*)

THELMA (*wearing a coat*). I still think I've got the wrong coat. . . . The buttons on mine was trimmed with silver and these has got gilt on them. (*Everyone ignores her.*)

NOW YOU'RE TALKIN' – ACT I SCENE II

VERONICA (*holding a handful of shells*). Anybody know what we're supposed to do with this stuff?

JACKIE. We're supposed to express ourselves with our materials.

THELMA (*sitting at the table, holding a twig*). You mean, like a wee ornament, or something?

COLLETTE. Do you feel like an ornament?

THELMA. What are you talkin' about? I'm going to make a wee coat-stand.

JACKIE. We're supposed to express how we felt on the beach.

VERONICA. How do you express your arse bein' freezin'? (*There is silence as they 'create'.*)

ISAAC (*rushes in carrying a tray with cups*). Right! Tea up, everybody! Before you start twenty bloody questions – I'm Isaac – I work here, so I do! (*Rhetorically.*) Oh really, Isaac? Aye, I live here on the premises. I'm goin' to be lookin' after yous . . . and I'm an anarchist – tea? (*The women take the tea from the tray.*) Sugar, love?

JACKIE. No, I'm on a diet.

ISAAC. What! With a lovely wee trim figure like that? You're kiddin' me? (*To* COLLETTE.) Sugar?

COLLETTE. No, it's okay, I've sweetners here.

ISAAC. Aye, you shud be on a diet right enough. . . . Are yous doin' the 'my life's nothin' but a bag a' dulse' caper? That's a good one. Right! Who's the Prods and who's the Taigs? (*There is silence.*) It's all right, you can tell me, I'm nothin', I'm an anarchist.

THELMA. You're far too young to be talkin' like that.

ISAAC. I know what you are, Missus, you're a Prod, aren't you? And, see her over there, (*indicating* VERONICA) she's a Taig so she is . . . (*Lifts* THELMA'S *twig.*) Hey, what's that?

THELMA. It's a wee coat. . . .

ISAAC. No, don't tell me . . . fifteenth century Chipperfields, am I right?

THELMA. It's a wee coat-stand.

ISAAC. Aye, it looks just like the thing. . . .

VERONICA. Isaac, there isn't a pub near here, by any chance?

ISAAC. Don't drink. (*Offstage* MADELINE *shouts.*)

MADELINE. Isaac! Isaac!

ISAAC (*covering his head with his hands – speaks to himself*). Oh God! It's not, is it? Oh no!

MADELINE (*entering – dishevelled*). Isaac! Ach, Isaac son! Am I glad to see you! Hi ya, everybody!

ISAAC. Mad . . . (*trying very unsuccessfully to interrupt.*)

MADELINE. This place is in the arsehole o' nowhere! I had to get a bloody bus – my feet are throbbin' . . . !

ISAAC. Madeline, you went home three weekends ago!

MADELINE. Aye, I know, son, but that scabby pig threw me out! Member I was tellin' you about him and his oul' shingles? Jesus, it's not my fault that he has them! Anyway, he's runnin' about like a beatin' bear at the minute . . . like, just because he can't go out, doesn't mean I have to sit in like a friggin' nun! . . . Oh, no offence to anybody here. . . .

ISAAC. Madeline! Watch my lips! Carter won't let you stay . . . you have for til go.

MADELINE. And where the hell am I going to go?

ISAAC. What about them battered shelters for weemin?

MADELINE. Aye, you're not geggin'. I was near bate to death only I made my escape in time. It was all over the head of nothin' too.

ISAAC. Come off it. A man doesn't bate a woman unless she's done somethin' to annoy him.

MADELINE. I swear t'God, Isaac! Me and big Irene, do you remember big Irene was here with me? Me and her went down to our Club for two or three wee drinks and we dandered home about two in the morning . . . like, half two. (*Addressing everyone now.*) I mean, the kids in our flats is still out playin' b'Jasus! Well, what he didn't call me! . . . 'Get out and don't come back, y'big' . . . even our wee budgie was shakin' with fear.

ISAAC. Why didn't you stick one on him? Give me his address and I'll get the boys on to him.

MADELINE. Oh, don't get me wrong, Isaac, I'm no softie, but there's no woman's a match for an angry man, so out I goes. I'd to sleep in the alleyway of our flats – all night. Like a bloody ice-cube, I was . . . big gaunch!

ISAAC. Here, my mate Snicker works for the Simon Community . . . you go down there – say Isaac sent you – you'll get a bed, no problem.

MADELINE. Sure, isn't this place supposed to be about peace, love and friendship?

ISAAC. Only if you've booked in advance.

MADELINE. Jesus, Isaac, you wouldn't turn a destitute woman out in the streets, would you?

ISAAC. Madeline, if it was up to me . . . but my orders come from down below.

MADELINE. Ach, go on, Isaac son. Sure you and Carter's right and thick. See what you can do for me.

NOW YOU'RE TALKIN' – ACT I SCENE II

ISAAC. You'll be gettin' me the boot, so you will. Last time (*addressing the other women*) it was her all-night parties, this time it's her oul' lad's shingles. . . . You know where I'll end up? In a home for destitute punks!

MADELINE (*waving a pound note in the air*). I'll give you a pound.

ISAAC. Ach no, Madeline, I cudn't take a poun' off ye.

MADELINE. Go on.

ISAAC. I couldn't take a pound off ye . . . have ye got one-fifty? (*Exits.*)

MADELINE (*addressing everyone*). Here, yous don't mind me stayin' with you this weekend?

JACKIE. No, not at all.

MADELINE. Like, I don't care what yous are, Catholics or Protestants, it's all the same to me. The way I look at it – you can't ate a Union Jack or a Tricolour for yer dinner.

JACKIE. That's a very interestin' way of putting it.

COLLETTE. If a bit simplistic.

MADELINE. All our problems are simple, love. It's people likes to complicate them so that when they've sorted them out it'll make them think they've done somethin'. Are you doin' your wee walk on the beach caper?

COLLETTE. Yeah, I'm just tryin' to finish this poem.

MADELINE. We'd a geg when we done that. Big Irene, my chum, found this used Durex on the beach. 'Here', she was to Carter, 'Carter, how do you express yourself w'that?' . . . I was cut to the bone.

COLLETTE. Seeing as you were here before, what do you think of the whole set-up?

MADELINE. You're not allowed to drink . . . (*producing a large bottle of vodka from her bag.*) Course I've my big bottle of Vodka with me. . . . Maybe we'll have a wee drink later on?

JACKIE. But sure, if you're not allowed . . . ?

MADELINE. Ach, you don't pay no heed to that. (*Spotting* THELMA'S *'creation'.*) What's that, love – a wee bit of Modern Art?

THELMA. It's a wee coat-stand.

VERONICA. Do they call you Madeline McNulty to your maiden name?

MADELINE. Aye.

VERONICA. I used to go out with your brother, Seamie.

MADELINE (*obviously not wishing to pursue this subject*). Ah . . . he's in . . . America.

VERONICA. Oh, is that right now?

CARTER (*entering*). Madeline?
MADELINE. Carter.
CARTER. Okay, Madeline, what's the problem?
MADELINE. Carter, love, he threw me out . . . down two flights of stairs. . . . If it hadn't been for our big labrador, Jimbo, breakin' my fall, I'd've been lyin' up in the Royal. . . .
CARTER. Isn't it possible to go home and talk it out?
MADELINE. Talk it out? Talk it out? Neither man nor beast is safe with him when he takes a wobbler! I couldn't!
CARTER. Madeline, we are completely booked up.
MADELINE. Sure, could I not double up with some of themins? (*There is silence.*)
CARTER. All right, I'll get Isaac to put a camp bed in the attic for you.
MADELINE. Thanks, Carter.
CARTER. Now ladies, let's examine our responses to our walk on the beach and see what we have discovered within ourselves . . . Collette, have you done something for me?
COLLETTE. Yes. I decided I'd like to express myself in poetic form.
CARTER (*obviously impressed*). Poetry – you write a lot of poetry?
COLLETTE. Well no, not a lot, but there are times when an event or a situation can move you so much that everyday language doesn't cover it – it has to be poetical.
CARTER. Good, can I have a look?
COLLETTE. Well, it's not quite finished yet . . . I'd like to refine it first before I show it to you.
CARTER. Good, keep it up. Veronica?
VERONICA. I haven't nothin' done.
CARTER. I see. Did you have difficulty in understanding?
VERONICA. No, not at all – I just didn't see the point of it. I mean, you can think the same things standin' in the middle of yer kitchen as you can out there, a handful of shells and a lump of seaweed doesn't change anything.
CARTER. Yes, but sometimes the beauty of nature helps us think more deeply, helps us appreciate our existence.
VERONICA. Maybe you're right, but there aren't too many beaches around our way.
CARTER. Jackie?
JACKIE. I've been doing an abstract drawing about my thoughts and feelings on the beach . . . but I'm not too happy about it. . . . Perhaps you and I could have a chat about it sometime?
CARTER. Yes, I'd like that.
THELMA (*jumping up*). Carter, I've done something . . . (*holds up her*

twig which is now decorated with coloured paper, shells, etc.)
CARTER (*pulling a chair into the centre of the room*). Okay everybody, gather round and Thelma will try to tell us how she has expressed herself in this morning's exercise. (*The women gather around* THELMA. *She is intimidated.*) Shall I interpret it?
THELMA. Yes.
CARTER (*holding up the twig*). I can see from this configuration....your spiritual self is rising in response to an environment in which you really want to express yourself.
THELMA. Yes?
CARTER. Now, I want us all to relax and really start talking to one another. (*There is a bewildered silence as* CARTER *produces a maypole and places it in the centre of the stage.*) Now, you all know what this is . . . it's a Maypole. Maypole dancing is a quaint old English custom. . . . You may think this is rather silly, but I find it useful for demonstrating the point of this exercise. The Maypole represents the world. Each coloured ribbon represents a different country. You can see now that they are separated but we are going to unite them. By intertwining the ribbon, we create a pattern of unity which demonstrates that it is possible. As we dance around the Maypole, we must try not to think of ourselves but to liberate our minds from the futility of personality. We must try to free our spirits but remain aware that we can never be free while the rest of humanity is divided. (*Hands the white ribbon to* JACKIE.)
COLLETTE. It's not really equal though, is it, Carter? You take your own country of America, just for an example. Did you know that in America they waste more food than they do anywhere else in the world and yet there are places in Africa where people are starving to death . . . it's never been equal. . . .
CARTER (*gently*). We are not talking about worldly wealth. We are talking about how we as people can find an inner peace so that we can understand and appreciate another way of life.
THELMA. Love thy neighbour as thyself. Is that what you mean?
CARTER. Before you can find peace with your neighbour you must find peace within yourself.
VERONICA. It's very hard to have peace with your neighbour if you're running about with not a stitch on your back and your neighbour's got a sixty pound coat!
THELMA. What's my coat got to do with anything?
VERONICA. I wasn't talking about your coat – I was just using it as an example.

CARTER (*placing his hands on* VERONICA'S *shoulders*). Veronica, when you find peace with yourself, envy will vanish because you will have everything you need within yourself.

THELMA (*to* VERONICA). Though, I think that people should get what they deserve . . . the Lord helps those that helps themselves.

MADELINE. Here, Carter, member the last time when big Irene was here? You wanna seen her. She's hands like J.C.B. diggers! Anyway, here she was, 'Give me the green one and I'll be Ireland!', and pulls it off the Maypole and wrecks the whole exercise. (CARTER *is not amused*, VERONICA *is*.)

JACKIE. Carter, did you think that that was symbolic?

CARTER. Jackie, without empathy, sharing and understanding, anything can fall apart.

COLLETTE. Sure I'll take the green and be Ireland. I don't mind. (*Lifts the green ribbon*.)

THELMA (*taking a blue ribbon*). I'll be Northern Ireland. (*They each take a ribbon*.)

CARTER. Now, we all spread evenly 'round the Maypole and we can begin . . . now, it's very simpl . . . over and under . . . over, under . . . (*Turns on a tape. SFX maypole music. The women are unsure of what to do and become confused.* CARTER *stops the music*). Well, it's not perfect, but it's a start. . . . You see now on the Maypole we have a pattern of unity and harmony. This is what we want to achieve. . . . I think we will take a break for some coffee now. You all deserve it.

COLLETTE. Thank you. That was very interesting. (*The women all leave the stage.* VERONICA *is last to go*.)

CARTER. Veronica! Could I have a word with you, please? (*Leads* VERONICA *gently into centre stage. Awkwardly*.) Veronica, is there something bothering you?

VERONICA. Me? No, there's nothin' wrong with me.

CARTER. I'm just sorry you're having difficulty. I want to help.

VERONICA. What are you talking about? I'm here, aren't I? I'm joinin' in. What more do you want me to do?

CARTER. But you're not enjoying it.

VERONICA. Oh, I'm sorry . . . is that one of the rules . . . you must enjoy yourself?

CARTER. There are no rules. I'm just sorry you don't get on with the other people here.

VERONICA. Ah, I get it . . . it's because I'm not throwin' myself at you like everybody else is. Listen, you've got enough on your plate without worryin' about me.

NOW YOU'RE TALKIN' – ACT I SCENE II

CARTER. But I am worried about you.
VERONICA. Why?
CARTER. Because I think you're unhappy and I want to help, whatever way I can.
VERONICA. Well there's nothin' you can do, so just forget it.
CARTER (*tentatively*). Maybe I could help you relax for a while . . . ?
VERONICA (*unsure of what he means*). How?
CARTER (*touches her gently. Softly*). I'll see you later, okay?
VERONICA. Okay.

SLOW FADE TO BLACKOUT.

ACT I – SCENE III

SCENE: *SFX pub music. LX up.* VERONICA, MADELINE, JACKIE *and* COLLETTE *are sitting around a table, drinking.*

MADELINE. Oh aye. I had a hysterectomy when I was only nineteen.
COLLETTE. God, that's awful!
MADELINE. Ach it wasn't so bad . . . the wee shop might be closed but the playpen's still open . . . !
VERONICA. So who are you married to now, Madeline?
MADELINE. How come you know so much about me?
VERONICA. Cause Belfast's a very small place. . . .
MADELINE. Anyway, I'm not married to this one. I'm waitin' on my second divorce comin' through. But, sure y'never know, third time lucky!
COLLETTE. Lucky? Sure he beat you up and threw you out!
MADELINE. I don't care about that, so long as he loves me. (*Produces a bottle of vodka from her bag.*) Here, keep dick somebody, for I'd swear that barman has eyes in his arse! (*Pours some vodka into each glass.*)
JACKIE. Collette, do you fancy a dance?
COLLETTE. Are you jokin', to that oul' fogey music? These country pubs is all the same.
VERONICA. Oh, sorry love. If we knew you were comin' we cud've hired the Ulster Orchestra!
JACKIE. Did you see Carter dancin'? He was brilliant!
VERONICA. Aye, he was a laugh a minute.
COLLETTE (*to* JACKIE). C'mon, you and me'll ask him up to dance.
MADELINE. It's manners to wait til you're asked, love.
COLLETTE. This is 1985.
JACKIE. Ach, he'll probably come over. He's tryin' to get round everybody.
COLLETTE (*sarcastically*). Aye, he's bein' what you might call 'circumspect'. Isn't that right, Jackie? (JACKIE *obviously does not know what this word means.*)
MADELINE (*to* COLLETTE). You've a couple of them wee 'O' levels, haven't you? . . . I've a B.A. 'Bugger All'! (CARTER *and* THELMA

NOW YOU'RE TALKIN' – ACT I SCENE III

enter. Both are out of breath.)

THELMA (*excitedly*). Oh Carter, that was brilliant! I'm outta breath! (*Addressing everyone, but especially* CARTER.) You know I don't get much chance to get away for the weekend like this, what with my husband having the wee window cleaning business and the two men under him. I hafta stay in and take his calls. He's very high up in the Masonic, y'know. He's that many functions to attend . . . he's thinkin' of gettin' one of those wee answerin' machines. . . .

CARTER. Now, don't be drinking too much, Madeline. We've got a big day ahead of us. . . .

THELMA. Our Olive would love this, so she would. (*No-one is interested in what* THELMA *is saying.*). . . . Only this is the weekend she bakes her Christmas cakes. They're beautiful, so they are. They'd melt in your mouth. . . .

MADELINE (*interrupting* THELMA). Carter, I hear you're being 'circumspect'?

CARTER. Well, how is Madeline? Are we getting over our little domestic crisis?

MADELINE. Aye well, sure you have t'keep your chin up. . . . I'll maybe talk to you about it later on?

CARTER. Yeah, well....

JACKIE. Carter, would you have five minutes to talk about the shell session . . . member you said . . . ?

CARTER. Sure. . . .

COLLETTE. Fancy a dance, Carter?

JACKIE (*abruptly*). Carter and I have some things to talk about!

MADELINE (*to* JACKIE). At least yours won't be men trouble, will it? No . . . you're happily married, aren't you? Two point five kids. What I need is a point forty-five to get rid of this oul' ghett I'm livin' with!

COLLETTE. God, it would put you off marriage for life, wouldn't it, Carter? Thank God I'm single. . . .

MADELINE. Here, I'm not married either, love. No, twice was enough for me – this time it's 'taste and try before you buy'.

CARTER. Right, who's dancing?

COLLETTE. Sure it's a fast one. I think we should all get up.

VERONICA. Oh? I thought you couldn't dance to this oul' fogey music?

COLLETTE. Oh, I don't mind. I can come down to anybody's level!
(JACKIE *and* COLLETTE *jump up to dance.*)

THELMA. No Carter, if I dance any more I'll have palpitations.

NOW YOU'RE TALKIN' – ACT I SCENE III

CARTER. Veronica?
VERONICA. No thanks, I'll skip this one.
JACKIE (*as they are leaving the room*). Carter, I have a few ideas myself for exercises. . . .
MADELINE (*looking after them as they leave*). Just cause she knows two or three big words she thinks she's Brains Trust....friggin' Liquorice Allsort!
THELMA. Well, there's no harm in tryin' to better yourself in this world. Wee Mrs Watson is with a lovely group of women. She's very impressed . . . one of them is married to a doctor!
MADELINE (*very impressed by this*). Ach away!
THELMA. Mrs Watson says she's a lovely woman. Very down to earth and takes time to talk to you. Very interested in reconciliation.
MADELINE. Aye sure, y'never know who you'll meet in a place like this . . . I remember big Irene fell in with a Minister's wife. Big Irene is the biggest heathen that ever walked. . . . I remember when them Christians came round the doors. D'y'know what she would have said? 'Jesus, I thought the lions ate all yousins . . . !' Now you couldn't bate her outta the Rectory. She's atin' tea and buns wi' them, and her a Catholic!
THELMA. Oh, it's marvellous what a wee bit of tolerance can do.
VERONICA. So, is your Seamie still in Long Kesh then, Madeline?
MADELINE (*nearly choking on her cigarette*). Yes, he is. He's two more years to do. Are you satisfied now? Why don't you go up and get yer man to announce it? The rest of the people mightn't've heard you!
THELMA. Well, I think I'll go and get my coat now. (*Stands.*) I'll see yous in the morning at the Ecumenical Service.
VERONICA. Oh, have you an armed guard out there watching it? Sit down, Thelma! Sit down and enjoy yourself!
MADELINE (*to* THELMA). Sorry about that, love. (*Pats* THELMA'S *leg. To* VERONICA.) Listen you, I don't need people like you throwin' things up in my face. I'm tryin' to forget all that. . . .
VERONICA. And do you think that Seamie will forget about it? Do you think he'll forget his own sister who turns her back on him? No, Madeline, he's in H-Block, not friggin' Butlins! He's there because of people like you and me. And what does his dear sister do while he's in there sufferin'? She bloody denies him! You're a Judas, you are! A bloody Judas!
MADELINE. Listen you! I never asked anyone to suffer for me! I don't want to see people dyin' for Ireland. I want to see people livin' for Ireland!

THELMA. You're right, Madeline. That's the oul' drink talkin' . . . (*To* VERONICA.) You leave Madeline alone.
VERONICA. You want to see people livin' for Ireland. What's livin' in your books, eh, being at the bottom of a heap, being made to feel like scroungers in our own country? No, Madeline. That's why Seamie's in there, so that some day we'll be able to live like people – not friggin' animals!
THELMA. Here you, there's a time and place for that kind of talk.
MADELINE. Listen you, this is a Protestant pub. Do you wanna be the next dead hero?
VERONICA. Oh there's a time and place is there, Thelma? Okay then, tell me where it is cause I'd like to know.
MADELINE. Here you, leave Thelma alone. Our fight's not with Thelma, it's with the Brits.
THELMA. I'm British.
VERONICA. C'mon, Thelma, where's the time and place?
THELMA. Look, Carter and the rest will be back soon, or will you not be satisfied til you've upset everybody?
VERONICA. Maybe it's in the City Council, is that it, where your friggin' D.U.P. headbangers spray us with air freshener and play 'The Sash' on toy bugles? No, Thelma, that's not it either. So you tell me where it is.
THELMA. Ach, is it any wonder nobody will listen til yous if you behave like this?
VERONICA. Ah well, this is my time and my place . . . (*Stands and sings loudly.*) 'A Nation Once Again, A Nation Once Again'!
MADELINE. For God's sake, sit down! Sit down, you! (CARTER *and the other women return.* CARTER *drags* VERONICA *out. They are all embarrassed.*)

LX BLACKOUT.

ACT I – SCENE IV

SCENE: *Blackout. LX up. The women are lying on the floor, yelling and screaming.* CARTER *is standing above them.*

CARTER. Yes, get it out, get it right out! (*Shakes the tambourine he is holding as a signal for the women to stop screaming.*) Good. Now just relax. Let yourself go. Primal screaming loosens the unreal self and broadens the real self. I want you to relive and fully experience those unreal feelings of hate. Remember, the more pain you feel, the less you actually suffer. Now, open your mouth as wide as you can. Open your mouth and keep it that way. Now, pull that feeling out of your belly, feel that scream all over your body. (*The women's screaming rises to a pitch. After a while* CARTER *stops them again.*) I know what you're feeling . . . it's like a lightning bolt that's felt all over your body. It has broken your unconscious control of your body. You are now relaxed and ready to build real feelings of love towards each other. Don't be tense and resist those feelings of love. There is no need to revert to your unreal behaviour. Don't allow yourself to become someone else again. There should not be residual tension. You are now feeling well. Now, reach out, reach out and touch. Reach out and touch as if for the first time, like a child without hate. Let's come together. Let's join our bodies together, our minds together, and come together as one, like the petals of a flower.

LX FADE OUT.

ACT I – SCENE V

SCENE: *Everyone is dancing to 'The Farmer and the Cow Man Should be Friends'. The music ends.*

CARTER. Well, did you all enjoy that? I know it's just a corny old musical but did you listen to the lyrics? 'The Farmer and the Cow Man Should Be Friends'. Even though they had their territorial feuds they still managed to live together in peace and harmony in Oklahoma. Now, we're going to see if we can do the same thing right here. Jackie, Thelma, you two are going to be the 'farmers', so you get on this side. The rest of you will be 'cow men', so you get on that side. Right now, let's see if we can make this work, okay?

THELMA. Excuse me, Carter, could I not be a 'cow man' instead? Cause I was reared on the Beerbridge Road and there's not many farms about there.

CARTER. Well, Thelma, I want you to be a 'farmer' just this time. Okay?

VERONICA. Why?

COLLETTE. Cause they're Protestants i.e. 'farmers' and we're Catholics i.e. 'cow men', right?

CARTER. Right. Jackie, take it away.

JACKIE. Hi there, Mr Cow man. I'm gonna darn well take down my fences and let you through my cotton pickin' fields!

COLLETTE. Well, what do you think of that, fellers?

MADELINE & THELMA. Yippee!

COLLETTE. In fact, I'm so pleased by that, I might even let you marry my daughter Ellie Mae!

CARTER. Come on now, keep it going.

THELMA. How do, Mr Cow man. How's about me letting you have some of my crops and you letting me have some of your cow dung?

MADELINE. No problem, Jethro, I'll even fork it over the fence for you.

CARTER. Veronica, why don't you join in, huh?

VERONICA. You want me to join in?

CARTER. Yeah, come on.
VERONICA (*moving seductively towards* CARTER *and lifting her skirt*). I've come for ma boy. How's about you comin' up to my haystack sometime? God, you're a dickhead!
CARTER (*pulling* VERONICA *to one side*). Now, Veronica, you've tried to be destructive in every exercise. You must not burden everyone with your negativity. For your own sake, try and co-operate. Things are possible. Please?
VERONICA. Okay.
CARTER. Well, I think it's time we had a celebration.
THELMA. I know, let's have a big picnic and I'll bake a blueberry pie!
COLLETTE. To hell with your blueberry pie. I think we should have a goddamn barn dance!
CARTER. Okay, let's have a barn dance. (CARTER *switches on the music and everyone dances.*)
THELMA. Oh, Carter, this is great! Oh, our Olive would love this! How do you like my American accent? Oh, if she could see me now....
VERONICA. Why don't you just shut up? Hee, hee, hee. Ho, ho, ho – dancing around there like a stupid bloody cow! (CARTER *stops the music.*)
JACKIE. Veronica, I don't think this is the time or place....
VERONICA. Shut you up, too. (*To* THELMA.) You know what I'd like to do with you? I'd like to get your sixty pound coat and ram it down your bloody throat!
THELMA. Jealous! Just jealous! If you want to run around like a rag picker, that's your own affair.
VERONICA. Listen you, do you want to know what I have to do with sixty pound?
COLLETTE. Aw come on now, Veronica....
VERONICA. You stay out of this! I have to feed five kids and pay rent on a house that's fallin' down round me!
THELMA. What's it to me if yous lot want to breed like rabbits?
MADELINE. Here, Thelma, who's breedin' like rabbits?
THELMA. You bloody Catholics! Aye, and it's my man has to pay his taxes to keep you Fenian hordes!
COLLETTE. Thelma, this is our country.
THELMA & JACKIE. It's our country too!
COLLETTE. We'd pay taxes too if we could get jobs, but there's no chance of that while your government is discriminating against forty per cent of it's own friggin' population!
THELMA. Is it any wonder when all yous want to do is blow it to

Kingdom come?

JACKIE. Aye, and what about your own politicians, nothin' but a pack of bloody murderers!

VERONICA. Listen love, if you're so fond of bein' British, why don't you fuck off back to England where you belong?

MADELINE. Aye, yous have no right to be here in the first place!

JACKIE. I have a right to be here, it's my bloody country! (*This argument develops into a full scale row.* JACKIE *and* VERONICA *come to blows.* CARTER *separates them.*)

CARTER. You, ask yourselves, why is our dear country devastated by violence? Shall I tell you why. Shall I? It is because we have stopped looking for the things that unite us. You have been given an opportunity to look for those things. You have been given an opportunity to get rid of these hateful, destructive, negative emotions. And yet you still feel hatred. Until you get rid of these emotions, you will keep fighting, you will keep hating, you will destroy yourselves, you will destroy your land. So, it's up to you. The choice is yours. You can have good emotions or you can feel bad emotions. You can feel hate or you can feel love. Now, which do you want?

JACKIE (*lunging at* CARTER). Yes! Yes! We do hate each other! We bloody well hate each other! That's what all this trouble is about. What the hell do you know about it? You don't know anything about it! Why don't you just fuck off? (CARTER *runs offstage.*)

LX BLACKOUT. END OF FIRST ACT.

ACT II – SCENE I

SCENE: *SFX 'The Farmer and the Cow Man Should be Friends.' LX up on same positions as at end of ACT ONE. There is a long pause.*

MADELINE. Pig! And to think I fancied him too!
COLLETTE. You fancied him?
MADELINE. Aye, why do you think I came here this weekend?
COLLETTE. I thought your man threw you out?
MADELINE. Yes, so he did, but I deliberately fought with him so I could come here . . . he gave me the impression he fancied me too the last time. I went and bought one of them wee black and gold nightdresses. (*Takes a nightdress from her bag.*) Look! Nine ninety nine that cost me!
COLLETTE. You're not going to believe this . . . I fancied him too!
VERONICA. He really turned me on. . . .
COLLETTE (*to* MADELINE). Would you really have slept with him?
MADELINE No problem!
JACKIE. Yous are all havin' a nice wee cosy chat now. You were callin' us 'Orange Bastards' a minute ago!
COLLETTE. What's wrong? Did you fancy him too?
JACKIE. You know I did, but not in the way you think . . . I thought he was really wonderful . . . !
THELMA. Well, you ought to be ashamed of yourself, and you a married woman!
JACKIE. It doesn't stop you havin' feelings.
THELMA. I think the half of yous came down here for a bloody orgy!
JACKIE. Thelma, there is a difference between having an orgy and really believing in someone so much you just want to be with them.
THELMA (*going to get her bag*). Bold tinkers, the lot of you! I'm going home!
JACKIE. Who's stoppin' you?
THELMA (*stopping*). I have as much right to be here as you!
JACKIE. The whole purpose of the thing is ruined anyway. You were the one who was screaming 'Fenian Bastards' at people!
THELMA. Well I wasn't going to let them get away with calling me

NOW YOU'RE TALKIN' – ACT II SCENE I

names. (*There is a loud knock at the door.*)
JACKIE. If that's Carter, tell him to piss off!
VERONICA (*shouts*). Piss off, Carter!
MADELINE. You're a waste of space, Carter! (CARTER *enters. There is an awkward tension.*)
CARTER. I think we should forget the recent ugly display and try to talk.
THELMA. Talk? Look what talkin' done the last time.
CARTER. Okay then, we won't talk. I'm afraid you'll have to leave.
COLLETTE. What are you scared of, Carter, afraid we might start shootin' each other?
CARTER. Please, your buses will be arriving.
JACKIE. Oh that's really great, that is! You make us expose our innermost feelings and then you ask us to leave! Well I'm sorry, I can't go feelin' the way I do.
CARTER. If you would just like to wait in the hallway . . . ?
JACKIE. I'm not going to be treated like a piece of baggage. I don't feel like going!
CARTER. Jackie, I have opened my home to you. . . .
JACKIE (*screaming*). I don't want to listen to you!
CARTER. Right then, we have to leave.
THELMA. Why do you keep sayin', 'we'? You mean 'us', us here. We have got nothin' to do with you!
CARTER. I see you've learned a lot this weekend, Thelma.
THELMA. Listen here, you, do you think I'm stupid? I could teach you a thing or two! You Yanks comin' over here . . . clear away back to your own country and sort out your own cowboys and Indians! Aye, and what about all them war stories, all them poor soldiers you got killed in Vietnam for nothin'!
CARTER. Are you proud of yourself now?
THELMA. No, I am not indeed. I feel very angry!
COLLETTE. That about sums it up.
CARTER. Well then, if you won't leave, I must ask you to remain in this room. I don't want these feelings spreading to the other people in the house, people who have enjoyed this weekend and used it very positively.
VERONICA. Carter, why don't you just fuck off? We have nothing to say to you, all right? (CARTER *leaves.*)
MADELINE (*shouting after* CARTER). I got threw out once this weekend. I'm not gettin' threw out again! (*Pauses.*) There goes nine ninety nine down the drain.
VERONICA. Well, it could have been worse. You could have got one

of those ones with fur round the bottom to keep your neck warm!

JACKIE. How can yous make jokes after what we've said to each other?

VERONICA. What do you want us to do?

JACKIE. We could sit down and face up to it. . . . Why don't you tell us why we're 'Orange Bastards', why you hate us?

COLLETTE (*sarcastically*). Oh, you mean sit down and analyse it, see what we come up with?

JACKIE. Can't you see? Carter didn't want us to do that. He wanted us to pretend we don't have the feelings we do.

COLLETTE. You were the one sayin' we should all crawl back over the 'Border'.

VERONICA. That's right! Listen you, this whole island happens to belong to us . . . (*They begin to argue again.*)

THELMA. Ah for God's sake, don't be startin' all that oul' carry on again. (*Pauses.*) Wait til you see these postcards I got down in the village. (*Produces some postcards from her bag. Walking to each of the women.*) Look at it . . . just look . . . do you not see? . . . It really is beautiful . . . Jackie, look . . . (THELMA *gets no response from the others. She walks to the window and stares out at the view.*) It is the exact same view as we've got here . . . You know, we really are very lucky, we do have a beautiful country.

JACKIE. Are we going to go home and pretend that this didn't happen? If we don't want to talk about it, what was the point in coming?

MADELINE (*sighing*). Talk. . . .

COLLETTE. All right, I'll talk, I'll listen. Where do you want to start?

MADELINE. Aww, fuck this for a game of soldiers. I'm away. Is anybody comin'?

VERONICA. Aye. (*Gets up to leave.*)

COLLETTE. That's right, crawl out like bloody wee lambs!

MADELINE. What are you slabberin' about?

COLLETTE. If you walk out that door now we might as well apologise, say we're sorry, we've been naughty girls and crawl on home.

MADELINE. Apologise?

VERONICA. I'd look sick!

MADELINE. Crawl? Me, crawl?

COLLETTE. Do you not think that's what it looks like?

MADELINE. We'll soon see about that . . . (*Starts to sing.*) 'Craigavon sent the 'Specials' out to shoot the people down . . . ' (VERONICA *joins in.* COLLETTE, THELMA *and* JACKIE *stand in silence.* VERONICA *and* MADELINE *sing with their arms raised defiantly*).

NOW YOU'RE TALKIN' – ACT II SCENE I

> He thought the IRA was dead
> In dear old Belfast town.
> He got a rude awakening
> With his rifle and grenade
> When he met the first battalion
> Of the Belfast Brigade.
>
> Oh, glory, glory to oul' Ireland
> Glory, glory to oul' Ireland.
> Glory to the memory
> Of the men who fought and died
> No surrender is the war cry
> Of the Belfast Brigade.

MADELINE. Now shove that up your nose, Carter! (THELMA *marches defiantly to the doorway, bringing* JACKIE *with her.* JACKIE *and* THELMA *begin to sing.*)

> It was old but it was beautiful
> And its colours they were fine.
> It was worn in Derry, Aughrim,
> Enniskillen and the Boyne.
> (MADELINE, COLLETTE *and* VERONICA *join in.*)
> Sure, me father wore it in his youth
> In the bygone days of yore. . . .

CARTER (*entering – screams*). Shut your fucking mouths! (*The women stop singing.*) Thelma, I hope you're satisfied . . . Mrs Watson has shit herself! . . . Right you lot, get the fuck out of here and don't come back! (*There is silence. No-one moves.*) Do I have to repeat myself? I said, get out!

MADELINE (*starting to sing. Quietly*). We won't go, we won't go, we won't go . . . (*The other women join in, one by one.* CARTER *tries to shout over the singing but fails. He exits in a fury* VERONICA *and* JACKIE *are lit frontstage for dialogue. The others change to bedroom scene still singing.*)

VERONICA (*leaning as out of a window*). Hey Lily! Get you on the bus and go home and if you see any of the kids in the street bring them in and give them a bowl of cornflakes til I get back!

JACKIE. Sandra! Tell my Bobby to get the uniforms ready for the mornin' and makes sure they do their homeworks.

VERONICA. Yes Lily. Send us up a big bottle of vodka and a clatter of

fish suppers!
JACKIE. Make sure they're in bed by eight o'clock and they take their 'Halibut Orange' tablets!
VERONICA. No Lily, you can't stay . . . just get on the bloody bus, will ya?
JACKIE. Look, that must be wee Mrs Watson they're carryin' onto the bus!
VERONICA. Lily, give my regards to Ballymurphy!
JACKIE. And mine to Ballybeen! (*The others stop singing. Full LX on bedroom.*)
THELMA. Is that the buses away then?
MADELINE. Aye.
THELMA. Oh my God, how am I goin' to get home?
COLLETTE. Jesus, you wouldn't go when you were told.
MADELINE. Frig it, I'll thumb a lift. It wouldn't be the first time.
VERONICA. What the frig am I doin' here?
ISAAC (*entering*). Here, what's goin' on? Are yous havin' an occupation? Are yous occupyin' the place? You wanna see Carter . . . (*imitates* CARTER) . . . 'Jesus, this dreadful moral crisis! Shit! Maybe they'll turn violent?' He doesn't know whether to send for the cops. . . .
THELMA. Police!
ISAAC. . . . cause he doesn't know if yous are havin' an occupation. . . .
THELMA. Here, I'm not mixed up in this carry on, so I'm not. I'm going home!
COLLETTE. How?
THELMA. Aww my God! This is shockin' altogether, and me bad with my nerves!
ISAAC. C'mon. Look, what are yous doin'? Jesus! Typical bloody women, can't make up your minds! You're as bad as my Ma!
MADELINE. Here, less of that, you.
THELMA. Here, Isaac, here's my husband's wee business card. Phone him and tell him to come immediately.
ISAAC. What's it worth?
THELMA. Here's fifty p.
ISAAC. Phone calls, fifty p a touch! Any more phone calls?
THELMA. Run on, hurry up!
ISAAC (*exiting*). Up the revolution!
THELMA. I came here to be friendly with Catholics, to make my peace.
COLLETTE. Oh, that was big of you, Thelma!
THELMA. I don't want to get involved in any of this Protestant/

Catholic carry on.
COLLETTE. That's what this place is supposed to be about!
THELMA. It's not supposed to be about people hating each other! Carter's a very nice man . . . I don't want. . . .
VERONICA. He's a wally and you know it!
THELMA. I've disgraced my Church. Wee Mrs Watson'll never speak to me again. I'll never be allowed back!
JACKIE. I wouldn't recommend this place to the Devil himself!
THELMA. Yous started it. I was all right til I came here.
JACKIE. What did you think you were comin' for, just to have a nice time and sing hymns?
THELMA. Yes! And what's wrong with that, or are you a bloody heathen as well?
JACKIE. Oh aye, have a nice time, meet people of different religions, let's all be happy and friendly together!
THELMA. Yes! Yes!
JACKIE. And what happened? You turned into a foul mouthed bigot! And now you're sayin' it's all right, it doesn't matter? That's what all this bloody trouble is about!
THELMA. Leave me alone! I'm too old for all this carry on! Sittin' down and talkin'. Am I not entitled to a wee bit of peace at my time of life?
JACKIE. Well, go on then. Go on home to your wee cosy house. This flippin' country's goin' to get nowhere because of people like you.
THELMA. Yous are just jealous of me, jealous that my man's not on the dole, jealous of my bit of a coat, jealous that my man was smart enough to start a wee business!
MADELINE. You don't have to be smart to clean windies!
VERONICA. Just to remind you, Thelma, you may be all right, but there's people in this country who aren't!
THELMA. I don't live in West Belfast. I don't live in no troubled area. . . . Thank God I live among decent people! (*The other women all start shouting together.*) All right, I'm sorry. I'm sorry. Would yous leave me alone? Anyway, my Ronnie'll be here soon and that'll be the end of it.
VERONICA. Good! I'm glad you're goin' cause I couldn't stick another word that comes out of your mouth!
THELMA. You can't stick it because I don't want to be involved!
COLLETTE. Thelma, you are involved, whether you like it or not.
MADELINE. Here, Thelma, do you think your Ronnie would give us all a wee joy ride home?
VERONICA. Don't be daft, Madeline, it's Sunday. Won't he be out at

Church? Letting all the people see that he's a good, decent, upright, Protestant windie cleaner!

MADELINE. Aye, sure why don't we all sing a wee hymn, just to make her feel at home? (*Begins to sing.*) 'Abide with me. . . .'

JACKIE. You take nothin' serious.

MADELINE. I'm still here, amn't I? I didn't go.

ISAAC (*entering*). Thelma! Ronnie's not in!

VERONICA. Aww frig! Ronnie's not in. He gets her away for the weekend and he's out flyin' his kite!

ISAAC. But Carter has managed to get a minibus to take yous all home.

THELMA. Oh, tell Carter 'thanks very much'! Tell him we'll all be down . . . (*Falters.*) Tell him we'll all be down shortly.

ISAAC. Aww, yous aren't goin', are yous?

THELMA. Aye, in a few minutes. (ISAAC *exits.*)

MADELINE. Well, why don't you go on, Thelma?

THELMA. Well, I think we should all face the music together!

MADELINE. What bloody music? I'm not afraid of 'bucky beard'!

JACKIE. Go on, Thelma, go and lick Carter's arse! Tell him what a wonderful time you had meetin' Catholics! Tell him how wonderful reconciliation is!

THELMA. What are yous all doin'?

JACKIE. What's it to do with you? You're not involved, remember?

THELMA. I need to go to the toilet!

COLLETTE. Go on then. Nobody's stoppin' you!

THELMA. This isn't fair! You started all this and now I have to go out there on my own and Carter might be out there!

VERONICA. Stick your sixty pound coat over your head and he won't know who it is!

THELMA. I can't help it. It's my nerves makes me like this. I can't hold my water since I had my gall-bladder out! Oh, these oul' tablets the doctor has me on, I could be goin' every five minutes! Oh my God, this night! Oh, I'm goin' to wet myself! I haven't another clean pair of knickers either! Wait til you get to my age, you'll know all about it. . . .

MADELINE (*having lifted the wastepaper bin and brought it over to* THELMA). Here, stick your arse on that!

THELMA. Oh, I couldn't go fornenced yousins!

JACKIE (*grabbing a blanket off the bed*). Here, we'll cover you with a blanket! C'mon! (JACKIE *and* MADELINE *hold the blanket in front of* THELMA *while she squats. There is a pause.*)

THELMA. I can't go because yous are listenin'!

MADELINE and JACKIE (*sing*). 'Raindrops keep fallin' on my head....'
MADELINE. Are you finished?
THELMA. No. (MADELINE *and* JACKIE *continue to sing.*) It's all right, I'm done. (ISAAC *enters before* THELMA *has time to adjust her dress. He looks shocked.* THELMA *screams while* JACKIE *and* MADELINE *try to cover her with the blanket.* THELMA *is mortified and covers her face with her hands.*)
ISAAC. I think my cockatoo's wilted! Here, Thelma, do they not have toilets on the Beersbridge Road? Oh, that was a lovely wee G-string you had on there ! Let's have another look....All right now, important news! Pay attention! The minibus has arrived at the front door, right. Carter has made up a big pile of sandwiches for yous. Does he want yous outta here? He's even made the cleanin' ladies wait behind so they can get the place ready for the new lot comin' in the mornin'. He's goin' buck mad!
MADELINE. Is there a new lot comin' the morra? 'S'pose they're all from the Shankill and the Falls? Catholics on the right, Protestants on the left. Let's all hold hands and be sisters in love together. (MADELINE *and* ISAAC *dance together singing a parody of 'The Farmer and the Cow Man'*)
MADELINE & ISAAC. Oh, the Fenian and the Orangeman should be friends. Oh, the Fenian and the Orangeman should be friends.
JACKIE. There's no need to make a mockery out of it.
VERONICA. Don't tell me you take all that Maypole shite seriously?
COLLETTE. No. It didn't work, but he was tryin'. Thank God we're not depending on people like you to get people together!
VERONICA. Damn right you're not dependin' on me! I happen to live in the real friggin' world!
JACKIE. Well, now that we're here, why don't we try to find other ways? ... If we do it together....
VERONICA. Oh aye, we could all sit down and hold hands and sing and maybe the Brits'll disappear off the streets of Belfast!
JACKIE. Look, Veronica, we can't do anything overnight. We can't do anything at all until people sit down and talk. How can they talk if people like you won't listen?
VERONICA. Listen, love, people have been talkin' for seventeen years and got nowhere!
COLLETTE. You won't even listen to the Protestants. They might have a point of view too, y'know.
VERONICA. You know what you are, you're a bloody traitor!
COLLETTE. Here we go.

VERONICA. You're a Catholic. Your people are bein' mutilated on the street, and you want me to sit down and hold hands with Protestants! Go and talk to the British . . . government! That'd answer you better!

JACKIE. Look, we are prepared to meet you half way and you don't want to know. You're as bad as her over there (*indicating* THELMA). She doesn't want to know either. Oh, you know all right, you just don't want to listen.

COLLETTE. What the hell did you come here for in the first place?

VERONICA. To get away from my rotten, stinkin' life for two days. Is that a crime?

JACKIE. Are you sayin' you won't even listen to us?

VERONICA. All right! All right! Go ahead. Yous talk, I'll listen.

JACKIE. Veronica, I grew up in a Protestant housing estate. I still live there. We never had the opportunity to meet any Catholics. . . .

ISAAC. Hold it! I'll just go and tell Carter yous are stayin'. . . .

BLACKOUT. SFX *Kris Kristofferson 'Help Me Make it Through the Night'.*

ACT II – SCENE II

SCENE: *SFX Kris Kristofferson 'Help Me Make it Through the Night'. The women are seated in debate.*

JACKIE. You see, the problem is that all the community centres are in Catholic or Protestant areas. What we need is a central reconciliation centre so that both sides can use it.
VERONICA. Isaac, any chance of a drink around here?
MADELINE. Remember the last time, you knocked off a big bottle of vodka?
ISAAC. No, I don't.
COLLETTE. Look, are you listening, or what?
THELMA (*going towards her bed*). If you're goin' to turn this into a shebeen, I'm goin' to my bed.
MADELINE. For Chrissake, it's only a drink!
THELMA (*getting into her bed*). Isaac, would you please get me a glass of water til I take my Valium.
ISAAC. Aye, houl on a minute til I ate this here bap!
VERONICA. Go on, Jesus sake, we'll give you the money for it.
ISAAC. Bloody right, you'll give me the money for it! Here, don't be talkin' about your sex lives til I get back. Weemin always do that, I think it's weeker! (*Exits.*)
JACKIE. Look, are yous listenin'?
COLLETTE. Yes! I think it should be in the city centre too, that way there'd be no danger of anybody associatin' it with either side. . . .
JACKIE. It could be used for a whole lot of things. People could come and learn about history, our history. Y'see, when I was at school, I was taught about the Tudor Kings and Queens of England . . . I mean Irish history.
COLLETTE. Aye. You could have Irish dancing classes, poetry classes, play readings by the great Irish writers, people like Synge and O'Casey, and you could have Irish language classes, Irish music. . . .
JACKIE. That's stupid! The Irish language is dead. Nobody talks like that anymore. What's the point in wasting time with a lot of oul' fiddle music and a language that nobody understands?

COLLETTE. Jackie, that's our culture.
MADELINE. I love an oul' bit of Irish music. . . .
JACKIE. Not my culture.
VERONICA. You could have joy riding classes, an 'O' level in glue sniffin', wine makin' for alcoholics. . . .
COLLETTE. So what you're sayin' now is you're not Irish . . . ?
JACKIE. Are you listenin' to me? What I'm sayin' is that there are more important things that we should be concerned about, like ending discrimination, religious and sexual.
MADELINE. Yes, I'll go along with that, cause you see if I was a man I wouldn't be here, I would've threw him out instead! That's what we should have, self-defence classes for women!
COLLETTE. But we are the ones that's going to have to put the emphasis on culture cause we are the ones who are going to have to preserve it.
JACKIE. What culture? We haven't really got one . . . we build one . . . a new culture for everybody!
VERONICA. Lambeg Drum batin' classes . . . !
JACKIE. Look, Collette, I know that the majority of people won't accept that culture. You can't ram it down people's throats.
VERONICA. Jesus, you can just see it! Thousands of Orangemen, all tied to chairs, tin whistles stickin' outta their gobs. (*Imitating a male voice.*) 'You must enjoy Seamie O'Houlihan's Celidh Band!'
JACKIE. All right, we can't agree on that, we'll come back to it. Mixed schools . . . integrated education. . . .
MADELINE. Yes! I went to an all girls school, it wasn't fair! See when I left, I was really green. When it came to boys I hadn't a clue . . . mind you, I knew it wasn't for stirrin' your soup with. . . .
COLLETTE. Ach Madeline, be serious! Yes, integrated education, I think it's a priority.
JACKIE. And definitely not having religion as a subject.
COLLETTE. Oh no, Jackie, people's faith is very important to them, they need to know that their kids are bein' brought up with their own beliefs.
JACKIE. Sure their parents can do that in the house, I do it with my two.
COLLETTE. No. What you have is integrated education but you also have religious education classes where kids learn about each others' beliefs. That gets rid of fear and prejudice.
JACKIE. No, Collette, it's private, a personal thing.
COLLETTE. Okay, sure , we'll come back to it. . . .
VERONICA. Frig, this is great, yous haven't agreed on anything yet,

but that's about right for here.
ISAAC (*entering with a tray of drinks*). Cocktails, everybody!
JACKIE. I hope this isn't going to turn into a boozin' session.
ISAAC. I'm not going to get drunk . . . I'm going to roll a wee joint.
THELMA (*sitting up in bed*). Drugs! Ah, Holy God!
ISAAC. Sure aren't you on Valium?
COLLETTE. Please! Look, I just know that if everyone here was to listen and co-operate, we could come up with something really radical. . . .
VERONICA. Thank Christ for the drink, Isaac son, it's going to be a long, long night. . . .
ISAAC & MADELINE. Here! Here!

BLACKOUT. *SFX Kris Kristofferson 'Help Me Make it Through the Night'.*

ACT II – SCENE III

SCENE: *LX up. The women are drinking and singing.*

> Now the bomb and the bullet
> Has pierced every heart
> And a whole generation
> Has been torn apart.
>
> Robbed of the chances
> One had to be free
> Oh blitzed Belfast city
> God bless you today.

JACKIE. You know, given all the trouble, I really love Belfast. Me and my Bobby were thinkin' of goin' to Canada when things were getting too much here . . . but we just couldn't do it. I think it's because we're so consolidated with this place. . . .

ISAAC. I think I'll go to the Bahamas.

VERONICA. Aye, it's Belfast people, isn't it? Like, you see when they want to hate, they can really friggin' hate . . . but you see when they wanna love, they can do it in an even bigger way. Belfast . . . it's friggin' magic!

ISAAC. If I don't like the Bahamas, I might try the Seychelles.

COLLETTE. It's weird though, because it's like this city has decided – 'you are my people' – instead of the other way around. I've loads of aunts who live in England and they're always sayin' to me, 'why don't you just get away from it all and come over here', but I can't explain it . . . you see I belong here, you know what I mean?

ISAAC. I don't fancy the Middle East . . . cause you can't get Strongbow there.

MADELINE. You see we all belong here. . . . Now I remember when me and my Peter lived in Dublin for two years, Jesus, I hated it. Ach, I think I hated it cause I missed home. Ah, but I remember when I did come home . . . I went into a wee shop in Sandy Row . . . now, I know the wee girl was a Protestant. . . . I asked her for a wee bit of cheese . . . well, she cut me the best bit there was

NOW YOU'RE TALKIN' – ACT II SCENE III

and she rambled on about everything and anythin'. God, I could have cried! Like, I says to myself, 'these are my people', and yet, God I don't know. . . .

ISAAC. Yuck! Typical bloody weemin, always getting soppy! Hey, can I turn the radio on?

MADELINE. What do you want to turn that bloody thing on for?

ISAAC. Ah . . . there's a wee band I wanna hear.

MADELINE. There's only the oul' eleven o'clock news on.

ISAAC. If you don't let me put it on I'll start to sing.

VERONICA. Oh for Jesus' sake, stick it on! (*SFX radio*.)

THELMA (*awakens – startled by the music*). Where am I?

MADELINE. You're in the Vatican! (ISAAC *holds up the radio for all to hear. SFX radio news report*).

SFX. A group of women have barricaded themselves in a room in Atlantic View, a reconciliation centre in Portrock. The women were on a weekend organised by various community organisations and Church groups. The women, a mixed group of Catholic and Protestant housewives, have not yet stated their demands or, indeed, the reason for the protest. The women – Madeline Turk, Thelma Parker, Jackie Calderwood, Collette McGrath and Veronica Murray – are not believed to be armed. (*There are cries of alarm from some of the women.*)

THELMA. My name on the radio! This is terrible, awful!

MADELINE. It's brilliant! I've never heard my name on the radio before!

ISAAC. I wonder how they found out?

JACKIE. Don't you 'wonder how they found out' me! I bloody know how they found out! It was you! (*She thumps* ISAAC. VERONICA *and* MADELINE *hug* ISAAC, *both laughing.*)

THELMA. My husband's wee business! This is shockin', shockin'!

MADELINE (*sings*). 'I'm a star, superstar . . . !'

VERONICA (*sings*). 'I'm a rebel, super rebel . . . !'

JACKIE. I'm glad you can be so pleased about it! They mentioned arms!

MADELINE. Well, so we have, love . . . I've two, he's two, she's two . . . !

COLLETTE. Aye, you're a bloody comedian, Madeline.

VERONICA. The men round our way think they're the only ones to be rebels, that'll show them! (*Sings.*) 'I'm a rebel and I'll never be any good!' (*To* THELMA.) 'Oh with his shammy and his bucket, Oul' Ronnie says fuck it . . . !'

JACKIE. Isaac, you're an interferin' wee bugger!

NOW YOU'RE TALKIN' – ACT II SCENE III

ISAAC. That's the thanks I get for tryin' to make yous famous. (THELMA *slaps* ISAAC'S *face. There is silence.*)
THELMA. They mentioned Protestants and Catholics! My man's run is in East Belfast! The Protestants'll not want to know him.... You've ruined his wee business, you and your big mouth!
ISAAC (*placatingly*). Thelma....
THELMA. Don't you 'Thelma' me!
ISAAC (*leading* THELMA *gently to a chair, sitting her down and massaging her shoulders*). Sit down there, Thelma love, take the weight off your feet. . . . See, after that news today, the Protestants in East Belfast will be out throwin' dirt on their windies just to get your oul' fella t'clean them . . . !
THELMA. What are you talkin' about?
ISAAC. I'm tellin' ye. They'll be sayin', 'there's that windie cleaner, Thelma Parker's man'. They'll be queueing up to get refilling his buckets . . . !
THELMA. D'you think so?
ISAAC. He'll have that many windies t'clean. The hand'll be droppin' off him!
JACKIE. Isaac, these things they said weren't true. Why did you mention arms?
ISAAC. Sure I said yous didn't have any . . . yous are 'armless!
COLLETTE. Listen, wee lad, first thing you do tomorrow morning is to get back on to the phone to the radio station, tell them we don't have any arms, we're a group of women, we're here to talk about peace!
MADELINE For God's sake, this is Northern Ireland. If you didn't mention arms you wouldn't get a write up in the bloody Beano!
BLACKOUT.
ACT TWO
Scene Four
SCENE: *SFX early morning radio programme. LX up. The women are all asleep.*
ISAAC (*entering. Sees that everyone is asleep – shouts very loudly*). Cock a doodle doo!
MADELINE. Fuck off, Isaac.
ISAAC (*pulling* THELMA *out of bed*). Thelma, wake up, wake up, it's nine o'clock! . . . Thelma, Ronnie's just phoned . . . he says your phone hasn't stopped ringin' since ten to seven this morning for business. Everyone wants their windies cleaned by Thelma Parker's man! (*Addressing the others.*) See, I told you . . . (*Helps* THELMA *out of bed.*) He says, could you stick it out til tonight at

least . . . he's much too busy to come and collect you, anyway.

THELMA. Oh this is marvellous, marvellous! . . . I wonder is it in the Newsletter? . . . I wish we could get a copy. . . . See news like this, Madeline, it's better than puttin' an ad in the East Belfast Post!

MADELINE. That's great, Thelma. I knew some good would come out of all this . . . (ISAAC *coughs. He is looking for payment.*)

THELMA. Thanks very much, Isaac son. I'll see you right later on . . . (ISAAC *exits.*)

COLLETTE. Am I hearing right? Are you actually going to stay in here just for the sake of money? Have you no principles whatsoever?

MADELINE. People has a right to make a livin' any way they can!

COLLETTE. Oh aye. I suppose you're hopin' for a rake off from Mr Windowlene if it all works out okay? Have you no principles either?

MADELINE. Principles don't pay the rent, love.

THELMA. What are you gettin' on your high horse for. We're not doin' you no harm?

COLLETTE. I don't want my name associated with a money grabbing toad!

JACKIE (*to* THELMA *and* MADELINE). What about all those things we talked about last night? Can we not just carry on? Veronica, you didn't say much . . . shouldn't we all talk more? I mean, we're all here now, so why not? We have to carry on . . . we only got half way. . . .

CARTER (*entering*). Good morning, ladies. I hope you all had a good night's sleep?

THELMA. Yes, thank you.

CARTER. I'm sure you are all aware of the news report on the radio – the report which has been broadcast all night long. You are famous now!

MADELINE. What?

CARTER. Journalists and photographers have been ringing constantly since midnight wanting a statement. In fact, you will find that several of them have already arrived from Belfast.

MADELINE (*looking as if out of a window*). Jesus! You wanna see the big, flashy car out there!

COLLETTE. A statement? They want a statement from us?

JACKIE. What about?

CARTER. Exactly.

THELMA. Madeline, would you put a wee bit of make-up on me? I've never had my photograph in the paper.

CARTER. Ladies, it is not a good idea to keep the gentlemen of the

Press waiting, otherwise you'll find they'll write their own statement.

MADELINE. Here, you tell them to houl on til I do my face and I'll go!

COLLETTE. No . . . Carter, would you mind leaving us and we will discuss this between ourselves? (CARTER *exits*.)

JACKIE. Why don't we just tell them to go because we're not ready to talk?

THELMA. No. Did y'not hear Carter? We'd better talk or they might write a pack of lies.

COLLETTE. We must make a statement, otherwise they'll think we're just a pack of stupid women who can't speak for themselves.

JACKIE. I suppose we could say we're stayin' in here to talk about . . . well, just to talk.

THELMA. For God's sake hurry up, or they'll be away!

COLLETTE. We can't just say we're here just to talk . . . I mean . . . big deal . . . so what? The papers will want something positive. What about all those things we discussed last night . . . our ideas for reconciliation and compromise?

JACKIE. But we didn't all agree, really, on any of them. There was always something. . . .

COLLETTE. That's what compromise is about, love. If we just tell them there are points we are trying to come to an agreement on. . . .

MADELINE. Aye, that's the truth, like . . . come on, are we ready?

JACKIE. Veronica, what about you?

COLLETTE. I think Jackie and I should talk to the Press.

THELMA. What's wrong w'us?

JACKIE. No, Collette, I wouldn't know what to say.

COLLETTE. Look, Jackie, this may be the only chance we will ever have in our lives to tell them what we feel as women, not just women, but a mixed group of Protestant and Catholic women.

JACKIE. But we never really. . . .

THELMA. Hurry up, for God's sake!

MADELINE. Jesus sake, I could've been out there and back again!

JACKIE. I just think we need to discuss more. . . .

COLLETTE. No, we have to let the public know that we have found talking points. . . .

VERONICA. I have never heard such a load of crap! Are yous all blind, is that possible, or are yous all just thick? People starved themselves to death for this country, people are bein' murdered for this country, and you have the gall to walk out there and say the answer to it all is a fuckin' community centre in the middle of

High Street! Jesus Christ! Do you not see that nearly every atrocity that's happened in the last seventeen years, the last bloody seven hundred years, could have been avoided? People aren't sectarian for the good of their health. They're sectarian because of the situation they're in!

COLLETTE. Well, the killing's gettin' us nowhere.

VERONICA. Do you think I want to see anybody die? I feel for every Peeler, every British soldier that's shot. But yous answer me this question. Who is responsible?

THELMA. Aye, well, you would know all about that, wouldn't ye?

COLLETTE If you and your sort would just listen for a change instead of cryin' on about freedom for Ireland and martyrs! Listen, love . . . (VERONICA *turns away*.) . . . whether you like it or not, the Protestants do have rights too. It is their country too. And until you realise that, not you, or your children, or your children's children will ever have peace in this country.

VERONICA. Don't give me any of your oul' 'I'm goin' to be a reasonable Catholic' shite! Where are you goin' to find all the reasonable Protestants to talk to? (*Looks pointedly at* THELMA.) Maybe you should try George Seawright and his lot. He stands up and says that all Catholics should be put in an incinerator and burnt and he gets the biggest vote of any councillor in our elections! You should go and talk to him and see how reasonable he is. But make sure you've got your flame proof knickers on!

JACKIE. Veronica, we have got to find a peaceful solution. I'm prepared to talk and there's other people like me. There must be. . . .

THELMA. How the hell can you talk about peace with the like of her (*indicates* VERONICA) around?

COLLETTE. Shut up!

MADELINE. You keep out of it. Otherwise we'll never get this friggin' statement together!

THELMA. I don't have to listen to that oul' carry on!

JACKIE. You may just content yourself, Thelma. You're not gettin' your photo took the day. We're not going anywhere til we get all this sorted out.

COLLETTE. Jesus, that's why we never get anywhere in this country, because there are just not enough people who have the guts to say, 'yes, I'm prepared to start somewhere', cause they just don't bloody care!

JACKIE. I do bloody care!

MADELINE (*to* COLLETTE). Just cause you've a wee bit of education doesn't mean you're the only one can speak for us!

THELMA. I'm away to talk to them.
JACKIE (*pulling* THELMA *back*). You are not going anywhere.
MADELINE. Jesus, it's only the Belfast Telegraph! You'd think to hear yous it was News at Ten!
COLLETTE. Aye, you go and talk to them, Thelma. But don't forget to tell them how many windows your man has cleaned today!
THELMA. It would answer you better if you went on a diet, you big lump, ye!
JACKIE. Oh for cryin' out loud, can we not just stay here and ignore the papers?
MADELINE. Why should we? Just because you don't know what to be at. You're like a child's bloody arse!
VERONICA. Now, now, sisters. What about the 'peaceful solution' we were goin' to find? Peace, sisters! No quarrelling. now! (*Sings.*) 'We shall overcome, we shall overcome . . . '
MADELINE. Aww, shut up!
COLLETTE. Shut up, Veronica! If you'd just listen you'd know we've nothin' to do with the 'peace people'! What is it . . . are you incapable of shutting your mouth . . . ? (COLLETTE *tries to put her hand over* VERONICA'S *mouth.* VERONICA *keeps on singing.*)
VERONICA. Now, now, no violence, please . . . (*Sings.*) 'We shall overcome some day . . . '
COLLETTE. Don't push me . . . I'm warning you. . . .
JACKIE. Calm down, Collette. . . .
COLLETTE. Don't you tell me to calm down. Just keep out of this . . . !
JACKIE. You're just provoking her.
COLLETTE. Me provoking her!
MADELINE. Would yous all, shut up!
COLLETTE. You should be supporting me! (*To* VERONICA.) Will you shut up! Right, I'll shut you up! (COLLETTE *grabs a scarf from one of the beds and gags* VERONICA *with it.* MADELINE *and* JACKIE *tie* VERONICA'S *hands and legs to the end of one of the bunks.*)
COLLETTE (*while* VERONICA *is being tied up*). How's that feel, Veronica? Somebody's silenced you for a change! God, I'm so sick of people like you! You see, I know you and your sort, you don't fool me. I'm tired of the violence, of what people like you have done to this country. Why can't you leave well alone? Green, bloody fascists! You don't liberate anybody. You oppress them! (*Stops – shocked at what they have done.*) I'm sorry, it simply had to be done for your own good. . . .

BLACKOUT.

ACT II – SCENE V

SCENE: *SFX glass breaking. The women scream. LX up* – JACKIE *is onstage, holding a brick with a note attached to it*

JACKIE. Bastards!
THELMA. What is it, Jackie?
JACKIE (*reading the note aloud*). 'Dear Sisters, Don't be mislead by Catholic scum. This association with Jezebels will only lead to another feather in the cap of the IRA. Furthermore, remember your duties, remember your womanhood. What is to become of the future sons of Ulster when their mothers behave in such an unnatural, unwomanly way? Protestant women are not rebels. Continue this pact with evil and you will suffer the just desserts of a traitor! Signed: C.T.L.D. – Carson's True and Loyal Defenders – for the Protestant people of Ulster.' Bastards! Evil bastards! They don't speak for me! The Protestant people of Ulster, who is the Protestant people of Ulster? These maniacs are, until the ordinary Protestant has the courage to stand up and say, 'No! No! No! They are not going to speak poison in my name!' I wanted the Assembly! I wanted to have power sharing. But nobody's gonna listen to me! Look, I don't care if they fly the tricolour up in West Belfast. I don't care if they call Londonderry 'Derry'. If only they would be reasonable and listen to the Catholics! It's their country too! We can live together . . . we can . . . we can! This country would have been all right if they'd listened to the Catholics in 'sixty-nine. I agreed! I agreed with Civil Rights and nobody wanted to know!
THELMA (*who has being reading a magazine while all this has been going on*). Keep your voice down! The papers might be out there.
JACKIE. Thelma, those people don't speak for you, do they?
THELMA. They're a crowd of headbangers! I pay no heed.
JACKIE. You pay no heed to them, but other people do! This is our chance to speak out!
THELMA. Well, I'm not too happy about that tricolour flying over Whiterock Leisure Centre. . . .
JACKIE. All right, we'll scrap that one, but those people are speakin'

in our name. Can't you see that they are the reason we can't solve this problem in Northern Ireland?

THELMA. Ach, away on, Jackie, and give my head peace.

JACKIE (*looking around for support. Moves towards* MADELINE). Madeline, love?

MADELINE. Look, love, why don't you take a couple of Thelma's wee tablets and you won't give a shit what anybody says?

JACKIE. I know you're a Catholic, don't think I don't see your point of view. It's not me that won't listen.

THELMA (*trying to change the subject*). Madeline, did your Peter take the shingles all over him or did they just come out on his neck?

MADELINE. God, it was shockin', he was like a leper!

THELMA. Although I'm very bad with my nerves, thank God I never took the shingles.

MADELINE. Aye, but those Valium is like 'Smarties' to me. I like the Mogadon, myself. Two Mogadon and I sail about like the QE2.

JACKIE (*in despair now*). Is nobody listenin' to me?

COLLETTE. Just calm down, Jackie. We can't afford to lose our cool. That's exactly what the papers expect because we're women and I'm not going to let it happen.

JACKIE. Collette, I want to make a statement. I want to let everyone know what I think of those people!

COLLETTE. Now, Jackie, we mustn't be too extreme, otherwise we're just as bad as they are. It's up to us to show we're above that kind of ignorance.

JACKIE. But they have threatened us!

COLLETTE. I know! Just ignore them, they've nothing better to do. Okay? (JACKIE *looks around. Everyone is ignoring her. She catches* VERONICA'S *eye.*)

JACKIE. Veronica?

THELMA. Ah, for god's sake, don't be startin' that fishwife up again. I never met the like o' you, wee girl, for interferin'!

JACKIE. Look, I know that you find it hard to believe me after all you've been through in your life, but, Jesus, I didn't know that years ago they were throwin' Catholics into the water in the shipyard and stonin' them! It's not my fault! But, as a Protestant, I feel it is my responsibility to do something about it! I don't want you to feel alienated, I want you to feel part of this country. (*Pulls down the gag from* VERONICA'S *mouth.*)

VERONICA (*singing loudly*). 'God save Ireland cried the Provos . . . !'

JACKIE (*placing the gag back over* VERONICA'S *mouth*). Veronica, please! Is nobody listenin' to me?

NOW YOU'RE TALKIN' – ACT II SCENE V

MADELINE. There's wiser locked up, wee girl.
THELMA. I say, Madeline, you've got a lovely figure for your age.
MADELINE. Aye, sure you have to when you're still on the market, y'know what I mean?
THELMA. S'pose I've no need, for I've got a man.
ISAAC (*entering, carrying a big basket of flowers*). Special delivery for Mrs Thelma Parker! (*Spotting* VERONICA *tied up and gagged.*) Here, yous might have told me last night you were into bondage! Thelma, Ronnie says business is boomin' . . . he says can you stick it out til tonight at least and he'll send you up anything you require. Oh, by the way, there's a man here from 'The Guardian' and he wants to talk to (*looks at* VERONICA) all five of you.
THELMA. Ah, Holy Jesus!
JACKIE. You'd better tell him to houl on a minute.
ISAAC. Yous had better hurry up! (*Exits.*)
JACKIE. We'd better untie her.
COLLETTE. Bloody sure, we don't! She'll just rant and rave and we won't get a word in edgeways.
THELMA. Could we not stick her out on the fire escape til the papers go away?
JACKIE. Don't be daft, he's expectin' to speak to five of us!
MADELINE (*to* THELMA). How many Valium have you?
THELMA. Only a couple left.
MADELINE. Sure between us we've enough to dope her!
COLLETTE. It's 'The Guardian', a quality newspaper, they'd suspect.
ISAAC (*entering*). What's happening? Your man is expectin' five of yous. (COLLETTE *and* MADELINE *eye* ISAAC *up and down, obviously with the same thought.*)
MADELINE. Who got us into this bloody mess in the first place?
COLLETTE. C'mon, Isaac, you owe us a favour. . . .
ISAAC. What are you talkin' about?
MADELINE. Two ticks and I'll have his face made up!
ISAAC. On yer bike! I'll go and tell him yous don't want to talk. . . .
COLLETTE. We'll give you a fiver for it?
ISAAC (*flinging off his jacket*). Right! Where's the make-up?
JACKIE. He's not going to believe this....(MADELINE *starts to put lipstick on* ISAAC.)
COLLETTE. Shut up and help me put her (*indicating* VERONICA) on top of the bunk! (*The women cover* VERONICA *with a blanket, tie her up, and hoist her onto the top bunk.*)
THELMA (*looking at* ISAAC'S *hair*). And how the hell are we going to cover that monstrosity?

45

COLLETTE. Stick a towel over it and pretend he's....she's washed her hair!
ISAAC. Hope I don't look like a cissy....(*Puts on* THELMA'S *dressing gown.*)
MADELINE. Shut up, you're getting a fiver for it!
JACKIE. He's never going to believe it. We'll never get away with it . . . !
MADELINE. What about her (*indicates* VERONICA.) . . . ?
COLLETTE. Just say three Hail Mary's. . . . Right, go and tell him we're ready . . . (JACKIE *exits.*)

ACT II – SCENE VI

SCENE: *As at end of previous scene.* JACKIE *returns with the* REPORTER *from 'The Guardian'*

JACKIE. Oh, by the way, I'm Jackie Calderwood.
REPORTER. Hello, I'm Greg Havistock of 'The Guardian', pleased to meet you. (*Shakes* JACKIE'S *hand.*)
COLLETTE. Hello, my name's Collette McGrath, I'm very pleased to meet you.
REPORTER. Hello. (*Shakes* COLLETTE'S *hand.*)
MADELINE. I'm Madeline Turk.
REPORTER. Hello, Madeline. (*Shakes* MADELINE *by the hand.*)
THELMA. Thelma Parker. (*Shakes the* REPORTER'S *hand.*)
REPORTER (*checking his notebook*). And you must be Veronica? (ISAAC *giggles coyly.*)
COLLETTE. Ah, yes, take a seat.
REPORTER. I must say how grateful I am that you have allowed me to come here today and talk to you about your occupation.
JACKIE. And we're delighted, particularly at this time, to have an opportunity to speak to someone from 'The Guardian'.
REPORTER. Thank you very much. . . . Now, firstly, how did your occupation begin?
JACKIE. Well, we became very disillusioned with the methods of reconciliation here. . . .
MADELINE (*interrupting* JACKIE). Have you got your wee camera with you?
REPORTER. No – we don't intend to publish a picture with this story. . . .
MADELINE. Ach, that's a pity, isn't it?
REPORTER. As you were saying. . . .
COLLETTE. Well, we decided it just wasn't working. You see what we think is that, no matter how well intentioned outsiders may be, reconciliation is really going to have to begin with the people of Northern Ireland themselves.
JACKIE. But that isn't what we wanted to start the statement with. . . .
COLLETTE. Actually Jackie, it was.

JACKIE. Excuse me Collette, but. . . .
REPORTER (*interrupting* JACKIE). Sorry! If you could just bear with me a minute . . . (*Writes in his notebook.*) Have you discussed the possibility of an Anglo/Irish Agreement in the Autumn?
COLLETTE & JACKIE. No, we haven't, not really.
MADELINE. Yes we have! Now, we've been talking, Catholics and Protestants together, and we've decided, for the sake of our country, to stick it out . . . (*Everyone looks puzzled.*)
REPORTER. Yes . . . very good. . . . What do you think the possible reaction of the Protestant extremists will be . . . ?
JACKIE (*jumping up and grabbing the brick that had previously been thrown through the window*). I'll tell you about extremists! This brick was thrown through the window. We have been. . . .
COLLETTE (*interrupting* JACKIE). Jackie, Jackie, you know what'll happen. You'll get over excited and the gentleman won't understand. We feel, actually, that it is high time that the people of this Province stood up to the men of violence, on both sides. You see, we think that women can actually be a lot more tolerant than men. So, it's only logical that a feminine influence has got to be brought to bear upon the troubled waters of Belfast at this point in time.
REPORTER. I couldn't agree more. And I'm sure most of our readers would also agree that the feminine principle is more tolerant, less inclined to violence, more generous . . . (ISAAC *makes a retching sound. Everyone looks at him in horror.*)
ISAAC. Pre-mensual tension!
REPORTER (*placing his hand on* ISAAC'S *knee*). Please don't feel embarrassed. I thoroughly sympathise. . . . So, would you say in fact, you were a revival of the Peace People?
COLLETTE. No! Definitely not. We know we've got a much more realistic thrust than they ever had. We know nothing just happens overnight when you just say, 'let's have peace, please' . . . What we are prepared to do is to work very hard within the present constitutional framework . . . (VERONICA *has freed herself by this stage and now leaps from the bed onto the reporter.*)
VERONICA. What the fuck do you know about it? Bloody Brit . . . ! (*The women scream.* VERONICA *is pulled off the* REPORTER *and shoved onto a chair.* MADELINE *sits on her –* JACKIE *holds her arms.*)
REPORTER (*being helped to his feet*). Who is this woman?
THELMA. She's a bloody IRA woman! Now, Mr Reporter, there's the real truth! We have tried to talk, we have sacrificed our home

comforts, we have put our families at risk. We have tried everything with that woman. We are worn done tryin' to compromise, but there was no use talkin' to her. How are we going to get anywhere with stubborness like that going on? That's why we'll never get nowhere. Here you've got two decent Protestants and two Catholics tryin' their best to find an answer, but how can we with that bloody murderess about? Look out there. (*Leads the* REPORTER *as if to a window.*) What do you see? It's beautiful, isn't it? And it's ours....but with the likes of her tryin' to blow it to Kingdom Come. . . . Are you surprised we had to shut her up? Tell your paper that good, decent people are tryin' their best and ones like her is doin' their best to wreck it!

REPORTER. And you are Mrs Thelma . . . ?

THELMA. Mrs Ronnie Parker of the Beersbridge Road, East Belfast.

REPORTER. And can I quote you as the spokesperson of the group?

THELMA. You certainly can!

BLACKOUT.

ACT II – SCENE VII

SCENE: SFX *radio announcement.*

SFX. The occupation of Portrock Reconciliation Centre continues. In an exclusive interview with 'The Guardian' newspaper yesterday, the five women involved spoke about their search to find a lasting solution to the Northern Ireland problem. Their spokesperson, Mrs Ronnie Parker, stressed their determination to carry on until a solution had been found. There are four women involved in the talks, two Catholics and two Protestants. The fifth woman, the most extreme and intransigent of the Catholics has, sadly, refused to co-operate. (*LX up.* THELMA *is seated in the centre of the room, writing.* MADELINE *is styling* THELMA'S *hair while taking an occasional swig from a bottle of champagne.* JACKIE *is on the top bunk, depressed.* ISAAC *is seated at* THELMA'S *feet, reading holiday brochures.*)

MADELINE. Your Ronnie's a quare good man, all the same, Thelma. There's no oul' dirt about that champagne.

THELMA. Don't trouble yourself with trivialities, Madeline, there's plenty more where that came from.

MADELINE (*putting some lipstick on* THELMA). Now, you have to look your best today, Thelma, it is English TV.

THELMA. Yes, and no running in front of the cameras, Isaac. We are supposed to be promoting peace and understanding.

ISAAC. Sure I want to promote peace and understandin'. . . .

THELMA. Not with that hair-do, you'll not!

MADELINE. That's you, Thelma . . . God, you'll win the hearts of millions . . . !

THELMA (*reading her speech*). 'I appeal to all responsible and peace loving people of Northern Ireland to join me and my Roman Catholic colleagues – Madeline Turk and Collette McGrath – in our struggle' . . . oh, no – that's the word the IRA uses . . . 'in our fight' . . . no . . . (*thinks*) . . . I know, 'as we diligently strive for peace'.

MADELINE. God, Thelma, that's marvellous . . . ! It's a wonder nobody thought of that before. . . .

ISAAC. I want to talk about the unemployed youth when you've finished talkin' about peace.

THELMA. Isaac, if you want anybody to listen to you, you'll have to start lookin' like a human bein'.

COLLETTE (*running in – breathless*). Right, everybody, they are ready! They have cameras set up and everything and they want to speak to you, Thelma!

THELMA. Have you got your wee speech ready, Collette?

COLLETTE. What?

THELMA. Your speech?

COLLETTE. Oh yes . . . 'As a Catholic I fully endorse the statement just made by my Protestant Colleague, Mrs Thelma Parker. We are fully aware that here in Portrock we carry the hopes of every reasonable person, both Protestant and Catholic, in Northern Ireland today'.

THELMA. Good. Now, are we ready?

COLLETTE. Thelma, don't forget to mention what we were talking about last night about non-sectarian sports stadiums. Don't forget to tell them that we are the ones setting up negotiations between the Northern Ireland Football Association and the GAA.

THELMA. Oh yes, they will really love that!

ISAAC. What's the good in that? Sure the football in England isn't about Prods and Taigs and they still beat hell out of each other!

COLLETTE. Come on! Hurry up! And don't forget, Thelma, speak up nice and clear. It is English TV!

THELMA. Don't you worry yourself, I know my P's and Q's.

MADELINE. Hey, Thelma, the Queen might be watchin' this!

ISAAC. Her Majesty! I'm goin' to give the fingers in the background. (*They exit.*)

MADELINE (*looking back at* JACKIE). Ah well, you have to start somewhere, love, you know what I mean?

JACKIE (*to herself*). It's just a waste of time talking about peace. It's not going to happen, not that way. There is never going to be any compromise . . . not now . . . not ever. But what else is there? (*Looks at* VERONICA *who is still tied up.*) God, no, that can't be the only way! . . . (*Jumps off the bed and approaches* VERONICA.) Veronica, will you please listen to me . . . (*Kneels beside* VERONICA.) Do you realise how frightened I am of you? You see, when I was growing up, I was told about the evil of Popery and how awful things were going to happen to me if there was ever a United Ireland . . . there would be poverty, we'd all be ruled by priests and nuns, all the Catholics would gang up on us and treat us like dirt, they'd take away our religion....Just try not to blame me for everything that has happened to you. I don't run the country. The leaders, the politicians, blame them for feeding us hatred and lies about you....and now those same politicians are telling us we have to fight to the death to protect Ulster.... There are even ones who say they are going to burn it off the face of the earth rather than give it up. Is it any wonder, hearing that, that we've all been driven mad? I don't know where to turn . . . I'm scared . . . bloody scared! . . . And I hate having this fear of ordinary people like you, but what else do you expect? Fuck bein' proud to be British! . . . It's not pride, it's fear! . . . I'm Irish too, but I'm not allowed to be! (JACKIE *unties* VERONICA. VERONICA *takes her bag and coat and makes to leave. She stops when she gets to the doorway and returns.*)

ALTERNATIVE ENDING WOULD BEGIN HERE

VERONICA. Jackie, do you think the Brits want Northern Ireland? Do you not think they'd love to get rid of it and dump the whole bloody lot of us into the Irish Sea, if they could?

JACKIE. But, they can't do that. . . .

VERONICA. They can do whatever they want.

JACKIE. But, they can't . . . it's written into the constit. . . .

VERONICA. What law did they need when they came in and took it in the first place? Do you think they're gonna give a damn about any law if they want to get out? You see, Jackie, that is the real fear . . . if the Protestant people would only face up to it.

JACKIE. And what would happen would be too terrifying to think about . . . there would be a blood bath . . . we're, I mean, the Protestants are already behavin' like cornered rats.

VERONICA. Why does there have to be a blood bath? The Unions in the shipyards, the factories, the mills in Belfast. It was ordinary, workin' class Protestant people that created those Unions. Why can't they use those same guts and determination to help us all build a new Ireland? What the hell use will they be screamin' like frightened rats? Jesus, don't think that I want to live in a priest-ridden Ireland, but I can't live in one run by Union Jack waving bigots!

JACKIE. It is too late now . . . (*Grabs* VERONICA'S *hand and leads her to the window.*) Look out there . . . tell me what you

NOW YOU'RE TALKIN' – ACT II SCENE VII

see? . . . Thelma, the true-blue Brit . . . and there are thousands like her who are not going to give an inch . . . how are you going to make people like that come with you into a new Ireland?

VERONICA (*quietly*). What other choice do you have . . . 'Fight to defend Ulster until every drop of Protestant blood has been spilt'? (*SFX 'The Farmer and the Cow Man Should be Friends'.*)

LX FADE. END OF THE SECOND ACT.

ALTERNATIVE ENDING TO
NOW YOU'RE TALKIN'

(VERONICA *takes her bag and coat and makes to leave*).

JACKIE. Were you listening to me? Veronica.

VERONICA (*stops*). Jackie, I have nothing against you, nothing against the ordinary Protestant, but this is a war and we have every right to fight as an army.

JACKIE. Do you know who the people are you have to convince . . . that you want a New Ireland for? The majority of people in this country whose ancestors fought and died in the Somme for it. The people whose husbands are being shot dead because they join a police force to protect it. They believe in their cause just as much as you believe in yours cos there's 20 ordinary Joes standin' by to take the place of every policeman that's killed.

VERONICA. The cause that they are fighting for is the oppression of the Minority in their own country That's *their* cause.

JACKIE. But all they see are bombers and killers, not a poor minority!

VERONICA. What other choice was there, what happens to peaceful protest, eh? Nobody listens, nobody wants to know, cos we are nothing only dirty Catholic scum that your leaders don't give a damn about . . . I'm sorry, Jackie, but as I said it is a war . . . and people suffer. (VERONICA *about to leave again.*)

JACKIE (*takes* VERONICA *to window*). There's the people you have to convince! Thelma, the true blue Brit! How are you going to convince her and thousands like her that the IRA have the way forward for the New Ireland! Do you think they are going to believe that murderers, cos that's how they see you, want what's best for them?

VERONICA. How could we convince them before when nobody wanted to know?

JACKIE. What choice do you have now, bomb them into United Ireland?
VERONICA. What choice do you have. . . . 'Fight to defend Ulster until every drop of Protestant blood has been spilt'? (*They look at each other with nothing left to say.* VERONICA *leaves.* JACKIE *leaves slowly. SFX music 'The Farmer and the Cowman should be friends'.*)

END OF SECOND ACT.

GOLD IN THE STREETS

PLAY IN THREE ACTS

BY
MARIE JONES

DEVISED BY
CHARABANC THEATRE COMPANY

CHARACTERS

ACT I: Belfast, 1912

AGNES MULLAN – Catholic in her early thirties
MOLLY BAXTER – Agnes' sister
JOHN JOE MULLAN – Agnes' husband
GEORGE BAXTER – Molly's husband, Protestant
LADY GLENDINNING – Agnes' employer
COOK – employee of Lady Glendinning
SANDY – Protestant neighbor of the Baxters
JAMESIE – Protestant neighbor of the Baxters

ACT II: Belfast, 1950

MARY CONNOR – Irish widow of an English soldier, early thirties
TAXI DRIVER
MARY'S MOTHER
MRS. MURPHY – neighbor
KATHLEEN – an old friend of Mary's
TERESA – an old friend of Mary's
JOAN WILCOX – Mary's daughter
MRS. MOLLOY – neighbor
ELISH – neighbor child
BERNIE – neighbor child
SHOPKEEPER
MRS. O'LEARY – neighbor

ACT III: Belfast, 1985

SHARON MCALLISTER – married woman in her early thirties
MAUREEN – Sharon's coworker
DAVY MCALLISTER – Sharon's husband
BEEZER – friend of Davy's
SMICKER – friend of Davy's
JESSIE – woman friend of Davy's
SAM – Sharon's father-in-law
MAISIE – Sharon's mother-in-law
JOHN – Maureen's husband

GOLD IN THE STREETS
PRODUCTION NOTE
All characters are played by four actors who sit visible on chairs placed just outside main acting space. Actors, when it is specified, enter lighted centre stage to speak. At other times, lines are spoken from chairs. Scenes flow fluidly one into another. Actors arrange table and chairs on the stage when needed. Additional small roles are not specified in the play character lists.

GOLD IN THE STREETS

ACT I

The Agnes Mullan Story

SCENE: *SFX music 'Mountains of Mourne' and voice over Music hall – general.*

M.C. (*entering*). My Lords, Ladies, and Gentlemen . . . the Management of this beneficent . . . emporium of Thespianism . . . (*cheers*) . . . wish to present for your delectation . . . (*cheers*) . . . Belfast! (*Raucous cheers from crowd.*) Between 1901 and 1911, the population rose by 37,767 – fairly steady – equivalent numbers to Dublin . . . (*Booing*). . . . But Belfast . . . the heart of the Empire, had the industrial clout. Roll up, roll up and see the greatest Shipyard, Ropeworks, Tobacco Factory, Linen Spinning Mills, Dry Docks, and Tea Machinery Works in the world . . . and we built the biggest ship . . . the Titanic, launched May 31st, 1911.
MAN ONE. Sank, April 15th, 1912.
MAN TWO. Tragedy!
WOMAN ONE. Stunned the town!
WOMAN TWO. Cast a shadow on our reputation as shipbuilders. (*SFX music hall.*)
M.C. 1912! The Empire. The Grand Opera House. The Hippodrome. . . . The Alexandra Theatre.
ALL (*shouting*). What about work?
M.C. Work? Well, of course there was work . . . in the Heart of the Empire. (*SFX chanting first verse of 'Mountains of Mourne' up to 'gold in the streets'. In a round.*) Thousands migrated from rural Ulster . . . thousands of the poor and unskilled from West of the Bann . . . like Agnes there.
AGNES (*entering*). Poor and unskilled, that's what they call us. Well, let me tell you, for the past lot of months, I've been feeding myself, my John Joe and the two wains, payin' my rent, feeding three hens that will not lay an egg between them, layin' out tuppence a week for his baccy and porter *all* on six shillings a

week. And if that's not a skill, I don't know what is. Let me tell you, I'm at my wits end. Now I never was one for the moaning. I was always the girl for takin' things in my stride, but times is hard in Ballymacartin. My John Joe was the best handloom weaver this side of the Bann. In fact, he still is. Trouble is, nobody wants him now, what with all this fancy new machinery they have now for weaving. The poor creature is gettin' less and less work from the local mill, and there's less and less for me to put on the plates of a night. . . . See that. (*Holds up letter.*) That is a letter from my sister Molly in Belfast. She's doing very well for herself as well. She's away nine years now. Married a Belfast man.

MOLLY (*at table writing letter*). So, my George just says to me, 'Molly, my wee jube-jube, if you want that hat in Cleaver's window, you just get it'. Oh, our Agnes, it's great up here in the big city. My George is gettin' on the best at the docks. He's always gettin' picked cos he's a quare hard worker and he says soon there will be a day when his jube-jube will want for nothin'. Might even come the time when I can leave the spinning and stay in my own wee house with my very own gas light. Mammy would take a buckle if she saw it, it's the quare oul' mark. Sure maybe you and your John Joe should think of movin' up here some time. My Edwina has two teeth and young George got full marks for sums. My George says he's goin' to be the smart one. Love, from your sister Molly. P.S. Sorry to hear you're having troubles.

AGNES. There's one thing you can say about John Joe. You'll always know where to find him – in McGlene's pub, with a face like a Lurgan spade, starin' into a jug of porter. (*To* JOHN JOE.) If you don't buck up, boyo, and get the hell out of this place you may go and graze with Barney McCartin's cows from now on, for I've tholed enough of this. I'm packin' our wee bits and pieces and goin'. With you or without you. (*To audience.*) And I left him gawkin' at McGlene's door what I slammed in his face as I left. (AGNES *blows out candle and exits.*)

JOHN JOE. Belfast? Is she off her head? . . . She'll go with me or without me? . . . She wouldn't . . . she couldn't . . . Holy Jasus, she might! By God, I think she would. (*Shankill road.* WOMAN *enters – cleans windows.*)

AGNES (*enters*). Excuse me, would this be Forth Street?

WOMAN. Yes, love.

AGNES. Are all the evens down this side? I mean, would number six be here?

WOMAN. Number six, would that be wee Molly Baxter?

AGNES. Aye, do you know her?
WOMAN. Any connection?
AGNES. Yes, I'm her sister.
WOMAN. Well, you won't find too many chapels about here, love. (AGNES *crosses to house and knocks on door. Knocking door.* MOLLY *opens it.*)
MOLLY. Agnes! What are you doin' here? You'd better come in. Did you not get my letter?
AGNES. Aye, I got all your letters. That's why I'm here. Sure, I wrote and told you there was no future for John Joe, what with the mill getting mechanised.
MOLLY. Didn't I write back and tell you not to sell anything, to stay where you are, for my George near had rickets. What with the shortage of jobs, there's just enough room here for me and him and the three wains.
AGNES. I never got no letter, Molly. But you said everything was great up here and we should think of movin' some time.
MOLLY. Aye, but sure I never thought you would!
AGNES. John Joe and our ones will be arrivin' soon. They got a lift up in McIlwaine's cart with my wee sticks of furniture! What are we goin' to do?
MOLLY. Nothin' much you can do. I suppose you have to stay here, but I don't know what he's gonna say when he comes in from the docks.
AGNES. I don't want to get you into any trouble. We could always move somewhere else.
MOLLY. Are you daft, Agnes? Houses don't grow on trees up here, you know. There'd only be the workshouse....Jesus... I'm not lettin' my own flesh and blood be put in there. We'll manage someways. I'll move my three in with us, and you and John Joe and your two can take the two settle beds in the wee room.
AGNES. That's awful decent of you, Molly. (*Factory horn.*)
MOLLY. My George will be in soon. I'd better get a bit of dinner. Keep an eye on our wee Edwina there.
AGNES. Molly, would you speak up for me in the mill?
MOLLY. I couldn't do that, Agnes. I'm only in myself because his Mina spoke for me, and anyway there'd be an awful commotion if you got started and ones that lives here can't get in.
AGNES. Well, could I take in washin' or somethin'?
MOLLY. No! My George wouldn't hear tell of that. You only do that if you're really poor. You'd get an awful bad name.
AGNES. But I am really poor!

MOLLY. Aye, but you have to have a wee bit of pride, our Agnes. Look, you could maybe get into service in one of the big houses up the Antrim Road.

AGNES. Aye, anything!

MOLLY. I'll get my George to look in the paper for you . . . right . . . what age is your wee cub John Joe?

AGNES. Coming eight!

MOLLY. Aye, well, I'll let on he's eight and get him started as a Half Timer. It'll be easier to speak for him in the mill than a woman cos he'll be cheaper.

AGNES. Molly, he's only a child!

MOLLY. He may be only a child, but he's another mouth to feed. Talkin' of which, I may get on a wheen more pratties – Edwinas gurnin'! (MOLLY *exits*.)

GEORGE (*enters and shouts*). I'm home, Molly! (*Sees* AGNES.) Who are you?

AGNES. We've never met. I'm Molly's sister, Agnes.

GEORGE. Are you just up for the day?

AGNES. Ah, well, no, not exactly. You see, I wrote and said. . . .

GEORGE. Aye, and I instructed her to write back. Like it's no offence to you, but your man is not going to get a job up here, not with his background, love.

AGNES. What do you mean? My John Joe was a great hand-loom weaver.

GEORGE. I don't mean that. Hold on! Molly! I want a word with you.

AGNES. I'll see to the dinner. (AGNES *exits*.)

MOLLY (*enters*). Yes, luvvie?

GEORGE. What's this carry on?

MOLLY. They left before they got the letter. Nothin' we can do now. They've left their mill house. We can't turn them out onto the street.

GEORGE. He's not gonna get a job as a weaver up here. That's a skilled job. He's a Catholic.

MOLLY. Could you not speak for him? You know Ernie Simpson, he's the foreman in the factory. He's in your lodge. Could you not vouch for him even though he's a Catholic?

GEORGE. And get branded as a Fenian lover? I'd a hard enough time when I married you, love, but thank God you've done me proud. I'm not gettin' myself into trouble. Ulster Day is coming up soon and I want to be able to hold my head up in these streets.

MOLLY. Knowin' our Agnes she'll want to go til the Chapel!

GEORGE. Ah, Holy Jesus, is that the start of it? Tell her to go by the

back entry and up through the wee lane and don't let anyone see her. You're not going with her, you tell her that.
MOLLY. Of course, love, that was my promise to you.
GEORGE. It's not me, love, it's just the way things are. I could be in for a Button soon at the Docks, and we don't want nothin' standin' in the way. I'll tell him to go down to the Pens tomorrow, and he can take his chances.
MOLLY. George, would you look in the paper for our Agnes and see is there anybody lookin' for servants?
GEORGE. Not be many lookin' a Catholic unless they're recommended by our clergy. (*Looks at* MOLLY.) I know! You want me to get a letter from the Rev. Trimble.
MOLLY. Sure don't you do his garden for him an odd time. He'll do it for you, luvvie.
GEORGE. What are you?
MOLLY. A bloody nuisance! . . . Don't worry, George, once they get a few shillings past them, they get their own house.
GEORGE. Hey, guess who's on the Empire the night.
MOLLY. Who?
GEORGE. Marie Lloyd . . . what do you say?
MOLLY. Aye, sure our Agnes will look after the wains.
GEORGE. Aye. (MOLLY *and* GEORGE *exit. Voice singing 'My Old Man Said Follow the Van' until the line 'Lost our way and don't know where to roam'.*)

AGNES (*knocking at imaginary doors*). It's about the job as a kitchen maid.
VOICE ONE. Do you fit all the requirements in the advertisement?
AGNES. Well, yes, apart from one, but I have a letter of recommendation from the Rev. Trimble.
VOICE ONE. The Rev. who?
AGNES. The Rev. Tr. . . .
VOICE ONE. Sorry, don't know him! (*Knock, knock.*)
VOICE TWO. Oh, God, not another one with a letter of recommendation from the Rev. Trimble. I think the Rev. Trimble is becoming a one man Vincent de Paul. . . . Aren't there any Protestants living on the Shankill any more? (*Knock, knock.*)
COOK. The Lady is out. I'm the cook. What do you want?
AGNES. It's about the job. I have a letter from the Rev. Trimble.
COOK. That means you're one of the other sort.
AGNES. Yes, but it's a letter of recommendation.
COOK. You're wastin' your time, love. Even if King Billy and his army

were to vouch for you, she wouldn't start you.
AGNES. But why? It's not fair.
COOK. Don't be lookin' at me, love. I only make the dinner! (*Knock, knock.* AGNES *rips up letter and throws it away.*)
AGNES. I've come about the job in the paper.
VOICE THREE. Well, you know what we're lookin' for? A cook's help! Honest, clean, Protestant, good character, early 30's. . . . Do you fit the requirements?
AGNES. Yes.
MOLLY (*entering, puts on hat and shawl and, singing 'Daisy, Daisy' links* GEORGE'S *arm*).

> Daisy, Daisy, give me your answer do,
> I'm half crazy, all for the love of you
> It won't be a stylish marriage, I can't afford a carriage
> But we'll look sweet, upon the seat of a bicycle made for two.

Will there be a tram soon, George?
GEORGE. Shu'd be, luvvie.
MOLLY (*offers bag*). Will you have a wee jube jube?
GEORGE. No thanks, luvvie.
MOLLY. What's wrong, George? You usually laugh your leg off at the Sand Dancers, but you weren't at yourself the night.
GEORGE. Molly, you know this big Ulster day is coming up soon?
MOLLY. I do. Don't you worry, George. I've your Orange Sash pressed and in the dresser, and your Bowler hat dusted down. You know I won't let you down.
GEORGE. Oh, I know you'll do right by me. It's not that, love. . . . You see that is going to be one of the biggest days Ulster people will ever see. Lord Carson wants the whole country out to sign that Covenant as a pledge of allegiance to the union with Great Britain. You know what things are like round our way on the Twelfth of July. The blood's up. This is going to be worse, and I just think for their own sake Agnes and John Joe should stay well out of the road.
MOLLY. But where are they goin' to go, George?
GEORGE. Tell them to go into the wee room and lie low. Aren't they quick enough goin' in there when they want to say their mumbo jumbo and rattle rosary beads!
MOLLY. George!
GEORGE. Ah – no offence to you, luvvie.
MOLLY. Oh, here's a tram now.

(*LX cross fade to* AGNES *and* JOHN JOE *in* MOLLY'S *house.* AGNES *is kneeling, saying the end of the rosary.* JOHN JOE *is keeping watch out of window.*)

AGNES. In the name of the Father, the Son and the Holy Ghost, Amen. There, that's me.

JOHN JOE. You're all right. No sign of them yet.

AGNES. I'm sick of this, hidin' and jukin' in corners, sayin' the prayers in secret – it's like back to the dark ages this carry on.

JOHN JOE. Well, sure we don't want to offend George. He's been good to us.

AGNES. Oh, aye? Well, if he's that good why doesn't he get you a start down in the docks?

JOHN JOE. He can't, love. That's the Protestant pen. I have to go to the deep sea dock; the work's heavier and it doesn't come as often.

AGNES. It's not fair, you're standin' there in the freezin' cold for nothin', and there's George gets picked nearly every day.

JOHN JOE. Aye, but that's a different pen, and I know nobody there. It'll all take time. C'mon, we'll go til our beds.

AGNES. Sure, we can't talk in there with the weans sleepin'.

JOHN JOE. Oh now, what I have in mind doesn't need no talkin'.

AGNES. John Joe Mullan! Now sit down there – I need to talk to you serious like!

(*LX cross fade to* MOLLY *and* GEORGE *who are seated as if on tram.*)

MOLLY. Isn't our Agnes doing awful well in that big house on the Antrim Road? Maybe the Reverend Trimble could get John Joe a start as a gardener; he's awful good with his hands.

GEORGE. What? Two Catholics in the one household! Molly, enough is enough. Next thing you know the Pope of Rome will be makin' the Reverend Trimble a Saint for the Catholic church. (*They* BOTH *laugh.*)

MOLLY. I hope they're in bed when we get home.

GEORGE. Why?

MOLLY. We've had the childer sleepin' with us this past few weeks, so if the front room was free. . . .

GEORGE. Molly Baxter! Please stop talkin' like that! (*They* BOTH *laugh.*)

(*LX cross fade to* AGNES *and* JOHN JOE)

AGNES. It's just that, Lady Glendinning's been very good to me. I know in my heart that if I was to tell her that it's all been a pack of lies and I'm really a Catholic, she'd admire me for bein' so honest. I just hate deceivin' her, it's been three weeks now, and there's never been a cross word. She's awful decent.

JOHN JOE. I think you'd be better waitin' until this Covenant thing is over. The Protestants is all a wee bit excited just now.

AGNES. Aye, maybe you're right.

JOHN JOE. C'mon, we'll get to our beds. You know they like the place to themselves when they come in.

AGNES. John Joe, y'know I've never been to a theatre, and there's Molly and George goes maybe twice a week. Can we go sometime? I mean when you get a job?

JOHN JOE. D'ye want a wee concert? I'll give ye a wee concert. (*He exits to bedroom.*)

AGNES. John Joe, what are you at?

JOHN JOE (*offstage*). Never you mind. Just keep jukin' out that windy.

AGNES. Ach, John Joe, you're an awful oul' lig – there's the lamp lighter out – they'll be home soon. (JOHN JOE *enters wearing George's sash.*)

JOHN JOE. My name is Lord Carson, and this is my big white charger, King Billy the goat. I'm gonna put all you Catholics in a dungeon and throw away the key. (*Chasing her around the room.*) He's coming to get ye....(*Grabs rosary beads and puts them in his mouth.*) and he ates rosary beads for his diner.

AGNES. John Joe! God forgive you! (AGNES *is screaming and laughing.* JOHN JOE *jumps on to the table with rosary beads in his mouth. As this is happening* GEORGE *and* MOLLY *are approaching house. They enter. Stunned silence.*)

AGNES. We were just havin' a bit of fun.

JOHN JOE. Agnes was sayin' she'd never been to the Empire. . . .

AGNES. We didn't mean any harm, George.

GEORGE (*notices rosary beads on table*). Molly, get them things out of my house. (AGNES *grabs them.*)

MOLLY. I think you'd better go to bed, our Agnes.

AGNES. Aye, c'mon, John Joe. (*They exit.*)

MOLLY. George . . . (*She exits.*)

GEORGE. This sash has been in my family for one hundred years. My father wore it, and his father before him, and if God spares my wee children, they'll wear it. I never thought I'd see the day that it could be treated in such a manner under my very own roof. (*He leaves.*)

(LADY GLENDINNING *rings a little hand bell and* COOK *enters.*)

LADY GLENDINNING. Well, Cook, I believe you wanted a word with me.
COOK. If it's not wasting your time, Lady Glendinning.
LADY GLENDINNING. I can spare a few minutes. What is the problem?
COOK. Well, see the other day when Agnes and me was in the kitchen and you know the lovely blue china you have what sits up on the big cupboard? Anyway, you know the big, big plate that you could very near set a sheep on?
LADY GLENDINNING. Yes.
COOK. Well, anyway, 'Here dear', I says til Agnes. 'It's about time yon thing got a good cleaning for you know the way it's never used'. So I says til Agnes, 'Get up on thon stool and get that down, for it's only an old dust gatherer, and we'll give it a bit of a rub'. So here's Agnes, 'Aye alright', so then she gets the stool and gets up at it. I was kneading a big lump of dough at the time.
LADY GLENDINNING. Cook, I really can't spare that long. . . .
COOK. I'm near done now....anyhow, up she gets on the stool, like I wasn't payin' much heed of her for I was workin' . . . and then I hears a gasp of breath and 'Oh, Jesus,' for she near dropped it, and then just as I turned my head I caught her.
LADY GLENDINNING. Caught her? . . . Did she fall?
COOK. No . . . I caught her on.
LADY GLENDINNING. Sorry?
COOK. Doing this (*Crosses herself.*) God forgive me.
LADY GLENDINNING. Are you sure?
COOK. May I never bake another loaf, as large as life like this. (*Crosses herself.*) May God forgive me.
LADY GLENDINNING. Did you confront her about it?
COOK. I was too shocked . . . I tried to say to myself, 'Lila, you're seeing things,' but oh, no, as plain as the nose on your face, she did it. . . . She was an awful nice girl, too.
LADY GLENDINNING. I have been deceived.
COOK. I know, because you started her and all.
LADY GLENDINNING. Thank you, Cook. Would you please send Agnes up?
COOK. Like it's none of my business, but I thought it only right and proper you should know. . . . Oh, aye, and here's another thing . . . I asked her, just to catch her out like, would she be going to the Covenant Rally on Saturday, and she said she was busy. . . . Busy! . . . like you know yourself, Lady Glendinning, any Protestants who wouldn't be at that are either dead or in America.

GOLD IN THE STREETS – ACT I

LADY GLENDINNING. Thank you . . . just send Agnes up, please.
COOK. Yes, Lady Glendinning. (*Shouts.*) Agnes! (COOK *leaves and* AGNES *enters.*)
AGNES. You want to see me, Lady Glendinning?
LADY GLENDINNING. Agnes, I have some excellent news for you.
AGNES. For me?
LADY GLENDINNING. Yes, Agnes . . . I have been more than pleased with your services these past few weeks, and I have a very special task for you.
AGNES. Really?
LADY GLENDINNING. Well, we all know what day Saturday is.
AGNES (*unaware*). Am . . . no, Lady Glendinning.
LADY GLENDINNING. Agnes, I'm surprised at you . . . Saturday? Ulster Day? History in the making? No?
AGNES. Oh, yes . . . it . . . am . . . just slipped my mind.
LADY GLENDINNING. Well, Agnes, that day is going to be a special day for *you*. A day to remember as long as you live.
AGNES. Oh, really?
LADY GLENDINNING. Yes, Myself and the Cleaver Ladies Guild have been requested by Lord Carson himself to be on duty at the City Hall to arrange tea for the officials during the signing. What an honour, eh?
AGNES. Yes, a great honour for you, I'm sure.
LADY GLENDINNING. And I have chosen you from all my staff to come along and help with the festivities at the City Hall. Now what do you say to that, Agnes?
AGNES. Yes, I am very honoured, but you see, Lady Glendinning, I am busy that day. I can't come.
LADY GLENDINNING. Do I believe my ears? How can anyone be busy on the most important day we good loyal Ulster people will ever see? Everyone will be there apart from our enemies.
AGNES. I'm very, very sorry, but I can't come. Any other time, yes . . . I'm sure Cook would love to go.
LADY GLENDINNING. No, Agnes, I want you. In fact, I am ordering you to go, as your employer.
AGNES. Well, I can't.
LADY GLENDINNING. Agnes, if you are disobeying my order, you are not doing your job.
AGNES. I'll do anything else you need me to do. I'm a hard worker.
LADY GLENDINNING. No, I'm afraid if you can't carry out the job required, then I must ask you to leave.
AGNES. But Lady Glen. . . .

GOLD IN THE STREETS – ACT I

LADY GLENDINNING. You may collect your wages at the end of the day. Thank you, Agnes. (AGNES *leaves*. LADY GLENDINNING *follows*.)

(FOUR CHARACTERS *enter giving covenant details*.)
ACTOR ONE. Read all about it! Carson's Army marches on!
ACTOR TWO. Extra, extra! Insult to King George. Carson takes Royal Salute at Portadown.
ACTOR THREE. Ulster Day! Saturday 28th September 1912. Sir Edward Carson, a Dublin barrister, the man who helped prosecute Oscar Wilde, had found a new cause to fight – Asquith's 'Home Rule for Ireland' bill.
ACTOR ONE. Read all about it! Rallies held all over Ulster!
ACTOR TWO. Carson presented with orange lilies. Inspects forty Ulster Volunteers armed with toy rifles.
ACTOR FOUR. A covenant is to be signed by all Loyal Unionists on Ulster Day as a pledge of allegiance to the union with Great Britain.
ACTOR TWO. Read all about it. Covenant fever!
ACTOR ONE. All Unionists over sixteen years of age eligible to sign.
ACTOR THREE (*buys paper*). Now this was odd, because women, even over 21, still couldn't vote. There was speechifying here, there, and everywhere.
ACTOR FOUR (*reads from paper*). 'So long as I help to prevent the blighting curse of Romanism from crossing the Boyne, I will have spent my life in a good cause'.
ACTOR TWO. Get your Ulster Day buttons here. Crossed Union Jacks with the Red Hand of Ulster, only a penny apiece.
ACTOR THREE. It would be just as reasonable to give the convicts of Dartmoor the control of that institution as to give over a parliament controlled by the votes of Irish Nationalists. (*Bodhran beat. Song to the tune 'Mine Eyes Have Seen the Glory'. The* FOUR CHARACTERS *sing and mark time*.)

> Brothers shall we sever
> Never! Never! Never!
> But we'll cling forever
> To Union and to King.
>
> No Home Rule shall bind us,
> But traitors ever find us
> For Union and for King.

No surrender, this is Ulster.
No surrender, this is Ulster.

No surrender, we'll defend
With all our might
For Ulster will fight
And Ulster will be right!
(FOUR CHARACTERS *exit.*)

MOLLY (*entering with* AGNES). One day, our Agnes! That's all it is. One day! That's all you would have had to do! Just go down there! Close your eyes and ears and just get on with it. Aye, but you know better! You and your principles! Aye, well look where your principles got you! Now you've no job, and me and my man have to keep you and yours! And your wee children runnin' about without a shoe on their foot. You should be ashamed of yourselves. Excuse me! (MOLLY *lifts her hat and goes to mirror.* MOLLY *Sings 'Oh the Pope He Had a Pimple On His Bum'. She is trying on her new hat.*)

Oh the Pope he had a pimple on his bum
And it nipped, nipped, nipped so sore.
So he went to King Billy, and he rubbed it with a lily
And it nipped, nipped, nipped, no more.

AGNES. Molly, that's terrible for you to sing that. What has come over you?
MOLLY. That's a good wee tune. I love the old Orangey songs. (*Sings.*) 'We'll fight for no surrender'!
AGNES. Molly, our da would turn in his grave.
MOLLY. Our da wasn't married til an Orangeman livin' on the Shankill Road.
AGNES. Do you realise what you're doin'?
MOLLY. I'm signin' the ladies one. It has got nothin' to do with fightin'.
AGNES. Molly, you're a Catholic!
MOLLY. Didn't I turn?
AGNES. Turn? Does that mean you have to be more Protestant than the Prods?
MOLLY. Yes. Yes, it does, our Agnes. My George says if I want to get on up here I have to be against Home Rule. England has put a lot of money into Ireland.
AGNES. Our da spent his life hopin' we could all be free from the

English, Prods and all.

MOLLY. My freedom is to make sure my wains are fed and clothed, and that I'm accepted in these streets. And if it means me dressin' up as King Billy and ridin' a horse down to the City Hall, I'll do it, and if you and your John Joe had any sense you'd be down there signing and making sure everybody seen you and all. (GEORGE *enters.*)

GEORGE (*dressed in bowler and sash*). Well, Molly, how do I look? The day's the day when good Ulster men and women pledge to fight for Ulster. Your wee hat's lovely.

AGNES. Who are you pledgin' to fight, George?

GEORGE. Anybody that tries to take away our British heritage.

AGNES. Going to fight your fellow Irish men, maybe?

MOLLY. Oh, God, no, we won't go as far as that . . . (GEORGE *cuts her off.*)

GEORGE. Yes, Molly, afraid we will have to do that if it comes to the bit.

MOLLY. Whatever you say, George.

GEORGE. I am going down now to the City Hall with a couple of boys who are very high up in the Docks. Now, it would be a mark up for your John Joe if he was seen with me.

AGNES. No fear, George. We'll not sell ourselves out. We'll take our chances with all the other Catholics. You can't bate us all down.

GEORGE. I don't want to bate anybody down, love. I want to try and help you.

AGNES. We don't want no help from nobody that doesn't respect what we believe. Go you on, George. Tip your cap to all the big wigs, same boys that would walk over the top of you if they didn't need you. Go on, do their dirty work!

GEORGE. Don't say we didn't warn you. (*SFX of flute band.*)

MOLLY. There's Woodvale Flute Band. Will we follow it down to the City Hall?

GEORGE. Now, Molly, what have we to remember?

MOLLY. No Surrender. . . . Kick the Pope, am . . . Ulster will fight and Ulster will be right.

GEORGE. Good girl, I'm proud of you. . . . No Surrender! (GEORGE *and* MOLLY *exit.*)

AGNES (*at table writing*). Dear Annie, many thanks for your letter. I've a chance to write for the wains are in bed, George and Molly's away to the rally and John Joe's at the pub. We've had a wee bit of a dip in our fortunes since last I wrote. I no longer am in service to Lady Glendinning but, please God, something else will turn up soon. There's great excitement with this Ulster Day business the day. I suppose the Covenant's all signed, sealed and delivered by now. John Joe and me had a long talk last night. It's bad for him being here under another man's roof with no job. He's talking about uprooting ourselves again and going to England. It seems a long way away to me. But then it was me made him shift up here in the first place. I don't rightly know what we'll do. It's a wee bit uncomfortable between us and Molly's neighbours just now, this being such a big Protestant area. Ach, but it'll die down, we'll ride out the storm, sure we've weathered worse. They're good people at heart, just frightened and misled. (*Knock at door –* SANDY *and* JAMESIE.)

AGNES (*opening door*). George and Molly are out.
SANDY. We didn't come to see them, came to see you, love.
JAMESIE. Can we come in . . . cold out here, love.
AGNES. Aye . . . what can I do for you?
JAMESIE. Well, me and my mate Sandy are very concerned about you, love.
AGNES. About me? Sure I don't even know you.
SANDY. Look, love, for your own good and your wee children, you wanna think of getting out of here.
AGNES. Out? Look, who are you?
JAMESIE. Now, love, we don't have to spell it out for you, now, do we, eh?
AGNES. Yes, yes. I'm afraid you do have to spell it out.
SANDY. T-A-I-G-S.
AGNES. What?
SANDY. Taigs.
JAMESIE. Sandy, please . . . he's only showin' off because he can spell. Things is bad around here, love, and they are going to get worse . . . every man Jack in this district knows you and your man was not at the City Hall the day, now that's serious.
SANDY. Very serious.
AGNES. There were thousands that weren't at the City Hall the day.
SANDY. Aye, but they don't live in our street.
JAMESIE. Everybody's not like me and Sandy here, that's why we came before the hooligans get to you. They're calling it 'The Cleaning Up the District Campaign'.
AGNES. Cleaning up of what?
SANDY. T-A-I-G-S.
JAMESIE. Of them that would be against us.
AGNES. Me and my man want to do no harm.
SANDY. I'm sure you don't, love, but we can't make no exceptions. That wouldn't be fair on the rest.

AGNES. Fair? Fair? You are here to tell me I have to get out with no money, my man with no job, out on the streets with my wee children and you say that is fair. Jesus Christ, I wouldn't like to see what you would do if you were being unfair.

JAMESIE. Look, you don't understand.

SANDY. Why weren't you at the City Hall?

AGNES. Because I believe we should all be facing up to the British or we'll never have peace.

SANDY. What about being free of Popery?

AGNES. If you mean my religion, that is my own affair.

JAMESIE. Aye, well, if we get ruled by that shower in Dublin, it will be our affair because we'll have no choice. That's why we're against you, so, love, it's best for your own sake you get out now. That's all I'm sayin'.

SANDY. Come on, Jamesie, we have a few more calls to make – Alfie and Eithne MacNamara. Them ones with the thirteen children.

JAMESIE. Aye. We may send up Geordie Marley's cart to rid them out. Mind yourself, love. (*Starting to leave, they bump into* GEORGE *who has just arrived.* JAMESIE *grabs* GEORGE'S *arms and whispers.*) Get them the fuck out, alright? (SANDY *and* JAMESIE *exit.*)

AGNES. Where's John Joe? Is he not with you?

GEORGE. Were you daft letting him go down to the pub on a night like this?

AGNES. What's wrong, where is he?

GEORGE. Just a wee bit of a scrap. He's alright.

AGNES. Oh, dear God, where is he?

GEORGE. Alright, calm down, I got him up to the Mater. It's just his nose and a wee cut above his eye.

AGNES. Dear God in Heaven, why? Why him? He wouldn't hurt a fly . . . I'm going up.

GEORGE. Sit your ground. He's alright. Molly's with him . . . just a couple of the boys got carried away with the celebrations. Anyway he shouldn't have been there.

AGNES. He's every right to go out for a drink. Jesus Christ, are they all animals or what?

GEORGE. They got excited.

AGNES. Oh, no, George, it's more than that. They want us out and they mean business.

GEORGE. I have a few shillings put by I don't mind loanin' it to you, help you get out on the Falls in the Pound Loney.

AGNES. Did you not tell me earlier that was filth and squalor?

GEORGE. It's not. I was tryin' to make you see sense.
AGNES. Don't talk to me about sense, George. What sense is there in all this, eh?
GEORGE. You're just better off among your own sort.
AGNES. And according to you, my 'own sort' are lepers that have to be put in some place away from everybody else . . . you are not going to do it to me and my family. We're gettin' well out of this . . . to England.
GEORGE. Across the water?
AGNES. Don't look so shocked, George. Isn't that what you want, away where we won't bother nobody anymore?
GEORGE. Look, it's not me. . . .
AGNES. Go and tell your mates, George. They've won. Go on, laugh and have a pint with them because we won't dirty your street any more . . . At least over there nobody will care what we are, we can be what we want . . . my man will get a job for what he can do not what he believes. (GEORGE *exits*.) Nobody is goin' to bother us any more. (*Sings*.)

Mountains of Mourne

Oh, Mary, this London's a wonderful sight.
Where the people are workin' by day and by night.
They don't grow potatoes, nor barley, nor wheat,
But there's gangs of them diggin' for gold in the streets.
At least when I asked them that's what I was told,
So I just took a hand at this diggin' for gold.
But for all that I've found there, I might as well be
Back where the dark Mournes sweep down to the sea.

(AGNES *exits*.)
ACTOR ONE. What's the difference between an Irish man and a ham sandwich? Ham sandwich is only half as thick.
ACTOR TWO. Two Irish men are on an iceberg. Paddy says to Murphy. 'We're saved, we're saved'. Murphy said 'How do you know that?' Paddy says, 'Here comes the Titanic'.
ACTOR THREE. What's 25 miles long, got an IQ of 40? A Saint Patrick's Day march.
ACTOR TWO. What's the difference between a St. Patrick's Day march and one hundred ham sandwiches? Bugger all!
ACTOR FOUR. Bloody Irish. Why don't they go back to where they belong.

LX FADE. END OF ACT ONE

ACT II

The Mary Connor Story

SCENE: *Blackout. SFX air raid siren.*

VOICE (*shouting*). All Clear! (*LX up.*)

NARRATOR (*entering*). Right! That's World War II dealt with! Evening, everybody, I am brought to you courtesy of the 1947 Education Act. You got free school, so you got free teachers as well. Now pay attention and I'll bring you all up to date. It is now 1950. Well, we've still got rationing, of course, but then when you're only sixty miles away from another state that doesn't have restrictions, it's very easy to smuggle stuff across the border. Oh, aye, the border. Well, as you all know, Ireland got partitioned in 1922. Anyway, 1945 saw a Labour Government in Britain, and plans for a Welfare State. We didn't just get free education two years after the rest of the U.K., we got free health care as well. People were queueing up for free dentures, glasses and surgical stockings. Oh, aye, Belfast people have a great eye for the bargains, whether they need them or not. Now, the Blitz shocked a lot of people in more ways than one. They saw others sleeping under trees and in sheughs for anything up to ten weeks at a time. No houses built to replace those destroyed. So Belfast started building. Set up a Housing Trust, borrowing money from government to subsidise rent and rates. Great, isn't it? Everything was radical and new. Even the fashion, for a reaction to war time austerity came in the shape of the New Look. Women wore dozens of yards of material in just one skirt. So, new houses, new system – pity about the old beliefs. (*Enter* MARY CONNOR.) Mary Wilcox is just about to find out all about them – she was born Mary Connor in Ship Street, Belfast. She met a British serviceman in 1940 and married him, much to her parents' horror, in 1941. She moved across the water to live with Phil's parents – well, it was handier for whenever Phil was on leave. His ship was torpedoed in 1945, very near the end of the war in Europe, leaving Mary a widow with a young child. She's been away nine years. The exile returns. (*Exit*

GOLD IN THE STREETS – ACT II

NARRATOR. MARY *enters centre stage.* TAXI DRIVER *pulls up.*)
MARY. Taxi?
DRIVER. No, I'm Peter. Only geggin'. Jump in, love. Just off the overcoat?
MARY. What?
DRIVER. The boat.
MARY. Yes. I've been in England nine years. It's great to be home. Your accent is music to my ears. It's great to be home.
DRIVER. If it's that great, why did you leave it in the first place?
MARY. My husband was in the forces.
DRIVER. Englishman, is he?
MARY. He was killed.
DRIVER. Aye, well, us Belfast men did our bit back home too, y'know. I was in the A.R.P.
MARY. Really?
DRIVER. Aye, arsin' round pubs!
MARY. You're a geg.
DRIVER. No, t'tell you the truth, I was tryin' t'keep myself alive!
MARY. God, look at all the spare ground. Belfast did get a quare old knockin' in the Blitz.
DRIVER. Don't give a shite about that. A lot of old bricks and mortar. It's people you can't replace. Tho' mind you we had some quare old laughs when the siren went. All runnin' to the hills. There was solidarity then, love, I'm tellin' you. This particular night the siren goes and here's me and the wife runnin' like two men and a wee lad up to Divis Mountain. Halfway there the wife shouts 'Hey, Peter, I forgot my false teeth'. Says I, 'Holy Jesus, Mary, it's bombs the Krauts are droppin', not bloody Paris buns'. We couldn't run for laughin'.
MARY. Aye, typical Irish, get a laugh anywhere.
DRIVER. What! I'm not Irish, love, I'm British. . . . S'pose you'll be lookin' a job.
MARY. Oh, yes. That's the first thing. My wee girl's coming over next week . . . we're livin' with my mother, then I'm gonna get a house of my own . . . .are there many jobs about?
DRIVER. Aye, any God's amount, love, if you're prepared to work under the English. . . . Sure, they have all the top positions here. The bloody place is comin' down with them. That's why I took up the taxiing, just me and the oul' hackney. And she doesn't talk back to me neither.
MARY. English people don't bother me. Sure I've lived there for nine years.

75

DRIVER. That's it, love. They're all right when they're in England . . . but they come over here and treat us like dirt . . . bloody foreigners.
MARY. Well, that's what you get for wantin' to be British.
DRIVER. I am British 'cause I am for the King. He's a great fella . . . but I definitely disagree w'him plantin' a lot of his subjects on us. Subjects! That's a geg . . . rejects is more like. That's why they come here, cause they can't get good jobs over there . . . they get a wee bit o' power here and go back . . . jumped up hallions.
MARY. My husband was English. He was neither a hallion nor was he jumped up.
DRIVER. Ach, there's the odd one that's not bad . . . put them all together and it's friggin' dynamite . . . same as you get the odd Catholic or two, that's alright.
MARY. Where does that put me, then? I'm a Catholic and my husband was English.
DRIVER. If I were you, love, I'd go til Australia. (*Pause.*) Gonna tell you the honest t'God's truth, I would take you Taigs any day til the English . . . as I say, love, better the devil y'know.
MARY. Thanks.
DRIVER. Don't mention it! . . . Where abouts, love?
MARY. Just round the corner here.
DRIVER. Jesus, look at the state of this place.
MARY. What's wrong w'it?
DRIVER. If I were you, love, I'd get back to where I came from if you want to get on.
MARY. You just look after yourself, love. I'll be all right . . . much is that?
DRIVER. A shillin' to you, love . . . (*Shouts.*) Quit your climbin' or I'll put my toe in your arse . . . wee buggers . . . what are they doing? There's a hell of a lot o'breedin' going on here.
MARY. Have you any better suggestions, seeing as how you're so well informed?
DRIVER. What! I've been married 25 years and I've no childer . . . mind you, the wife's left 24 o'them.
MARY. Can't think why.
DRIVER. Only geggin' love!
MARY. Anyway, cheerio.
DRIVER. Don't be worrying, love, there's worse places . . . after all, there's people in graveyards would love to be like you. (DRIVER *exits.* MARY *looks around at the street.* TWO WOMEN *come*

GOLD IN THE STREETS – ACT II

forward with basins, start to scrub paving stones. MARY'S MOTHER *comes forward.* MARY *and her* MOTHER *exit. SFX 'My Aunt Jane')*

> My Aunt Jane, she called me in
> Gave me tea out of her wee tin
> Half a bap with sugar on the top
> Three black balls out of her wee shop.

(*Street scene. Two more* WOMEN *come forward with basins. All four actors scrub their stoops in following scene.*)

ACTOR ONE. Cobbled streets and Rag o'bone men.
ACTOR TWO. Harrished mothers yell for childern. Corncrake screams of. . . .
ACTOR THREE. 'You're a wantin''.
ACTOR ONE. Gurnin' weeins runnin' pantin',
 Snattery noses piggin' faces,
 Holey jumpers, untied laces,
 Poucey shawlies laughin', singin',
 Chapel bells and washin' wringin',
ACTOR FOUR. Money lenders faces trippin',
 Countin' cursin' tempers rippin',
ACTOR THREE. Josie's man who likes to punch her,
 Staggers home in oily duncher.
ACTOR TWO. Oul' d'Da Da Dempsey spits and stutters.
 His mangey dog lies in the gutters.
ACTOR ONE. White washed yards and outside lavies,
 Bookies crammed with greasy navvies.
ACTOR FOUR. Shufflin' home with batin' dockets,
 Tappin' mates with empty pockets.
ACTOR TWO. Mangle grinders,
ACTOR THREE. Wash-board scrubbers,
ACTOR FOUR. Spinners, doffers clad in rubbers,
ACTOR THREE. Body washers,
ACTOR FOUR. Tick collectors,
ACTOR TWO. Mitchers duke the school inspectors,
ACTOR ONE. Scrubbed half-moons and shiny knockers,
ACTOR TWO. The heavy trudge of weary dockers,
ACTOR ONE. Hapenny chews and pokes and sliders,
 Rusty prams make hand-made guiders.
ACTOR THREE. Cabbage, bacon, smelly nappies,
 The homely stench of Jean Mahaffies,

ACTOR ONE. Scores a childer gettin' fed,
ACTOR THREE. Sardine packed to five a bed.
ACTOR ONE. Willie wino hugs the lamp-post,
   Bares his soul and sings his utmost,
   Calls it 'Nelly,' slugs his Mundies,
   Damns the pubs for closin' Sundays.
ACTOR ONE. Yellowed blinds shut out the daylight.
ACTOR FOUR. Pawns are bunged til Friday pay-night.
ACTOR ONE. Pigeons clock in missin' slatins.
ACTOR THREE. Drunken men are givin' batins.
ACTOR TWO. Annie Sweeny's had her sixth son.
ACTOR FOUR. Wee Ma Black has breathed her last one.
ACTOR ONE. Front doors slam from til to shut,
   To the wantin' whine of an oul' stray mutt.
      (FOUR ACTORS *exit.*)

(LX on Mary's Mother's house.)
MOTHER *(entering, hearing the sound of bells)*. God. There's the eight o'clock chapel bells, and she's never here yet. Suppose the boat's been held up. Well, the place is all shipshape anyway. I'll just finish my wee brasses. (MARY *appears.*)
MARY. Well, I'm home.
MOTHER. God save us, you scared the heart out o'me, a thought it was somebody.
MARY. It is, it's me.
MOTHER. C'mon in . . . where's the child?
MARY. She wanted to stay with Phil's mother and father til her school finished . . . so I came first to sort things out.
MOTHER. Does she talk as if she has a mouthful o'caramels like her da?
MARY. She's English, Mummy. She was born there.
MOTHER. Aye, well, we'll soon see to that. (*Pause.*)
MARY. The place hasn't changed much.
MOTHER. I got a new wee bit o'oil cloth for you comin'.
MARY. It's lovely . . . the house seems smaller.
MOTHER. Awful sorry it's not what you're used til.
MARY. I didn't mean that . . . it's just when you're away and come back. . . .
MOTHER. Aye, nine years is a long time. It's easy to forget where ye come from.
MARY. Don't worry, Mummy, I never forget.
MOTHER. Course this family was never good enough for him.

GOLD IN THE STREETS – ACT II

MARY. Please don't be rakin' up the past . . . Phil's dead. I don't want to talk about him.
MOTHER. S'pose Phil's Ma's house was big.
MARY. Well . . . bit bigger than this.
MOTHER. Bathroom?
MARY. Yes.
MOTHER. S'pose you didn't tell them you had to sit fornenst the hearth in a tin one to wash yourself down.
MARY. Course I did.
MOTHER. Did they all get a good laugh at us? (*Pause.*)
MARY. How have you been?
MOTHER. Sure, I could h'been dead and buried, not a sinner hardly puts a foot over the door since the Lord took your poor father and you skipped off t'England.
MARY. I went to be with my husband, I didn't just skip off.
MOTHER. Have you been going to Mass?
MARY. Would you like a cup of tea?
MOTHER. I said have you been going to your church? (*Pause. Door knock.*)
MRS MURPHY (*entering*). Saw you arrivin' Mary . . . God, you're lookin' well . . . how are ye? She's lookin' quare and well, isn't she, Mrs. Connor? Awful sorry to hear about your man gettin' kilt. All the same your mother will be glad to get you home for the company, won't you, Mrs. Connor?
MOTHER. Aye, if God spares me and she doesn't start her gallivantin'.
MRS MURPHY. Sure there's nobody round here to go gallivantin' with now. Your chums are all well and truly married now, more interested in gettin' their wee childer up. My Teresa has just had her fourth and Bernie Riley has five.
MARY. They've only bin married same time as me.
MRS MURPHY. Well, like how long does it take? You'd just the one, Mary?
MARY. She's coming eight.
MRS MURPHY. Funny world, like, when you think about it. If I so much as looked at my man, I was booked intil the maternity and there's your poor mother could only have you. Maybe you've the same wee bit a trouble. God. people's innards are funny, aren't they? If I've any more Doctor McGee says I'll have to have all down there took away and there's you and your poor. . . .
MARY. It's nice to see you again, Mrs. Murphy, but if you don't mind my mother and I haven't seen each other for a long time. . . .
MRS MURPHY (*very disgruntled*). I'll go on, then, Mrs. Connor. I'll be

seeing you about, Mary, I'm sure. (*She leaves.*)
MARY. Nosy bloody cow.
MOTHER. Less a that kind a talk. You're not in England now. Your father never used that language in front of me, and nither will you, my girl. (*Pause.*)
MARY. Maybe I should call round and see Kathleen and Teresa.
MOTHER. You needin' be lookin' them to go traipsin' about with ye. They've enough on their plates. (*Pause.*) Are you comin' to Mass with me?
MARY. I'm only in. I'm tired.
MOTHER. Put your head down for half an hour. We'll go to eight o'clock Mass.
MARY. No!
MOTHER. What?
MARY. I'm away for a walk. (MARY *leaves.*)
MOTHER (*shouting*). I'll meet you outside the chapel . . . 8 o'clock. (MOTHER *exits.*)

(*LX on Kathleen's house. SFX sound of baby crying.* KATHLEEN *puts bottle in baby's mouth.*)
KATHLEEN. Here, shove that in your gub you gurnin' wee ghett . . . ack, where's a doodie woodie . . . you're the picture of your dadsie wadsie. God help you, you're scarred for life.
TERESA (*entering*). Kathleen, it's only me . . . guess who has just hopped out of a taxi with a big suitcase.
KATHLEEN. Don't tell me – Errol Flynn.
TERESA. No! Mary Connor.
KATHLEEN. Honest t'God.
TERESA. I'm after seein' her goin' til her ma's.
KATHLEEN. S'pose she's all Englified.
TERESA. I wasn't speakin' to her. I was jukin' out the curtains at the time.
KATHLEEN. I'm surprised her ma ever speakin' to her again for marrying that soldier. Protestant too.
TERESA. Sure, he's dead now.
KATHLEEN. S'pose that's why she's home.
TERESA. No, sure that was ages ago. He got killed in the war.
KATHLEEN. Well, for some. Wish to hell somebody would a took that ghett a mine off til the war and lost him.
TERESA. Will you and me and Mary have a wee get-together?
KATHLEEN. Who's gonna mind our squad?

TERESA. My ma.
KATHLEEN. Sure she has eight of her own.
TERESA. Sure your five and my four will not make that much difference.
KATHLEEN. That's seventeen childer.
TERESA. So?
KATHLEEN (*looks at baby*). Right, m'boy'o, you're the last. For, by frig, I'll sleep in the coal hole if I have til. (TERESA *and* KATHLEEN *exit.*)

(*LX on* MOTHER'S *house.* MOTHER *enters and sits at table, humming and darning a sock.*)
JOAN (*off*). N-I-P spells Nip. P-O-N-D spells Pond. F-L-U-F-F spells Fluff. (MARY *enters from outside.*)
MOTHER. You took your time, didn't you?
MARY. There was a whole lot in for the job. I had to take my turn.
MOTHER. How did the interview go anyway?
MARY. It's hard to tell. I'll just have to wait and see.
MOTHER. Huh. I wouldn't like to be holding my breath.
JOAN (*entering*). Mam, wait til you hear this. (*Kneels and reads from book.*) Nip goes to the pond. Nip and Dick go to the pond. Nip and Dora go to the pond. Nip and Fluff go to the pond. Nip and Dick and Dora and Fluff all go to the pond. Nip has a. . . .
MOTHER. Wish somebody'd drown that bloody dog!
MARY. Mummy, please. She is trying to learn.
MOTHER. Aye, well, if you'd send it to school like the rest of the childer.
MARY. I'm going up to get changed.
MOTHER. Where are you going now?
MARY. Out!
MOTHER. Sure you're only in. Oh, I suppose you're leaving it with me again. Poor wee crater hasn't a friend in the world seeing as how you didn't see fit to provide it with a brother or a sister when you had your man. Aye, and I'd like to know why.
MARY. Joan, go on out and play in the entry.
JOAN. All right, Mam. (*Exits.*)
MARY. Because I didn't want to breed like a rabbit like half of them round here.
MOTHER. Who do you think you are, Lady Muck? Here's me who prayed to St. Gerard day and night to bear another child and God knows what evil you've been up to, deprivin' your man of his rights and your wee child of a family.

MARY. Phil didn't want a big family either. We wanted to get on our feet.
MOTHER. I suppose you weren't even churched after that wee one.
MARY. No, I wasn't bloody going to the Chapel to be forgiven for making love to my husband and havin' a child.
MOTHER. Don't you speak that filth in my house.
MARY. Ah, is making love to your husband filthy, mother?
MOTHER. I'm not listening to this. I'm going out. You can mind her yourself for a change. I don't know what them English heathens have put into your mind.
MARY. Mammy, I just want to be left alone. I don't want to work in a factory. I never wanted a squad akids, and I have no desire to ate the alter rails.
MOTHER. If your father was here now to listen to this, and his own wee grandchild has never seen the inside of a chapel. The day that Englishman put his foot over that door this family was cursed.
MARY. He was the best thing ever happened to me, getting me out of this priest-ridden hole! (MOTHER *looks at her and leaves.* MARY *exits.*)

(*LX on the street.* MRS. MOLLOY *and* MOTHER *enter.*)
MRS MOLLOY (*brushing*). Away to Mass, Mrs. Connor?
MOTHER. Aye.
MRS MOLLOY. Anything wrong?
MOTHER. Mrs. Molloy, I'm at my wits' end. That Mary one hasn't put her foot inside the chapel since the day and hour she came home, and that wee child doesn't know the meaning of the word catechism.
MRS MOLLOY. You're a saint, Mrs. Connor, a saint. It's not your fault. You're at the church more than the priest himself, I always say.
MOTHER. That wee child is not even christened, I'm ashamed to say it.
MRS MOLLOY. Oh, dear God in Heaven! That is a sin! What if something was to happen to her? She would be in limbo, make no mistake!
MOTHER. That Mary one took the Statue of the Blessed Virgin out of her bedroom and the Sacred Heart off the bedroom wall!
MRS MOLLOY. It's out of your hands now, Mrs. Connor. I think Father McCabe is the only answer. Look at the time my Kathleen wanted to marry that Protestant from the Shankill Road. Her father went straight to Father McCabe, and Kathleen never mentioned the fella again.

GOLD IN THE STREETS – ACT II

MOTHER. Father McCabe, do you think so?
MRS MOLLOY. What else can you do?
MOTHER . I'll speak to him after Mass. (MOTHER *exits.*)
MRS MOLLOY (*calling after her*). I'll pray for you, Mrs. Connor. (*Exits.*)

(*LX on* CHILDREN *playing ball in street, while singing*).

>Up the long ladder and down the short rope,
>Away with King Billy and God Bless the Pope.
>The Pope is a gentleman who wears a watch and chain.
>King Billy is a beggarman who lives down the lane.
>What shall we do?
>Tear him up in two,
>And send him down to hell
>With his red, white and blue.

JOAN (*entering*). Hello there, can I have a go?
ELISH. What's your name?
JOAN. My name's Joan Wilcox.
BERNIE. Where d'you live?
JOAN. I live in Number 17.
BERNIE. You're a liar. Nobody talks like that in our street.
ELISH. What school do you go to?
JOAN. Well, I don't know yet. I expect Mam will tell me after the school holidays are over.
BERNIE. Mam! Do you hear her, Elish? She can't even talk proper!
ELISH. Bernie, make her say "Bout ye' – Go on. (*They laugh.*)
BERNIE. Right, we'll let you play if you say "Bout ye'! (*Pause.*)
BERNIE. Come on!
ELISH. Flip sake, hurry up!
JOAN. 'Bout yoou!
BERNIE. Oooh!
ELISH. 'Bout you! That's wick!
JOAN. Can I play now?
BERNIE. No! 'Cause you didn't do it right.
ELISH. Bernie, maybe she's a spy. A German spy.
BERNIE. Elish, you're stupid. They don't have spies that young. Anyway, my daddy says we killed them all in the war.
JOAN. I'm not German really, I'm English. Cross my heart and hope to die.
BERNIE. My da says he hates the English cause they think they're

brilliant cause they talk posh.

ELISH. During the war when m'da was at sea, we used to have a pretend uncle who talked funny. He was brilliant. Brought us chocolate and everythin'. My da hated him too. One day when he came back from the sea he told Uncle John if he ever set foot in our house again, he'd blow his Willie John off! (*They laugh.*)

JOAN. My daddy's English too. He was nice, but he got killed.

BERNIE. Aw! Aw! She's an orphan.

ELISH. You should be in a Home.

JOAN. I'm not. I live with my Mummy.

BERNIE. Liar! Liar!

ELISH. Orphan! Orphan!

BERNIE & ELISH (*sing-song*). Joan has no Daddy; they'll lock you in a prison.

BERNIE. My mummy says yous don't believe in God and are goin' to burn in the fires of hell like the black babies in Africa.

JOAN. No. My daddy is in heaven. The man in church said.

ELISH. He couldn't be; he's a Prod.

BERNIE. Even worse than that, he's English. He's bound to be in hell.

JOAN. But we believe in God as well.

ELISH. Well, God doesn't believe in you, so there!

BERNIE. Sure my mummy had a friend who went into a Protestant church by mistake and her hair all fell out, and she went all funny, and they had to put her in a home, and that proves it.

ELISH. Bernie, d'you know what my mummy says about her mummy? That she's only an old soldier doll, that she hung around the Plaza tryin' to pick up soldiers. Then she went off to England and married an English one!

BERNIE. Oh, I'm not playin' with her!

JOAN. That's lies! It's all lies! I'm not playing with you bad girls. I'm away home to tell my mam. (*Exits.*)

BERNIE & ELISH.
Tell tale tit,
Your mummy can't knit.
Your daddy can't go to bed,
Without a dummy tit. (*Running after* JOAN.)

(*LX on Plaza Ballroom.* KATHLEEN *and* TERESA *enter, sit at table.*)

KATHLEEN. I'm bored.

TERESA. So am I.

KATHLEEN. This oul' dance is dead borin'.

GOLD IN THE STREETS – ACT II

TERESA. So it is.
KATHLEEN. Mary doesn't look bored.
TERESA. No.
KATHLEEN. Y'know she's far, far. . . .
TERESA. Too old for him?
KATHLEEN. Aye.
TERESA. She's changed.
KATHLEEN. Since she came home from England?
TERESA. Can't talk til her now.
KATHLEEN. She'd dance with oul' Nick!
TERESA. Ssh. She's comin' over. (MARY *arrives and sits.*)
MARY. Hey, girls, that was smashin'! Did you see thon fella jivin' w'me? He was brilliant!
KATHLEEN. I'm bored!
TERESA. So am I. Is it not near over yet? I haft to get home.
MARY. Just relax and you'll enjoy yourself.
KATHLEEN. Oh aye, that's what my Seamus said to me once, and I ended up with a squad of kids!
MARY. Hey, look at that fella over there in the white suit! Isn't he gorgeous? (MARY *waves.*)
TERESA. Mary Connor!
KATHLEEN. Hope my wee ones is all right with your ma; our Peter's a wee bit colicky and he might play her up.
TERESA. Kathleen, you should use that gripe water; it's great. Our wee Anne-Marie was the same. . . .
KATHLEEN. Oh aye, I know, but the wee fellas is more inclined that way, y'know what I mean?
TERESA. Y'know, I think I'm away with it again, an' our Patrick's only three months!
KATHLEEN. My God! Y'never get a break, do ye?
MARY. Well, why don't you use somethin'?
KATHLEEN. What? Iron knickers?
MARY. No, it's somethin' your husband uses.
KATHLEEN. Well, frig me! I know what mine uses and I have five kids to prove it! What did yours have? A magic wand!
TERESA. Oh, my God! Here's two fellas comin' over. Just keep talkin'; never let on you see them!
MARY (*as two imaginary men pass by*). Hello! (KATHLEEN *and* TERESA *are disgusted.*)
TERESA. You think we're green, Mary Connor! I know rightly what you're talkin' about, and it's a sin!
KATHLEEN. It's dead disgustin', that's what it is!

MARY. It's not a sin. It just means you can relax and enjoy yourself.
TERESA. Ssh!
KATHLEEN. If I came round the morra, would you loan us some of the gripe water? Like just til he gets his pay.
TERESA. Aye, and if he doesn't like it, just stick it on his dummy, shove it in his mouth. That's what I do!
KATHLEEN. Who? Seamus or the child? (*They laugh.*)
MARY. Flip's sake! Can yous two not give over about friggin' kids?
KATHLEEN. Mary, you may not care about your child, but we do care about ours!
MARY. I do care about her, but this is supposed to be our night out away from them.
TERESA. Kathleen, don't you be gettin' up to dance without me.
KATHLEEN. Oh no fear, Teresa.
MARY. The fellas aren't goin' to take a bite out of ye!
TERESA. Look, if we wanna get up, we'll get up, and if we wanna talk about our children, we will!
KATHLEEN. I mean, it doesn't stop you throwin' yourself at every sailor you clap eyes on!
TERESA. Maybe in England you do that sort of thing, but we are married women!
MARY. Alls I'm saying is this is our night out without the kids. We came here to enjoy ourselves!
KATHLEEN. Oh, I am enjoyin' myself! Are you enjoyin' yourself, Teresa?
TERESA. Oh, I'm enjoyin' myself. I just don't need a fella eyein' me up and down to enjoy myself!
KATHLEEN. There! Y'see we are enjoyin' ourselves!
MARY. Hey, girls, do y'remember the nights we used to come here? All the Greek sailors? They were great oul' nights, weren't they?
TERESA. Aye, they were all right! Anyway, as I was sayin', Kathleen, I have to get home, to do the nappies before I go to bed.
KATHLEEN. I've me mammy doin' mine! She gets them that white and I don't know how she does it!
TERESA. Ach, that's great, Kathleen. . . .
MARY. Seriously! Have yous nothin' better to talk about than nappies and babies and friggin' gripe water?
KATHLEEN. Who the friggin' hell do you think you are? You're not comin' over here to make little of us, just cos you spent a couple of years in England!
TERESA. Lookin' a job in an office! If you're lookin' to pick up another sailor, you're not trailin' us through the muck with ye!

C'mon Teresa! (*They stand up.*)
TERESA. Yes. C'mon Kathleen. Some of us have families to see to!
MARY. Just go on, I'll pick up a few soldiers, maybe a sailor and the odd Chinaman and take them all up the entry at the same time!
KATHLEEN. Suits ye! (KATHLEEN *and* TERESA *exit.*)
MARY (*shouting after them*). You won't make me feel guilty, either. (MARY *sits alone. To herself.*) This looks awful bad. I'm sitting here on my own. . . . Oh, no, that fella in the white suit is lookin' over! If he comes over, what am I gonna say? Can't say I'm a widow, or he'll think I'm out lookin' for a man! If I say I'm not married he'll think 'What's wrong with her?' God, here he's comin'! What am I gonna do? Frig't, I'm going home! (MARY *runs out as lights fade.*)

(*LX on the local corner shop.* SHOPKEEPER *enters.*)
SHOPKEEPER. 100 Parkdrive, 10 of Ovaltine, 10 of H.P., two boxes of broken biscuits.
MARY (*entering*). Evening.
SHOPKEEPER. Evening, Mary. You've just caught me. I was about to close up. Stocktakin', y'know.
MARY. Y'couldn't give us two loose cigarettes and a box of matches?
SHOPKEEPER. Can indeed, love! Now d'ya want t'pay for those, or will I put it on your ma's slate?
MARY. Could y'put it on her slate?
SHOPKEEPER. Will indeed, love. Any sign of a job?
MARY. No, not yet.

SHOPKEEPER. Aye, well, night, night, love! (SHOPKEEPER *exits.*) (MARY *enters street and smokes. She stands alone on corner of street.*)
MOTHER (*passing*). What are you doing standin' at the corner on your own like a hussy?
MARY. I wanted out to get a bit of air.
MOTHER. Could y'not find nobody to go gallivantin' with?
MARY. Aye, there's a regiment a soldiers meetin' me here in half an hour. (MOTHER *goes to smack her face.* MARY *grabs her arm.*)
MARY. Don't dare! I'm a grown woman!
MOTHER. You're a disgrace. Teasie Murphy's mother was round and told me the carry on a you at the Plaza the other night . . . the filth that was coming out of your mouth.
MARY. Go away and leave me alone.
MOTHER. Father McCabe is coming round the night to see you. It's

terrible when I have to go to the Holy Father to try and save your name.

MARY. Mammy, please listen to me. Please. I'm not evil. I'm not ashamed of my background. Why do you think I came home? I just want to live my own life. I won't do no harm. I came back to be with you, to be in Belfast where I belong. Why are you makin' it so hard for me? What in God's name have I done to people? Tell me!

MOTHER. You haven't come back to your people, you've turned your back on them. I brought you up a good decent Catholic girl. How do ya pay me? You marry a bloody Englishman and then the two a yous debase yourselves for your own pleasures. You're not even married cause he's a Protestant . . . you're not married in the eyes a your church. . . . What does that make ye? You disgust me . . . Father McCabe . . . at 6 o'clock . . . be there! (*Exits.* MARY *looks after her, then exits.*)

(*LX on neighbors cleaning stoops.*)
MRS MOLLOY. Mornin' Mrs. Murphy!
MRS MURPHY. Mornin' Mrs. Molloy!
MRS MOLLOY. Cribbin's piggin'!
MRS MURPHY. Bloody dogs!
MRS MOLLOY. Awful warm!
MRS MURPHY. I'm sweatin'!
MRS MOLLOY. Mornin' Mrs. O'Leary!
MRS O'LEARY. Mornin' Mrs. Molloy, Mrs. Murphy!
MRS MURPHY. Mornin' Mrs. O'Leary!
MRS MOLLOY. Lovely mornin'!
MRS MURPHY. Beautiful mornin'!
MRS O'LEARY. Gorgeous mornin'! (MARY *enters.*)
MARY. Good morning!
MRS MOLLOY. S'pose it is.
MARY. It is a lovely day, the sun is shining.
MRS MOLLOY. How is your poor mother's nerves?
MARY. My mother's nerves are fine as far as I can see.
MRS MOLLOY. It's an awful pity of her, for she was a fine wee woman.
MARY. What do you mean 'was'? She's still alive and kicking.
MRS MOLLOY. Aye . . . well, she hasn't had her sorrows to seek.
MARY. Good morning . . . bloody hypocrites! (MARY *walks across the stage past the* THREE WOMEN. *They are on their hands and knees, each washing the ground in the traditional half-moon shape. When they speak, it is not to each other, but to themselves or the*

audience.)
MRS MURPHY. Bloody cheek!
MRS MOLLOY. Walks past here with her head in the air!
MRS O'LEARY. Just because she's lived in England.
MRS MURPHY. Near broke her mother's heart.
MRS MOLLOY. And her brought up a good decent Catholic.
MRS O'LEARY. We're s'posed to feel sorry for her because her man got killed.
MRS MURPHY. Germans should have blew a hell of a lot more of them to smithereens.
MRS MOLLOY. The brass neck to go lookin' a house.
MRS O'LEARY. There's poor wee creatures had theirs blown to hell in the Blitz and still waitin'.
MRS MURPHY. While she was gallivantin' round England.
MRS MOLLOY. No conscience.
MRS O'LEARY. Back here because they don't want her over there because she had no man.
MRS MURPHY. Serves her right.
MRS MOLLOY. Not be long afore she takes up with another one.
MRS O'LEARY. And the last one hardly cold.
MRS MURPHY. Oh, she'll be round knockin' about the dance halls.
MRS MOLLOY. Pickin' up anything in a uniform.
MRS O'LEARY. She's well and truly sullied now.
MRS MURPHY. She may as well go the whole hog.
MRS MOLLOY. Hussy.
MRS O'LEARY. Traitor.
MRS MURPHY. Turncoat.
MRS MOLLOY. Aye, them ones that take up with the English.
MRS O'LEARY. Never have any luck.
MRS MURPHY. Deserve all they get.
MRS MOLLOY. My back's near broke.
MRS O'LEARY. Could be doin' with another surgical stockin'.
MRS MURPHY. Must get another clatter of them wee nerve tablets.
MRS MOLLOY. I'll get a dose of them for him for his old nerves are going too.
MRS MOLLOY. Sure we might as well.
MRS O'LEARY. Seeing it's free.
MRS MURPHY. Thank God for it.
MRS MOLLOY. Quare oul' mark.
MRS O'LEARY. This free health.
MRS MURPHY. Best thing the British ever done. (WOMEN *throw water out of basins and exit.*)

SFX (*voice off*). Mrs. Molloy, have y'heard the latest about Mary Wilcox?
MRS MOLLOY (*comes back*). Aye, runnin' t'the Plaza! . . . I know! The mother's for Graham's home . . . Protestant school, tut, tut! . . . Sailors! . . . Father McCabe! . . . Sacred heart o'Jesus!!! (*She exits.*)

(*LX on* MOTHER, *cutting newspaper in squares for toilet paper, and* MARY *sit in silence.* JOAN *enters.*)
JOAN. Mam, can I have some sweetie coupons?
MARY. I have none left.
JOAN. Gran, I want to go to shop on t'corner.
MOTHER. I told you before it's not gran, it's granny.
JOAN. What's the difference?
MOTHER. Plenty! And it's not 'shop on t'corner' it's the 'shap' at the bottom of the street. Say it! 'Shap'!
JOAN. 'Shap'!
MOTHER. That's better. I don't know what hell way they taught you to speak over there, but you're not doing it here.
JOAN. My dad spoke like that and so do all my friends.
MOTHER. Sure didn't you come in gurnin' the other day cos the childer laughed at the way you spoke?
JOAN. Gurning? What's that?
MOTHER. Holy God, am I going to have to educate you for-by. (*Hands* JOAN *squares of newspaper on string.*) Here, put those in the toilet.
JOAN. Gran, I've learnt some Irish words.
MOTHER. (*Impressed.*) Irish . . . that's great, let me hear them.
JOAN. What about ye?
MOTHER. That's not Irish, that's English.
JOAN. We didn't say it in England.
MOTHER. Irish is a different language.
JOAN. Then why don't you speak it?
MOTHER. Cause I speak English.
JOAN. Then why are you shouting at me for speaking English?
MOTHER (*irritated*). Get your coat. I'm taking you out.
JOAN. That means that everybody in Ireland speaks English even though they're Irish. Why is that, Gran?
MOTHER (*glares at* MARY). Why don't you ask the English? Come on, I'm taking you round to St. Mary's school to get you started.
JOAN. Mam says I'm not allowed to go to the Catholic school.
MOTHER. Your mother has enough sins to answer for. She'll have no

more if I have anything to do with it.
MARY. Joan, go and play in the entry.
JOAN. I'm sick of playing in the entry.
MARY. Do as you're told. (JOAN *leaves.*)
MARY. Mother, she is going to a state school.
MOTHER. You mean a Protestant school, don't you? Why don't you say it?
MARY. All right . . . a Protestant school.
MOTHER. Mother of God. (*Breaks down.*) What did I do to deserve all this? My only child sending me to my maker with a sin on my soul. You don't know what you're doing to me, nor do you care.
MARY. Mummy, I do care, but you're going to have to let me live my own life. Phil and I agreed. . . .
MOTHER. Ach, Phil this, Phil that. Did he not give a damn about you and your faith? But then he's English, he's like the rest of them. It has to be their way all the time, to hell with the destruction they cause.
MARY. It's because he gave a damn about you and me that he's not here now but lying at the bottom of a bloody ocean somewhere.
MOTHER. He's not lying at the bottom of no ocean on my account. I never asked the English to fight no war for me. Haven't they caused enough of a bloody war here for the last 100 years?
MARY. Why do you have to blame every English person for something that happened before they were even born?
MOTHER. Ach, what do you mean before they were even born? Sure aren't they still here? Why do you think you can't get a job? Why do you think you're finding it so hard to get a house? It's because you're a Catholic, and you'll never get a chance while they're still here. All they've ever done for you is turn you into a heathen. There's nothing left for me to do only pray for you. (MOTHER *leaves.* MARY *follows.*)

(TAXI DRIVER *places two orange boxes for taxi. Goes to house and knocks.*)
JOAN (*entering*). Mam, I'll get it! (*Runs and opens door.*) Taxi?
DRIVER. No, I'm Peter! (*No response.*) Wee joke, love!
JOAN (*running to pick up case*). Hurry up, mam. (JOAN *runs to car with case.*)
DRIVER. Just hop in the back, love! (*He gets into taxi and waits.* MARY *enters and sits in taxi.*) Where abouts, love?
MARY. The boat! (TAXI DRIVER *looks in mirror.*)
DRIVER. Have I seen you somewhere before?

MARY. Don't think so!
DRIVER. Holidays?
MARY. No!
DRIVER. Goin' away t'live?
MARY. Yes!
DRIVER. Aye, you're just as well, love. No future here now!
MARY. Well, if this place is so bad why don't you go and live in England?
DRIVER. And live among a lot of bloody Englishmen? I'd take the Catholics any day, love. Better the devil y'know! (*SFX of whistling, tune 'O Mary, This London's a Wonderful Sight'. Taxi stops.*)
DRIVER. God, there's desperate crowds here the night. Is half o'Belfast emigratin', or has somethin' happened and nobody's told me about it?
MARY. Aye, did y'not hear? They're diggin' for gold in the streets over there!

BLACKOUT ON TAXI. FADE SFX. END OF ACT II.

ACT III

The Sharon McAllister Story

SCENE: *LX up. Bare stage. SFX news jingle.*

NARRATOR (*enters*). And now the news in brief. Trouble flared today in the black township of Soweto when police clashed with mourners at the funeral of two of the youths killed in last Wednesday's riots. Bishop Tu-Tu has condemned the police action as direct incitement to violence and called for the immediate suspension of Martial Law. Actress Joan Collins arrived in Heathrow today to a barrage of questions from journalists about rumours of a possible reconciliation between Alexis and Blake Carrington in the spring series. Ms. Collins refused to comment. Lawyers acting on behalf of Mr. Alf Roberts are to sue the *Sun* newspaper for defamation of character after allegations of corruption were made in last Wednesday's issue. Mr. Roberts refutes the suggestion that, as a local councillor, he is privy to information on redevelopment in the area, and that this constitutes a conflict with his private business concerns such as buying Hildas house to provide space for extension to the corner shop. And there are rumours of a co-operative link between Coronation Street and the Crossroads Motel. George and Ivy are considering one of the hotel's attractively priced week-away schemes as a honeymoon. The search continued in Dallas today for the murderer of Bobby Ewing. The Police Department are carrying out skyscraper to skyscraper enquiries and the Cattlebarons' Ball has been postponed as a mark of respect. Former motor magnate John De Lorean, convicted last year on a narcotics charge, who has since become a born-again Christian, has applied to the Northern Ireland Office for funding to convert the former DeLorean car plant in Dunmurry, West Belfast, into a multi-million pound drive-in worship complex. It is estimated the scheme could provide up to fifteen hundred new jobs with government backing. A spokesman for the Northern Ireland Office said they were giving the proposal serious consideration. A

GOLD IN THE STREETS – ACT III

representative of C.U. – the Christian Union – said they were willing to enter into negotiations with the management. And still with DeLorean, 23 year old mother of two, Sharon McAllister of Suffolk estate in West Belfast, whose husband Davy has been unemployed since the firm's collapse four years ago, has been watching this bulletin. (*Exits.*)

(SHARON *enters. Sits on seat centre stage.*)
SHARON (*recites*). Four years now he's out of work
    Don't ask me how we've got through
    We don't grin but we have to bear it
    What the hell else can we do?
    It's the kids that help keep my sanity
    Like I know that sounds a bit odd
    But sure you can't take it out on themins
    Is it their fault their da's no job?
    He used to work for your man DeLorean
    He was foreman, if you please, in there
    Damn you, Mr. Big Shot DeLorean
    And your castles in the air.
    Ach, he was bringing home smashing wages then
    We were doin' great for a while
    The children wanted for nothin'
    And I'd the best of style.
    Never once did it even strike me
    That one day it would come to a stop
    No more Marks and Spencer's
    Now it's the Oxfam shop.
    And my ma helps us out a fair wee bit
    God love her, she's been like a rock
    If he knew she paid the TV every month
    He'd go off his friggin' block.
    Cause he wouldn't ask nobody for nothin'
    Never has and he never will
    But I can't afford to be as proud as that
    When the wains' bellies are needin' filled.
    It's destroyin' him slowly but surely
    His pride's took a hell of a blow
    I keep sayin', 'Look, love, there's thousands like you'
    But he just doesn't want t'know.
    I'm tryin' to keep up for all our sakes
    But it's puttin' m'head away

For we're just another pile on the scrap heap
And it's gettin' bigger every day.
I got m'self a wee job cleanin'
Sure it keeps the old wolf from the door
Sneaking up entries to get there
You can't trust nobody no more.
There's dole snoopers round every corner
Just tryin' to call your bluff
As soon as he gets a job legally
I'll tell them to go and get stuffed.
   (SHARON *gets up and begins to clean chair with cloth.*)
MAUREEN (*enters cleaning*). I says to him this morning 'I can't drive the Rolls Royce, it's too big', so he's going to get me a wee sports car, you know, just for running back and forward to the Spar.
SHARON. I wouldn't bother myself with driving, Maureen. The chauffeur takes me everywhere. Like why put yourself out when you don't have to?
MAUREEN. I like my independence. We had an awful row when he bought the private jet. I says, 'No, John, no way'. I want to be able to travel to the Bahamas with ordinary people. I am not going to get airs above my station.
SHARON. Davy and I are having champagne and caviar beside the pool the night. Would you like to join us?
MAUREEN. Sharon, I can't tell you how bored I am with rich food. Me and him's gonna run round to Bettie's for fish and chips. See how the poor live. (*Phone rings.*)
SHARON. Hello . . . yes . . . what? . . . no . . . you're a liar . . . honest to God . . . no geggin' . . . swear on our Gary and Jill's life . . . really? . . . on the level . . . what is it? . . . aye . . . alright . . . see you . . . aye . . . cheerio.
MAUREEN. Well?
SHARON. That was my Davy on the phone. I knew it must be important cause he wouldn't phone me at work if it was nothin'. Only if it was something up with the kids like. You'd better sit down. (MAUREEN *gets chair and sits beside* SHARON.)
SHARON. He's got a job!
MAUREEN. Liar!
SHARON. Honest to God, he's over the moon. He's away out now with a couple of his mates to celebrate. Borrowed the money off my ma, of course, and then me and him is goin' to the Four Winds Supper Dance the morrow night.
MAUREEN. I'm delighted for you. What is it?

GOLD IN THE STREETS – ACT III

SHARON. What's what?

MAUREEN. The job.

SHARON. He said he didn't want to talk about it over the phone. He'd tell me later.

MAUREEN. Probably MI5.

SHARON. Are you jokin'? My Davy in the Secret Service? He couldn't hold his own water, for God's sake!

MAUREEN. I got a lovely new dress in Tina's Club book I'll loan you it for the do.

SHARON. Thanks, Maureen, but I've waited for this minute for so long, I'll borrow the money and buy myself something.

MAUREEN. You're just right, love. Suppose that's you givin' up your wee job here now. There'll be no stoppin' you.

SHARON. Tell the truth, Maureen, would it be anybody's choice to clean up after people?

MAUREEN. Frig, I'll be cleaning this school til Nelson gets his eye back.

SHARON. What about your John? No sign of work?

MAUREEN. That's a good joke. That one of Frank Carson's?

SHARON. Awful, isn't it?

MAUREEN. Ach, it could be worse, as my ma always says. She's a great comfort, 'Cheer up, our Maureen, there's people in hospital with no legs'.

SHARON. I get the same from my ma. 'Yous don't know yous are livin'. I never had a pair of shoes til I was seven years of age'.

MAUREEN. Know what? I've a wee quarter bottle of vodka up in the cleaner's cupboard. I was hidin' it from him for my wee Saturday night drink. Sure this is a celebration.

SHARON. I'll pay you back, Maureen. You and me's gonna hit the town some night, no expense spared. (*They start singing 'If They Could See Me Now'.*)

> Now isn't that stickin' out
> The bold Davy with a job
> He'll be round at my ma's
> For the loan of a few bob.
> Now I can pay all my debts off
> God, we'll be atin' like Queens
> For I'm sick, sore and tired
> Of tea, an' toast and baked beans.
> His mates will be sleggin' him
> For they're all on the Bru

And they're all doin' the double
Sure what the hell else can you do?
He'll be out celebratin'
Y'know like actin' the lad
Buyin' drinks all round him
Bummin' and blowin' like mad

(SHARON *and* MAUREEN *exit.*)

(*SFX of Northern Ireland soccer match.* BEEZER, DAVY *and* SMICKER *are watching.* JESSIE *is drinking and ignoring them.* BEEZER, DAVY *and* SMICKER *cheer and shout 'Mexico, here we go, here we go'.*)

BEEZER. Northern Ireland. No problem
DAVY. Hey, it should be written into the Constitution that Pat Jennings isn't allowed to leave the team.
SMICKER. I'm going to put in for a discretionary grant to the DHSS to get me to Mexico.
BEEZER. What do you think of that, Jessie. . . . No problem
JESSIE. Will it put the price of drink up?
DAVY. What are you on about?
JESSIE. Then it won't make a hapworth of difference, will it?
BEEZER. Jesus sake, Jessie, there could be nuclear war declared and all you'd care about is the price of vodka.
SMICKER. Right . . . down to the serious stuff . . . three pints, Charlie!
DAVY. It's alright, Charlie. These are on me.
JESSIE. Wee vodka, Charlie.
BEEZER. Where did you get the spondulicks?
ALL (*sing*). Davy's been doing the double.
SMICKER. Last time you bought a drink, Davy, Rubens cut his left ear off.
DAVY. I got a job.
BEEZER. Aye . . . have you been knockin' up that wee blonde dole clerk?
JESSIE. Less of that filthy talk, yousens, there's a lady present.
DAVY. Where?
SMICKER. Don't tell us, Davy. You're goin' to dress up as a fairy and you're goin' to be a singin' telegram.
BEEZER. Hey, Smicker, he doesn't have to dress up.
DAVY. Watch it, yousens. No, this is all above board – job for life – good pensions – help with the mortgage.
SMICKER. A vicar?

JESSIE. Got a job, Davy? That's a quare pity, like.
DAVY. What?
JESSIE. Means you'll not be able to afford to drink no more.
DAVY. Not atall, these are decent wages....because I am goin' to be a public servant.
BEEZER. Shite house clerk for the City Hall toilets.
JESSIE. Mind you, I've never been as well off since he knocked off work.
DAVY. How come?
JESSIE. Well, his old lung collapsed last year and he took til his bed. I collect his sick every Wednesday, and seeing he's not able to spend it, I get it all. (*Lifts the glass.*) Here's to ye, ye baldy old ghett. (*Laughs uproariously.*)
DAVY. All I'm telling yis is, when I start this job, I want a wee bit of respect from yous hallions.
SMICKER. I've got it – brain surgeon.
DAVY. Well, if I was, I'd have a hard job operatin' on you.
BEEZER. You'd have to operate on his arse.
DAVY. Right, yous have had three guesses. I, David Reginald McAllister, will wish to be known in future as Constable McAllister.
SMICKER. Fuck me, he's goin' to be a peeler!
BEEZER. Who'd have thought it? Davy McAllister a peeler. How the hell did you get through the interview? I thought you had to be smart.
SMICKER. Not atall, sure you let anybody in these days.
JESSIE. Jesus' sake, they'd take my Tommy and he's only half a lung.
DAVY. What? You have to have a very high IQ.
BEEZER. Aye, very high Eejit Quality. (JESSIE *roars with laughter.*)
DAVY. Right, Beezer, your car out there, is it taxed? Furthermore, you've had three pints and two halfins, may I have your keys or I'm going to have to report you. (*Again* JESSIE *roars with laughter.*) And you, Madam, I believe you are claiming benefit for attending your sick mother who died twenty-five years ago. What about that, Jessie?
BEEZER. And we knew him when he was nothin'.
JESSIE. And me that used to wipe the snatters from his nose when he was two hands higher than a duck's arse.
DAVY. Only geggin'.
SMICKER. All jokin' aside, mate . . . put it there.
BEEZER. Aye, all the best, big lad.
DAVY. Aye, sure what else is there these days?

BEEZER. Not what it used to be. 'Member when we were nippers, all the peelers did was stop you playin' football and carry home drunks?
DAVY, SMICKER, BEEZER. Aye, your da. (ALL *three point to someone else.*)
JESSIE. Me.
SMICKER. Dangerous old game now, Davy.
DAVY. Jesus' sake, give over. We're here to celebrate. Charlie! All the drinks are on me.
BEEZER. Fuck me, if Rubens was here he'd cut his other ear off.
JESIE. What's the wife say about it, Davy?
DAVY. She was at work when I got the letter. Frig sake, she'll be over the moon. No more borrowin' off her ma, she'll have Marks and Spencers bought out.
BEEZER. Hey, Davy, why are the police the police?
DAVY. What are you talkin' about?
BEEZER. Because they are. . . .
SMICKER. You see! . . . Hey, Jessie, help the RUC, bate yourself up!
JESSIE. It'll be the last one I'll buy you. Peelers aren't what they used to be. In my day, you could trust them. Ah, it's a different kettle of fish now. (JESSIE *exits singing Z cars theme.* DAVY, SMICKER, *and* BEEZER *exit.*)

    (SHARON *enters.*)
SHARON (*recites*). Six months he's been with the police
    And it's taken six years off my life
    But they don't tell them that in the trainin'
    The hell you go through as a wife
    Here I'm sittin' with all m'orders
    He says 'Love, just buy what y'want'
    But night after night it's the worry
    God, I wish I could stop, but I can't.

    (SHARON *is sitting on a chair asleep. SFX door slamming.*)
DAVY (*enters and sees her. Sings the following line*). Moonlight becomes you. . . .
SHARON (*wakes up*). What bloody time is this?
DAVY (*continues singing the line*). It goes with your hair. . . .
SHARON. If you weren't coming home why didn't you bloody phone?
DAVY. I forgot my address book . . . sorry, love . . . just a wee joke. (*Sobers up.*) No, wee Jim Armstrong's wife just had a wee boy, and Jesus, he's as pleased as punch, so we all went to wet the

baby's head.

SHARON. Well, you needn't be lookin' for any dinner, cause it's in the bin.

DAVY. No, don't you bother, love. I had a Hoo Flung Dung on the way home. A wee drop of tea will do. God, we'd a laugh the day. Big Ginger Donaldson – member I told you about him 'the Michelin Man', gets his uniform made by Alec Simmons, me and him and George McBride were all in the Landrover goin' up by Divis Flats when there was the sound of gunfire. So we stopped, all jumped out and took cover. Suddenly there was another burst of machine gun fire, silence, and then a psshh sound. As quick as a flash George pipes up 'Hey, Ginger, was that you or the back wheel?' We near pissed ourselves laughing.

SHARON. Oh Christ! Are you stupid or what? You were supposed to be home at seven o'clock....it's now eleven. For all I knew you could have been lying in an alley with your brains splattered all over the ground. If you don't care about your own bloody life, you can spare a thought for me sitting here night after night, wondering if I'll have a husband to wake up beside in the morning....and when you're not working, you're out getting plastered with your mates, having a laugh and cracking jokes about who was nearly shot at, nearly committing suicide. It's not a bloody game. People are getting killed, so you'll forgive me if I don't join in the joke.

DAVY (*angry*). What do you want me to do? Be frightened goin' out my front door in the mornin' in case somebody takes a pot shot? Be terrified answerin' the door in case it's not the postman? No. It's not a joke, but it's part of the job, and you and me both have to accept that that's the way things are.

SHARON. I don't have to accept you bein' downright careless and to be perfectly honest – just plain selfish. It just happens to be our anniversary today and I'd bought us a couple of steaks and a bottle of wine to celebrate. But I suppose that's too normal for you. Hardly a high risk situation. Course we could always swap the steaks for two piranha fish!

DAVY. Sorry, love, you're just tired.

SHARON. Tired! I'm a nervous bloody wreck.

DAVY. Jesus, love, it's the job. What can I do? It's not my fault. I only went for a wee dr. . . .

SHARON. No, it's not your fault. Nothing's ever anybody's fault, is it? But why the hell should I have to live with it?

DAVY. Jesus Christ! Would y'rather I was on the dole again?

SHARON. Yes. Yes. At least I could sleep in my bed at night. Look at me. I haven't had a full night's sleep in months. Bloody nightmares. Every car that stops, every noise, sleeping with a gun beside m'bed. It's a bloody nightmare.
DAVY. We'll get used to it, love. Soon it'll just be a way of life. Just as normal as any other job.
SHARON. Normal! A way of life! I'd rather be in a bloody madhouse! (SHARON *exits.*)
DAVY. I hope she's not going to go on like this to my ma and da the morra night! (*Exits.*)

(*LX on* SHARON *and* DAVY'S *kitchen. Following night.* SHARON *and* MAISIE *enter.*)

SAM (*entering*). Isn't the kitchen lookin' great, Maisie? Sharon, love, you're a lucky girl that has a man who can turn his hand to anything.
MAISIE. I'm sure the microwave is handy when Davy's on lates.
SHARON. Well, we haven't had a chance to use it yet. He just brought it home yesterday. Said someone in the station could get them cheap. We haven't even paid for the stereo and the computer he bought for himself yet, but Davy says that he'll do all the worrying over the bank balance.
SAM. Don't you fuss yourself. Our Davy has his head screwed on. He just wants the best for you and the wains.
SHARON. I know he does. It just depends what you mean by best.
SAM. What does any man want for his family – a nice home, money comin' in, a good holiday once a year.
MAISIE. Is there anything wrong, dear?
SAM. Ach . . . stop buttin' in, woman, and away in and make a drop of tea. Sharon, love, you're a great wee girl, but I've never met the like of you for worrying.
SHARON. Do you think Davy's changed since he's joined the police?
SAM. Changed! What – he's a different man!
SHARON. I was beginning to think it was just me.
SAM. For a while after he lost his job in De Lorean, I'll have to admit to you, I was worried about him. Nothing interested him. The fella was lost for something to do. You know – maybe he never mentioned it to you, but he was even considering emigratin' to Australia. But I'm tellin' you the police has been the makin' of that fella. You should be proud of him. Your husband and men like him are keepin' this country from goin' to wreck and ruin.
SHARON. Yes, but in the carryin' out of that job, what's it doin' to

them inside?

SAM. I would say givin' them pride and dignity. They're what's stoppin' this country from civil war. They're keepin' the peace. (MAISIE *comes in.*)

MAISIE. Ach, Sam, you're not talking about old politics again.

SAM. Actually, Maisie, I was talking about how proud I am of our son – or is a father not allowed to show affection for his own flesh and blood?

MAISIE. Sharon, did you pick the wallpaper? It matches the tiles perfectly!

SHARON. I'm sure you are very proud, but what's pride compared to living every day in fear?

SAM. That's the name of the game, love!

SHARON. Excuse me, but with all due respect, you're sitting over there in East Belfast well out of it. You can switch off the news when you don't want to listen any more. Me and my family are in the front line. You can't switch off your brain like your television set. Davy may be doing his job and doin' it very well, but it's making him hard. Oh, he's different alright. He's a man with a gun and a target for more guns. So, no, Mr. McAllister, we are not on the pig's back, we're in a bloody nightmare only we can't wake up.

MAISIE. You look a wee bit peaky. Maybe you need a tonic, dear.

SHARON. Only yesterday your grandchild of five years of age was playing cowboys and indians with his daddy. Except it wasn't bows and arrows, it was a .45.

MAISIE. Tut tut! Our Davy has no sense.

SHARON. And when I screamed at Davy he said 'Sorry, love, it was just a game'. So what happens the next time our Gary wants to play some games? He might just want to pick up a loaded one and. . . .

SAM. Sharon, love . . . look (*Embarrassed.*) I didn't mean to upset you. Maisie, is that tea near brewed yet? (*Silence.*) Sure I'll go and bring it in.

MAISIE. Look, Sharon, take that wee tablet. (*Hands it to her.*) It'll calm you down. They're lovely curtains, love. Are they them Sanderson's or what? (SHARON *leaves.*)

SAM. Know what's wrong with that wee girl? She's too much time on her hands to worry.

MAISIE. She'll be alright. I gave her one of my wee green and red nerve tablets.

SAM. What age is her youngest wee buck now?

MAISIE. Coming five.
SAM. Aye, well, it's about time she had another one, that should give her something to think about. I'll speak to our Davy about it.
MAISIE. She should never have given up that wee cleaning job. She has too much time to think.
SAM. And too many bloody gadgets in thon kitchen.
MAISIE. Well now, I was admiring the wee tumbler dryer.
SAM. Aye, well, you can get it out of your head. What is wrong with God's good fresh bloody air . . . 'mon home, woman, and get my tea on. (*Exits.*)
MAISIE. Aye, I'll light an oul' fire in the garden and throw on a boar's head. (*Exits.*)
SHARON (*recites*). His da's a bloody comedian
Oh, that's the name of the game
But if his son gets blown to pieces
He'll soon find somebody to blame.
I'm tryin' dead hard to be normal
But our marriage has fallen apart
I'll try to be strong for his sake
But I can't hide the fear in my heart. (*Switches on radio.*)
SFX (*news jingle*). Belfast Councillor the Rev. Stanley Foster has made a statement condemning the policing of a DUP Rally in Dundonald last night. The Rev. Foster claims to have inside information that Loyalist gatherings are policed by a hand-picked squad of Roman Catholic RUC men bearing grudges. Trouble flared between police and mourners today in Co. Tyrone at the funeral of the two youths shot dead on Monday when their car refused to stop at an RUC checkpoint. The incident happened when police refused to allow the cortege through a Protestant housing estate. Local Sinn Fein Councillor Damien Heaney, who has already called for an enquiry into the shooting, was arrested when scuffles broke out. The funeral took place today of Reserve Constable Ivan Little, who was shot dead as he drove to work in his North Belfast butcher's shop last Tuesday. Constable Little leaves a wife and three small children. The Secretary of State, Mr. Tom King. . . .
SHARON (*switches off radio. Sings*).

Valium

Doctor, doctor, my old nerves is bad
He's on patrol again and I'm going mad
The kids are going to drive me round the bend

GOLD IN THE STREETS – ACT III

And I need them pills, I'm at my wit's end
Just write the prescription with your usual smile
And I promise you won't see me for a while
I know you're too busy to listen to me
But I won't bother you if you give them to me.
Give me a V
Give me an A
Give me an L-I-U-M
Oh valium
Put it in my mouth and then I count from one to ten
Oh valium
The world seems to fade away by the time that I reach five
And then it's almost like being alive
The kids are screamin' but they don't fuss me
I got a little bottle and my mind is free
When the bills come round and I forget to pay my debts
I don't give a damn when the pills take effect
And if I use them all and there is no more
Sure it's easy to borrow from Lily next door
If Lily runs out she doesn't really care
She can borrow from Jean next door to her
Give me a V
Give me an A
Give me an L-I-U-M
Oh valium
Put it in my mouth and count from one to ten
Oh valium
The world seems to fade away by the time I reach seven
Almost like being in heaven
Oh doctor, doctor, I don't mean to moan
I'm so afraid and I feel alone
I don't need to talk, I won't waste your time
'Cause nobody listens, these fears are all mine
God, I wish I could stop but I don't have the will
What use is your life when you depend on a pill?
Why didn't you tell me it's a dangerous game?
Now I'm a bloody junkie but who's to blame?
Give me a V
Give me an A
Give me an L-I-U-M
Oh valium
Put it in my mouth and then count from one to ten

Oh valium
The world seems to fade away by the time that I reach seven
And then it's almost like being in heaven.

SHARON (*reciting*). We're havin' Maureen and John for supper.
Now if only my Davy could see
Tho' he's no job and she works like a slave
I wish t'God it was me
(SHARON *sits.* MAUREEN, DAVY *and* JOHN *sit down as if coming from another room.*)
DAVY (*to* JOHN). So I says to the UDR fella, 'Aye, you'd be a bloody comedian if you hadn't a gun in your hand'. You know like, we're the professionals, John. Frig sake, they make Dad's Army look like the KGB.
MAUREEN. The supper was beautiful, Sharon.
DAVY. Then we get a call to this house, you know like, disturbing the peace after twelve. Flies round in the squad car, four of us. A lot of social workers having a hooley! Place coming down with drink. We goes in – dead embarrassing, you feel like an old killjoy – we left two hours later near full. Apart from the driver, like ... bloody geg ... we call it perks of the job.
SHARON. Get enough to eat, John?
JOHN. Aye, smashing, love.
DAVY. Aye, the old microwave is cracker – just bung in the stuff and Bob's your uncle.
SHARON. They're alright, Maureen. Sure the cooker does the same job.
JOHN (*looking round the wall*). What's this, Davy? The Encyclopedia Britannica?
DAVY. Aye, looks great, doesn't it? A thousand smackers....but I get it taken out of the bank every month. Sure you don't miss it and they last a lifetime.
SHARON. That's cause he never opens them.
JOHN. I wouldn't mind them, love.
MAUREEN. Aye, I'll get you two sets ... one for each wall!
DAVY. Anything you want to know, John, just phone me and I'll look it up ... Sharon love, they could be doing with a wee bit of a dust.
SHARON. Will yous have a wee drink for the road?
JOHN. No, Sharon ... we wanna catch the last bus.
DAVY. Yous are getting no last bus, I'll run you home.
SHARON. No, Davy, you've been drinking.
DAVY. It's alright. I'm a peeler. That wee turbo runs like a wee

dream, John. Do you wanna spin in it?
SHARON. No!
MAUREEN. No, we should be going. Let the babysitter away. (*Loud knock on door.* DAVY *gets up.*)
SHARON (*jumps up*). Davy, don't answer . . . don't go near it!
DAVY. Jesus' sake, love . . . see what I've to put up with. Who is it? (*Voices. 'Halloween is coming'.*) Away to hell – Halloween's not for another couple of weeks. (*Whispers.*) Knock it off, will you, love?
MAUREEN. We'll get the bus.
JOHN. Anyway, it's Maureen's treat the night, she's paying the fares. (*They get up.*)
DAVY. Now, I'll get you them forms if you like.
SHARON. What forms?
JOHN. The ah . . . football pools. It's a wee formula ah . . . Davy has. He's goin' to give me them.
MAUREEN. What forms has he asked you for?
JOHN. Nothin' . . . come on.
SHARON. Davy, I wish you'd mind your own business.
DAVY. What did I do?
MAUREEN. Davy, the only thing he is joining is the snooker club. (*Sound of car pulling up.*)
SHARON. Who's that? We're expecting nobody at this time.
DAVY. It's not for us.
SHARON. Ssh! (*She listens.*)
JOHN. We'll go on now, Davy.
SHARON. No, wait a minute. (DAVY *goes to window.*) Come you away from that window!
DAVY. Sharon, you're puttin' years on me! (*Goes to door off.*)
SHARON (*calls*). Davy!
DAVY (*voice off*). Alright, Stevie, brass monkey weather the night. (*Enters.*) It's the next door bloody neighbor!
MAUREEN. I'll give you a ring, Sharon.
JOHN. Come on, you.
DAVY. Hey, do you fancy takin' her with you? (MAUREEN *and* JOHN *leave.*)
DAVY. You're doing well, Sharon . . . ten out of ten for driving friends away . . . top of the class, love. (*He leaves.*)
ACTORS (*recite*). That's the name of the game, love.
  You're lookin' a wee bit peaky.
  Sharon, you are puttin' years on me.
  Your husband is keepin' this country from wreck and ruin.
  Ten out of ten for driving our friends away.

Now You're Talkin', 1985, (above left to right), Carol Scanlan, Marie Jones, Eleanor Methven, Aidan McCann, Ann Forsythe; below Marie Jones, Aidan McCann, Ann Forsythe, Carol Scanlan. Photos: Rod Tuach. (For Equity reasons Marie Jones used the name Sarah Jones as an actress. To avoid confusion her true name is used throughout the captions in this illustration section. In 1993 Carol Scanlan began using her maiden name Carol Moore, but as in all the Charabanc plays she performed under her married name, Scanlan has been retained.)

Gold in the Streets, 1986. Left to right, above: Marie Jones, Eleanor Methven, Carol Scanlan; below Eleanor Methven, Carol Scanlan, and Marie Jones.

Gold in the Streets. Above: Carol Scanlan, Eleanor Methven, Marie Jones, Rosena Brown; below: Marie Jones.

The Girls in the Big Picture, 1986 . Above: Ian McElhinney and Eleanor Methven; below, left to right: Carol Scanlan, Marie Jones and Eleanor Methven.

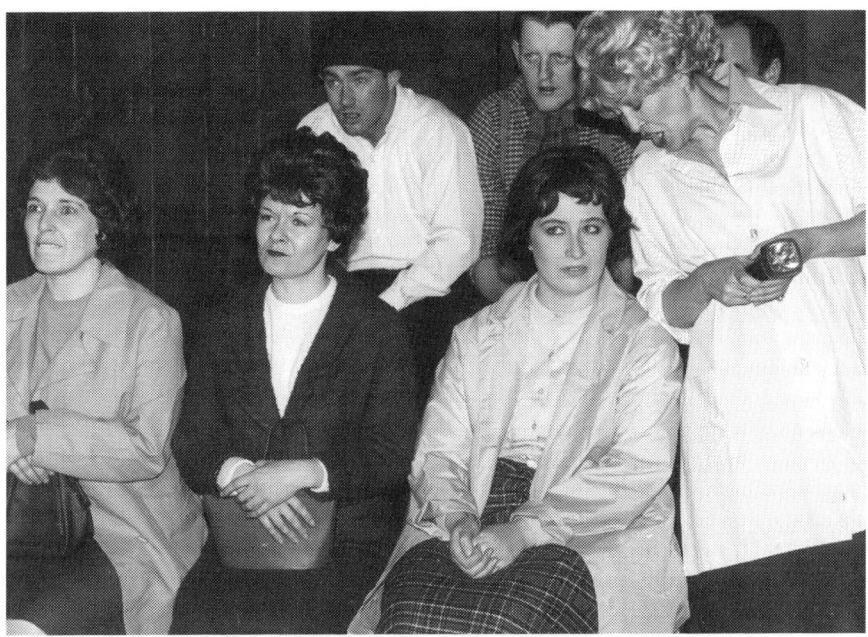

The Girls in the Big Picture. Above, left to right: Marie Jones, Carol Scanlan, Martin Maguire, Sean Kearns, Eleanor Methven, Rosena Brown, Toby Byrne (almost entirely obscured); below: Sarah Jones and Eleanor Methven.

Somewhere over the Balcony, 1987. Left to right: Carol Scanlan, Marie Jones, Eleanor Methven. Photo: Sheila Burnett. Below: Sean Kearns and Marie Jones in *Weddins Wee'ins & Wakes*, 1989. Photo: Henrietta Butler.

Lay up Your Ends, 1983. Left to right: Carol Scanlan, Marie Jones, Maureen Macauley, and Brenda Winter. Below: Carol Scanlan, Marie Jones, Aidan McCann, Brenda Winter, Eleanor Methven.

Above: *The House of Bernarda Alba*, 1993. Left to right: Emma Jordan, Eleanor Methven, Catherine Walsh, Niamh Linehan and Andrea Irvine. Below: Mark O'Regan (left) and Michael James Ford in *Me and My Friend*, 1991.

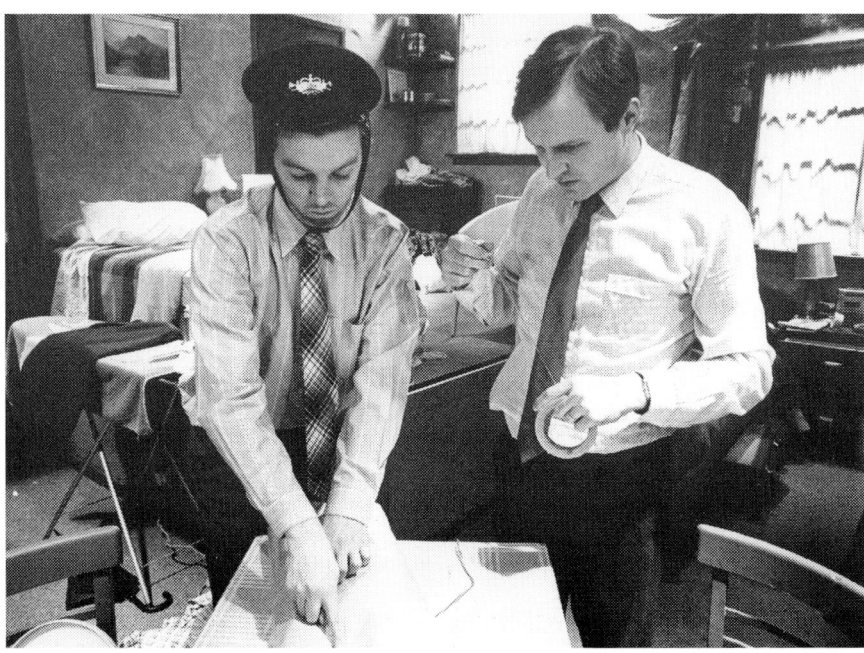

Probably MI5 or somethin'.
Somethin' wrong, dear?
It's just the way things are, Sharon, and you and me just have to accept that.
I'm sure that microwave's quare and handy for when Davy's on the lates.
Jesus Christ, would you rather I was back on the dole again?
Why don't you take one of them wee tablets love, they're good for your nerves.
That's the name of the game, love.
Sharon, you are puttin' years on me.
Somethin' wrong, dear? (*lines are repeated in a jumble until* SHARON *silences the voices by screaming.*)
SHARON. Shut up! (*Exits.*)

(*LX on* MAUREEN'S *house.* MAUREEN *answering door.*)
MAUREEN. Ach, Sharon.
SHARON (*entering*). Brought you this. (*Hands her vodka bottle.*)
MAUREEN. Don't be daft, Sharon. Sure my John's buying Bushmills cause I'm sick of vodka....you'd no call doing that.
SHARON. Just came up to say, cheerio.
MAUREEN. Oh? Holidays?
SHARON. I'm going to our Pat's in London.
MAUREEN. Ach, that's great, Sharon.
SHARON. For good.
MAUREEN. What!
SHARON. Can't stick it any more, Maureen . . . I've had enough. . . .
MAUREEN. I don't know, love, maybe you are as well. Like I give out to him all the time about being under my feet, but I'd rather that than not knowing if he was coming home at night.
SHARON. Davy's not going. Just me and the kids.
MAUREEN. Jesus! (*Produces two paper cups and pours vodka.*)
SHARON. He won't listen to me. I've pleaded with him but he says I'm just being selfish. Like living with a stranger. Even his old mates have no time for him. But the police is his whole life now . . . .they can do nothing wrong. If only he would stand outside of it, for a day and look what it is doing to his family. All I can say to him is 'Sorry I'm not strong enough'. I know I'm letting him down, but it's just me. I can't help it.
MAUREEN. Ach, Sharon love, maybe it will become just normal.
SHARON. But let's face up to it, Maureen, it's not normal. I have to send his shirts to the laundry. I can't advertise the fact that he's a

policeman by hangin' them on the line. He can't walk out the door in his uniform in case anybody sees him. So, no, it's not normal. Not me and Davy, our marriage, our children or this bloody country. And I'm tired of it! That's why I want to get away. Things will be better over there. I left him a letter beggin' him to reconsider and come with me and the kids. I hope he will, for all our sakes. Anyway, see you, Maureen.

MAUREEN. Aye, well, mind yourself!

SHARON. Oh, Davy's on lates. Would you call round to my house the morrow and tell him to collect his shirts from the laundry?

MAUREEN. Aye, well, cheer up. Whenever you get there maybe your Pat's man will get him a start at whatever he does.

SHARON. He was made redundant last year.

MAUREEN. Ah well, sure, you never know. (BOTH *exit.*)

ALL (*singing*).

>Strangers to our country
>Try hard to understand
>Why does an Irish man
>Fear his fellow Irish men?
>They weren't born with bigotry
>So why this bloody war?
>They'll get this simple answer –
>It's just the way things are.

CHORUS.
>So don't blame me, sister
>It's the way it's always been
>I mean no ill to no-one
>It's just the way it is
>I don't want no trouble
>I've never done no harm
>It's not discrimination
>It's just the way things are.
>Just keep your head down
>Deeper in the sand
>Never look for answers
>Even tho' they're close at hand
>Yes, there will be reason
>They might cause us shame
>Only when we find the answers
>We'll find out who's to blame.

BLACKOUT. END OF ACT THREE.

THE GIRLS IN THE BIG PICTURE

PLAY IN TWO ACTS

BY
MARIE JONES
DEVISED BY
THE COMPANY

CHARACTERS

JEAN MOORE – plump, plain, unmarried woman of thirty-two
MARY JO MCILFATRICK – a single woman in her early thirties, cheerful and open
MARGARET ANDERSON – friend of Mary Jo, thirty-five years old, runs a café, caustic

VIOLET TUCKEY – a spinster in her late sixties
EEDIE TUCKEY – Violet's sister, same age, spinster
ESTHER MOORE – widowed mother of Jean and Sidney
TOBIAS STINTON – forty-year-old suitor to Jean, newphew of the Tuckeys
SIDNEY MOORE – nineteen-year-old brother of Jean
PAT SHARKEY – wide boy wheeler-dealer from Belfast
PATSY FLOOD – gossipy cinema usherette, about thirty-five
THE MCKERNANS – village louts
MICKY MATTHEWS – village lout
SALLY SPENCE – seventeen-year-old girlfriend
WILLIE JOE FERGUSON – neighboring farmer to Moores, recently lost his mother
AMBROSE FITZCADDEN – uncouth farmer

ACT I

Scene One:	Moores' farm kitchen
Scene Two:	Back kitchen of Anderson's café
Scene Three:	Movie theatre
Scene Four:	Moores' farm kitchen
Scene Five:	Moores' farm kitchen
Scene Six:	Basket Tea auction

ACT II

Scene One:	Back kitchen of Anderson's café, two weeks later
Scene Two:	Moores' farm kitchen
Scene Three:	Back kitchen of Anderson's café
Scene Four:	Movie theatre, one year later

THE GIRLS IN THE BIG PICTURE
ACT I – SCENE I

SCENE: *LX up on Moores' farm kitchen. SFX of cows being herded in to evening milking.* VIOLET *and* EEDIE TUCKEY *are seated at kitchen table, teacups, etc. on a tray in front of them.* EEDIE *is counting buttons into a tin silently. Throughout the scene a clock ticks steadily.*

VIOLET. How many have ye?
EEDIE. Thirteen.
VIOLET. Throw one out!
EEDIE. I will not. I'll get another one and make it fourteen.
VIOLET. It's safer to get rid of one!
EEDIE. Better to find one.
VIOLET. And where are ye goin' to find one? (*Pause as* EEDIE *examines her layers of clothing.*)
EEDIE. There's one hangin' off my cardigan.
VIOLET. How many will that leave on ye? (EEDIE *counts buttons on her cardigan silently.*)
EEDIE. Thirteen! (VIOLET *triumphantly snatches a button from the tin and gives it back to* EEDIE. EEDIE *examines button.*) Oh, look see, Violet. That one was off a wee button boot of mine. I wore it when I met Edward Brolly in 1911. That was almost half a century ago.
VIOLET (*referring to the imaginary window*). Sure, you'd need eyes like a hawk to see in here with thon oul' tree!
EEDIE (*examining another button*). D'you mind that one, Violet?
VIOLET. I do! It's mine, and you've no call havin' it!
EEDIE. For sake of Jean and the competition!
VIOLET. Ach, I never heard the like! Seein' how many buttons ye can sew on a postcard for a competition!
EEDIE. D'you want Jean to lose it?
VIOLET. Hasn't leafed in two year!
EEDIE (*another button*). Don't mind that one. Don't mind that one at all. (JEAN *enters with bowl of eggs, dressed in working clothes and wellies which she proceeds to take off by the dresser where she*

THE GIRLS IN THE BIG PICTURE – ACT I SCENE I

leaves the eggs.)
JEAN. Evenin'. There's your eggs.
EEDIE. Five double yolkers last time, Jean.
JEAN. Sure don't I tell them who they're layin' for, and they lay them special! (TUCKEYS *chuckle at her little joke.*)
VIOLET. We have the buttons for ye, Jean. (JEAN *takes the buttons proffered.*)
EEDIE. That one's off a wee button boot of mine I wore when I met Edward Brolly. . . . D'you mind I told ye? The Lord took him at the Somme.
VIOLET. What concern is that of Jean's?
JEAN (*returning button*). That's very kindly of ye, Miss Tuckey. You too, Miss Tuckey. I've been muckin' out. Sidney was in a hurry, said he'd put away the hens. He's still at it out there. (*She goes to leave.*)
VIOLET. Has he thon rooster locked up?
JEAN. Oh, he can be wilful, that Rusty! You'll excuse me, for Tobias will be here for me soon. (*Exits.*)
VIOLET. She's a nice big girl.
EEDIE. Kindly.
VIOLET. Time our Tobias got a move on.
EEDIE. He's forty years of age.
VIOLET. And she's no spring chicken! (VIOLET *lifts a copy of a sixties film magazine from the table.*)
EEDIE. What's that?
VIOLET (*leafing through it*). It's about the Pictures.
EEDIE. Whose pictures?
VIOLET. Lordy days! There's a photie of Princess Margaret!
EEDIE. A bad egg!
VIOLET (*reads*). 'The fairy Princess of show business. She got with it as the beatnik would say'. . . .
EEDIE. The what?
VIOLET. Will ye let me finish! (*Reading on.*) 'And she was so keen on the wild rock music, that she took off her shoes and beat time in her stockinged feet' .
EEDIE. Makin' a holy show of her mother in her stockin' soles!
VIOLET (*putting down magazine*). I never took to her or the sister! (ESTHER *enters, also in wellingtons, carrying a small billy can. She too deposits it on the dresser and removes her boots.*)
ESTHER. There's your milk, and there's none better in the five townlands. (*SFX chickens squawking.*)
EEDIE. Sidney havin' bother?

ESTHER (*exiting to the kitchen*). Oh, they're wilful.
EEDIE (*whispering to* VIOLET). It's eleven year tomorrow.
VIOLET. He was a fine big man.
EEDIE. Big Silas Moore. (ESTHER *enters with a black coat and sits to darn it.*) Are you for the grave tomorrow, Esther?
ESTHER. If God spares me, Miss Tuckey. Did yous get enough tea?
TUCKEYS. Aye. (*SFX eight chimes of grandfather clock.*)
VIOLET (*when they've finished*). There's eight o'clock!
EEDIE. Our Tobias will be here for Jean soon; you could put the kettle on for him, he's that regular!
VIOLET. Esther, do ye mind we would've heard Silas windin' her up? D'ye mind that, Eedie?
EEDIE. I do, for he was that regular too.
VIOLET. In he'd come from the byre, and you'd have heard the rattle....
EEDIE. And the shufflin' of the big boots.
VIOLET. He'd have took them off by thon dresser, and he'd have said, 'All's well for another day'.
EEDIE. And then, 'Brave night, Miss Tuckey'.
VIOLET. 'And you too, Miss Tuckey'. A quare fine big man.
EEDIE. None better. (TOBIAS *enters.*)
TOBIAS. It was on the latch. Evening, Mrs. Moore, Aunt Eedie, Aunt Violet. Your brack.
VIOLET. Jean's changin', she was muckin' out.
TOBIAS. Aye, I seen Sidney runnin' about cursin' at the chickens.
EEDIE. Nessie Banner's gettin' worse, Tobias.
TOBIAS. Were you by the day?
VIOLET. We never miss these past few weeks.
EEDIE. I've been makin' her up a good feed of parsnip and bakin' soda for her stomach's leavin' her now.
VIOLET. Them Andersons'll be glad to see the tail end of her goin' out the door til they get their claws in!
EEDIE. That's two shops they'll have in Cloughmartin now, and they don't belong here.
VIOLET. Blow-ins!
TOBIAS. Are you for grave the morra, Mrs. Moore?
ESTHER. I am.
TOBIAS. I was that road the day. Reverend Bennett says he's havin' trouble with them McKernan boys, runnin' amok through it of a night. He caught the bigger one last night with some young cutty!
VIOLET. In the graveyard?
TOBIAS. Oh, we have a bad element, all right. Constable Baird's

keepin' an eye on them ... that Rio Grande Cafe...the carry on of them of a night, for I can watch them through the bedroom windy.

VIOLET. Oh, the town's not like it was.

TOBIAS. 'Tis not. Are you needin' anything off the van, Mrs. Moore?

ESTHER. I have what'll do me.

TOBIAS. Aye. . . . Can't please that oul' Willie Joe Ferguson since the mother died; says he, 'That's not what my mother bought'. Says I, 'It's the way you cook it'!

VIOLET. An oul' gurner!

EEDIE. Lookin' hisself a wife!

VIOLET. Who would have him?

EEDIE. He's gormless!

TOBIAS. And an oul' miser! (*Pause while* TOBIAS *and the* TUCKEYS *share their joke.*)

ESTHER. He has ninety acre! (SIDNEY *Enters in a hurry, pulling off boots by the dresser.*)

SIDNEY. That's the last time I'm puttin' them chickens away!

VIOLET. Have ye thon rooster locked up?

SIDNEY (*exiting to stairs off*). He's lucky he's not in Willie Masterson's butcher's shop!

EEDIE. Brave night for a walk tonight, Tobias.

VIOLET. You and Jean are right and fond of the walkin'.

EEDIE. You've covered a lot of loanins in four year!

VIOLET. Is it that long?

EEDIE. Aye.

VIOLET. My! My! (TOBIAS *moves to the window to change the subject quickly.*)

TOBIAS. I seed on the road up here the first digger has arrived for the electricity station. . . . Can't rightly see it from here.

VIOLET. Young Maguire sold the land for it.

EEDIE. Oul' James would never have done it.

VIOLET. Oul' James wouldn've died first.

TOBIAS. Oh now, I think it'll be a good thing, for as soon as the workmen move in, I'll be up there to see if I can get a regular run out of it.

EEDIE. The townlands would be lost without your wee van, Tobias.

VIOLET. They would. (JEAN *enters, changed, with a mac over her arm.*)

JEAN. There, that's me.

VIOLET. Aye, we'll head on too now, Esther. (JEAN *leaves down coat and gets eggs and milk. She hands these to the* TUCKEYS *and receives the button tin in exchange.*)

EEDIE. Full moon the night, the wee folk'll be out and about.
JEAN. Aye, the fairy thorn'll busy the night.
VIOLET. We'll take a wee juke as we pass . . . goodnight all!
ESTHER. Goodnight and God Bless. (*They exit.* TOBIAS *Helps* JEAN *on with coat. SFX hens squawking, turkeys yelling.*)
JEAN. Oh no, Sidney hasn't locked Rusty up! (JEAN *seizes wellingtons and runs out. Awkward silence*)
ESTHER (*continuing to sew*). She might be needin' a hand.
TOBIAS. Oh, that rooster'd pay no heed to me. . . . It's took an awful umbrage agin my aunts . . . funny that. Afore Rusty came, Carson was the same way with them. . . . Thompson Courtney's selling his mobile shop; I'm thinkin' of gettin' it.
ESTHER. What's he goin' to do?
TOBIAS. Openin' a Mace shop in Lisnacloughey. That wee van of mine's gettin' too small. I'm doin' more lines, so. . . . Mrs. Moore, it's been four years now that Jean and I. . . .
ESTHER. Your aunts might need a lift home. That Rusty upsets them.
TOBIAS. Aye. (*Exits.* SIDNEY *enters carrying a guitar – tries to exit behind his mother quickly so she won't see it.*)
SIDNEY. See you later, Mammy.
ESTHER. Sidney!
SIDNEY. Yes? (ESTHER *continues to sew without looking up.* SIDNEY *has to come down to her. He bumps the guitar while trying to place it gently against the dresser.*)
ESTHER. Where are you going with that?
SIDNEY. The Hop.
ESTHER. First time I've seen you takin' that.
SIDNEY. Just at the break, I'm goin' to sing.
ESTHER. Don't hear you singin' much in the Lord's House of a Sunday.
SIDNEY. They asked me.
ESTHER. What would your father have thought? (SIDNEY *doesn't answer.*) Them McKernan boys were in trouble with Constable Baird. He caught the bigger one in the graveyard. . . . Will the wee Spence cutty be there?
SIDNEY. We're only playin' records.
ESTHER. Are ye fond of her?
SIDNEY. Only see her once a week.
ESTHER. You're a good lad, son, and I know you would never do anything to shame me, or your father, but. . . . there are times when the Lord takes it upon himself to test our will, and our faith.
SIDNEY. Don't worry.

THE GIRLS IN THE BIG PICTURE – ACT I SCENE I

ESTHER. Promise me.
SIDNEY. I promise.
ESTHER. You promised me you'd fix the henhouse roof and what happened?
SIDNEY. I forgot. (JEAN *enters and deposits boots by the dresser.*)
JEAN. The Tuckeys are still a bit windy; Tobias offered to leave them home. (*Sits at the table*)
ESTHER (*dismissing* SIDNEY). Wind the clock up on your way out. (MARY JO *runs in.*)
MARY JO. Only me for my milk. Hello, Sidney, where are you goin' with that?
SIDNEY (*desperate to get out*). The Hop.
MARY JO. Are you playin' it?
SIDNEY. Maybe.
MARY JO. Gonna sing?
SIDNEY. Don't know.
MARY JO. What are ye gonna sing?
SIDNEY. Nothin'!
MARY JO. Must be somethin'!
SIDNEY. I made it up myself.
MARY JO. Let's hear it.
SIDNEY. I will not!
MARY JO. Mrs. Moore, have you heard it?
ESTHER. No!
SIDNEY. Goodnight! (*Exits.*)
MARY JO. God, he's gettin' that big! Me that used to change his nappy.
JEAN (*hands her the magazine*). Here y'are, Mary Jo, I've done with that. You can have it.
MARY JO. Oh smashin'. 'Picture Goer'. I can't stay long, Jean. Me daddy's barkin' for his milk. Do you know the latest with him? He's losin' his memory! Says I, 'You can mind rightly how to get to Eugene O'Connor's pub, for yer boots would go themselves they know it that well'!
ESTHER. He never was the same since your mother died.
MARY JO. Ach, I'm sick of that excuse! (*Reading*) 'Irene Papas will be sensational in *The Guns of Navarone* says Gregory Peck'. I hate oul' cowboys, all they ever do is shoot. We like the good gurney ones, don't we, Jean? I'll go on, for he'll be growlin'. (*She goes to leave, stops beside* ESTHER.) Doin' a wee bit of mendin'?
ESTHER. For grave the morra.
MARY JO. Oh, God, that's right! And here's me blabbin' on, too. Me

daddy sends his respects on account of it bein' the anniversary.
ESTHER. Thank him.
MARY JO. I can still see him standin' here as plain. . . . Your Sidney's the spit of him too. Anyway, goodnight, Jean, goodnight Mrs. Moore. (*Goes to leave again.*) Oh, Jean, did ye hear? Nessie Banner's pickin' at the wallpaper! Means her mind's goin' now, too.
ESTHER. It comes to us all.
MARY JO. Ach, well, she's eighty-five, she can't complain. Jean, God forgive me for wishin' death on anybody, but the sooner Nessie dies and Margaret gets her shop and I get runnin' that cafe fulltime the better!
JEAN. Your daddy's not goin' to like bein' left on his lone all day.
MARY JO. He's just goin' to have to lump it, for I'll pay no heed! Jean, you know them pigs is not enough to keep us, so he'll just have to thole it!
ESTHER. He is your father.
MARY JO. We have to live, Mrs. Moore. . . . Goodnight.
JEAN. Goodnight, Mary Jo. (TOBIAS *enters as she reaches the door.*)
MARY JO. Oh, Tobias, my daddy owes you money for tobacco, but don't worry, I'll give it to ye.
TOBIAS. Aye. (MARY JO *leaves.*)
ESTHER. I'm away to my bed.
JEAN. Mammy, are you not well?
MARY JO (*off*). Mrs. Moore, your clock's stopped!
ESTHER. It's all right, Sidney forgot. (*She gathers sewing and goes to exit to bed.*) No, I'm just a bit tired, that's all.
TOBIAS. Goodnight, Mrs. Moore, God Bless. (ESTHER *leaves, making no reply.*)
JEAN (*rising and buttoning mac.* TOBIAS *goes to chair*). Well, did you get them settled?
TOBIAS. Aye, they're away to make a cup of sweet tea with docken juice. Claim it's good for the nerves!
JEAN (*ready to go*). Well, are ye comin'?
TOBIAS. Sure there's no need to go out now your mother's away to bed.
JEAN. It's a lovely night. I wanted to go for a walk.
TOBIAS. Felt a few spits of rain.
JEAN. Aye . . . all right. (*Takes off mac.*)
TOBIAS. Your mother was very quiet tonight.
JEAN. It's just me daddy.
TOBIAS. Would she have never thought of maybe marryin' again?

THE GIRLS IN THE BIG PICTURE – ACT I SCENE I

JEAN. Are ye away in the head! Now there wouldn't be a man on this planet could touch my father in her book. Oh, he had his faults like anybody else, but you'd never think it to hear people talkin', 'Big Silas Moore', when you're dead, you become perfect, don't ye? (*Pause*) Did you bring my Merrymaids?

TOBIAS. Aye. (*He brings a quarter pound of sweets from his pocket and hands them to* JEAN.)

JEAN. Thank you. (*She takes one and puts the bag down. Sits.*)

TOBIAS. Let's see, there's ten in a quarter, a bag a week for four years, come next Tuesday, that's two thousand and eighty Merrymaids you've ate!

JEAN. I like them!

TOBIAS. Jean?

JEAN. Umm? (*Her mouth is full with the sweet.*)

TOBIAS. Thompson Courtney's sellin' his mobile shop. I'm thinkin' of gettin' it.

JEAN. Well, it's what you've always wanted.

TOBIAS. I'd be better able to provide for you. . . . Well, what do you say? (*Pause*)

JEAN. Well....I'll think about it, and I'll let you know soon.

TOBIAS. Aye. (*He too takes a sweet and starts unwrapping it.*)

JEAN. C'mon, you and me'll go down to the Hop!

TOBIAS. Us!

JEAN. Aye.

TOBIAS. I think them Merrymaids have gone to your head! (JEAN *smiles.* TOBIAS *puts a sweet in his mouth.*)

LX FADE

ACT I – SCENE II

SCENE: *The back kitchen of Anderson's Cafe.* MARY JO *dressed in overall looks through the serving hatch from off stage as* MARGARET *enters from the street.*

MARY JO. Sneakin' in the back door! What time's this?
MARGARET. About half one.
MARY JO. Why'd you not come in?
MARGARET. What for?
MARY JO. Margaret, look at the crack on thon plate. People's goin' to get the mange!
MARGARET. Tell them to eat somewhere else.
MARY JO. There's only been four people in the day for their dinner. (*Whispers.*) Nan Flood's cookin's gettin' worse.
MARGARET. Ach, don't annoy me.
MARY JO. Margaret, d'ye want to hear the news?
MARGARET. No, I want a cup of tea. (MARY JO *comes into back kitchen.*)
MARY JO (*shouts*). Nan, go you on home now, Margaret's here. Margaret, Nessie Banner's pickin' the wallpaper!
MARGAET. Oul' cow's supposed to be dyin' not doin' the place up!
MARY JO. No! Pickin', like this. (*Demonstrates.*) That means her mind is goin' now, too.
MARGARET. Who toul ye?
MARY JO. Everybody knows. Doctor Small was there until late.
MARGARET. Mary Jo! That's cheered me up no end! I think we should have a wee drink to celebrate! (*She gets a bottle of 'Babycham' and two glasses and pours them each a drink during the following dialogue.*)
MARY JO. Oh, Margaret, God forgive you!
MARGARET. 'Anderson's Fashions' and my own wee flat!
MARY JO. Margaret, d'you think I could ask your daddy, you know, when I take over here, if I could make changes? Home bakin', do the place up, make it a really nice place again.
MARGARET. Mary Jo, I don't care if you turn it into a zoo, as long as I get out of it! At last, my own wee shop. Hope she hurries up!

MARY JO. Margaret, that's awful!
MARGARET. Y'know, she's that carnaptious she'll hang on til the bitter.
MARY JO. Wouldn't you, if you were dyin'?
MARGARET. Well, here's to the future. Good health and long life, Mary Jo. (*They toast and drink.*)
MARY JO. Well, how did it go last night?
MARGARET. How did what go?
MARY JO. Brian, the teacher. Did he walk ye home?
MARGARET. Oh, Brian Thompson. I think he walked home with Pearl Pinkerton.
MARY JO. You didn't do what I told ye!
MARGARET. You mean throw myself at him! No, I did not!
MARY JO. I didn't. I said, 'Invite him back here for a cup of tea to explain something in the class ye didn't understand' .
MARGARET. Exactly. Throw myself at him!
MARY JO. But he made the first move.
MARGARET. He read out my essay on 'Banshees'; that is hardly the first move!
MARY JO. And then?
MARGARET. Nothin'.
MARY JO. And then he said so's the rest of the class couldn't hear, 'Wish they were all as smart as you, Margaret' . He used your first name!
MARGARET. What did you want him to call me? Clara Bow?
MARY JO. And then he said, 'What's your perfume? It's nice' .
MARGARET. So, he likes a bit of scent!
MARY JO. Margaret, why are you crabbin' at me? It was you who told me all this in the first place!
MARGARET. I'll keep my mouth shut in future!
MARY JO. What happened when the class was over? Did ye not get a chance to speak to him?
MARGARET. He asked me what way I was walkin' and I said. 'With one foot in front of the other'!
MARY JO. Y'see! Why could you not just have said 'Same way as you, Brian' ?
MARGARET. Just leave me alone. (SIDNEY *enters breathless and in working clothes.*)
SIDNEY. Hey, yous, did a fella come in here lookin' me?
MARY JO. No, not the day, Sidney.
SIDNEY. Are ye sure, now?
MARGARET. All your mates have been barred.

THE GIRLS IN THE BIG PICTURE – ACT I SCENE II

SIDNEY. He's not a mate, he's an older fella.
MARY JO. Well, who is he and what's he want with you?
SIDNEY. D'ye mind last night I went to the Hop? Well, this fella was there and heard me singin'. He says I should go in for the first heat of a talent competition that's goin' to be held in Ballymena.
MARGARET. What's a fella from Belfast doin' at a Young Farmers Hop?
SIDNEY. He was tryin' to flog Ernie Simpson a new record player for the hall cos his is done, but he heard me singin'!
MARGARET. Is he runnin' this contest?
SIDNEY. Naw. See all the ones that's goin' in for it has to do like a test before a panel. Well, I'm too late for that, so Pat, that's his name, he says he can pull strings.
MARY JO. Is he here sellin' record players?
SIDNEY. Says he's here buyin' up oul' furniture and things.
MARGARET. I've heard of them boyos.
MARY JO. I wonder would he be interested in my da? (SIDNEY *takes a bottle of coke and opens it.*)
MARGARET. He sounds like a fly customer to me . . . sixpence!
SIDNEY. Mary Jo's payin'.
MARY JO. Ach, Sidney, that's my only tip! (*Pays.*) Sidney, I don't think your mammy'd be too happy about all this.
SIDNEY. Swear you won't tell her, Mary Jo! (PAT SHARKEY *appears at hatch.*)
PAT. Where's the new Ricky Nelson, then?
SIDNEY. You came, Pat!
PAT. Would I let you down, kid? (*Leaves hatch and enters back kitchen.* SIDNEY *proffers hand.*) Phew, you're Abraham Lincoln, Sidney!
SIDNEY. I've been muck-spreadin'.
PAT. Hi, chicks!
MARGARET. This is a private area. Can I help you?
PAT. Milkshake, love.
MARY JO. We don't do milkshakes.
PAT. What? What is this? 'Ye Olde Tea Shoppe'?
MARGARET. The Rio Grande opens at six o'clock.
PAT. Right, I'll remember that.
MARY JO. Would ye like a coke?
PAT. Right, love. (MARY JO *scuttles over to get him a coke.*) Well, the deal's done, kid. You're on!
SIDNEY. Really! Aw, thanks, Pat!
PAT. Don't thank me, kid. You're the star!

SIDNEY. Do you still think I should sing 'Jenny'?
PAT. Aye, we'll give it a whirl, might need souped up a bit.
MARY JO. Is that the song you wrote yourself?
PAT. This your big sister?
SIDNEY. Naw, these are my sister's mates; this is Mary Jo, and that's Margaret. (MARGARET *is studiously ignoring him.*) Pat, I have to go. The tractor's blockin' Willie Masterson's.
PAT. Right. Two weeks from now, same time, same place, to get ye decked out, okay? See ya later, alligator?
SIDNEY. Aye . . . right ye be! (*Exits.*)
PAT. Livin' in the dark ages here. No milkshake machine!
MARGARET. No desire for one!
PAT. Y'have to keep up with the times, love.
MARGARET. Why?
PAT. If you wanna get on.
MARGARET. Oh? That's odd comin' from you. Sidney's been tellin' us you're buyin' up oul' furniture!
MARY JO. Are you buyin' up oul' stuff?
PAT. Buyin', sellin', whatever you dig!
MARY JO. We live, me and my da, just out the road, second loanin' on the left, Moore's lane. My da's got a right load of oul' stuff.
PAT. I might just try and get a run up to you, then.
MARGARET. Oh, you're honoured, Mary Jo!
PAT. This your mate?
MARY JO. Ach, that's just her way.
PAT. Anyway, about this milkshake machine. I know a fella that gets them cheap.
MARY JO. The customers that come in here wouldn't know a milkshake if it hit them over the head.
PAT. You don't get the young ones, then?
MARGARET. We have a very respectable clientele!
MARY JO. What brings you to Cloughmartin, of all places?
PAT. Just passin' through. Thought I'd give you a look over.
MARY JO. Where are you from?
PAT. Here and there.
MARY JO. No, really?
PAT. Really!
MARGARET. Well, we're closed now. Mary Jo, it's two o'clock. You're finished.
MARY JO (*exiting to get coat*). Oh, right.
PAT. So. This is the hotspot, then? (MARGARET *begins closing up the cafe, ignoring him.*) What d'ye call an Ulster woman with a frog

THE GIRLS IN THE BIG PICTURE – ACT I SCENE II

on her head?
MARGARET. Stupid!
PAT. Lily! (*No response as* MARGARET *stands beside the door waiting for him to leave.*) Aye, well . . . I'll head on then.
MARGARET (*as he opens door*). If anybody needs any advice on what to charge for their old furniture, tell them to call in. I know a bit about antiques.
PAT (*chucking her under the chin and exiting*). See ya!
MARY JO (*entering having pased him*). Cheerio!
PAT (*off*). Bye, love!
MARY JO. He was nice, wasn't he?
MARGARET. Are you half wise?
MARY JO. Well, I thought he was.
MARGARET. He's probably married with a squad of youngsters and ones he doesn't know he has.
MARY JO. Ach, d'ye think so?
MARGARET. Rob ye blind as quick as look at ye.
MARY JO. Ach now, you don't know that.
MARGARET. Well, think what you like, I'm only givin' my opinion.
MARY JO. Flip sake, I thought he was nice too! Anyway, I'll see ye the night. (*Goes to exit.*)
MARGARET. What's on?
MARY JO. Don't know. I never looked to see.
MARGARET. If it's cowboys I'm not goin'.
MARY JO. Hope it's a good gurney one. (*Goes to exit again.*)
MARGARET. Are you goin' into the Chapel?
MARY JO. Sure, I always go on my way home.
MARGARET. Say a wee prayer for Nessie Banner.

LX BLACKOUT

ACT I – SCENE III

SCENE: *Under the film soundtrack three chairs are set up downstage for* MARGARET, MARY JO *and* JEAN. *The* MCKERNAN BOYS *and* MICKY MATTHEWS *are seated behind them on the table to give the appearance of a cinema rise.* PATSY FLOOD *enters with usherette's tray and torch. During dialogue the film soundtrack fades down and comes to prominence again in the pauses. The full soundtrack used is printed at the end of the script. Film SFX.*

> JOHNNY. No use denying it, Angela. Malcolm saw you. You were in his arms, dancing in the Blue Lagoon.
> ANGELA. No, Johnny, you don't understand!
> JOHNNY. My God, it's a double betrayal, you and Ronald. Hah! He was going to be our best man. Good God, he belongs to my Club! (*Film sound down.*)

PATSY. Less of that carry on in the back row there. Are you not a wee bit young for that? I'll tell your mother. Hello, Derek. Hello Valerie. Oh, Derek, mammy says since she got the cat done it's been great. (*She comes down to sit beside* JEAN.) Isn't Stewart Travers a hunk? Here yer woman Grogan's up there with her ma. She came round to our house the other night, yellin' and screamin' all over the street! From Belfast, wouldn't ye know! She accused our Jason of hittin' her wee Collette, so I just says 'Wash yer mouth out with soap and water.' Here, whisper to Mary Jo, for she loves our wee Jason. (JEAN, *who has been trying to watch the film throughout this diatribe, has to lean across* MARGARET *to tap* MARY JO.)
JEAN. Patsy says. . . .
MARY JO. Aye, tell her that's shockin'!
JEAN. She says, 'That's shockin''. (*Film SFX.*)

> ANGELA. Johnny, why are you looking at me like that? What are you going to do? Johnny, answer me! (*Pause with tense music and dog barking. Film sound down. SFX slap as the three turn to* PATSY *and miss it.*)

THE GIRLS IN THE BIG PICTURE – ACT I SCENE III

MCKERNAN ONE (*as they rain popcorn down on the girls*). Go ye, girl, ye!

PATSY. Quit it, yousins! I told Jamesie Bickerstaff not to let them in. That's them McKernan boys, they can't get a wee girl so they take it out on everybody else. Oh, the wee dog gets killed here. There's a wee job for you, Derek! (*Film SFX.*)

> DRIVER. Cor, strike a light! She came at me so sudden like!
> POLICE CONSTABLE. Stand back! Stand back!
> JOHNNY. Oh my God, Angela. I've killed her!
> WOMAN. Looks dead enough to me, all right. Nasty thump!
> DRIVER. Wasn't my fault, guv'nor. She came at me out of nowhere. This woman's my witness!
> WOMAN. I didn't see nuffink, mate! (*SFX film sound down.*)

PATSY (*shines her torch in* MARY JO'S *face*). Here, the other day this boyo was round at my ma's lookin' to buy up her oul' ornaments. Quare good looker too! Says I, 'Are you married'? Says he, 'No'. Says I, 'Well go down to Anderson's Cafe for I knows someone there would do ye'. So there ye are now!

MARGARET. I'm trying to watch this, Patsy Flood!

PATSY. I wasn't speakin' to you, Margaret Anderson! (*To* JEAN.) Oul' sour gub! Any wonder she never got herself a man. Here, what about you and Tobias? It's been four years now, hasn't it? He'll get that oul' Tuckey house when they go, won't he? Ye have to pity them bein' on their own, don't ye? Still, they've got each other.

MCKERNAN TWO. C'mon up here and we'll feel yer diddies!

PATSY. Yous are warned! (MARGARET *rises and turns to reply.* MARY JO *pulls her down.*)

MARY JO. Don't! You'll only start them off again.

PATSY. They're bad boyos!

MATTHEWS. Hey, girls, see the night. Don't forget to put the candles out! (*They make a popping noise with fingers in cheeks.*)

PATSY. Ach, who rears them? (*Film SFX.*)

> JOHNNY. I say, Doctor, can't you give me any news? I've been here all night and most of the morning.
> DOCTOR. Ah, yes, Mr. Carlton. Are you some relation? She asked for you when she regained consciousness.
> JOHNNY. Did she? Did she really? Yes, well, I was her fiance.
> DOCTOR. Was?

THE GIRLS IN THE BIG PICTURE – ACT I SCENE III

JOHNNY. Oh, Doctor, you've got to tell me the truth. Is she going to be all right?
DOCTOR. Well, she's in a coma, Carlton. Maybe you're just the ace we can play. Perhaps if she were to hear your voice. (*SFX film sound down.*)

MCKERNAN ONE. Hey, frosty bake! C'mon up here and we'll warm ye up!
PATSY. Last chance!
MCKERNAN TWO. We're not that desperate!
PATSY. Ach, they're just bored.
MATTHEWS. Hey, big arse!
PATSY. Ignore them, Jean.
MATTHEWS. Loan us yer knickers. We're goin' campin'!
MARGARET (*on her feet, facing them*). Micky Matthews! I am going to speak to your mother in the morning!
MATTHEWS. Hey, Margaret! D'ye want the cobwebs blown off yer fanny? (*The lads dissolve into whoops and giggles.* MARGARET *sits down.*)
PATSY (*goes up to them*). Right, yous, that's enough. You never leave them poor women alone! Yous have had yer orders – now, out!
LADS. Ssh, Patsy!
PATSY. I said get out of here! Who do yous think yous are coddin'? Out!
LADS. Patsy! We're trying to watch the film!
PATSY. If you don't get out I'm goin' to get Jamesie for you! (*General protestations of innocence from the lads.*) Jamesie! (*She signals to him by shining her torch in the direction of the projectionist, then at the lads, then to the exit. the lads scramble out, followed by* PATSY.) And don't come back! (*A piece of popcorn hits her thrown from offstage. Exiting.*) Ye cheeky monkey! (*Film SFX.*)

ANGELA. Oh, Johnny, your mother's ring! Put it on my finger, darling.
JOHNNY. Angela! You can move your hand! Oh, my precious darling. Hold my hand through all eternity! Angela – well-named, my Angel!
ANGELA. I must be an Angel. I think I'm in Heaven!
JOHNNY. Darling girl!

PATSY (*enters as 'God Save the Queen' begins*). Right, I'll fill ye in on what ye missed! (*The three friends gather up scarves, bags, etc. and*

THE GIRLS IN THE BIG PICTURE – ACT I SCENE III

step downstage and, therefore, outside as the furniture is cleared upstage.)
MARGARET. Well, that was just great, wasn't it?
MARY JO. Margaret, just you pay no heed to them.
MARGARET. Did you enjoy it, Jean?
JEAN *(looking off)*. We'll walk you down as far as the corner.
MARGARET. Why?
JEAN. The McKernans and their mates. They're all gathered there.
MARY JO. Is your Sidney not there?
JEAN. No, he is not! He's at a practice.
MARGARET *(moving off)*. Goodnight!
JEAN *(following with* MARY JO*)*. Aye, we'll walk with ye.
MARGARET. No! It's all right! I'll manage on my own.
MARY JO. But look at them. They're like a pack of wolves!
MARGARET. They don't bother me! *(She is thoroughly irritated now.)*
MARY JO. I'll see ye in work tomorrow, then?
MARGARET. Oh, I just can't wait! I can smell the grease now! See yis!
   *(*MARGARET *exits. They watch her progress off.)*
JEAN. There, she's past them.
MARY JO. Y'know, Jean, she asks for it.
MATTHEWS *(off)*. Hey 'sexy', j'ya buck? *(Whoops and wolf whistles.)*
JEAN. Nobody deserves that!
MARY JO. When they come into the Café, I'm as nice as ninepence, but she shouts at them and that eggs them on. *(They link arms and walk round to the front stage. The scene is now the country outside Cloughmartin on a customary walk home. SFX river.* JEAN *and* MARY JO *stop and look down upon the town in the valley below.)*
MARY JO. Now let's see who's up and who's in bed.
JEAN. Phyllis Pinkerton'll be glued to her window, watching all the young ones comin' out of the Hop. Seein' who's with who.
MARY JO. Margaret's upstairs light's on. She'll be reading her legends.
JEAN. She's awful crusty this weather.
MARY JO. The Cafe's drivin' her mad, and she can't wait to get out of it! Since her mammy and daddy retired, she just feels like a lodger.
JEAN. D'ye mind? When we were wee we used to sit here.
MARY JO. That's right! I was goin' to open up an ice cream shop and have three sets of twins.
JEAN. And I was either goin' to be Vera Lynn or be a missionary and convert Africa! What happened to the ice cream shop?
MARY JO. What happened the twins?
JEAN. I wonder where you'd be now if you hadn't have had to come back?

THE GIRLS IN THE BIG PICTURE – ACT I SCENE III

MARY JO. A Nursing Sister maybe.
JEAN. Married?
MARY JO. Frig my da!
JEAN. Mrs. Hagan?
MARY JO. Oh, Jean, he was gorgeous!
JEAN. What do you think of Mrs. Stinton?
MARY JO. Has Tobias asked you to marry him?
JEAN. I said I'd let him know soon.
MARY JO. D'ye want to?
JEAN. I don't know. I'm just sick of people askin' 'When's the big day'.
MARY JO. I'm sick of people askin' 'Is there ever goin' to be a big day'.
JEAN. It's worse when they stop askin' – like Margaret.
MARY JO. The young ones call her 'Miss Anderson' now. Jean, what is your mammy goin' to say?
JEAN. She doesn't know. Don't be lettin' on, Mary Jo. I just need to be left alone to make up my own mind.
MARY JO. D'ye love him?
JEAN. I've never loved anybody.
MARY JO. Well, I wouldn't hang around here waitin' for it.
JEAN. Suppose I'll have to go someday.
MARY JO. When Sidney gets married.
JEAN. Did you love Arnold?
MARY JO. I still get a wee flutter when I think about him.
JEAN. See ya at the W.I. (*She moves off.*)
MARY JO. Aye, goodnight, Jean. Might be somebody at the Basket Tea!
JEAN. Ye never know!
MARY JO (*moving off opposite direction*). There never is!
JEAN. There's Willie Joe Ferguson!
MARY JO. He's got ninety acre!
JEAN. Mammy's boy!
MARY JO. He's a big drip!
JEAN. He's a big nose!
MARY JO. He's a big wallet!
JEAN. What?
MARY JO. Wallet! Jean Moore!

LX BLACKOUT

ACT I – SCENE IV

SCENE: *Moores' farm. Sunday evening.* JEAN *is dozing by the table.* SIDNEY *has his coat on and is restless.* ESTHER *can be heard drying cutlery in the kitchen. SFX clock ticking.* ESTHER *can be heard humming 'Abide With Me'.*

SIDNEY (*in a half whisper*). Jean. . . . Jean! (*He goes forward and shakes her.*)
JEAN (*waking*). What?
SIDNEY. What am I goin' to say?
JEAN. I toul ye earlier on, I don't know!
SIDNEY. Aw, thanks. I don't know what I'd do without ye!
ESTHER (*off*). Grand sermon the night!
SIDNEY. Aye!
JEAN. What did ye make a date with her on a Sunday for? You know what mammy's like!
SIDNEY. I hate Sundays! I hate, hate, hate them!
JEAN. It's no pantomime for me either, Sidney.
SIDNEY. You're just goin' to be sittin' there, anyway.
JEAN. Well, why don't you tell her where you're goin' and be done with it!
SIDNEY. She hates Sally Spence! She hates the whole family!
JEAN. She's worried for ye, Sidney.
SIDNEY. Ach, you're just like her!
JEAN. Sidney! Who's been tellin' lies for ye this last lot of months when ye forget to do somethin'? (*SFX clock strikes eight.*)
SIDNEY. I said I'd meet Sally at the bottom of our loanin' at a quarter to. There's eight o'clock. (ESTHER *enters.*)
ESTHER. Time to wind the clock. (*She goes off.*)
SIDNEY. Mammy, Stanley Semple's sick. I said I'd take a run down with a couple of books for him.
ESTHER (*off*). How sick is sick?
SIDNEY. Brave and sick. Bokin' and that.
ESTHER. Well, what happened him?
SIDNEY. Dunno. A bug, I think.
ESTHER (*re-entering and crossing back into kitchen*). Aw, well, it's

best not to go near him, son, for fear it's smittal.
SIDNEY. She can't have Stanley either!
JEAN. You do run about with some wild boyos.
SIDNEY. Ach, anybody that enjoys themself is a wild bad boyo! Jean, d'ye see if I get out of here, will you tell her about the talent competition?
JEAN. You tell her!
SIDNEY. No, you're good with her. You could put it much better than I could.
JEAN. She's likely heard.
SIDNEY. D'ye think so? (ESTHER *comes in with book and sits down.*)
ESTHER. Well, that's all cleared up. Are ye not sittin' down, son?
SIDNEY. Aye. (*He does so.*)
ESTHER. And what are ye doin' with your jacket on ye in the house?
SIDNEY (*taking it off resignedly*). I was out checkin' on the heifer.
ESTHER. Aye, keep a close eye on her, son. She'd trouble the last time. Will yous be able to deliver without me?
JEAN. Why?
ESTHER. Well, the dear knows I'm gettin' no younger; can't manage like I used to.
SIDNEY. Is there somethin' wrong with you, mammy?
ESTHER. Just said I was gettin' no younger, that's all. There will come a time when I won't be able no more, and Jean won't always be with us.
SIDNEY. Why, where are you goin'? Jean! Are you goin' to marry Tobias?
JEAN. Naw.
ESTHER. You've turned into a fine young man, son. Your father would be proud of you if he were here now. I'm thinkin' of increasing the herd, gettin' a few fresians.
SIDNEY. Fresians!
ESTHER. Thinkin' on it.
SIDNEY. But you always said you wouldn't have them about; ye said they were only half a beast, you wouldn't wash the floor with their milk!
ESTHER. Well, I've changed my mind, haven't I? Willie Joe Ferguson's goin' to advise us; says there's a grand lot of yearlings comin' up next Derry market. (SALLY *spence bursts in.*)
SALLY. It was on the latch. Sidney, you comin' or not? Hello, Mrs. Moore. Awful dark in here, Mrs. Moore. Are ye for gettin' the electric? Sure that'll be great for yous. Quarter to eight, Sidney, is a quarter to eight, not five past! Look at the state of my feet

comin' up that yard! Flippin' oul' cows! (SIDNEY *goes to her, trying to shut her up.* ESTHER *goes back to her book.*)
SIDNEY. Sally, I can't make it tonight.
SALLY. What? You keep me standin' twenty minutes, and you've the cheek to say you can't make it since you're doin' nothin!
SIDNEY. I'm busy.
SALLY. Busy doin' what?
SIDNEY. Sally, please. I'll see you on Tuesday night.
SALLY. Oh, you think so? (*She goes to leave.* SIDNEY *grabs her and talks in an urgent whisper.*)
SIDNEY. Give me five minutes, five minutes!
SALLY. One minute! (*Exits.*)
SIDNEY. Sorry about that! (*He sits down again.*)
ESTHER. He says there's a grand lot of yearlings comin' up next Derry market day. And he's prepared to lease us the field that borders our top one, for next to nothin'.
SIDNEY. How did you manage that? Sure oul' Willie Joe wouldn't give ye the time of day!
ESTHER. Aw now, son, there's folk round here would see you doin' well for sake of your father.
SIDNEY. Jean, did you know about all this? (JEAN *shakes her head.*)
ESTHER. Mind you thon top field'll need fencin' off; there won't be much time for gaddin' about!
SALLY (*off*). Sidney. . . . Sidney Moore!
SIDNEY. I'll not be late, mammy. (*He puts on his jacket and sheepishly leaves.*)
ESTHER. If you'd have been a son, I wouldn't have had the worry of him.
JEAN. He's young. He's pleased. He's always wanted to increase the herd.
ESTHER. Only want what's best for yous.
JEAN. He loves this farm. He'll be all right.
ESTHER. Willie Joe Ferguson has expressed interest in you.
JEAN. I don't like him!
ESTHER. D'ye know him? Willie Joe could provide for ye.

LX FADE

ACT I – SCENE V

SCENE: *SFX clock continues over scene change. Lights up on* MARGARET *sitting in a shawl in Moores' kitchen. An old gramophone is playing 'You'll Never Miss a Mother's Love Til She's Buried Beneath the Clay'.*

MARGARET. Go! (JEAN *enters in man's jacket and cap with a clay pipe.*)
JEAN. Howdy, sis. It has been many years!
MARGARET. It has been forty long ones, Robert. Will ye have a sup with me? (*Proffers one of two enamel mugs.*)
JEAN (*taking mug*). Was mother's death slow?
MARGARET. When death comes it's as quick as the beat of a heart!
JEAN. Poor, poor mother!
MARGARET. How is Americkay?
JEAN. I miss the old country.
MARGARET. You never wrote. You never answered our plea for money to help live!
JEAN. I thought of you both.
MARGARET. We were almost starvin'; we are very poor; you are a wealthy man!
JEAN. True, true. I made my fortune.
MARGARET. And kept it all to yourself!
JEAN. If I'd given it away, I would have had no fortune!
MARGARET. You never married?
JEAN. I didn't.
MARGARET. We are your only kin. I mean, I am your only kin!
JEAN. That's true, Mary.
MARGARET. And you only came back to sell the house and the possessions since you are first born.
JEAN. They are rightfully mine since mother is dead!
MARGARET. I too am not wed! Where will I go?
JEAN. Get work, you're not too old! (*Aside.*) Ah, this is a fine house. I will make many dollars to bring back to Americkay!
MARGARET. Strange to think, if you had died mother would have inherited your fortune!

JEAN. But alas, she is dead. Poor, poor mother. Pity I had not seen her in forty year! (*Drinks.* MARY JO *dressed in shawl comes out of upstage entrance.*)
MARY JO. I am not dead, Robert!
JEAN. But the wire to come home from Americkay?
MARY JO. You would put your sister on the street for greed!
JEAN. I am the rightful owner of this house! Aaagh! (*She clutches her throat.*) Poisoned. I've been poisoned! (*Falls to the floor dead.*)
MARY JO. And now Mary and me are the rightful owners of all your fortune. You see, Mary, he was repaid for his greed!
MARGARET. And then we get up and bow, right? (*They do so and take off costumes, turn off record, and tidy away props during the following dialogue.*)
JEAN. That was great, Margaret!
MARY JO. Betty Bothwell is determined to win; she's got nine in her cast!
MARGARET. Size of cast is no guarantee of literary ability!
JEAN. Right. I'll make the tea. (*She exits upstage to kitchen.*)
MARY JO. I think I'll write one, about an oul' lad who wouldn't give his daughter an earthly; and she poisons him! You should've heard him when I was doin' the 'Basket Tea' poster; 'makin' a holy show of yourself, lookin' for a man at your age'. Are you hopin' Brian'll be at the Basket Tea?
MARGARET. Brian who?
MARY JO. You said the day!
MARGARET. I said I wonder will he be.
MARY JO. Sure that's the same thing!
MARGARET. Wonder does not mean hope. It means your imagination is asking a question.
MARY JO. Why would your imagination want to ask a question in the first place?
MARGARET. All right then. You said to me today, 'I wonder will Willie Joe Ferguson be there, seein' as how he's on the market'? Does that mean you're hopin' he will?
MARY JO. Ach, away and chase yerself!
MARGARET. Point made!
MARY JO. Jean! (JEAN *enters with tea tray.*) I'm goin' to fight you for Willie Joe Ferguson!
JEAN (*putting down tray and pouring tea*). Don't remind me! She hasn't mentioned him since; I think she must've caught herself on.
MARGARET. Oh, now, Jean, you could do worse!
MARY JO. Tobias is Elvis Presley compared with him! Well, you know

what I mean.
JEAN. Elvis! He's Errol Flynn!
MARGARET. Oh! You kept that a secret!
MARY JO. Jean Moore!
JEAN. Quit it, yousins! Here, Margaret, maybe Brian'll be at the Basket Tea.
MARGARET. Flip sake! Now you're at it! See when I get my own wee flat, I won't care if Yul Brynner himsel' comes knockin' at the door!
JEAN. How are yous goin' to get rid of Nan Flood? Have yous thought?
MARGARET. Well, she's droppin' plenty of hints! – 'Nan, you shouldn't be standin' all day with your legs. Nan, I knew a woman who took clots with standin' too long, and she died with thrombosis'!
MARY JO. Aye, well, it's not funny! It's me's goin' to have to do it! (PAT'S *and* SIDNEY'S *voices are heard off.*)
MARGARET. Who's that?
JEAN. It might be our Sidney; he's away to meet thon fella about the singin'.
MARY JO. Oh? Ach, he's a good laugh, him.
MARGARET. He's a creep!
JEAN. He has our Sidney's head turned! (PAT *enters.*)
PAT. Well, look who it is! The 'Girls in the Big Picture'! Ladies, I present for you, all the way from Cloughmartin, Si Stevens! (SIDNEY *walks in and puts guitar against dresser.*)
MARY JO. Ach, it's only Sidney!
SIDNEY. His head's away. That's our Jean, Pat.
PAT. Hello, love.
JEAN. Hello. (PAT *takes cider bottle from his coat.*)
PAT. Want a slug, girls?
MARY JO. What is it?
JEAN. Is it drink?
PAT. Naw, it's cider. (MARY JO *takes a drink.*)
MARY JO. Oh, Jean, you should have some!
JEAN. No, mammy'd smell it!
PAT. Well, sis, what d'you think of him now? Through to the semi-final!
JEAN. Great. It's his mother.
PAT. That's why I'm here. Don't you worry; I'll sort it out!
SIDNEY. Pat said he'd speak to mammy!
MARGARET. When are we goin' to hear this masterpiece, then?

MARY JO. God, Sidney! I'd love to hear it!
SIDNEY. Naw. Sure my ma'll be back any minute!
PAT. Go on, kid! Give them what you gave the birds last night!
SIDNEY. Naw. I'm too embarrassed!
PAT. What? You had two hundred people eatin' outta your hands! C'mon, kid, don't let me down! Got full marks from all three judges!
MARGARET. Do as your manager says, Sidney!
SIDNEY. Aye, well, all right, I'll do a wee bit for ye. (*He takes up guitar and begins introduction.*)
PAT. Take it away, Si!
MARY JO. 'Si Stevens'. Your ma will lay an egg!
SIDNEY (*sings*).

>Jenny is with me every night
>She holds me when I sleep
>Her lips are soft and gentle
>She comforts me when I weep
>And when I see her lovin' smile
>To tell me I'm a man
>I wake up in my lonely room
>And my darlin' she has gone.
>
>Oh, oh, oh, Jenny,
>Why aren't you real?
>Oh, oh, oh, Jenny,
>Why can't I feel
>Your soft white skin,
>Your golden hair,
>Your lovin' eyes so full of care?
>But, Jenny, you're not there.
>But, Jenny, oh you're not there.
>But, Jenny, oh you're not there.

(*All applaud.*)
MARY JO. Marvelous, Sidney!
PAT. Signed autographs, two bob each! Just gettin' into practice.
SIDNEY. I've only won a heat. The semi-final's goin' to be real hard!
PAT. Look, I toul ye, I got big hopes for ye, kid!
MARY JO. Sidney, you are marvelous. You're goin' to walk it!
PAT. Hey, love. I can get a milkshake machine.
MARGARET. Isn't that nice for ye?

PAT. Good salesman never gives up!
SIDNEY (*leaving guitar down*). How's the heifer, Jean? Did ye check?
JEAN. Aye, take another look at her before she comes back, for she'll ask.
SIDNEY (*leaving*). Aye, she'll drop soon!
MARY JO. How's the furniture buyin' business goin'?
JEAN. Excuse me, I'll just go and have a wee word with our Sidney. (JEAN *exits.*)
PAT. I could maybe lay me hands on a juke-box for ye.
MARGARET. Funny, that! I went to bed last night and I prayed to God that a man would come into my room and say, 'Margaret, I can get a juke-box for ye'!
PAT. A whipped ice-cream maker?
MARGARET. That was the other dream!
MARY JO. Margaret won't be runnin' the cafe. She's leavin'. I will!
PAT. Emigratin'?
MARGARET. No, more's the pity!
MARY JO. This woman, Nessie Banner's dyin'. Margaret's goin' to get her shop.
PAT. Another cafe?
MARGARET. Fashions.
PAT. I'll get you a load of clothes from a warehouse; I know a fella.
MARGARET. For people with expensive tastes!
PAT. Here?
MARGARET. Yes!
PAT. This girls' night out, then?
MARY JO. Ach, we just swop houses every now and again!
PAT. Where's the men then?
MARY JO. Oh, now!
MARGARET. Further away the better!
MARY JO (*goes to where* PAT *is examining the dresser*). Pat, have you ever been to a Basket Tea?
PAT. A what?
MARY JO. We have them here every three months. You see, the women all make up big baskets of food, right? And Jamesie Bickerstaff, he's the auctioneer, he auctions them, and the men all bid for them. Whoever basket you buy you stay with that person all night and you dance.
PAT. When's this?
MARY JO. Next Friday. Will you be here?
PAT. Maybe, maybe not.
MARY JO. Aw, you goin' home?

THE GIRLS IN THE BIG PICTURE – ACT I SCENE V

PAT. So, who'll be at this basket hooley, then?

MARY JO. We wouldn't miss it! Would you like to come?

PAT. Well, Mary Jo, if the basket I buy doesn't have a nice fresh apple tart in it, there'll be trouble. (MARY JO *giggles and sits down again.* PAT *lifts an oil lamp from the dresser.*) Called into two oul'dolls the day, about half a mile up that lane. You wanna see the stuff. They don't know what they have!

MARY JO. That's the Tuckey sisters; they've been here hundreds of years!

PAT. They looked it! Y'know when the electric comes into these parts, these things are goin' to go like hot cakes for ornaments. (JEAN *and* SIDNEY *come in.*)

JEAN. Excuse me, Mr. Sharkey, but you don't know my mother, and I think that bringing up this business about Sidney will only make things worse!

SIDNEY. Jean, leave off!

PAT. Give the kid a break. You were young once, love!

JEAN. You want to increase the herd? Well, you can't do that til you fence off that top field!

SIDNEY. I'll do it!

JEAN. When? You never leave thon thing down! (*Indicating the guitar.*)

SIDNEY. Jean, it's my life!

MARY JO. Jean, he should. This farm is going to be here when he's not!

PAT. Houl on a second! If fencing off a couple of fields is goin' to stand in the way of Sid's future, I'll get a couple of mates to do it!

JEAN. You don't understand! Once my mother decides on something, nothin'll shift her! (ESTHER'*s voice is heard off.* JEAN *grabs lamp and replaces it on dresser.* MARGARET *quickly hides bottle below table.*)

PAT. Look, I know what I'm doin'! (ESTHER *comes in.*)

MARY JO. Evenin', Mrs. Moore.

MARGARET. Evenin', Mrs. Moore.

PAT. Evening.

ESTHER. Ach, are yous here, girls? Come in, Willie Jo. (WILLIE JOE FERGUSON *enters nervously.*) Sidney, pull over a chair for Willie Joe. (SIDNEY *takes another chair to the table.*)

WILLIE JOE. Evenin'. (*He pulls off his cap and sits.*)

JEAN. Evenin', Mr. Ferguson.

WILLIE JOE. It's a brave night.

MARY JO. Aye.

THE GIRLS IN THE BIG PICTURE – ACT I SCENE V

PAT. Mrs. Moore, I'm Sidney's manager. Could I have a word?
ESTHER. His what?
SIDNEY. Pat wants a wee word, mammy.
ESTHER. Aye, well, I'm goin' into the parlour. He may speak to me there. Jean! Wet the tea for Willie Joe.
WILLIE JOE. Thankin' you kindly.
ESTHER. C'mon, Sidney.
PAT. If I could have a word alone, Mrs. Moore?
ESTHER. You'll be in in a minute, Sidney?
SIDNEY. Yes, mammy. (ESTHER *exits to parlour.* PAT *signals okay to* SIDNEY *and follows.*)
JEAN. Tea, then?
WILLIE JOE. Aye, it would be nice. (JEAN *goes to kitchen door.*)
SIDNEY. I'm away to wait for Pat at the 'Rio'.
JEAN. Where am I goin' to say y'are?
SIDNEY. Think of somethin', will ye? (*He goes.* JEAN *exits to kitchen. The situation between the three at the table is awkward.*)
WILLIE JOE. Is your da for gettin' the electric?
MARY JO. Chair! (*No response from* WILLIE JOE.) Only Jokin'!
WILLIE JOE. Oh. Aye. . . . Still, it's brave night all the same.
JEAN (*entering with teapot and extra cup, squeezes in to set cup beside* WILLIE JOE). Excuse me; there y'are, Mr. Ferguson. (MARGARET *and* MARY JO *are desperately stifling giggles as* JEAN *shuts kitchen door and comes back to pour milk and tea.*)
WILLIE JOE. Thankin' you kindly. Willie Joe – I never was one for the formalities, Jean.
JEAN. Right. Sugar? (*She indicates the bowl and goes to sit down in her own place.*)
WILLIE JOE. Aye, two and a bit. (JEAN *gives him the sugar and sits.*)
MARY JO. Are ye for Basket Tea, Mr. Ferguson?
WILLIE JOE. Naw, I never was one for the dancin'.
JEAN. Tobias and me will be there, as usual.
MARY JO. We wouldn't miss it.
JEAN. Were you just passin', Mr. Ferguson?
WILLIE JOE. No, your mother and me was discussin' business and she invited me over for a drop of tea.
JEAN. Oh, I see.
MARGARET (*rising*). Well, we'll go sure. . . .
JEAN. Not at all, take your hurry! (MARGARET *sits again.*)
MARGARET. You must miss your mother, Mr. Ferguson.
WILLIE JOE. Aye.
MARY JO. She had a right good innin's, though.

138

THE GIRLS IN THE BIG PICTURE – ACT I SCENE V

WILLIE JOE. She had indeed, aye. (*The pauses are getting longer and more awkward.*)

MARGARET. Well, we really *must* be goin' now, Jean. (MARGARET *and* MARY JO *both rise.*)

JEAN. Margaret, I'm really not very happy with that sketch. I think we ought to have another rehearsal.

WILLIE JOE. Oh. Am I in the road? (*He puts down his cup and it looks as though they're going to get rid of him.*)

JEAN. Well, it's just we have to rehearse this wee sketch for the Women's Institute.

WILLIE JOE. Oh, your mother didn't say anything about havin' visitors. My apologies.

JEAN (*rising in preparation to seeing him out*). She must've forgot just.

WILLIE JOE. Aye. (WILLIE JOE *lifts his cup again; he's not moving. they're going to have to go through with the sketch. Props and costumes are hastily gathered up and the record put on.* MARGARET *sits down.* MARY JO *retires to the kitchen and* JEAN *leaves the room.* WILLIE JOE *turns to watch.*)

JEAN. Howdy, Sis. It's been many years!

MARGARET (*rising and giving her the cup*). It has been forty long ones, Robert. Will ye have a sup with me?

JEAN. Was mother's death slow? (*They both realise with horror the significance of departed mothers in the present company.*)

MARGARET. When death comes it's as quick as the beat of a heart!

JEAN. Poor, poor mother! (*It is obvious they can't continue.* MARGARET *signals with the mug to* JEAN; *the light dawns and she drinks.*) Aagh! Poisoned, I've been poisoned! (*She dies.* MARY JO *runs in from the kitchen, totally bewildered.*)

MARY JO. Now Mary and me will inherit all your fortune! You see, Mary, he was repaid for his greed! (*They all bow.*)

MARGARET. Did you enjoy that, Mr. Ferguson?

WILLIE JOE. Aye. (*They frantically collect up props and costumes. As they finish* ESTHER *enters with* PAT.)

ESTHER. Well, goodnight now, Mr. Sharkey.

PAT. Goodnight. (*He exits.*)

ESTHER. Where's Sidney?

JEAN. Oh, he must be still out at the byre.

ESTHER. Still here, girls?

MARY JO & MARGARET. No, we're just goin'. Goodnight. (*They exit.*)

ESTHER. When Sidney comes in, tell him to come up to me. I want a word with him. Goodnight, Willie Joe.

WILLIE JOE. Goodnight, Mrs. Moore. (ESTHER *exits.* JEAN *moves to*

THE GIRLS IN THE BIG PICTURE – ACT I SCENE V

the table and begins clearing the tea things.)
JEAN. Well, I'll just red these things up for the mornin'.
WILLIE JOE. Are ye not sittin' down, Jean?
JEAN. Aye. (*She sits. There's an awkward pause.* PAT *enters.*)
PAT. Excuse me.
JEAN (*rising*). Mr. Sharkey!
PAT. Is your mother away to bed?
JEAN. Yes.
PAT. Think I forgot something. (*Crosses and picks up bottle of cider. He is almost out the door again when* JEAN *speaks.*)
JEAN. Mr. Sharkey! I really think you and me ought to have a discussion about Sidney's future! (*The penny drops with* PAT *that she needs help.*)
PAT. Aye, surely.
JEAN. Family business, Willie Joe. You do understand?
WILLIE JOE. Oh! Aye, well, I'll take my leave, Jean.
JEAN. It was awful nice seein' you again.
PAT. Goodnight.
WILLIE JOE. Goodnight. (*He leaves.*)
JEAN. Safe home, now. Thank you.
PAT. Is your mother match-makin'?
JEAN. Seems likely.
PAT. I don't blame you for wantin' rid of him. D'you want a slug of cider?
JEAN. Naw. I'll just red these things up and away to bed.
PAT. Here.
JEAN. Naw, you're all right.
PAT. Lovely and sweet.
JEAN (*giving in and taking the bottle, drinks*). Bitter. (*She gives the bottle back to* PAT.)
PAT. Gets sweeter.
JEAN. Well, did you talk to my mother?
PAT. Aye. Don't know what all the fuss was about.
JEAN. What? Dear God in Heaven, what got into her?
PAT. My charm!
JEAN. Must be something.
PAT. Never fails! I just let on he'd no chance of winning. Is Sid gone down to the 'Rio'?
JEAN. Aye.
PAT. Can't keep him away from the women! Here. (*He proffers bottle.*)
JEAN. Naw.

PAT. Go on. Have another drop. (JEAN *takes another drink*.) You look like him.
JEAN. Our Sidney?
PAT. Same eyes. Feeling a bit more relaxed now?
JEAN. A bit.
PAT. If he comes lookin' for you, I'll challenge him to a duel. (JEAN *laughs*.) Lovely smile.
JEAN. Spose I'll have to face him sometime. (*She drinks*.)
PAT. You'll be gettin' tipsy.
JEAN. Sure you could be dead tomorrow.
PAT. Let me feel your shoulders. (JEAN *starts in alarm*.) Just to see if you're really relaxed or if you're only lettin' on. (*He moves behind her and starts to massage her shoulders.* JEAN *is being completely seduced. As he finishes he moves round again to her side*.) Not bad.
JEAN. Must be the cider.
PAT. Aye. The cider, the night and the music.
JEAN. There's no music. (*He leans forward. So does* JEAN, *as though to kiss, but* PAT *simply grabs the cider bottle*.)
PAT. Now you keep practisin' that smile, d'ya hear? (*He goes to leave*.) Oh and hey! Look after my protege! (PAT *goes, leaving* JEAN *alone. She exits*.)

LX FADE

ACT I – SCENE VI

SCENE: *SFX Basket Tea auctioneer* JAMESIE BICKERSTAFF *over scene change. Cheap garlands, cloth on table, centrepiece and cups.* PATSY FLOOD *and* MARGARET *are dressed in their best, listening to* JAMESIE. PATSY *has a money tin and raffle tickets.*

SFX JAMESIE. Is this thing on? Testing, testing, one, two. Mary had a little lamb. Ladies and Gentlemen. I would like to draw your attention to Mr. Tobias Stinton and Mrs. Patsy Flood, who are making their way round the hall with the ballot tickets. We would appreciate your partaking in the purchase of these to assist the Scouts in the purchasing of a piano. Six pence for two and one shilling for five. The prizes consist of, first, five pounds of new spring lamb by Willie Masterson; second, a beautiful set of tumblers kindly donated by Jackie Peoples; third, a tin of Jacob's Assorted by Tobias Stinton. Thankin' you, and it's back to Ernie McAfee and his merry men. (*SFX country band continues under dialogue.* PATSY *and* MARGARET *join in applause.*)
PATSY. Ach, Margaret, your wee outfit's beautiful! You're a quare one for the style, aren't you?
MARGARET. Thanks, Patsy. (JEAN *enters with two bottles of orange and straws.*)
PATSY. I say, she's a quare one for the style. Here, Jean, will you take a ticket off me or are you buying yours off Tobias?
JEAN. No, you give me five. (*She gets change from her purse.*)
PATSY (*giving her tickets*). God, there's a quare dearth of men the night. Must be because the Drumshandy boys haven't showed.
VOICES (*off*). Patsy Flood!
PATSY. All right, I'll be over in a minute! That's them McClure sisters from Ballyskilly. Here y'want to see their baskets, they'd feed the five thousand! (*She exits.*)
JEAN. It's not like Mary Jo to be late; she's always heart scared of missing something.
MARGARET. Her da's probably playin' her up. Selfish oul' brute!
JEAN. Ach, he's ould.
MARGARET. He wants to hold on to her. Your mother's lookin' to rid

of you, and my ma says there was never anybody good enough for me. If they'd leave us alone, we'd be all right. Does Tobias know Willie Joe was round?

JEAN. No! Don't remind me!

MARGARET. Ach, I felt sorry for the big lump.

JEAN. Go you and marry him, then!

MARGARET. She couldn't be serious about you and him!

JEAN. You think not?

MARGARET (*laughs*). Away and chase yerself!

JEAN. Margaret, it's not funny!

MARGARET. Just tell her...(MARY JO *enters with basket.*) What kept ye?

JEAN (*rising and taking her basket.* MARY JO *sits in her seat*). Put your basket in quick afore it starts!

MARY JO. I'm not walkin' away up there on me own.

MARGARET (*waving*). Patsy!

JEAN (*getting another chair*). You all right?

MARY JO. He hid my basket on me. Like a big child, he is!

MARGARET. Was he drunk?

MARY JO. No, he was not! Look at the state of me! I'm out without handbag nor nothin'!

MARGARET (*gives her a compact*). Here, away into the toilet. Put a bit of powder on ye.

PATSY (*entering*). Oh, here, Mary Jo, you haven't bought a ticket off me! That Tobias Stinton one's stealin' all my customers! Like, no offence, Jean, but cute as get out, gets me talkin' then he's away harin' up my side. Here! D'ye want one?

MARY JO (*leaving*). No!

PATSY. What's ailin' her? Suppose it's because the Drumshandy ones hasn't showed?

MARGARET (*indicating basket*). Just take that up, Patsy, will ye?

PATSY (*peeking under cloth*). Mmm! Apple tart! Y'know I met my Norman at one of these, fifteen years ago. Y'know what his very first words were? 'What have ye in your basket? For I could eat a horse, jockey and all'. He's still the same! (*She exits, laughing.*)

JEAN. Poor Mary Jo!

MARGARET. You don't know the half of what she goes through with that oul' lad! She doesn't say, but I know. Sooner Nessie dies the better!

JEAN. That cafe'll be the makin' of her. She's never had nothin' since she came back to fend for him.

MARGARET. She never shuts up about it; she's my head turned!

JEAN. Is the loan cleared with the bank?
MARGARET. Mervyn Bradshaw was great! He says, 'Margaret, I know it's very important to you, this loan, but I had to make sure you weren't swimmin' out of your depth'. Ach, it'll be hard the first year or two with the repayments, so there'll be no discount.
JEAN. Your own wee flat, too!
MARGARET. I know, I can't wait! (MARY JO *enters and gives compact to* MARGARET.)
JEAN. Do ye want a mineral?
MARY JO. Sure I've no money with me!
JEAN. Sure I'll get it! (*She exits.*)
MARGARET. Why don't you get your Francine to take him for a while, give you a wee break?
MARY JO. Francine would take him, but he'll not go! If I mention it, he accuses me of tryin' to get rid of him, and then he cries!
MARGARET. When did he start this?
MARY JO. Ach, he's always done it! Imagine a man cryin', I hate it!
MARGARET. Mary Jo, I don't think you should suggest him goin'. Why don't you write to your Francine, tell her to ask him up to Derry, to make him feel really wanted? Try it.
MARY JO. All right, I'll try it! (*SFX* JAMESIE BICKERSTAFF. JEAN *enters, gives* MARY JO *a mineral, sits.*)
SFX JAMESIE. Attention, Ladies and Gentlemen. As I'm sure you are already aware, due to a bit of commotion at the door, we are pleased to announce the arrival of the Drumshandy minibuses. And I have been informed by a spokesman from the Drumshandy Social Committee that the late arrival was caused by the two drivers, the O'Neill brothers, who were called suddenly to attend to the birth of two bull calves! Congratulations to the proud fathers!
MARY JO. Is Pat Sharkey here?
MARGARET. I wouldn't depend on it.
MARY JO. I was only askin!
JEAN. Ach, he could turn up yet, though I thought I heard Sidney say he was givin' him a practice.
MARY JO. Margaret! Brian's here, the teacher. Now, don't shout at me, I'm only sayin'.
MARGARET. Mary Jo will you please sit down and stop embarrassing me!
MARY JO. Margaret, he's lookin' over!
MARGARET. Mary Jo! Look at them Drumshandy ones, they're like big vultures!
MARY JO. It's all right for you, Jean. Tobias is going to buy your

basket. You'll not be made an eejit out of.
JEAN. There's more men than baskets. You'll be all right.
MARY JO. Huh! Look at them! They're big geeks!
MARGARET. Be a good setting for a horror picture!
JEAN. Well, why do you come, then?
MARGARET. There's no need to get shirty, Jean!
JEAN. I'm just sick of people sayin', 'You're all right, you've got Tobias'!
MARY JO. It's well for ye to have somebody to be sick about! (PATSY *enters as* SFX JAMESIE *starts up again.*)
SFX JAMESIE. Gentlemen, we have many tasty snacks about the place tonight.
PATSY. He's a bad 'un, that Jamesie Bickerstaff!
SFX JAMESIE. I'm sure yous are all waitin' in anticipation to get your teeth into them! My wife informed me the night that when I first laid my hands on her basket I ate every pick and gave her back five buns for the oven!
MARGARET. He has a mind like a midden!
SFX JAMESIE. I hope the night, Gentlemen, that the buns'll go into your mouths and not into the oven! (PATSY *nearly dies laughing and comes over to the three women.*)
PATSY. 'I hope that tonight the buns'll go into your mouths and not into the oven!' (*Screeches with laughter.*)
MARY JO. All right, Patsy, we're not deaf!
SFX JAMESIE. So without further ado, what am I bid for Martha Moorhead's beautifully done up basket? (*During the following bidding* JEAN *first rises to her feet.* MARY JO *stands on her chair, and* MARGARET *stands to get a better view of the auction.* SFX JAMESIE *voiceover interspersed with various male voices bidding.*)
SFX. Five shillin's!
SFX JAMESIE. Any advance on five shillin's?
SFX. Seven and six!
SFX JAMESIE. Any advance on seven and six? C'mon now, gentlemen!
SFX. Eight shillin's!
SFX JAMESIE. C'mon, now!
SFX. Nine and six!
SFX JAMESIE. C'mon now, boys, any advance on nine and six?
SFX. Nine and nine pence!
SFX JAMESIE. Oh, now!
SFX. Ten shillin's!
SFX JAMESIE. Ten shillin's! Any advance on ten shillin's? Gone! Sold to the Gentleman in the green pullover! (*Applause.*) Now, if you'll

THE GIRLS IN THE BIG PICTURE – ACT I SCENE VI

collect your basket and pay the secretary, Bertie Bradshaw, in the corner. (*Applause.*) And now we come to Pearl Pinkerton's basket. What am I bid for young Pearl's beautifully turned out basket?
SFX. Five shillin's!
SFX. Five and thruppence!
SFX. Six shillin's!
SFX. Six and sixpence!
SFX. Seven shillin's!
SFX. Seven and sixpence!
SFX. Ten shillin's!
SFX. Twelve shillin's!
SFX JAMESIE. Any advance on twelve shillin's? Any advance on twelve shillin's now?
SFX. Thirteen shillin's!
SFX JAMESIE. Thirteen shillin's, c'mon now, boys. Any advance on thirteen shillin's? We're goin' well here now!
SFX. Fourteen!
SFX JAMESIE. Any advance on fourteen? Fourteen, fourteen, fourteen! No? Sold to the Gentleman in the grey suit next to Geordie McClurg there!
PATSY. Hey, Margaret, that's that teacher boyo, Brian!
MARY JO. Never worry yourself, Margaret!
SFX JAMESIE. And here we have Mary Jo McIlFatrick's basket! And what am I bid?
SFX. Six shillin's!
MARY JO. Oh, my God!
SFX JAMESIE. Any advance on six shillin's?
MARY JO. It's that eejit that brings the pig meal! I hate him!
SFX JAMESIE. Ach, now, c'mon Gentlemen, dig deep! Six shillin's isn't a lot to pay for the best baker in the five townlands! C'mon now.... No advance on six shillin's?.... No? Ah, that's a shame! Sold to Ambrose Fitzcadden! What's that, Archie?'
MARY JO (*getting down off chair*). God, that's all I need the night! (AMBROSE *enters, already tucking into a scone, speaking with his mouth full, a thoroughly uncouth character.*)
AMBROSE. Hey, girl! Did ye bake all this yourself? (PATSY *runs out in a fit of laughter.* AMBROSE *comes over.*)
MARY JO. Sit down over there! I'll be over in a minute! (AMBROSE *takes* JEAN's *vacant chair and moves it to the table, where he sits on another chair, leering and chewing.*)
SFX JAMESIE. Ladies and Gentlemen! I would like to interrupt the auction for a moment. Unfortunately I have to be the bearer of bad

news. A very old and respected member of our community here in Cloughmartin, I have just been informed by Nan Flood, has passed away to a better place, about five minutes ago. Mrs. Agnes Elizabeth Banner, better known to her friends as Nessie Banner. And I think we should observe a few minutes silence and then return to the festivities as I'm sure Nessie would have wanted'. (*The* THREE WOMEN *sit,* JEAN *and* MARY JO *sharing a seat.*)

MARGARET. I can't believe it!
MARY JO. Margaret, this is really it! (PATSY *enters, followed by* TOBIAS.)
PATSY. Ach, poor oul' Nessie!
JEAN. Aye, God rest her!
PATSY. I knew, I heard the banshee last night! Well, she's dead now. Life goes on! I suppose that'll be you movin' in soon, Margaret?
MARGARET. Well, that's the last thing on my mind just now, Patsy!
TOBIAS. There'll be a quare turn-out at that funeral!
JEAN. Aye, she'll be missed!
AMBROSE. Ye not sittin' down, girl?
MARY JO. In a minute! . . . Sorry!
PATSY. I'll just go on and see how Nessie went at the last. (*She exits.*)
SFX JAMESIE. Is this thing back on again? . . . Ladies and Gentlemen, I think we can begin again and hope that Nessie Banner's soul is on its way to a better place. The next treat in store is Jean Moore's basket. Are ye there, Mr. Stinton? Aye, and what am I bid for Jean's basket?'
TOBIAS. Five shillin's!
SFX JAMESIE. Can I presume there's no man brave enough to outbid Mr. Stinton?
WILLIE JOE (*off*). Ten shillin's!
MARY JO. God! It's Willie Joe Ferguson!
TOBIAS. Ten and six!
WILLIE JOE. Twelve shillin's!
SFX JAMESIE. Twelve shillin's! A very popular lady in the house tonight! Any advance on twelve shillin's?
TOBIAS. Fifteen bob!
WILLIE JOE. One pound!
TOBIAS. One pound, two shillin's!
WILLIE JOE. Two pound notes!
TOBIAS (*pulling all the money out of his pocket*). Two pounds, two and seven!
WILLIE JOE. Five pounds!
SFX JAMESIE. Five pounds! Lordy days! Am I bid no more? I say,

Jean Moore must have a quare special treat in that basket of hers the night! No advancers on five pounds, then? . . . Done! Sold to Willie Joe Ferguson! (JEAN *grabs her handbag and runs out.*)
MARY JO. Jean! Jean! . . . That is one crafty oul' brute!
MARGARET. It's not him. This is Esther Moore's doin'! Tobias! C'mon, we'll give Nan Flood a hand with the teas! (*She rises and takes* TOBIAS *by the arm.*)
TOBIAS. What?
MARGARET. You know what she's like! We'll give those cups a bit of a scrub or we'll be gettin' the mange!
TOBIAS. Oh, aye. (*They start leaving.*)
MARY JO. Margaret, what if somebody buys your basket?
MARGARET. Unless it's Robert Mitchum, don't come lookin' me! (*They exit.* MARY JO *is left looking at her suitor.* JAMESIE'S *auctioneering continues underneath the scene.*)
SFX JAMESIE. Lordy days! I can hardly lift this one; just as I suspected, Daphne McClatchey! What am I bid for Daphne's basket?
AMBROSE. C'mon and sit down, girl! (MARY JO *does so.*)
SFX JAMESIE. C'mon now, for Daphne's the girl with the golden brown touch!
SFX. Five shillin's! (*Bidding begins again.* WILLIE JOE *enters, carrying* JEAN'S *basket.*)
WILLIE JOE. Where did Jean go?
MARY JO. Oh, she went away up to the top table in the corner; she wants to be romantic!
AMBROSE. She didn't! She went out!
MARY JO (*elbowing him sharply*). No! That was Patsy Flood went out! She's up there in the corner!
WILLIE JOE. Aye. (*He goes back the way he came.*)
AMBROSE. What did I do?
MARY JO. Ach, eat that and shut up! . . . Sorry! (AMBROSE *is stuffing food into himself at great speed.*) You fond of your grub?
AMBROSE. I love it!
MARY JO. Does yer mammy feed ye well?
AMBROSE. Aye.
MARY JO. Did ye not get no dinner?
AMBROSE. Aye.
MARY JO. You still hungry?
AMBROSE. Aye. . . . Hey, girl, I'm surprised no one's snapped you up by now. This apple tart would melt in your mouth!

LX BLACKOUT. END OF ACT I

ACT II – SCENE I

SCENE: *Two weeks later in the cafe back kitchen.* MARGARET, *in overall, comes out of the centre door with a pile of plates which she stacks on the shelf. She sits down on the chair, very dispirited. Knocking is heard off.*

MARGARET. Closed!
MARY JO (*off*). It's only me! (MARGARET *goes off to let* MARY JO *in.* MARY JO *enters, followed by* MARGARET. *There is a huge difference in their moods.*) Margaret! Me da got our Francine's letter this morning! He's goin' to Derry for a fortnight! God, you want to've heard the lies she told, 'I really miss you, daddy. The children never see you; they think they have no granda'! And then she writes, God, I nearly had to laugh, 'It's not fair, our Mary Jo has you all the time and I never see you'! Then, here's the best bit, 'After all, she's goin' to have you the rest of her days'! (MARY JO *enjoys the joke.*) Course, I told our Francine to write that! . . . Anyway, it's worked!
MARGARET. I'm pleased for ye.
MARY JO. Margaret, would you not try and pull yourself together? Look, Tobias Stinton has bought that shop, and you're just goin' to have to accept it. It's no bed of roses for me either. . . . You've hardly spoke to anybody in two weeks now.
MARGARET. I've had a long day. I smell like a bucket of lard. What do you want me to do? Dance on the tables?
MARY JO. You think this whole town is laughing at you.
MARGARET. Aren't they?
MARY JO. No! That's your imagination. All they're concerned about is where somebody like Tobias Stinton got the money to buy Nessie's.
MARGARET. He did it deliberately!
MARY JO. Ach, he wouldn't have the gumption to do nothin' deliberately! But I'd love to know where the hell he got that money! Him that never had nothin' but an oul' beat-up van! Even Jean can't get it out of him! He's one sleeked wee skitter! (MARGARET *buries her head in her hands, crying.*)

THE GIRLS IN THE BIG PICTURE – ACT II SCENE I

MARY JO. You probably needed this, you know. (JEAN *enters*.)
JEAN. Door was open.
MARY JO. Hello, Jean.
JEAN. Sorry I haven't been in to see yous. I was too cut up about everything. Margaret, it had nothing to do with me. I nearly died when he told me!
MARGARET (*controlled again*). Not to worry!
MARY JO. Jean, where did he get that money?
JEAN. I swear on the Holy Bible, I don't know! All's I know is he sold young Kevin O'Connor the van, he had a bit put past him, and the rest I don't know!
MARGARET. Well, he wasn't long gettin' the sign up!
JEAN. Margaret, I'm sorry. I really am. If I'd have known what he was goin' to do I would've said!
MARY JO. Sure, it wouldn't have made any difference. Margaret couldn't have offered cash anyway.
MARGARET. When's the wedding, then?
JEAN. Is no wedding!
MARY JO. Margaret! Jean is sorry!
MARGARET. Well, I have to lock up now.
JEAN (*turning to go*). Right.
MARY JO. Will we not all have a wee cup of tea?
JEAN. Well, if Margaret's not too busy?
MARGARET. I've had enough of that flippin' kitchen!
MARY JO. I'll make it! (*She exits.* JEAN *awkwardly sits down with* MARGARET *at the table.*)
JEAN. She was sayin' she's got her daddy to go away to Derry. That was a good idea! Wish to God he'd take my mammy with him! She's hardly speakin' since the Basket Tea cos I insulted Willie Jo by running out!
MARGARET. I hear Willie Joe's been coming round?
JEAN. He lets on it's Sidney he wants to see about the fresians. I just excuse meself. She be's ragin'! I don't know what the next move's goin' to be!
MARGARET. Are ye goin' to marry Tobias?
JEAN. I didn't know before he got the shop and I still don't!
MARGARET. That's why he got it.
JEAN. No, it's more than just me. He's always been desperate to be respected. He'll maybe move up a pew in the Church now!
MARGARET. I'm sorry for eatin' the nose off ye earlier. I know you and Mary Jo have your own troubles.
JEAN. Why don't the two of you really get the place goin' here?

You. . . .

MARGARET. Ach, Jean, I have no interest. I wanted to be independent. I am thirty-five years of age, still living with my parents! Even they're startin' to take pity on me now!

JEAN. Buy this place off your mammy and daddy. Would they hear tell of that?

MARGARET. And be blow-ins with nothin'? They'd have to sit outside the Church and look in!

JEAN. What about another town?

MARGARET. Where I know nobody! I know what they think of me here. I'm just a crabbed oul' spinster! But at least I've learned how to handle them.

JEAN. Well, what are you goin' to do? (MARY JO *enters with the tea.*)

MARY JO. I've been thinking.

MARGARET. You'll hurt yerself!

MARY JO. When my daddy goes to Derry for a fortnight I could move in here full-time. Oh, now I know there's not enough money to pay me, but sure you know me, I'm all heart! I could stay in the kitchen with Nan Flood, try and improve things a bit! By the end of a fortnight she'll be that sick of me, she'll leave!

MARGARET. So will I!

MARY JO. She's never happy unless she's insultin' me!

JEAN. Nan's been here for years. She'll die over that stove!

MARGARET. Mary Jo, if you were to work here full-time, who could we get to look after your da? (JEAN *laughs. Slowly the light dawns on* MARY JO.)

MARY JO. Nan Flood! Ach, away you and chase yourself! (*Knocking on door, off.*)

TOBIAS (*off*). Jean! You said you'd give me a hand! I need you to hold up the other end of these shelves!

JEAN. Right! I'll be there in a minute! . . . Have yous your carrots dressed yet for the W.I.?

MARY JO. Yes, I've mine done up like a Canadian Mountie! It's great, isn't it, Margaret?

MARGARET. Aye.

JEAN. I seen Betty Bothwell there in the street. She's been hunting the place for a big one. She's determined because the sketch won the last time.

MARY JO. I'll get you a good big one out of my da's plot!

MARGARET. I told ye, I'm not goin'!

MARY JO. You're cuttin' off your nose to spite your face!

MARGARET. Ach, you know what they're like, Mary Jo! 'Awful sorry,

Margaret, about the shop, Margaret. Sure we'd no need of a dress shop anyway, Margaret'!

JEAN. Aye, well, me too! 'Oh, that'll be you and Tobias well set now, Jean. When's the big day, Jean'?

MARY JO. Betty Bothwell is near dead to get the ins and outs about Pat Sharkey. I just say, 'How would I know'?

MARGARET. Would ye want to know?

JEAN. You really can't stand him, Margaret!

MARGARET. If he was any port in a storm, I think I'd rather drown!

MARY JO. Margaret! Do you think he fancies you?

MARGARET. What? He's a big child!

MARY JO. I think he's a really nice man!

JEAN. Well, I may go and give him a hand with these shelves. I'll see yis! (*She exits.*)

MARGARET & MARY JO. Night, Jean!

MARY JO. Margaret, why don't we get stuck in here? Close the Cafe for a week, scrub the place down, decorate it, bring the customers back again?

MARGARET. Maybe.

MARY JO. Honest to God, and you're not jokin'? (PAT SHARKEY *is heard singing 'Jenny'. He comes in bearing a record player and a grip.*)

MARGARET. Pat Sharkey, have you not heard of knocking?

MARY JO. Hi ya, Pat!

PAT. Hello, love.

MARY JO. Congratulations! I hear Sidney won the semi-final!

PAT (*setting down record player and bag*). Aw, you should've been there, Mary Jo. He was a knock-out! You heard of Dave Glover? He takes me to one side afterwards and says, ' Pat, mate, you have a star on your hands'!

MARY JO. Dave Glover that has the showband?

PAT. A very big name!

MARY JO. And you know him?

PAT. Know him! Dave and me go way back! You see this? (*Holds up record.*) Sidney's first disc!

MARY JO. Margaret! (MARGARET *is derisively clearing up cups onto tray.*)

PAT. Where is he, by the way? He's supposed to be here to try these clothes on.

MARGARET. This is not the powder room of the Plaza!

MARY JO. Sidney's disc?

PAT. Aye, you go into a wee booth, sing away, and Bob's your uncle!

THE GIRLS IN THE BIG PICTURE – ACT II SCENE I

MARGARET. Mary Jo, since I'm here I'll go and step the peas for the morning. (MARGARET *exits*. PAT *begins setting up record player*.)
MARY JO. I didn't see you at the Basket Tea.
PAT. Aw, sorry, love. Got a big haul of stuff that day, had to take it back to my dealer. So who bought your basket?
MARY JO. Ambrose Fitzcadden! You want to see him, he's a slimy big geek! But I had no choice; he just kept biddin' and biddin' and nobody else got a look in!
PAT. Never mind, always next time!
MARY JO. Pat, I would really love to go up to the final to hear Sidney!
PAT. Mary Jo, I would really love to take you, but there isn't goin' to be any room. The car's goin' to be full of gear. (SIDNEY *enters*.)
SIDNEY. Did ye get them, Pat? Hi, Mary Jo!
PAT. Aye, try them on.
MARY JO. Sidney, congratulations. I hear you won the semi-final!
SIDNEY. Aye.
PAT. Mary Jo, would you excuse us for a minute?
MARY JO. Oh, aye! (MARY JO *goes and* SIDNEY *takes off his clothes during this scene and puts on tight trousers, tee-shirt and drape jacket*.)
PAT. What happened you last night?
SIDNEY. I just wanted to go home.
PAT. Could you not get it up or something?
SIDNEY. Just didn't want to!
PAT. Sally's ma and da away for the night, golden opportunity and you blow it! Me and the sister were just gettin' goin' and you come harin' through the room!
SIDNEY. There is no other way to get out!
PAT. Sid, son, what was the problem?
SIDNEY. Pat, she was cryin'!
PAT. They all cry! Doesn't mean they don't want to!
SIDNEY. But she's never before; I'm her first steady!
PAT. Some steady! Look, Sidney, I've had virgins, loads of them! You just have to have a wee bit more patience. S'pose you were like a bull at a gate!
SIDNEY. I just don't want to talk about it now!
PAT. And who's the one always goin' on about not gettin' it?
SIDNEY. Well, so what if I decided not to? Have I committed a crime?
PAT. Ah! Grow up, Sidney! That wee girl was up there with her teeth sunk in the tallboy! I nearly had to throw a bucket of water round her! . . . Sidney, I want you to talk to your mother tonight.
SIDNEY. Tonight! Pat, I can't speak to her tonight! You should've

seen the look on her face last night when I told her I'd won the semi-final! Pat, can we not just do this and let on we have a contract?

PAT. I'd get lynched! You're still a minor!

SIDNEY. I won't let anybody else manage me, I promise!

PAT. It's not just for me. It has to be signed for both our sakes!

SIDNEY. She'll not do it!

PAT. Play it down! Tell her a couple of spots here and there, a few bob to help with the upkeep of these cows you're gettin'.

SIDNEY. You told her I wouldn't win! That's why she let me do it in the first place!

PAT. A wee white lie.

SIDNEY. I hate tellin' her lies!

PAT. Well, all right then, away you go now, and we'll forget the whole thing! That's the best, isn't it?

SIDNEY. Forget about it?

PAT. No choice! Look, you have to play your part of the bargain.

SIDNEY. But she's oul' fashioned, Pat.

PAT. Well, it's an awful pity, Sid. You'd a great future.

SIDNEY. But I love the farm. I want to do both!

PAT. Well, then, will you listen to me? Do like I'm tellin' ye!

SIDNEY. I'll try.

PAT. Put that big sad face on and she'll not resist ye! All she's scared about is losin' ye. Convince her that's not the case and we're for the tops!

SIDNEY. Is Sally's sister not a wee bit young for you?

PAT. What's age got to do with it? Some chicks prefer older men – more experience.

SIDNEY. Would you not rather find someone more your own age?

PAT. No fear, one date and it's weddin' bells!

SIDNEY. Do you fancy Margaret?

PAT. What? I know what she needs!

SIDNEY. Then why do you always arrange to meet me here?

PAT. Is nowhere else.

SIDNEY (*having finished dressing*). Pat, away and get Mary Jo. I want her to see me.

PAT. Pity that wee woman wasn't your ma; he'd have no problems! (*He opens door into shop.*) Mary Jo!

MARY JO (*off*). Is he ready? (*Comes in.*) Sidney Moore!

PAT. Sid, kid, you're goin' to cause heart failure!

SIDNEY (*putting on shades and striking a pose*). I don't look stupid, do I?

TTHE GIRLS IN THE BIG PICTURE – ACT II SCENE I

MARY JO. No, you look gorgeous!
PAT (*putting on disc*). Wait'll you hear this, Mary Jo!
MARY JO. And your very own disc!
SIDNEY. It's only a wee plastic thing!
MARY JO. Sure that's what they're all like! (*SFX recording of 'Jenny'*.)
PAT. Want to dance, love?
MARY JO. Aye. (*As they dance,* PAT *manoevers her round to finally land her on* SIDNEY'S *lap*.)
PAT. What a voice! Wonder who it is? Sounds like Fabian. Funny we should mention Fabian, there he is! Hey, Fabe, baby, wanna partner? (*He deposits her roughly. She and* SIDNEY *laugh embarrassedly and get up to dance.* PAT *goes into the cafe*.)
SIDNEY. He's a big eejit, isn't he?
MARY JO. Aye. . . . That sounds different.
SIDNEY. Aye, Pat thought I should soup it up a bit.
MARY JO. I liked it better the other way.
SIDNEY. Pat said that was just for oul' dolls!
MARY JO. Oh. (*She breaks away and sits down.* SIDNEY *joins her*.)
SIDNEY. Not in good twist, Mary Jo?
MARY JO. Aye, I'm in great twist. You look gorgeous so you do.
SIDNEY. I won't be makin' an eejit out of myself, will I?
MARY JO. No, you will not. You have a beautiful voice.
SIDNEY. Pat sometimes gets carried away.
MARY JO. No, he's right. He really knows what he's doin'. . . . Does he have a girlfriend or anything?
SIDNEY. I don't know, Mary Jo.
MARY JO. You take your chance, Sidney. Don't you make the same mistake I did.
SIDNEY. Aye, you were a nurse in Belfast. . . .
MARY JO. Not for long. All our ones went off and got themselves married. I had to come back and nurse my Mammy. You were only a wee buck at the time.
SIDNEY. What age were you?
MARY JO. I was the same age as you, nineteen.
SIDNEY. Why'd you not go back – whenever she died?
MARY JO. Ach, it was too late, sure my Da couldn't manage on his own. I got landed with 'Grumpy Joe from Ballysloe'! That's what I call him.
SIDNEY. I wish you were my Mother. . . .
MARY JO. I wish I was nineteen!
SIDNEY. Why? What would you do?
MARY JO. Oh, now. . . . (*Looks at him coyly.* PAT *and* MARGARET

155

MARGARET. You never heard such silly oul' jokes, Mary Jo!
PAT. You laughed at the one about the Irish trick, remember?
MARGARET. I smiled, not laughed.
PAT. Mary Jo, pick a thumb. (*He holds out his hands.* MARY JO *picks his right hand. He puts them behind his back as though to jumble them and then holds out his clenched fists.*) Right, what hand's it in? (MARY JO *squeals with laughter.*)
MARGARET (*to* SIDNEY). Oh, looks good. Yes, I have to admit you have taste.
PAT. I pick them!
MARY JO. Pat, what do you call an Ulster woman with a frog on her head?
PAT. Stupid?
MARY JO. No! Lily! Margaret told me that! (MARGARET *hurries out with the tray.*)
SIDNEY. I'll head on. I want to go and show Sally.
PAT. Well, get you to bed early! No hanky panky!
SIDNEY. Goodnight. (*He leaves with the bag of clothes.*)
MARY JO. Good luck, Sidney! (PAT *turns and follows* MARGARET *out. He shuts the door and leaves* MARY JO *alone as the lights fade.* MARY JO *exits.* PAT *and* MARGARET *re-enter.*)
MARGARET. Pat Sharkey, do you think I've nothing to do all day but listen to you tellin' jokes?
PAT. The place is closed. What else would you be doin'?
MARGARET. I do have a life outside this cafe, y'know.
PAT. Aye, do y'change into a pumpkin?
MARGARET. I am studying legends for my night class.
PAT. Legends! Like what?
MARGARET. Why? Do you know any?
PAT. Aye. James Dean, he's a legend.
MARGARET. Irish legends!
PAT. Naw! He was American. . . . Are y'daft?
MARGARET. You can't just talk, can ye?
PAT. What?
MARGARET. Nothin' (*Beat.*)
PAT. I'm going to miss this wee place.
MARGARET. Why?
PAT. Had it with the buyin' and sellin' round here.
MARGARET. What about Sidney?
PAT. He'll not be doing many more spots round here. It'll be Belfast, Ballymena, Dublin. And if he wins the final, it'll be a full-time job

for me as manager.

MARGARET. Bored now with buyin' junk?

PAT. Makin' money isn't borin', love. I had one of my biggest hauls round here.

MARGARET. Who?

PAT. Two oul' dolls.

MARGARET. The Tuckeys?

PAT. That's top secret. They told me not to say. You wanna see their place. It was like a museum!

MARGARET. How much did they get? (PAT *points to his nose.*) Did they say what they wanted the money for?

PAT. God knows! Probably to stick under their mattresses. (*Beat.*) See y'later. (MARGARET *realises it is the Tuckeys' money. Beat.*)

MARGARET. Cheerio! (*Beat.* PAT *exits. Beat.*)

LX FADE

ACT II – SCENE II

SCENE: *SFX 'Jenny' covers scene change fading into clock ticking. Moores' farmhouse.* ESTHER *is setting the table with two cups, saucers, and a plate of barn brack. The door knocks and she quickly gets an old photo album from the dresser, opens it, and searches for a particular page.*

ESTHER. It's on the latch! (WILLIE JOE *enters carrying a box of chocolates.*) Ach. Willie Joe, punctual as ever.
WILLIE JOE. Aye, it was my mother that drummed that into me. Whoever we used to go out to visit, we would have stood outside their door for five minutes for we were always early. 'Neither be early nor late, for neither is polite', she would've said. (*He sits.*)
ESTHER. Sure, Silas was the same. Just before you came I was looking over the old scrapbook of all the newspaper cuttings of him.
WILLIE JOE. Aye, sure, there wasn't a week went by that he wasn't in it, for he was in everything!
ESTHER (*pointing at photos*). That's him and Sidney with the pride of the herd. 2000 gallons in three lactations. We got her as a calf off old Maguire.
WILLIE JOE. Aye, she looks a right beast.
ESTHER. Jean – when she was thirteen – in her Girl Guides uniform getting presented with her cookery badge.
WILLIE JOE. Aye....
ESTHER. When she was twenty-one – poultry farming at Loughrey.
WILLIE JOE. I'm a wee bit on the nervous side, Mrs. Moore.
ESTHER. Why's that?
WILLIE JOE. I've no experience at this courtin' lark on account of my mother being that strict with me. I mind about 15 years ago – I was walkin' out with a wee girl from Lisnacloughey, and if my mother had known I was goin' out with her she would've hid my suit! I would be sat yonder and sulked, but she paid no heed to me. She was right too, for the same wee cutty got herself into trouble, and she had to go away to England!
ESTHER. Tut! Tut! Don't you worry about a thing . . . will you take a wee sherry?

THE GIRLS IN THE BIG PICTURE – ACT II SCENE II

WILLIE JOE. Aye, I would. (ESTHER *goes to the dresser for sherry and a glass.*)

ESTHER. Jean'll not be long. She's down helping Tobias Stinton in his wee shop. Ach, awful kindly is Jean, always willin' to lend a helping hand. (*Pours sherry.*)

WILLIE JOE. Thon Tobias one didn't look too well-pleased at the Basket Tea.

ESTHER. Ach, that's because he usually buys her basket – I wouldn't worry about it.

WILLIE JOE. I brought her this wee box of sweets.

ESTHER. Oh, she'll love them. (WILLIE JOE *gulps down the sherry in one go.*) Another wee sherry?

WILLIE JOE. Aye. (JEAN *enters.*)

JEAN. Evenin', Mr. Ferguson!

WILLIE JOE. Evenin', Jean. (TOBIAS *enters.*)

ESTHER. Tobias! I wasn't expecting you.

TOBIAS. Evening.

JEAN. Sure you know Tobias always leaves me home.

ESTHER. Well, I thought on account of him working to the wee small hours and with him havin' no van. . . .

TOBIAS. Well, I wouldn't let Jean walk home on her own.

ESTHER. That's very kind of you, Tobias.

JEAN. Pull over a seat, Tobias.

ESTHER. Yes, Tobias, pull over a seat. Get your breath back. (TOBIAS *pulls over a chair to the end of the table and sits in the one beside it, leaving it for* JEAN *and putting himself between her and the empty chair beside* WILLIE JOE *at the other end of the table. As* JEAN *crosses to sit,* ESTHER *stands in front of the chair, forcing her to sit between* TOBIAS *and* WILLIE JOE.) Well, I'm sure that's the tea wet now.

JEAN (*going to the dresser*). I'll get two extra cups for me and Tobias.

ESTHER. Just the one – I wasn't having any. (JEAN *brings over a cup and saucer from the dresser, sets them down for* TOBIAS, *and pours them all milk.*)

WILLIE JOE. Congratulations, Tobias, on getting the Mace franchise.

TOBIAS. Indeed!

JEAN. Tobias is thinkin' of extending out the back. (ESTHER *re-enters with the teapot and pours the tea.*)

ESTHER. Sure, everybody's at the extendin' these days. (*To* WILLIE JOE.) How many more acre did ye get off young Maguire?

WILLIE JOE. I got twenty off him – just in the nick of time – for the Lord knows who he would've sold them to next.

THE GIRLS IN THE BIG PICTURE – ACT II SCENE II

ESTHER. Cloughmartin has a lot to thank you for, Willie Joe, for since old Maguire died that young buck's been goin' mad selling off land for electricity stations, council houses, and the dear knows what.

WILLIE JOE. Aye, well, it's grand land for wheat, and what's not so good, I'll grow rye.

ESTHER. Jean and her father often talked about goin' into wheat. Didn't ye, Jean? (JEAN *smiles awkwardly*.)

WILLIE JOE. It must be all go now, Tobias, with gettin' the shop ready for the opening?

TOBIAS. Aye.

JEAN. It's lookin' great so far. Tobias has done a lot of work.

ESTHER. Oh, well, if it's anything like that wee house of his it'll be a credit to him . . . no woman could keep it better.

JEAN (*lying*). Tobias was talkin' bout supplying the lemonade for the harvest picnic.

TOBIAS. Oh, aye, no bother. Well, Nessie always contributed, so I. . . .

ESTHER. Oh, it's a great day for Moore's farm. Silas's way of thankin' the Lord for a good yield. You'll be there, Willie Joe?

WILLIE JOE. Aye – God granted.

ESTHER. I was hoping you'd bring along the fiddle?

WILLIE JOE. Sure, I haven't laid my hands on that thing this years.

ESTHER. On account of Isaac Strain being bad with the arthritis.

JEAN. He was bravely last time I seen him.

ESTHER. Oh, and indeed he was complaining. Sure the day wouldn't be same without the music.

WILLIE JOE. Aye, well, I'll try my best, Mrs. Moore.

ESTHER. Tell you what, why don't you bring it round here some night and give us all a bit of a turn? Aw, Jean, Willie Joe and I were havin' a quare laugh at you there in your Girl Guides uniform!

WILLIE JOE. Aye, it was a nice wee picture. . . . (TOBIAS *pushes back his chair and heads for the door*.)

ESTHER. Are ye not finishing your tea, Tobias?

TOBIAS. I've had enough! (*He exits*.)

JEAN. Tobias! (*She goes to follow*. ESTHER *stops her by speaking*.)

ESTHER. Jean – look what Willie Joe brought. (*Indicates chocolates*.)

JEAN. Thanks very much, Mr. Ferguson.

WILLIE JOE. Is Mr. Stinton alright?

JEAN. I think he's a bit upset. (*As she is about to exit,* TOBIAS *returns*.)

TOBIAS. I've changed my mind. (*Sits*.) I will have that cup of tea. (JEAN *also sits. The situation is very awkward. Even* ESTHER *is flustered*.)

THE GIRLS IN THE BIG PICTURE – ACT II SCENE II

WILLIE JOE. You must be awful tired, Tobias, with getting the shop ready for opening?

ESTHER. He's a quare hard worker. Sure that wee van of his is never off the road. (SIDNEY *enters*.)

SIDNEY. Evenin' – what's this – the good crockery? Is the Queen comin'? (*He gets another cup and pours his tea*.) You here to see me, Willie Joe?

WILLIE JOE. Em. . . .

ESTHER. Willie Joe's here to talk about the yearlings.

SIDNEY. You toul' me yesterday, sure.

WILLIE JOE. Aye . . . I did . . . it must've slipped my mind. (SIDNEY *has no chair, so sits between* JEAN *and* TOBIAS.)

SIDNEY. Shove up, Jean. Are ye dotin'? I hear you got 20 acre off Maguire; you're not dotin' that much. Wish it was me, wheat crop is up forty-seven per cent.

WILLIE JOE. Aye, and the rye's up thirteen. I'll use it for feed and then plough it under so's that next year I'll have a good pasture.

SIDNEY. There's no flies on you, Willie Joe.

WILLIE JOE. Sure now you could have the same if you put your mind to it and stopped your gallivantin'! I was a quare hand at the fiddle when I was your age, but II didn't get carried away with it. I heard you won the semi-final. What does this mean?

SIDNEY. Nothin'!

WILLIE JOE. Nothin' indeed! If it was nothin', it wasn't worth missin' a good herd from Derry!

ESTHER. When was that? (WILLIE JOE *has enough gumption to realise he's put his foot in it*.)

SIDNEY. Last week. I was far too busy.

WILLIE JOE. Not to worry. We'll get him fixed up soon. I've big Matthew Overend lookin' out for him. Sure there's none better.

SIDNEY (*trying to deflect attention away from himself*). Can I have a sweet? 'Quality Street'! My, my Tobias, we are coming up in the world now we're a business man!

TOBIAS. I didn't bring them!

SIDNEY. Can I have one, Mammy? Whose are they? Jean's?

TOBIAS. I think they belong to Mr. Ferguson!

SIDNEY. Are they yours, Willie Joe?

WILLIE JOE. They belong to Jean

SIDNEY. Jean, are these. . . .

TOBIAS. Mr. Ferguson brought them for Jean, I would imagine! Am I right?

WILLIE JOE. Aye, I did.

TOBIAS. I hoped you thanked Mr. Ferguson, Jean?
JEAN. Yes.
TOBIAS. I think there's been a misunderstanding on Mr. Ferguson's part!
WILLIE JOE. Aye, aye, there has been a wee bit of a misunderstanding, Tobias!
ESTHER. Well, I think it's time we called it a day. We're all for up early in the mornin'. (JEAN, WILLIE JOE, *and* ESTHER *rise*.)
WILLIE JOE. Aye, well, thank you kindly for your hospitality, Mrs. Moore. (*He makes his way towards the door.*)
TOBIAS. Oh, aye! Skulk off now, cos you're too sleeked to admit you're in the wrong!
ESTHER. I'll see ye out, Willie Joe.
WILLIE JOE. I'm skulking nowhere, Tobias! I'm leaving as any decent gentleman would under the circumstances!
TOBIAS. Did you or did you not know that Jean and I are walking out, and that I have asked her to be my wife? (*As* WILLIE JOE *tries to leave,* TOBIAS *blocks his path.*) A gentleman indeed! What about the Basket Tea? Making little of me?
JEAN. Tobias, leave it!
TOBIAS. And you come here courtin' tonight, knowing full well I would be here with Jean!
WILLIE JOE. I knew nothing of the sort, Tobias!
TOBIAS. So, you were doin' it behind my back?
ESTHER. Willie Joe's as welcome here as you are, Tobias!
TOBIAS. You were in on this too! The two of yous, thinking I wouldn't be able to stand up to yous and leave! Him and his ninety acre! That's all you want, isn't it?
WILLIE JOE. I planned nothin', Tobias!
TOBIAS. You're a liar!
JEAN. That's enough!
TOBIAS. I want the truth, and I'm not leaving here until that liar tells. . . .
WILLIE JOE. Don't call me a liar, Stinton!
ESTHER. I want this stopped now!
TOBIAS. And you're a liar too!
SIDNEY. Tobias, I think you'd better leave!
TOBIAS. Not til I get to the bottom of this! (*He fetches Bible from the dresser and holds it out to* ESTHER.) Put your hands on this Bible, the pair of yis, and swear on the Lord's name that this wasn't planned behind my back!
ESTHER. I don't think there's any call for this exhibition!

THE GIRLS IN THE BIG PICTURE – ACT II SCENE II

TOBIAS. I know you've always hated me, but I'm goin' to show you up for what you are! Swear on that Bible!
ESTHER. What is it you want me to swear?
TOBIAS. That this wasn't planned behind my back!
ESTHER. What?
TOBIAS. Him, coming here with a box of chocolates!
ESTHER. Sure I knew Willie Joe was comin'. I invited him.
TOBIAS. Swear you didn't know that I would be here with Jean! (ESTHER *stares at the Bible.* TOBIAS *is triumphant, convinced he has her now. After a long pause....*)
ESTHER. Has my daughter given you an answer, Tobias?
TOBIAS. No, not yet. But she will soon, she said.
ESTHER. Is this so, Jean?
JEAN. Yes.
ESTHER. You have put us all in a very difficult position!
JEAN (*flabbergasted*). I'm very sorry, Mr. Ferguson. I'm sure you didn't realise, but yes, Tobias has asked me.
ESTHER. I knew nothin' of this intended marriage.
JEAN. There is no intended marriage!
ESTHER. I'm sorry you had to find out this way, Tobias.
JEAN. I never said I would never marry him!
ESTHER. Are you saying that at no time had you ever decided not to marry Tobias?
JEAN. That's right.
ESTHER. But not a month ago at that very table Sidney asked you if you were goin' to marry Tobias and you said 'no'. And Sidney will bear me out in that.... Sidney?
SIDNEY. I – I can't remember.
JEAN. I didn't mean never, I meant 'no' at that time.
ESTHER. And now? (JEAN *simply cannot answer.* TOBIAS *leaves the Bible back on the dresser and silently leaves.*)
JEAN. Tobias!
ESTHER. I'm very sorry you had to be put in this position, Willie Joe. I'll speak to you later. Sidney, see Willie Joe out. (SIDNEY *and* WILLIE JOE *leave.*) So, he asked you to marry him. And you hadn't even the guts to say no! You don't love him. You don't respect him as a man.
JEAN (*almost in tears*). He's a good man!
ESTHER. Jean, you are a big lump of thirty-two years of age! You should be down on your hands and knees thankin' the Lord that a man like Willie Joe Ferguson wants ye!
JEAN (*now completely broken*). I don't want none of them!

ESTHER. And who else d'ye think's ever goin' to ask ye? D'ye want to be an oul' maid the rest of your days?

JEAN. Then I'll marry Tobias!

ESTHER. That insignificant wee runt of a man! And what standin' would that give ye? You'd look a right picture standin' at the altar with him! Laughing stock of the whole town!

JEAN. Ach, Mammy, just leave me alone! (JEAN *runs out.* ESTHER *begins to clear up tea things.* SIDNEY *enters and tries to sneak across behind her.*)

ESTHER. Sidney! Sit down, son, I want a word with ye. (*They both sit.*) It's not easy for me to be a mother and a father. And I know there were times when you might've thought I stood in your way, but if I have it's because I love you and I'm worried for you. I know how much the singing meant to you. I should've been proud of you; the whole town's talkin' about how marvelous it all is. But I will never stand in your way again!

SIDNEY. Mammy, I need you to sign a contract. It's nothin', it's just so Pat can manage me, cos I'm under age. (*He gets contract from his jacket and gives it to his mother.*)

ESTHER. Well, get me a pen. (SIDNEY *does so, from the dresser.*) I suppose that'll be you going places now, as Patsy Flood informed me?

SIDNEY. Naw, it's just one or two spots a week.

ESTHER. They'll be wanting you across the water next.

SIDNEY. Not at all!

ESTHER. Oh, now, I'm sure Mr. Sharkey has big plans for you!

SIDNEY. Dunno.

ESTHER. What do you think we could arrange here, if you're too busy with all these spots?

SIDNEY. Mammy, I've been thinkin'. If I were to earn a few shillin's, we could hire people in. The money's good; Stanley Semple and the boys, they have no work. They'd be glad of it!

ESTHER. You've been thinking ahead.

SIDNEY. But sure, it'll never come to that.

ESTHER. Your father loved this land. Handed down by Lord Kilbarron to Edward Moore, for services to the crown, during the ninety-eight rebellion, when that big tree there was first planted. 'Moore's farm or nothin'', he'd have said. 'No stranger will ever tend my land'; those were his famous words. It'll be an awful shame to see it all go!

SIDNEY. Mammy, you're not sayin' you'd think about sellin' the land!

ESTHER. Ach, don't you worry about me, son. I'll be all right. And anyway, as you say, I'm sure it'll never come to that. (*She sighs.*) I'm tired now. Goodnight, son. (ESTHER *exits leaving* SIDNEY *with the contract.*)

LX FADE

ACT II – SCENE III

SCENE: *LX on cafe back kitchen.* MARGARET *is sitting on table with a cup of tea, reading a book.* MARY JO *enters from cafe with old exercise book.*

MARY JO. You enjoyin' yourself?
MARGARET. I'm havin' a break, am I allowed?
MARY JO. Margaret! You wanna see behind that big cupboard in there! It would turn ye! There's peas and carrots lyin' there about fifty years old! Look what I found! (*Gives* MARGARET *book.*)
MARGARET. Ach, for God's sake! First class! (*Reads.*) 'I live in a big red house. My mammy and daddy own a cafe. When I grow up I will own a cafe with forty tables. I will drink lemonade all day and eat buns. And then I will clean and clean and clean'!
MARY JO. What happened ye?
MARGARET. Ach, sit down. You haven't stopped since the scrake!
MARY JO. No, I'm too excited waitin' on my chip fryers comin'.
MARGARET. Pity we didn't have a bottle of champagne to celebrate!
MARY JO. Margaret, this was your idea too!
MARGARET. I wasn't jokin'!
MARY JO. Oh, it's goin' to be gorgeous, isn't it?
MARGARET. Think of the fun Nan Flood would've had with the chip fryers! There would've been grease from here to Drumshandy!
MARY JO. She has our house covered in it instead!
MARGARET. Did he eat the cabbage?
MARY JO. He ate every pick and licked his lips forby. The two of them, sittin' there yarnin' away all night about nothin'. There's me gave up everything to look after him, and now he couldn't care less if I never came home!
MARGARET. Isn't that what you always wanted?
MARY JO. It's a bit late in the day, isn't it? I'll go and get Jean and get more Flash.
MARGARET. Ach, leave her.
MARY JO. No, she said she'd help me kill them clocks!
MARGARET. He has her slavin' over there. This'll be the last thing she'll want.

MARY JO. She may as well be married to him. Y'know it's only two weeks to the time when she said she'd let him know.
MARGARET. She should marry him and get away from her ma.
MARY JO. She doesn't want to marry him!
MARGARET. I can see her point! (PAT *enters carrying milkshake machine.*)
PAT (*giving milkshake machine to* MARGARET). There ye go! Latest model on the market!
MARGARET. I don't believe it! You really brought it!
PAT. Would I let you down?
MARY JO. Pat? Did ye get my chip fryers?
PAT. The fella says he's havin' trouble gettin' them at that price. He's still beatin' the other fella down. Should be here by the weekend, though.
MARY JO (*very disappointed*). When are they comin'? Like what day? Cos I would need to know!
PAT. That's what Churchill used to say about the Germans, Love, but he just had to wait.
MARY JO. What's that got to do with my chip fryers?
PAT. German made!
MARGARET. Pat Sharkey! You're getting worse!
MARY JO. Margaret, you know we can't open until they come!
PAT. Trust me, Mary Jo! C'mon now, give us a big smile! Have ye heard Sid's new song? (*He puts his arms around her from behind and sways in time to the singing.* MARY JO *is pleased and embarrassed at his attention.*) 'After all we shared, she only left a letter'. . . .
MARY JO (*breaking loose and giggling*). We'd better get stuck in here, Margaret. This place is minging!
MARGARET. Aye . . . well, you go and get the Flash! (MARY JO *exits. Silence.*)
PAT. How are you, then?
MARGARET. Pardon?
PAT. I'm asking how you're keepin'?
MARGARET. Why?
PAT. You said . . . I can't just talk . . . well, no jokes . . . I'm talkin'. . . . So, how are ye, then?
MARGARET. I'm all right.
PAT. Good.
MARGARET. And yourself?
PAT. Not too craigavad. (*Catches himself.*) I mean . . . I'm not bad.
MARGARET. Good. (*Silence.*)

PAT. Margaret . . . fancy comin' up to the final with me at the weekend? (MARY JO *bursts in with a packet of Flash, followed by JEAN.*)
MARY JO. Right! All hands on deck! Jean's here!
JEAN (*heading for a chair*). Aw great, tea break.
MARY JO. Ach, for flip's sake!
PAT. I'm away to see stingy Stinton here. . . . (*He realises what he has said.*)
JEAN. He's still waitin' on his wood.
PAT. There's a wee bit of a problem there.
MARGARET. Is it German too?
PAT. I can't work miracles! I'll be back. (*Exits.*)
JEAN (*taking off shoes*). I'm exhausted. I feel like a herd of cows has sat on me!
MARY JO. Suppose we could always open the cafe next year!
JEAN. Ach, I'm sorry, Mary Jo, but I've been at it all day over there!
MARGARET. How's it look?
JEAN. Great. He's a real hard worker.
MARY JO. Well, I'm not goin' in to face them clocks myself, and you know what I'm like about mice!
MARGARET. In a minute!
JEAN. Soon as the shop shuts he's straight in to decorating the upstairs!
MARY JO. Oh, my God!
MARGARET. It'll be hard to say no.
MARY JO. I suppose the oul' Tuckeys have their fox furs dusted down ready for the big day!
JEAN. They stopped visiting when they heard about that night with Tobias and Willie Joe! God love them! My mammy was all they had once Nessie died!
MARGARET. Well, they're that picky who they visit, any road!
MARY JO. Ach, y'know them two. If you haven't lived in Cloughmartin since the flood, you're a stranger! Sure they can't have Margaret's mammy and daddy, say they're blow-ins! And they've been here forty-five years! Jean, you want to see our new menu; home-made mince tart, selection from the sweet trolley, cloth napkins, the lot! It's great, isn't it, Margaret?
MARGARET. Yes.
MARY JO. Margaret, y'know I've been thinkin'; we should get stuck in to them two oul' storerooms up the stairs and red them all out.
MARGARET. What for?
MARY JO. Do you realise them's two good sized rooms up there, only

been goin' to waste?

MARGARET. I am not movin' in here with you! (*Realizes she's been a bit sharp.*) I'd rather live with your da! (*They laugh weakly.* JEAN *tries to save the situation.*)

JEAN. C'mon, Mary Jo, you and me'll get stuck in to these clocks! (*Rising and putting on shoes.*)

MARY JO. Oh, you want to see them, Jean. They're the size of big pot lids! (SIDNEY *comes in.*) Oh, great! Here's another helper! Sidney, you're a man, kill these clocks for us!

SIDNEY. Naw, I want to speak to Jean.

MARY JO. Oh, just cos he sang in Ballymena last night he thinks he is somebody!

MARGARET. Mary Jo, c'mon!

MARY JO. I'll go and get a mouse-trap! (*They exit to cafe.*)

JEAN. Well, what's wrong?

SIDNEY. It's Sally; she's goin' to have a baby!

JEAN. Yours?

SIDNEY. Aye.

JEAN. Are you goin' to marry her?

SIDNEY. She wants to.

JEAN. What's mammy goin' to say? . . . Ach, Sidney! How could you have been so stupid? She'll be livin' in our house?

SIDNEY. Jean, will you tell her?

JEAN. Why me?

SIDNEY. Cos I can't face her!

JEAN. Ach, Sidney! You're goin' to be a father! Now go up there and face her like a man!

PAT (*entering with a bag of clothes*). There y'are! You should be practisin' your finger-pickin'! Sally's over there with a face like a wet fortnight waitin'. . . .

JEAN (*exiting*). I'll be over in Tobias's.

PAT. Tight jeans, leather jacket, knock-out! Try them on! (*Dumps bag on chair.*)

SIDNEY. Pat, Sally's goin' to have a baby!

PAT. You mean you've been. . . .? Jesus Christ! Did you not use nothin'. . . ? I hope this isn't goin' to affect your performance – three days to go!

SIDNEY. I'm goin' to marry her!

PAT. Catch a grip!

SIDNEY. I'm goin' to marry her!

PAT. Sid, there's no place for a woman and a kid in our plans!

SIDNEY. And I can't do it – the final.

THE GIRLS IN THE BIG PICTURE – ACT II SCENE III

PAT. Look, you're upset. There's no need to get carried away!
SIDNEY. No, Sally and I have to get married and that means givin' it all up!
PAT. After all the time and money I've spent on ye! Groomin' ye, dressin' ye! You're yella, is that it? Can't face the big world. Is that it?
SIDNEY. Ach, Pat, it's just not possible!
PAT. You're goin' to ruin your future for some stupid bitch!
SIDNEY. It was me too!
PAT. Sid, you're only nineteen! Did you read what it said in the Tyrone Constitution? 'Si Stevens, the face that's up for the tops'!
SIDNEY. Sorry, Pat.
PAT. And what about the spots we've lined up? I suppose you've forgotten we've a contract!
SIDNEY. There's plenty of people singin'.
PAT. Sidney! Look, away home and think about it, get your head clear! I'll talk to you later.
SIDNEY. No, my mind's made up!
PAT. I could sue you!
SIDNEY. Well, then, I can't stop ye!
PAT. Did that wee bitch over there put you up to this?
SIDNEY. Naw!
PAT. Well, you'd better get back over there! Tell her it's not a pop star she's marrying! She might change her mind!
SIDNEY. Sally loves me!
PAT. And how do you know it's yours? Have you asked her that?
SIDNEY. What're you talkin' about?
PAT. Just make sure you're not ruinin' your life for someone else's brat!
SIDNEY. Knock it off, Pat!
PAT. Don't be surprised if it looks like me! (SIDNEY *punches him. He falls to the ground.* MARGARET *runs in to see* SIDNEY *standing over him, ready to continue. Instead he lifts the bag of clothes, throws it at the cowering* PAT, *and runs out.*)
MARGARET (*kneeling beside* PAT). What happened?
PAT. He's got the Spence one up the spout, took it out on me!
MARGARET. Oh, Sidney!
PAT. And he's not goin' to do the final! That's me definitely leavin' this place!
MARGARET. I'll go and get a damp cloth. (*She exits to cafe and comes back with a bowl of water and a cloth. Holds the cloth to* PAT'S *eye.*) What are you goin' to do next?

THE GIRLS IN THE BIG PICTURE – ACT II SCENE III

PAT. Belfast for a while. Then, God knows!
MARGARET. Must be great, not to know!
PAT. You'd hate that!
MARGARET. You don't know!
PAT. What is there to know?
MARGARET. No family?
PAT. Somewhere.
MARGARET (*rising, putting the bowl on the table*). I'm glad you said that. I thought you were goin' to be dead borin' and tell me all about your aunts and uncles and second cousins. (PAT *rises, turns her and they kiss.*)
PAT. I'll be back on Friday; can I see you? (*They kiss again.* MARGARET *begins to respond more and more intensely until she is clinging to him, holding his face. He has to break away roughly. The situation is now embarrassing.* PAT *lifts the bag and goes to leave.*)
MARGARET. Til Friday, then?
PAT. Aye. (*Exits.* MARY JO *bursts in, carrying a mousetrap.*)
MARY JO. Right! That mouse is on my death list! Did I see Pat Sharkey coming in?
MARGARET. He's away.
MARY JO. Is he callin' back?
MARGARET. I think he's had enough of this place!
MARY JO. Is he away for good? (MARGARET *nods.*) What about Sidney and the singing?
MARGARET. I don't think he's goin' to do the competition.
MARY JO. What? But why? That's stupid! (JEAN *comes in.*) Hello, Jean.
JEAN. Hello. I thought that you two really ought to be the first to know that Tobias and me's gettin' married.
MARY JO. Honest to God?
JEAN. Yes. Harvest.
MARGARET. Congratulations!
MARY JO. Did ye change your mind?
JEAN. I never said I would never marry him. Well, one down, two to go! Wonder who'll catch the bouquet?
MARGARET. That's marvellous, Jean! I hope the two of yous'll be very happy!
JEAN. Thank you!
MARY JO. Congratulations!
JEAN. Mrs. Stinton!
MARY JO. Did you just decide the day?

JEAN. No, I knew. I just had to be sure, y'know.
MARGARET. Well, that's great, isn't it?
MARY JO. Aye!
JEAN. Well, I'll go on up and tell my mother. I'll see yis!
MARY JO & MARGARET. Night, Jean. (JEAN *leaves*.)
MARY JO. Margaret! I can't credit that! She never wanted to marry him!
MARGARET. She's no choice! Sidney and Sally Spence *have* to get married!
MARY JO. Oh, dear God in Heaven! That's awful! Sally Spence and Mrs. Moore in the one house! I wonder who'll win that battle?
MARGARET. What battle? Mrs. Moore'll be as nice as ninepence!
MARY JO. God! Jean married!
MARGARET. Just you and me left!
MARY JO. Maybe there'll be somebody at the Basket Tea!
MARGARET (*moves to door*). Come on and we'll get this kitchen cleared up. (*They exit*.)

LX BLACKOUT

ACT II – SCENE IV

SCENE: *SFX soundtrack of American bubblegum movie covers scene change and continues under the dialogue.* THE MCKERNANS *(two of them only) sit as usual on the rise. Three chairs are set downstage, but only* MARGARET *and* MARY JO *are there. It is a year later.* PATSY FLOOD *is there as usual.*

PATSY. Micky Matthews! If you don't want to watch the picture – O-U- T! Hello, Pearl! Hello, Brian! Should you be here, Pearl? You're very near your time! You're all out front! It's goin' to be a wee girl! (*Comes downstage.*) How's the girls? This is a load of oul' rubbish, isn't it?
MARGARET. I would like to see it, Patsy!
MARY JO. She likes it for the style!
PATSY. Oh here! Will yous babysit for me on Saturday night? It's my Norman's meat processin' dinner dance!
MARY JO. Aye.
PATSY. Our wee Jason says to me, 'Does Mary Jo have to bring oul' sour gub'? (MARGARET *moves up a seat in disgust.*)
MARY JO. Don't mind her!
PATSY. He likes you but! (*They watch the movie for a while.*)
MARY JO. You're right, Patsy, this is a load of oul' rubbish!
PATSY. The young ones'll watch anything these days!
MCKERNAN ONE. Would yous stop gossipin', we're watchin'?
PATSY. Aye, when it suits ye! . . . I'd rather be at home watching 'Quartermass', it's great! Have yous got the television yet?
MARY JO. No, but we're savin' for one.
PATSY. Ach, sure everybody has one nowadays, with the electric. Even oul' Willie Joe Ferguson – I believe he never leaves the house! Come to think of it, I haven't seen him in over a year, since thon Basket Tea, d'ye remember? It was a bloody turn! Oh here, I saw your da and Nan Flood at the Bingo last night; he was like a two year old! My Norman's all delighted; he says his ma has a new lease of life! So am I. It means I'm not tortured by her! It's great at their age, isn't it?
MCKERNAN TWO. See if ye don't shut up, we're goin' to get Jamesie

THE GIRLS IN THE BIG PICTURE – ACT II SCENE IV

for ye!
PATSY. Now y'know what we stick! (MARY JO *nudges* MARGARET, *who hands over a bag of sweets. She takes one, offers them to* PATSY, *and hands them back.*)
PATSY. Is it her turn?
MARY JO. Aye.
PATSY. Did you get them in stingy Stinton's?
MARY JO. Aye.
PATSY. I went over the other day for a quarter of jelly babies for our wee Jason. Well, it was a wee bit over, so he took one out, then it was a wee bit under. Says I, 'Tobias, d'ye want a knife to cut its head off'?
MARGARET. C'mon, we're goin'!
MARY JO. It's not finished yet!
MARGARET. I can't hear it!
PATSY. Sure, that's only for the young ones, that rubbish!
MARY JO. It'll be over in a wee minute!
PATSY. Does she get on like that in the house?
MARY JO. Only when she gets annoyed.
PATSY. Do yous fight?
MARY JO. I just ignore her.
PATSY. I'm the same with my Norman. (*SFX end of film soundtrack up.*)

SFX. I was born on the wrong side of the tracks for you, girl.
SFX. Oh, Marty, this is a new age. It's a world for our generation, and there ain't no room for any tracks to be on one side or the other of.
SFX (*song*). 'God Save the Queen.'

PATSY. Thank God that's over! (*Set is cleared as before.* JEAN *and* TOBIAS *enter to meet* MARGARET *and* MARY JO.)
JEAN. Well, how was the pictures?
MARGARET. The usual, starring Patsy Flood with her torch!
JEAN. She never stops, does she?
MARY JO. Where were yous?
TOBIAS. Visitin' the Bradshaws.
MARGARET. Mervyn and Beattie Bradshaw?
TOBIAS. Aye.
MARY JO. Oh! Comin' up in the world, gettin' invited to the bank manager's house!
JEAN. Naw. Betty just has a wee supper every once in a while.

THE GIRLS IN THE BIG PICTURE – ACT II SCENE IV

TOBIAS. Aye. Six couples, so Jean only has to cater twice a year, and we get for free the other ten times. It's a good system.
JEAN. Are yous for the Basket Tea come Saturday?
MARY JO. Ach no, Margaret doesn't want to go!
JEAN. Ach, Margaret! It's a wee night out for yous! (PATSY *enters on her way home.*)
PATSY. Are yous not away home yet? Ach, look who it is, Mr. and Mrs. Mace!
JEAN. Hello, Patsy.
PATSY. Here, that was a load of oul' rubbish, wasn't it?
MARGARET. Mary Jo, c'mon. Goodnight all!
MARY JO. Goodnight!
JEAN. See yous soon! (MARY JO *and* MARGARET *leave.*)
PATSY. Oul' sour gub! She has that wee one ruled! Would you look at the two of them, arm in arm! Awful shame, isn't it? Well, I must go on! My Norman hates me being out late! Shockin' jealous! (*Exits.*)
JEAN. Goodnight, Patsy!
TOBIAS. Oul' gossip that woman!
JEAN. There's our Sidney comin' out of Spences'.
TOBIAS. Have that pair got themselves sorted out yet?
JEAN. He was goin' to have another talk with her tonight, but I don't think she'll change her mind.
TOBIAS. C'mon we'll go.
JEAN. No, houl on a minute. (SIDNEY *enters.*) Well. How's wee Beverly?
SIDNEY. She was sleepin'. Sally's ma wouldn't waken her for me. She said, 'The child has a routine and she should stick to it'.
JEAN. Did you talk to Sally?
SIDNEY. She's down at the Rio with her sister.
JEAN. Give her time.
SIDNEY. Mrs. Spence laid into me the night. She said if I'd been any sort of a man, I'd have stuck up more for my wife instead of my mother!
JEAN. What did you say?
SIDNEY. Jean! You know my mammy couldn't have been any nicer to her!
JEAN. She just wanted you to herself!
SIDNEY. That's not true! She couldn't have done enough for Sally, and she loved the baby!
JEAN. No, it was Sally I meant that wanted you to herself!
SIDNEY. Aye, well. She can't see any further than the child.

THE GIRLS IN THE BIG PICTURE – ACT II SCENE IV

JEAN. Oh?

SIDNEY. Aye, well. I'll go on. Mammy waits up for me.

JEAN. I'll be up to see to the chickens as usual in the mornin'!

TOBIAS. Look after them turkeys, Sidney. That's the Christmas stock!

SIDNEY. Aye, goodnight! (*He leaves.*)

JEAN. Poor Sidney!

TOBIAS. Aye, well, you make your bed, and you lie in it! C'mon, I want to get those egg boxes filled up for the mornin'!

JEAN. Right! (*They leave. The table is placed downstage with the two chairs. We are in* MARGARET *and* MARY JO's *flat.* MARGARET *is sitting in her dressing gown reading. There's a pot of cold cream on the table.* MARY JO *comes in wearing dressing gown with two glasses of sherry and sits down.*)

MARY JO. There y'are. If that doesn't knock ye out, nothin' will! If business keeps goin' the way it's goin' we should be able to afford the TV set by next month....I can't wait!

MARGARET. Maybe we could find someone to get us one cheap? (*They laugh.*)

MARY JO. I wonder where he is now? . . . Think I'll go down and put my jellies in the refrigerator for the mornin'. And then I'm goin' to stay up and write my poem for the W.I.

MARGARET. I'll do it; you're useless at them!

MARY JO. No! Only means you'll write yourself a better one!

MARGARET. That's what you get for not bein' intelligent!

MARY JO. Who won the best recipe for a chocolate cake? (MARY JO *reaches for the cream.* MARGARET *gets there first. They both rub cream on their faces in silence.*)

MARGARET. Don't you leave the lights on the night!

MARY JO. Don't you snore!

LX FADE. HOUSE MUSIC UP. END OF ACT II.

FILM SOUNDTRACK

JOHNNY. No use denying it, Angela. Malcolm saw you. You were in his arms, dancing in the Blue Lagoon.

ANGELA. No, Johnny, you don't understand!

JOHNNY. My God, it's a double betrayal, you and Ronald. Hah! He was going to be our best man. Good God, he belongs to my Club!

ANGELA. Please, Johnny, let me explain. . . .

THE GIRLS IN THE BIG PICTURE – ACT II SCENE IV

JOHNNY. Don't try to bewitch me again!
ANGELA. Think what you please about me. I won't have you estrange your best friend!
JOHNNY. Aha! You try to defend him. You know how to twist the knife, Angela!
ANGELA. You're not being reasonable, darling!
JOHNNY. You'll push me too far, Angela!
ANGELA. Oh, Johnny, Johnny. He's wrong. Malcolm's so very wrong!
JOHNNY. Malcolm's the only true friend I have!
ANGELA. He's a liar and a coward! He'll say anything to turn you against Ronald!
JOHNNY. I think I truly see you for the first time, Angela; you'll blacken anyone's name to save Ronald Metcalf's skin!
ANGELA. You're not in your right mind, Johnny, to say things like that!
JOHNNY. No, I'm not in my senses! You've bewitched me and torn my life into little shreds and blown them away!
ANGELA. Oh, darling, please listen!
JOHNNY (*throws her across the room*). You slut!!!
ANGELA. Oh, Johnny, don't look at me like that! Those eyes were always filled with love and tenderness; now I see hate! Johnny, you're frightening me! Johnny, why are you looking at me like that? What are you going to do? Johnny, answer me! (JOHNNY *hits her.*) Aaagh! (*Sound of* JOHNNY *leaving. Door slams. Dog barks.*) Oh, Johnny, Johnny, don't leave me, don't go, don't go. I love you so much! Johnny! Johnny! (*Opens door and slams on dog in frenzy. Screech of brakes. Screams. Yelp.*)
WOMAN (*screams*). Oh, my Gawd, you've killed the lady!
DRIVER. Cor, strike a light! She came at me so sudden like!
POLICE CONSTABLE. Stand back! Stand back!
JOHNNY. Oh, my God, Angela. I've killed her!
WOMAN. Looks dead enough to me, all right. Nasty thump!
DRIVER. Wasn't my fault, guv'nor. She came at me out of nowhere; this woman's my witness!
WOMAN. I didn't see nuffink, mate!
JOHNNY. Oh, Angela, darling, I'm sorry!
POLICE CONSTABLE. Call an ambulance!
JOHNNY. I'll go with her! (*Ambulance sounds through streets.*)
JOHNNY. Oh, my God, Angela; her lips have turned blue!
AMBULANCE MAN. Looks like an emergency 'ere mate. Doc says up to feater at the double!
MALCOLM (*bursts into Club. To* RONALD). Oh, God, Metcalf.

THE GIRLS IN THE BIG PICTURE – ACT II SCENE IV

Something awful has happened! Angela's been seriously injured; she ran out into the road outside her flat!

RONALD. What are you jabbering about, Duckworth?

MALCOLM. Look, there's no time to lose! Johnny was in a bit of a state when he phoned me from St. George's Hospital; apparently it's touch and go with Angela. You've got to tell him why you were with her last night at the Blue Lagoon. Tell him about the inheritance. Oh, God! It's all my bally fault.

RONALD. For Heaven's sake, pull yourself together, man. You can explain that last cryptic remark later; just give me the number.

MALCOLM. Sloane 348.

RONALD. I say, Porter, be a good chap and get me Sloane 348 and step on it!

JOHNNY (*at the Hospital*). I say, Doctor, can't you give me any news? I've been here all night and most of the morning.

DOCTOR. Ah, yes, Mr. Carlton. Are you some relation? She asked for you when she regained consciousness.

JOHNNY. Did she? Did she really? Yes, well, I was her fiance.

DOCTOR. Was?

JOHNNY. Oh, Doctor, you've got to tell me the truth. Is she going to be all right?

DOCTOR. Well, she's in a coma, Carlton. Maybe you're just the ace we can play. Perhaps if she were to hear your voice.

JOHNNY. I'll give it a try, Doc. What are her chances of pulling through?

DOCTOR. The spine may be badly affected; she could be paralysed from the neck down. She'll need a lot of love and strength. Go in and see if you can bring her round. Gently now, no excitement. (*Door opens.*) Okay, Nurse, you can leave them alone. No change?

NURSE. None, Doctor.

JOHNNY. Angela, darling – can you hear me? It's me, Johnny. (*He kisses her.*)

ANGELA. What is it? Where am I? What happened? Johnny, is that you?

JOHNNY. Yes, darling, do you mind? You woke like 'Sleeping Beauty' from a kiss.

ANGELA. Mind? Darling, why on earth should I mind?

JOHNNY. Oh, Angela! I'm such a heel! Can you forgive me? I know I was wrong about you and Ronald. He told me all about the inheritance. Oh, darling, you were doing it all for me.

ANGELA. Oh, Johnny, let's put it all behind us. (*Dog barks.*)

JOHNNY. I brought you this puppy to replace 'Boo Boo'.
ANGELA. Oh, darling, he'll be a symbol of our love!
JOHNNY. I've an even better symbol!
ANGELA. Oh, Johnny, your mother's ring! Put it on my finger, darling.
JOHNNY. Angela, you can move your hand! Oh, my precious darling, hold my hand through all eternity! Angela – well-named, my Angela!
ANGELA. I must be an Angel. I think I'm in Heaven!
JOHNNY. Darling girl! (*Swells of music.*)

SOMEWHERE OVER THE BALCONY

PLAY IN TWO ACTS

BY
MARIE JONES

CHARACTERS

KATE TIDY – devout Catholic in her thirties

CEELY CASH – irreverent widow in her thirties

ROSE MARIE NOBLE – mother of twins in her thirties

ACT I

Balconies in tower block council flats, Belfast

ACT II

The balconies, next morning

SOMEWHERE OVER THE BALCONY

ACT I

SCENE: *LX up on* KATE TIDY, *a woman in her early thirties. She is wearing a nightdress and dressing gown and carrying a large tin rubbish bin. It is the first light of the day.* KATE *plonks the bin down and sits on it. Pause as she takes in the silence of the morning.*

KATE. See this bin . . . thank God for it . . . (*Pause.*). . . .Oh, the peace and quiet. If it wasn't for me having my own bin I wouldn't know peace and quiet. They all used to laugh at me in these flats. . . . 'Look at Kate Tidy with a bin . . . we have rubbish chutes, nobody needs bins in here'. But I was one step ahead of them. I get up at five o'clock every morning when there is not a sinner about and I can carry my bin out til my heart's content. (*Listens.*) There's not a peep . . . it's Heaven. (*Stares up as if looking at something.*). . . .I remember when the British army moved into that Tower block . . . my Frank says 'Them bastards are watching us'. Says me, 'They must have bloody good eyesight cos they are fifteen floors up'. 'Cameras', he says. 'Close-up cameras'. And lo and behold that very next day there was five soldiers hokin' around in my rubbish . . . but I don't care cos I never done nothin' as long I can't see them looking at me. So I just bought myself a new dressing gown and ignored them. . . . 'They can hear you too', says he. . . . Big deal (*Realizes she might be heard and changes to whisper.*) 'Big deal,' says me, 'sure them English can't understand us anyway'. (*Pause.*) I wonder how many videos they have of me and my bin? Hundreds over the years. I'll be doing them a favour cos they'll have nothin' to look at soon. . . . Only four families left in that block. (*Indicates block of flats opposite.*) The army won't be wasting any more film on them. We're the next block to be demolished, I can't wait. Wonder what they'll do then. Ach, I suppose they'll just get bored and go back to England. (*SFX early morning sounds – e.g. traffic, helicopters, the signature tune of an early morning soap, 'Neighbours'. LX up on balcony.* CEELY *and* ROSE *enter and climb onto balcony. They are women in their thirties. They are also dressed in nightwear.* KATE *joins them. SFX of helicopter becomes the dominant sound. The* THREE WOMEN *look up at the sky.* KATE *and* ROSE *pretend to bring chopper to land with hand signals while* CEELY *looks at it through army binoculars. SFX of chopper landing on roof of nearby tower block. SFX of music fades in for song. The following dialogue takes place during intro. Music for song.*)

ROSE (*tongue in cheek*). British Airways. They'll be staying in the Penthouse.

CEELY. I love it when a new tour comes in. All them wee pink arses not tanned yet.

KATE (*mock horror*). Ceely Cash, have you been fraternising with the foreigners?

CEELY. No, just spying on them from the balcony with my binoculars.

KATE. It's them that spy on us from the tower with their cameras. (*Said loud as if army can hear. They all wave and give rude sign.*)

ROSE. They'll be on the rampage tonight. Rape and pillage, getting drunk, projectile vomiting from the fifteenth floor, drowning the natives.

CEELY (*sarcastically*). Sure, aren't they on their holidays?

ROSE. Oh, is that what they call it?

KATE. Sure, everybody deserves a wee holiday.

CEELY. Everybody has a right to a holiday.

ROSE (*ironically*). Aye, holidays. (*They all sing.*)

Why Do You Watch Me

Why do you watch me? I won't do no wrong.
Just mind my own business and wish you were gone.
You hear every whisper, each move that I make.
And when I am sleeping you're always awake.
You're there every morning every hour of the day.
I've nowhere to run to, I just gotta stay.
I can't see your eyes, I can't look on your face,
but I feel your presence all over this place.
Why don't you leave me for my peace of mind?
You don't belong here, not one of my kind.
Go back where you came from, I'm tired of your stare.
I can't have my freedom until you're not there.

(KATE *climbs off balcony and sits on her bin while* ROSE *and*

SOMEWHERE OVER THE BALCONY – ACT I

CEELY *stare up at look-out post on tower block.*)
ROSE (*holding a child's bath sponge, e.g. animal shape. To* CEELY). The water is off again.
CEELY (*to tower*). The water's off again. Plenty of water in 1969 when yous all came lookin' cups of tea. (*To* ROSE.) To think I made a cup of tea for a Brit.
ROSE. My ma made tea for the Brits.
CEELY. Everybody's ma made tea for the Brits. Except for Mena Mackle over there. (*Reciting the rest of dialogue as if it is a well-known poem.*) She never gave in. She stood all alone and rattled her bin. When biscuits and cream buns were passed all around, she warned us beware of that vile English crown. . . . She is famous. That's the only woman on the Falls Road never to make a cup of tea for a Brit.
ROSE. How am I going to wash my twins?
CEELY (*shouts up at army*). How is she going to wash her twins? (*To* ROSE.) Sure you washed them yesterday. (*SFX helicopter as if flying just above them.*)
ROSE (*screams*). I wish the British Army would get silencers for their choppers.
CEELY (*looking through binoculars*). Jesus look at that soldier. (*Incredulously.*) He's joggin'!
ROSE (*grabs binoculars*). On the Tower block? They'll be sunbathing up there next.
CEELY (*fantasizing*). Oh, he'll be all hot and sweaty now.
ROSE (*shocked*). Ceely Cash, he's a Brit.
CEELY (*snaps at her*). He'll still be all hot and sweaty. (KATE *appears on balcony. She has a clothes peg on her nose.*)
KATE. Flies, wasps, ants, cockroaches . . . It is like a flippin' jungle in my flat. (*Shouts up at army post.*) My toilet is blocked up again. (*To* ROSE *and* KATE *as she takes clothes peg from her nose.*) I'm nearly stunk out. Has anybody seen my Pepe?
CEELY (*shouts up at army post*). Have yousins got her dog?
ROSE. I seen him playing over in the empty flats with Rambo McGlinchy. He'll get the mange, that dog is stinkin'.
KATE. Pepe! Pepe! (*SFX dogs barking as if they come running past them on the balcony and almost knock them over as they* ALL *shout* 'Get down,' 'Go away,' *etc. Then they look across to the flats opposite as if that is where the dogs have gone.*) Pepe, come you down out of those empty flats. That place is dangerous.
ROSE. I don't know why they just don't pull that oul' buildin down and chuck all them squatters out. Look, you can see right through

to the other side.

CEELY. Aye, but you get a lovely view of the City Hall.

ROSE. It is an eyesore. A death trap.

KATE (*calling to Pepe*). No, just stay where you are. Ach, he'll be safe with Rambo. That dog has ate more soliders' legs than dog biscuits, and him with only one eye.

CEELY. He used to be the quietest wee dog in Belfast, until he got shot with a plastic bullet.

ROSE (*sarcastically*). Oh aye, a legend in his own lifetime. (CEELY *begins to sing very reverently and* KATE *joins in to the tune of 'The Holy Ground'*).

> Oh, Rambo McGlinchey, you've only got one eye.
> You're fearless and brave, and for us you would die.
> You take on the British with no weapons at all,
> And when Ireland at last is free you will hang on our wall.

CEELY. Fine dog y'are.

KATE. Ceely, you know the state of my walls. Would you tell your ones to keep their ghetto blasters down a wee bit? I don't mind one, but four going at the same time. . . .

CEELY. Why don't you tell them to keep their choppers down?

KATE. It's terrible. Every time I sneeze they tremble. I've done four novenas, lit six candles, Mena Mackle is doing a daily vigil to St. Teresa, the two wee holy sisters from the Tower block are doin' 'Our Lady of Perpetual Succour', and wee Bridie is coming round the night to rub them down with Padre Pio's mitt.

ROSE. Padre Pio's mitt! For your walls?

CEELY. It's good for getting phlegm up.

ROSE. My Tucker could maybe do something. He is awful good with his hands.

KATE. Just ask him to put them together in prayer for me.

ROSE. Plasterin'? He has a quare straight eye, great snooker player.

KATE (*indignantly*). My walls are in the hands of St. Jude. (KATE *exits to ground level and sits on her bin.*)

ROSE. She cud start a Fruit Shop with fungus growing out o' her walls.

CEELY. Fungus! . . . Jesus, that reminds me. (*Shouts over balcony.*) Tucker! Come on up here and shave your granda. Tucker! Tucker! (*Sees* JOLENE.) Jolene, go and find our wee Tucker for me. Tell him to come up and shave his granda before these foreign visitors come. (*To* ROSE.) Can you take two of the 'Troops Out' movement?

SOMEWHERE OVER THE BALCONY – ACT I

ROSE. I can't take no movements . . . the twins are too strange . . . they wouldn't like it.
CEELY. No wonder they're strange . . . sure the poor wee articles never see the light of day.
ROSE. Anyway, you can't talk to them people . . . .going on about Marx and Lenin and all.
CEELY. You don't haft to talk to them. All you have to do is give them a fry.
ROSE. Aye, then they think we're stupid. Anyway, I've no room.
CEELY (*shouts over balcony*). Jolene! (*Sees* CRYSTAL.) Crystal, love, will you go and find our wee Tucker for me? (*To* ROSE.) No room, my arse. I've my seven kids, old granda Tucker, and I can find room.
ROSE. Where are you goin' to find the water to shave granda Tucker?
CEELY. Our Lady of Lourdes . . . (ROSE *is disgusted.*) I'll pour it back when the water comes on.
ROSE. I'll bet you their water never gets turned off.
CEELY. Hey, Rose Marie, that is a new batallion that just flew in. They haven't got our photographs yet. Come on . . . (*Strikes provocative pose for army.*)
ROSE. I will not.
CEELY (*striking several sexy poses*). Come on. They're free.
ROSE. Don't. That gives me the creeps. (*Tries to leave.*) I'm away to help my Tucker wash our car.
CEELY. There's no water.
ROSE. We'll polish it, then.
CEELY. The two of yous isn't right in the head.
ROSE. What is wrong with keeping it clean?
CEELY. It doesn't go.
ROSE. It is temporarily broke down.
CEELY. Then get it fixed.
ROSE. So the Hoods can drive it away and wreck it?
CEELY. Hoods!
ROSE. Hoods.
CEELY. That happens to be my children you are talking about.
ROSE. I know. (*SFX fire brigade, getting louder.* KATE *gasps as she looks on from her bin. The* OTHERS *rush to the end of the balcony.*)
CEELY & ROSE. Who is it?
KATE. One of the wee O'Neills has fallen out of the empty flats. But as luck would have it, he has landed on Mena Mackle's aerial.
CEELY & ROSE. Thanks be to God!

KATE. He's swingin' two floors up and them ones is taking photos of him.

CEELY & ROSE. Who?

KATE. Some big German photographers that are here for the internment anniversary. They asked wee O'Neill to hang out of an empty flat with a tricolour in his hand while they took his photo.

ROSE (*sarcastically*). Oh, aye, right enough. All them empty flats was missin' was a tricolour.

CEELY. And a child swinging from an aerial.

KATE. The worst of it is, Mena Mackle was sittin' watchin' the last ten minutes of 'Neighbours' and her reception went.

CEELY & ROSE (*sympathetically*). Ach, no.

KATE. She is now climbing up that German photographer's back trying to get at wee O'Neill . . . I think his name is Tucker.

ROSE. Aye, that is big Tucker's wee one.

KATE (*laughing*). He is telling the Fire Brigade not to rescue him, and Mena is raging. She's the only flat not blocked up on her row, and he goes and lands on her aerial.

CEELY (*shouts*). That will put the climbing out of you. (*Suddenly.*) Kate, here's one of them big Germans coming over here.

ROSE. He's got his cheque book out.

KATE (*jumps off her bin suspiciously*). Ten pounds? For my bin lid? Away and feel your head. That was the first bin lid ever banged on internment morning . . . was handed down from my granny. It is a collector's item. It's worth. . . .

CEELY & ROSE (*whisper to her*). Two hundred pounds.

KATE. Two hundred pounds.

CEELY & ROSE (*to German*). A bargain.

KATE (*looks disappointed*). A photograph? Aye, alright. (*Poses with bin lid.*)

ROSE. The authorities should issue the kids round here with parachutes.

CEELY. The least they could do is have twenty-four hour security on that oul' building. God knows there is enough lazy bastards out of work. (*To* ROSE.) Look at your Tucker.

ROSE. Aye, just look at my Tucker.

CEELY. Aye, just look at your Tucker.

ROSE. Look at your Tucker.

CEELY. My wee Tucker?

ROSE. No, your big Tucker.

CEELY. He's dead.

ROSE. I know he's dead. I'm just sayin'. . . .

SOMEWHERE OVER THE BALCONY – ACT I

CEELY. Just sayin' what?
ROSE. Just sayin' he's dead . . . Look at him when he was livin'.
CEELY. What about my Big Tucker when he was livin'?
ROSE. He was a lazy bastard, too.
CEELY. He wouldn't stand and wash a car that didn' go.
ROSE. Oh, is that right?
CEELY. Aye, just right.
ROSE. Aye, right.
CEELY. Right!
ROSE. Right! (*Shouts.*) Tucker! Go you over and guard them empty flats. You polished the car this mornin' . . . you hung your ma's curtains yesterday. . . . You don't need a crash helmet. . . . Just go. (*Changing her tone.*) I'll bring your sandwiches over to the site, pet.
(*SFX ambulance driving off at top speed.*)
KATE. Ach, no . . . I told that ambulance man not to get out of the driver seat.
ROSE (*over the balcony*). Mister, stop moaning. They'll burn it; you'll get your insurance.
CEELY (*shocked*). Look at my Tucker.
ROSE (*proudly*). Look at my Tucker.
KATE. Where is my Pepe?
CEELY. That's him with Tucker.
ROSE. My Tucker?
CEELY. No, my Tucker.
KATE. Your Tucker?
CEELY. Aye, my wee Tucker.
ROSE. Just look at my Tucker.
KATE. My wee Pepe with your wee Tucker?
CEELY. Aye, my wee Tucker.
KATE. Where?
CEELY. In the passenger seat of the ambulance.
KATE (*horrified*). Oh no! My Pepe is a joyrider!
CEELY. It's okay. My Tucker is experienced.
ROSE. Look at the tourists.
KATE (*delighted*). They are taking photos of my Pepe.
ROSE. No, my big Tucker.
CEELY. No, my wee Tucker.
KATE. My wee Pepe with your wee Tucker.
CEELY. Action pictures . . . five pounds a shot.
KATE. Joyriding poodles . . . we've the lot.
ROSE. Snap my Tucker guardin' the flats.

CEELY. My wee Marty chasin' rats.
KATE. Want a bin lid, two hundred pound?
ROSE. A photo of my Tucker marchin' round?
CEELY. Rubber bullets? Antique collection?
ROSE. Avoid the chutes, you'll get infection.
CEELY. Plastic ones are going cheap.
ROSE. Chase them kids from the rubbish heap.
CEELY. Gas masks, riot gear, souvenirs.
KATE. Collected with care over many years.
ROSE. Take them home for your living room wall.
KATE. And if you want a riot, just give us a call. (*Pause.*)
CEELY (*calling over balcony*). Tucker, give that wee man back his ambulance . . . give it back immediately, or I will break your two legs. And come up here now and shave your granda . . . (*To* KATE *and* ROSE.) That wee fella of mine is getting out of control.
KATE. Pepe! Put your paws up for the cameras. Dat a good boy.
ROSE (*exasperated*). No, Tucker . . . stay there. It is not a part-time job; you have to stay all day.
(*SFX helicopter hovering above through scene change.*)
CEELY (*as lights fade*). Quiet sort of a morning. (*They come off the balcony, which pushes back and becomes part of the set, and the rest of the play is played on floor level apart from three high chairs which they use for monologues. LX up on* CEELY, *who is perched on high chair with large headphones on her head with an attached mouthpiece.*)
CEELY. Good morning, are yis all tuned in?
ROSE & KATE (*from darkness*). Aye.
CEELY. After me . . . I promise.
ROSE & KATE. I promise.
CEELY. Not to tell.
ROSE & KATE. Not to tell.
CEELY. On Ceely Cash.
ROSE & KATE. On Ceely Cash.
CEELY. Who has an illegal pirate radio station. Oh, God, I am all eights and nines this mornin'. What with all these foreign visitors here and then I couldn't get oul' granda Tucker to ate his boiled eggs . . . he never ates on internment anniversary. Says I, 'I'll put you back into gaol and you will be glad of a boiled egg'. Ach, you wanna see him, he is sitting here wrapped up in my good pink duvet waitin' for a foreign photographer to take his picture. The laugh is he was never 'On the Blanket'; they threw him out of Long Kesh before that. Oul' eejit. Well, how are yis all this

morning? Did you see the big limousine arriving for Charlene MacAldooney's wedding? Dead exciting, isn't it? Oh, frig, I hope somebody told the driver not to get out of it. So how is everybody over in the MacAldooneys'? Up to high doh, eh? Mona, take you a wee brandy, love. It is not every day your daughter gets married. Oh, yes, that was the first thing that cured me at my big Tucker's funeral. Of course, I drank the whole bottle and near fell in the grave beside him. Right! I have a whole heap of good luck wishes for Charlene and Danny, but before I do that I have a bone to pick with all yousins out there and yis are sittin' on your arses and you know rightly what it is, and you are not lettin' on. There are sixteen visitors sittin' over in that chapel, two arms the one length, and they still haven't bin claimed yet. Now, I want them out and lodged before Charlene and Danny get married, right? Two 'Troops Out' from Manchester, men, middle aged. Two Basque Separatists; they want to stay together. Six of the Communist Party of Great Britain; they don't mind what way they're split up. Two Nicaraguan freedom strugglers and two 'Rock Against Racism' for anybody that likes a wee bit of a dance. And guess who? Valadimir the wee 'Solidarity' Polish fella. That cratur has been coming here on internment anniversary for three years, and he hasn't got a bed yet. Are yis gonna let the poor soul kip in the chapel again this year? I know nobody can be bothered because he doesn't speak any English, but still he is very, very interesting. By the way, where the hell is Nicaragua? I'll bet you oul' Granda Tucker knows. He knows everything. He's very deep. Jesus, that reminds me, Annie McGorman's flat was flooded out again last night.

ROSE & KATE. Ach, no. Not again.
CEELY. Yes, again. Can anybody take Annie and the kids in? They are over in the chapel too. Ach, look on the bright side, Annie. You are the only one in this complex that has any water. Now, I'd like to play a wee request for Charlene and Danny. It's very sentimental to me. Reminds me of my wedding. (*SFX intro. music for song.*) This is for you, Charlene and Danny, and please God and his holy Mother your wee reception goes off without any oul' trouble. (*They sing in turn.*)

The Wedding Song
CEELY. I had a beautiful wedding
My bridesmaids were all in pink.
(*LX up on* ROSE *and* KATE.)
We spent our honeymoon night

>   With him throwin' up in the sink.
> KATE. He was on the rebound
>   It was my mate he wanted instead.
> ROSE. You'll never want for nothing,
>   That's what my poor Tucker said.
> CHORUS. For poorer, not richer
>   In sickness, not health
>   It might have been better
>   To stay on the shelf.
>   Good times are coming,
>   Is that what they say?
>   Well, I'm still here waiting
>   Since my white wedding day.

(KATE *and* ROSE *climb off chairs and go to* CEELY. *The following lines are spoken.*)

ROSE. Right!
KATE. Right!
CEELY. I've changed my mind.
ROSE. Altar!
KATE. Altar!
CEELY. I'm too scared!
KATE. Too bad.
CEELY. I'm too young.
ROSE. Too late. (THEY *pull* CEELY *off of chair and put her white dressing gown over her head in the shape of a veil. The following lines are sung.*)
>   KATE. She never loved him like I did,
>   I jumped to his every whim.
>   ROSE. Married bliss and a lovely new semi
>   With babies, a freezer and him.
>   CEELY. A waster and a drunkard,
>   But eyes that could break my heart.
>   KATE. I cooked, I cleaned and I mended,
>   Perfect at playing my part.
>   CHORUS (*repeated*).

(*During chorus and intro to next verse the following dialogue and action take place.* KATE *and* ROSE *go behind a screen which is set at the back of the high chairs. The screen has several shelves on it containing several pieces of military equipment and a sign which reads 'Ceely's Souvenir Shop'.*

SOMEWHERE OVER THE BALCONY – ACT I

KATE *and* ROSE *change into their day clothes.* CEELY *remains in her nightwear throughout the play. The following lines are spoken.*)

CEELY. Hey! Did yous keep the light on?
KATE. I kept my nightie on.
ROSE. He kept his socks on. (*The following lines are sung.*)

> CEELY. I said every night I was leaving
>   Then he'd smile and my legs would go weak.
> ROSE. As the bills got higher and higher
>   The business went right down the creek.
> KATE. I did all the things I should,
>   But he still walked out the door.
> CEELY. Then he died, the ungrateful bugger,
>   But I yearn for his touch once more.
> CHORUS (*repeated.* KATE *and* ROSE *are now dressed and seated on chairs also*).
> CEELY. Then we moved to this luxury apartment,
>   It was just like a villa in Spain.
> ROSE. Kitchen, bathroom, and through-lounge,
> KATE. Designed to soak up the rain.
> ALL. I still have my lucky black cat,
>   My horseshoe and my rabbit's foot,
>   That little piece of blue ribbon
>   That was sewn to my stocking foot.
>   I had all the things you need,
>   So I'd never have bad luck again,
>   But I thought they might be suspect,
>   As I stood there in the rain.
> CHORUS (*repeated. End of song*).

CEELY (*speaking into headset*). Anybody that has seen the last 10 minutes of 'Neighbours' for Jesus sake go over and tell Mena Mackle what happened. She's ready to kill dead things. (*LX out on* CEELY *and a spot on* KATE, *who is speaking into a tape recorder. She is enveloped in a cloud of smoke.*)
KATE (*to tape and wearing a gas mask*). To whom it may concern. This is Kate Tidy of 19 . . . I repeat, 19 Dooney Row. My damp dividing walls are about to crumble before my two eyes, and my flat will soon be aptly named. (*Switches off tape and removes mask.*) Wee 'Child of Prague,' you've been hangin' there for 12 years, if that wall goes you've had it. Your wee head will fall off

and I will have bad luck til the day I die. I know what you are sayin'. . . . 'Why don't you take me down, then?' I can't. If I do I'll have bad luck til the day I die. You helped my wee Dustin pass his eleven plus, don't let them fall, wee child. (*Switches on tape.*) Will this walless pot hole be fit for a young boy with a bad chest who is going to grammar school? (*Switches off tape.*) Wee St. Francis, don't let my wee Pepe run away from me. He won't stay in the house. The children put the rubbish chute beside me on fire and it smokes us out. He doesn't like it. Wee St. Jude (*whispers*), you are my wee favourite. You won't let my walls fall. Ceely Cash and Rose Marie Noble still have theirs, and they don't even have you hangin' up. Is it just me? (*Switches on tape.*) It's a plot to get me out, cos I have no man and only one child left. (*Switches off tape. To herself.*) My wee Coreen gone. It wasn't my fault she slipped. (*Not allowing herself to get morbid.*) Maybe I should have bought bigger statues. I don't think these ones are workin'. Wee Bridie has a five foot St. Brigid in her hall, and she always wins at bingo. Yes, that's what's wrong, but then Ceely Cash uses Our Lady of Lourdes water when ours gets turned off, she never goes to the chapel, and nothing ever happens to her. God! Will you speak to the British army for me? Every time an army chopper flies by, my heart is in my mouth for my wee walls just can't take it. God! My wee Dustin is an altar boy in the chapel, servin' at his first weddin', I'm never out of the confession box. I collect for the missions in Africa. I even have my neighbours prayin', ones that don't usually, so that's good isn't it? Don't let them fall on me, God. (*Switches on tape.*) My Dustin will be going to grammar school in a wet uniform and mouldy books smelling of smoke. Is this to be the life of a young boy who could well be the brains of the future? Yours, Kate Tidy. (*Switches off tape as LX fade on her and up on* ROSE, *who is attaching the heads of plastic flowers to their stems. She talks as if she is speaking to the twins.*)

ROSE. Here, stop that! Give it back to him . . . I said give it back. You think I'm not wise? You wore that saucepan the whole way through 'Playschool'. It's his turn now. . . . He looks no more stupid in it than you did. . . . Anyway, it's not a saucepan, it's a helmet. . . . It was a helmet the whole way through 'Playschool'. . . . You can just leave them other pots alone. . . . Only one helmet allowed at a time, you haft to learn to share. . . . Hey, boy, turn that T.V. down a bit. My head is opening. . . . Down! . . . You will feel the back of my hand. . . . Well, if you can't hear it you'll haft to take your saucepan off. . . . No, you can't. . . . It's still his helmet even when he is not wearin' it. . . . Stop that. . . . It's not nice. . . . Turn that T.V. back. . . . Turn it back. . . . Stop playin' with the knobs; you will break the bloody thing. . . . Right! . . . Right! Off. . . . Off! If yous can't learn to watch it without fightin' it goes off. NO . . . that is not on until five past two. . . . No, I'm busy. . . . Plenty of stories in your books. . . . Then make one up out of your heads; that's what your imaginations are there for. . . . Leave me in peace. . . . Alright . . . alright . . . (*Agressively.*) . . . sit down the two of yis. . . . Once upon a time there were two wee Princes . . . who lived in a lovely wee blue castle on a hill . . . with their ma and da, the King and Queen. Now outside this castle was a cave with a big, big, ugly, slimy, pig of a dragon livin' in it. . . . Now up above the castle on a mountain lived thousands of knights who were supposed to be brave and protect the wee princes . . . yella big ghetts . . . they stayed up on the mountain gawkin' down. No, the King wasn't afraid of the Dragon; he just stayed out of his road. . . . He is brave . . . cos he's a King and Kings are brave. . . . He just made sure the Queen and the wee princes had nothin' to do with the dragon....Oh, yes, he could knock his shite in if he wanted to . . . because he just doesn't. . . . That's why. . . . Because. . . . A coward? . . . He is not a coward. . . . For frig sake, doesn't he protect the Queen and the wee princes? . . . What more do yis want? . . . He stands guard at the mouth of the dragon's cave. . . . Yes, just like they have in the big shops down the town . . . a security guard. . . . Well, he is a security King. . . . The villagers all love the King, cos he makes sure their children never go near the dragon. . . . Cos the King just stands there and here he is, 'Right, I am the King around here. . . . Get it? And all you cheeky wee skitters can just frig off home. . . .' Yes, there are children who get through security . . . and the big dragon just goes 'Whoosh! . . . Whoosh!' and burns the arses off them, and nobody sheds one tear for them dirty wee ghetts. Then the King becomes so famous for fightin' dragons that he gets invitations from all over the world to come and fight other people's dragons . . . with his own two bare hands. . . . Then one day the King comes to the Queen and the two wee princes and he says, 'Now that I am a brave and noble fightin' warrior we are gonna shift . . . get a new wee castle on the ground . . . with a wee bit of a garden and real roses for your ma the Queen'. . . . Yes, the two wee princes would be on the ground. . . . Aye, they could play in the stree. . . . No, that wouldn't matter, they could fall all they wanted to; they would be on the ground. . . . We'll be goin'

SOMEWHERE OVER THE BALCONY – ACT I

soon. . . . No, not tomorrow. . . . Because. . . . Just because. . . . That's why. . . . Ask your da. . . . Because he is a King. . . . He is a King if I say he's a King. . . . I am a Queen, his wife. He is a King, my husband, and yous are two gurny wee shites! (*SFX helicopter overhead. Sharply.*) No! . . . You can't go out and play on the balcony.

CEELY (*on balcony looking up at army post on the tower and talking to herself*). Just look at him, joggin' round and round. I wonder is he the same one that looked at me on the stairs yesterday. He did . . . cos I would know that chemistry anywhere. Must be cold up there . . . he'll be warm, though. . . . I wonder what he is like out of that uniform. Sexy . . . Jesus, Ceely Cash, he is a Brit. I would get done. Ach, sure I'm just lookin', that's all, just lookin'. (CEELY *stares dreamily over the balcony, then suddenly notices something. She runs to her chair and puts head set on.*) May Day! May Day! Slimy the Dole Snooper is in the area. He is disguised as a paratrooper with a foot patrol. Anybody doing the 'double', down tools immediately. The foot patrol are just passing the empty flats. (*Comes out onto balcony.*) Tucker, tell that snooper over there that you are voluntary . . . he **is** a snooper. He's a snooper dressed as a soldier so he doesn't look like a snooper. . . . Just tell the snooper dressed as a soldier. . . . No, not that soldier, that is a real soldier. The other soldier who is not a soldier, who is a snooper dressed as a soldier. Hey! Hey! Mister, don't you report him; he's not workin'. . . . He is just doin a job that's not a job. It's just a job until he gets a job, but it's not a job like a real job, like your job. Tucker! Punch that snooper in the gub. I said punch him in the gub. No! Not that gub, that is the wrong gub. That is a real soldier's gub. Mister, don't arrest him. He thought you were a soldier dressed as a snooper. No, I mean a snooper dressed as a soldier. Mister, don't arrest him. (KATE *has appeared and is watching her getting excited and tongue tied. She is holding her bin lid in her hand.*)

KATE. Stand back! (*Shouts.*) Rambo! Pepe! Brits! (*SFX dogs growling and barking.* KATE *and* ROSE *watch as soldiers scatter.*)

ROSE. Tucker, stay at your post!

KATE (*holding her bin lid in front of her like a riot shield*). Your Tucker is gonna need protection. I saw a dark figure trying to drop a toilet bowl on his head!

ROSE. Dark figure?

KATE. From the whisperin' stairs in the empty flats.

ROSE. Whisperin' stairs?

KATE. Two soldiers. . . . Glorious Gloucesters . . . got blew up over there. They come back to haunt. . . . I've seen them many a night floating about on the balcony and whisperin' like this. (*Mimics ghostly whisper.*) Then they throw things out and try to land them on people's heads. You think wee Tucker O'Neill fell? (*Nods her head mysteriously.*) They are getting their own back. Ever hear them chitterin'?
ROSE. I wouldn't go near them empty flats.
KATE. Won't be content until the whole buildin' is demolished . . . then they are going to come over here.
ROSE. Are they responsible for your walls?
KATE. Oh, no! I got Father McCann to exorcise me. (*Offers bin lid to* KATE.) Yes, your Tucker is gonna need protection.
ROSE. Ceely's souvenirs?
KATE. Yes.
ROSE. I don't like going into her flat. That oul' man. . . .
KATE. Oul' Tucker! Ach, he is a harmless oul' crater.
ROSE. Never talks. Just stares.
KATE. He used to talk years and years ago. Then he got interned. He was 56 years of age . . . too old . . . nearly killed him. Threw him out of the helicopter and it only six inches off the ground. Hasn't spoke from the day and hour they let him out. (*Looks up at army post.*) Cos of themmins listening. Claims they even bug the confession boxes.
ROSE. Are they bugged?
KATE. Oh, no doubt. I used to whisper to Father McCann, but he couldn't hear me, so I write it down on a wee piece of paper, slip it through the grill. He forgives me, and then I take it home and burn it. (*They turn to walk into* CEELY'S. *They go behind screen.* CEELY *is on her chair and speaking into radio.*)
CEELY. Has anybody seen my wee Johnny? I hear he is throwing toilet bowls out of them empty flats. Attention! Anybody going to the MacAldooney wedding, watch yourselves passing them empty flats. You are liable to get flattened. Oh, Mona, all this reminds me of my wedding day. Mickey McGlade our best man, he forgot what day it was and we had to go and get him out of bed, so my Tucker goes to the hire shop for the suits . . . suits that must have been for bloody big rugby players, and Mickey McGlade and my Tucker no bigger than two ducks' arses. Mona, I was affronted. It took me an hour and a half to find my Tucker's hand in the chapel. The chapel, never let me forget it til God calls me. My da goes without his teeth, so my ma sends him out in the middle of

the Mass to go and get them. Then the priest asks, 'Who is bringing up the offerings?' And isn't it my da who is on his way over here like two men and a wee lad for his choppers. . . . Silence. . . . The Priest asks again, 'Who is bringing up the offerings?' Well, my wee aunt Beattie (*whispers*), she is a wee Protestant, she didn't know what he was talking about. . . . She walked the whole length of that aisle and handed the Priest a set of pink His & Hers towels, and the whole congregation bursted out laughin'. I was cut to the bone. . . . Then we get to the community centre for the reception and get a phone call. The caterer's van has been highjacked and a bomb put in it, so we wait. Then BOOM! 45 roast turkey dinners and sherry trifles splattered all over Great Victoria Street. So my da, the toothless wonder, phones up the Chinese on the Falls Road, 45 curry chips . . . two hours later . . . 45 stone cold curried chips, the van was held up in a bomb scare. Of course by this time Mona, everybody is paralytic drunk and they ate them anyway. My da tries to touch up Tucker's aunt from America, so my ma beats her up, throws his teeth down the toilet and flushes it. As for me and my big Tucker, my weddin' dress was covered in curry sauce where he threw up on me. Then we couldn't get home here cos the Brits had this place cordoned off while they were raidin'. Oh, a day to remember all right, Mona. But then, that was in 1969 durin' all the trouble. . . . It's all different now. (ROSE *and* KATE *come from behind the screen.* ROSE *is wearing a police riot helmet and carrying the bin lid and a baton.* KATE *is wearing an army shirt.*)

ROSE (*takes off riot helmet and sniffs it*). Jesus, Ceely, where did you get this?

CEELY. It fell off the back of a police Landrover.

ROSE. Was the policeman still in it at the time?

CEELY. It will cost you fifteen pounds. (*To* KATE.) And two pounds for that shirt. (*To* ROSE.) I'll throw in the baton as well.

ROSE. Will you take it out of my Christmas club?

CEELY. Aye. (*SFX of clapping and cheering.*) Quick, everybody out on the balcony for the wedding.

KATE. Look at the size of Mona's hat. They'll haft to put her in the lift sideways.

CEELY. Charlene, your dress is a wee beauty. (*To* KATE.) You wouldn't know she was pregnant.

KATE. Look at all the wee flower girls. Ach, and the wee page boys. They are like wee. . . .

ALL. Angels!
KATE. That's why there's nobody about to stone the army.
ROSE. Yoo hoo, Danny!
CEELY. Ach, Danny and his wee best man.
ROSE. That is not right . . . the groom seeing the bride before the wedding. Bad luck.
KATE. Look at their wee blonde tips and their walkin' sticks.
ROSE. Ach, now, that is not right. Charlene, the Ma, the Da . . . bridesmaid, flowergirls, pageboys, all comin' out at the one time with their walkin sticks. . . .
ALL (*baffled*). Walkin' sticks?
KATE (*also baffled . . . then realising there is something afoot*). The whole family is lame.
CEELY. Curse put on them since birth. (*SFX of helicopters hovering above.*)
ROSE. That best man does not look like himself.
CEELY (*realises what is going on*). That's because he is not himself.
ROSE. Who is he?
CEELY. Tootsie O'Hare disguised as big Tucker O'Neill.
KATE. On the run disguised as Tootsie O'Hare.
ROSE. Why?
CEELY. Cos Danny didn't want Tucker, he wanted Tootsie.
KATE. So Tootsie is Tucker, and Tucker is Tootsie!
ROSE. Why is Tucker 'on the run' when he didn't do nothin'?
KATE & CEELY. So Tootsie could do best man . . . .For fuck sake!
ROSE. Right!
KATE & CEELY. Aye, right!
ROSE. Aye, just right!
KATE & CEELY. Aye, just right indeed. (*SFX helicopter as if hovering right above.*)
ROSE (*realises*). They're gonna arrest the whole weddin' party.
KATE. Quick, Mona, get into the lift.
CEELY. Mona, take your bloody hat off.
KATE. She is that stubborn.
CEELY. Push her, Charlene.
ROSE. Put her in feet first and set her at an angle.
CEELY. They're in!
ROSE. The Brits will get them when they come out of the lift.
KATE (*confidently*). That lift will never open.
CEELY. Who has been jammin' lifts in these flats since he was in nappies?
ROSE. Tucker O'Neill.

KATE. No....Tootsie O'Hare.
ROSE. And then what?
ALL (*singing*).

>Hear, see, and speak no evil
>Cos that's the jungle law.
>We just swing from our branches
>And see what we never saw.
>Hear, see, and speak no evil
>Don't think we monkeys are scared,
>We're too busy gathering berries
>To see what we never saw
>Or hear what we never heard.
>
>Hear, see, and speak no evil
>We eat bananas instead.
>The lion's never gonna catch us
>We don't say what was never said
>Or see what was never saw
>Or hear what was never heard
>Or see what was never seen
>Or heard what we never did hear
>Or see what we never heard
>Or say what was never seen
>Or hear what we never said
>Or see what was never heard.

CEELY (*during musical break*). Marty! Marty! Run away over to the chapel. Shimmy up the drainpipe and borrow a priest's vestments for Tootsie . . . quick!
(*Last section of song is repeated. End of song. They are looking across to the lift.*)
ALL. And the lift doors are opening. . . . ood morning, Father! (*SFX army jeeps etc. driving into square.*)
KATE. Look at all the flowers.
ROSE. Beautiful.
CEELY. Charlene, you're like a scene out of 'Dynasty'. (*SFX shouting and screaming.*) They're liftin' Tootsie. (*Shouts.*) Imagine arrestin' a man of the cloth.
ROSE. Pack of heathens.
KATE. Yous will all be struck by lightnin'.
CEELY. Shove your walkin' sticks up their. . . .

ROSE. Mona is fit to kill. She is beatin' the head of that paratrooper with her handbag.
CEELY. Charlene! Stuff your bouquet down his throat.
ROSE. Best dressed riot I have ever seen.
KATE. Mona! Beat him with your stilettoes. Pepe! You stay out of it.
CEELY. Knock his shite in, Father.
ROSE. Look at that big pig of a soldier. He's throwin' him into the saracen. Made a lovely wee priest, too.
KATE. Danny, pull Mona off the saracen. She'll be piggin'. (*SFX saracen driving off.*)
CEELY. They got him! Pigs!
ROSE. That's not right. Mona, the Da, Charlene, Danny, bridesmaids, pageboys, flowergirls, all barricadin' themselves in the chap....
ALL. In the chapel!
KATE. Look at your Tucker.
ROSE. My Tucker?
KATE. No, her wee Tucker.
CEELY. My wee Tucker?
KATE. With my wee Pepe!
CEELY. Where?
KATE. Drivin' the saracen.
CEELY (*delighted*). With Tootsie O'Hare disguised as a priest and my Tucker dressed as a soldier ... (*Shocked.*).... My Tucker dressed as a soldier?
KATE. I know where he got the uniform from. See that dole snooper who was disguised as a para! He's hiding in the empty flats disguised as a streaker.
ROSE. Tucker! Knock his shite in.
KATE. Hold on tight, Pepe!
CEELY. Tucker, put your seatbelt on ... you'll get arrested! (CEELY *and* KATE *exit Stage Right.* ROSE *follows, then runs back into her flat for a can of polish. As she reemerges onto the balcony she stops short. The following is played as in a conversation with an invisible soldier.*)
ROSE. What?....offensive weapon ... no it's only polish ... look. (*Sprays at the imaginary soldier.*) Ooops sorry.... It's just to clean this riot helmet ... no not a real riot, it's in case of toilet bowls ... oh, you mean *this* is an offensive weapon? (*Indicating truncheon.*).... No it's not ... well, I suppose it would be if you were to be offensive with it, but it's not goin' to offend nobody ... it's just for my man's work, and this is only an oul' bin lid I brought out to polish ... for I'm sure you know yourself how

dirty your bin can get . . . it's just for my husband's head and the soldiers on the whispering stairs – no, not live soldiers, dead soldiers. No, not real dead soldiers . . . ghosts . . . throwin' toilet bowls . . . Here mister, you can't arrest me. I've two wee twins not washed yet. (*She exits.*)

KATE. What a day, Ceely. My Dustin under seige. My wee Pepe on the run. And now Rose Marie arrested.

CEELY. And it's only half eleven.

KATE. It wasn't fair arrestin' her. All she was doin' was polishing Tucker's riot gear.

CEELY. What do you expect? She was standin' out in the middle of the balcony with a can of Pledge and half the British Army lukin' up at her.

KATE. She's never been interrogated before.

CEELY. She'll love it. Isn't it getting her out of the house for a couple of hours? (*SFX jeeps and saracens.*)

KATE. Oh, no! Here's more reinforcements.

CEELY. Poor Mona. Surrounded.

KATE. She'll be goin' daft. It's near time for your radio bingo, and she's only waitin on two numbers for the jackpot.

CEELY. Wish we could make contact.

KATE. Oul' Paddy! He's in the chapel. Mona's da. He's deaf. (CEELY *doesn't know what she is on about.*) Isn't he always complaining cos he can pick your station up on his hearing aid.

CEELY (*delighted*). Kate, if you had brains you would be dangerous. (*She runs to the head set.*) Calling Mona MacAldooney in the chapel. Calling Mona MacAldooney in the chapel. Mona, are you tuned in? (*SFX chapel bell tolls four times.*)

KATE. She can hear you.

CEELY. Mona, love, don't you take them barricades down. Don't you give in to them pigs. You didn't know when you got up this mornin' that you had all this excitement in front of you. We've had our own excitement here on the balcony. They have arrested Rose Marie Noble in her riot gear. It's all go. Mona, I hope you brought that brandy in with you in your beg. Take a double. You're surrounded by three hundred Brits and fifteen army jeeps, but don't you take them barricades down til they give in to your demands. God knows! A Priest, thirty chicken salads and a three piece band is no big deal.

KATE (*grabs mouthpiece*). Thirty chicken salads? Mona MacAldooney, are you gonna keep them visitors hostage and not feedin' them? What happened to the Ireland of the Welcomes?

(*Jumps off chair and runs to the balcony.*) Calling the British Army! Hey you! Mona MacAldooney changes that to forty-five chicken salads. (*SFX of continuous dog crying.*)

CEELY (*into headset*). The residents of the area would like to apologise to all our foreign guests held hostage in the chapel. We hope that you look on it as doing your 'Wee Bit for Ireland' . . . (*Comes out to balcony to* KATE, *who appears to be looking down at Rambo crying.*)

KATE. Poor Rambo. He is crying his eyes out cos he didn't get goin' on the run with my Pepe.

CEELY (*getting emotional*). I'm all excited so I am. . . . When wee Tucker was wee, big Tucker used to say to him, 'Son, what do you wanna be when you grow up?' And he would a said, 'On the run, Daddy'. Oh, God, if he was alive the day he would be so proud of him. He's probably in Bundoran by now.

KATE (*upset*). It's well for you, Ceely. My Dustin wants to become a computer scientist.

CEELY. My wee Tucker was just lucky, Kate. He happened to be in the right place at the right time.

KATE. I don't know if my wee Dustin can make it what with me and my walls.

CEELY. You should have let him go to America with my Dwane and the rest of The Deprived Children of the Area.

KATE. I was too scared. I heard if them Americans like your children they don't allow them to come home. They try and adopt them.

CEELY. Not my Dwane. Dr. Barnardo wouldn't even take him. (*SFX voice over. 'Attention! Attention! Would all residents please leave their homes immediately as we have reason to believe there is a suspect device in the area'. The same message is repeated through the following dialogue.*) Bloody army. That is gonna interfere with my radio bingo!

KATE. It's just an excuse so they can raid the place.

CEELY (*on headset*). Mona, pay no heed to themmins. Only tryin' to get you to surrender.

KATE. The Brits have surrounded Rose Marie Noble's car.

CEELY (*casually*). Probably gonna blow it up.

KATE. And it only polished, too.

CEELY (*shouts*). Tucker! The Brits are gonna blow up your wee motor. Tucker! . . . Doesn't realize . . . too busy guardin' that naked snooper. (CEELY *goes to exit and* KATE *stops her.*)

KATE. What about Granda Tucker and the tortoise?

CEELY. It would take more than a bomb to shift them two. But if they get blew up, they needn't come runnin' crying to me. (*To* KATE.) Come on. (*They are about to exit and* CEELY *Stops. Calls back.*) And if my good duvet gets smoke damaged it's comin' out of your pension. (*Stops and looks over balcony. To* KATE.) That is a beautiful day, isn't it?

KATE. That's what I love about this place. On a day like today you could be anywhere. (*They pause and take in the sun.*) Ceely, are you not going to put some clothes on?

CEELY. Sure, I'm only going to the Falls Road. (*As they exit....SFX bomb exploding.*)

BLACKOUT. END OF ACT ONE.

ACT II

SCENE: ROSE *and* CEELY *are seated on their chairs.* KATE *is seated on a toilet bowl covered by a sheet. She is dressed in army boots, trousers, and camouflage jacket. She speaks through the sheet.*

KATE. When I woke up this mornin', had my fry, give Dustin his syrup of figs, I thought to myself something is afoot, for when I looked to my walls my wee St. Francis was slanty. An omen, I says to myself, an omen. And I was right. (*Throws off sheet to reveal toilet bowl.*) Today, August 8th, 1987, as a result of a controlled explosion, my walls finally crumbled. (*LX up on* CEELY.)

CEELY. I am never speaking to Granda Tucker ever again. He is one miserable oul' decrepit pig. He sat there and let the Brits lift my wee pirate radio station, my only pleasure. And he just let them walk in and take it away. Jealousy, that's what it is, not gettin' enough attention. I wouldn't mind if he was company. Company! I'd rather live with Psycho . . . at least he spoke. (*LX up on* ROSE.)

ROSE. Arrested! Me arrested! And I never even done nothin'. I says, 'Mister, I'm only polishing my man's workin' clothes, his riot shield and his helmet'. Next thing I'm led down to a saracen with my two wee twins dukin' over the balcony shoutin', 'Is that the brave knights come to rescue us'? Says I, 'Mister, will you tell them wee boys that you're the King's messengers and we're away to look for a new wee castle?' They thought I was mental, and the shape of me tryin' to get up into the saracen. I couldn't. Kate's bin lid kept jammin' in the door, and then I hear Kate shouting from the balcony at the soldiers, 'Hey, watch my bin lid. That is a family heirloom'. I was prayin' they would ignore her, cos I heard if you complained you got beat up....And my poor Tucker and the twins were wavin' like I was goin' on a bus trip to Portrush. Oh, God, and the thing was speedin' . . . and this big ginger one . . . she was a woman, too . . . she kept sayin', 'Hold on tight'. But I was fallin' on top of her, and then fallin' back on the wee soldier with his rifle. Here's me, 'If that thing goes off I am dead,' but I couldn't help it.

KATE. Welcome to my through lounge and open plan kitchen and bathroom. Frank said when he left me, 'Kate, no good will come of you'. And he was right, for all I am left with is Ceely Cash's walls on one side and a rubbish chute on the other. How could any of that be my fault? (*Suddenly realises.*) Dustin can watch me goin' to the toilet. I won't be able to.

CEELY. Twelve miserable years I put up with him sittin' there. (*Sarcastically.*) 'Commandant Cash, Communications Officer' . . . communications my arse. No wonder the 50s campaign was a friggin' disaster. Our wee Dwane's tortoise has more sense. The people were dependin' on that radio the day for the bingo. Big jackpot game. Mona and everybody under seige in the chapel were going to play. The whole district. What has he left them with? (*Produces empty bean can, with string attached to one end, from back of her chair. Pulls string and speaks.*) Kate, Rose, will yous play bingo? (KATE *and* ROSE *also produce bean cans, but are more concerned with their own dilemmas and are half-hearted about it.*)

CEELY. Testing! Testing! One . . . two . . . single line across and then a full house.

ROSE. They get me into this room. 'Do you wanna be examined by a doctor?' I says, 'Mister, I'm not sick'. 'Take your belt off', they said. Here's me, 'What for'? 'In case you hang yourself'. 'Oh no, I wouldn't do that', I says, 'I am heart scared of heights'. Then here they were, 'We know your twins throw stones at us. Who taught them to do that? Your husband'? Here's me, 'What'? I near had a fit, but I had to calm myself cos the big ginger woman was dead scary lookin'. I said, 'My husband and my children would not lift their hand to nobody, we just mind our own business. Anyway,' I says, 'my husband has a bad chest'. Here they are, 'We know'!

CEELY (*through bean can*). Eyes down . . . look in . . . two rubber bullets, 22; armalite gun, 21; on the run, 31; unlucky for some, 13.

KATE. That's it, some people are lucky with their walls and some aren't. Ceely and Rose still have theirs, so it must be my fault. I get the blame for everything. That was Frank, everything was my fault. (*Mimics Frank.*) 'Kate, keep the kids in; there's rats'. Not my fault. 'Kate, them sewers is stinkin'. My fault? 'Kate, it's rainin'. (*Looks bemused.*) He still thinks our wee Coreen slipped on my rubbish on the stairs, but I've always had my own bin, handed down from my granny. I brought it here when we got married, carried it up these stairs myself cos the lift was broke. That was the start of things going wrong. 'Kate, the lift is broke'.

CEELY. I could have been somebody if it hadn't been for lookin after him. . . . Not even my friggin' da . . . could be calling out numbers in the Silver Star Bingo, makin' somethin' of myself. Oh, but I know what he wants . . . me to shut up too so that nobody says nothin' to nobody ever again, but I am a communicator, that's me, that's my way. My big Tucker used to say that to me when he had a few drinks in him. 'Ceely, you're a 'Mouth'. Two weeks before he died he went out and he stole me all my radio equipment. He knew he was dying and he wanted to leave me somebody to take his place. Not many men will leave you a whole district. (*Indicates Granda Tucker.*) I wish to hell they had never let him out of gaol. (*Back to bean can.*) Doin' time, 29; two dead men, number 10; two fat peelers, 88.

ROSE. Two big skinny ones they were, and one of them had a stutter. I tried to finish his sentences for him, but the big ginger one was givin' me dirty looks, so I just sat there. 'Do you wi wi want a' (*mimes cigarette*). But I wouldn't in case they were drugged, and they made me sign a statement. Then the other one says, 'What does your husband use his car for'? Here's me, 'Nothin'. It doesn't go. It is temporarily broke down'. Then the other one says, 'What would his poor invalid mother think of him bein' mixed up with smugglin' wea . . . wea'. I finished it for him cos I was gettin' annoyed. Here's me 'Weapons'! And the big ginger one, her glasses started steamin up. Here's me 'Weapons! Me and him is not like that. We have two wee children. . . . Wea Wea Weapons'!

CEELY (*in bean can*). Prods are dirty, number 30; Pope's a Mick, 26.

KATE. Check! 'Not my rubbish,' I says to Frank. 'Correen did not slip on my rubbish. Cos I examined it afterwards'. I said, 'Frank, none of us eat custard. We don't even like it'. But it was still my fault. I shud have watched her, but I've only got two eyes. Not my fault. 'No excuse,' he said. 'But you will be repaid,' and he was right. But what he doesn't know, my Dustin is going to grammar school. (*Switches on tape and speaks into it.*) This is the last will and testament of Kate Tidy of 19 Dooney Row. In the event of my ceiling falling down and suffocating me, Don't! I repeat, Don't! let Granny Tidy take my Dustin or my poodle Pepe. (*Switches off tape.*) My wee altar boy under seige. Ach, he'll be safe in the chapel, not even the soldiers would storm it, cos I'm sure most of them are Catholics too, and the statues barricading the chapel doors are their statues too. God, my wee statues are all broken, and the chapel statues are being used as a barricade. It's not their day either.

CEELY. Eyes down for your full house. Two big choppers, 77; sniffin' glue, 62; Brits are thick, 36.

ROSE. Then they says to me, 'Why do you want to move, Mrs. Noble? Is it not safe there anymore because your man is doin' a bit of toutin'?' Here's me, 'How did yous know? I mean, how did you know that we wanted to move, not that he's a tout, cos he's not a tout. I meant how did yous know?' Oh, God, and then I got all confused and I started stutterin' too. The big ginger one thought I was makin fun of the big skinny one. It was awful. Here's me, 'I want my Mammy', and they says, 'She is in the Isle of Man for her holidays'. I couldn't believe it.

CEELY. Oh, God, if I knew then what I know now I wouldn't be here, but then he knows what he did to me, and what's more he knows I know he knows what he did to me. But he doesn't know I know what to do. He just thinks he knows I don't know what to do, but he doesn't know that I know what to do. But then he who thinks he knows best . . . knows fuck all! (*Into bean can.*) Valium Heaven, number 7; blocked up loo, 42; great position, 69.

KATE. Yes, Frank, you would be havin' a quare laugh if you were here now. You're probably up there with them army ones, laughin' at me. That's where you have been all these years. Now yous can all sit and watch me goin' to the toilet. But my wee Pepe doesn't blame me, so there! He'll come back, so he will, and my Dustin will be all right in the chapel, so you see, you didn't get it all your own way. I still have them. You thought the rats would get me. (*Points to her army boots smugly.*) You thought I would freeze to death because of my walls. (*Points to Battle jacket.*) Oh, yes, you can laugh, Frank Tidy, but it is my brains that Dustin has. You couldn't even do the Daily Mirror crossword, but I could. I could do it. But don't think you can come runnin' back when Dustin is clever and I have walls, for you won't get in....But you're hopin' my walls don't get built cos it was my fault. But I had the nicest walls in our row. Beautiful pink with roses. All my wee saints. Four walls like in 'Playschool . . . one . . . two . . . '

CEELY (*through bean can*). Kate!

KATE. Aye.

CEELY (*has taken her slipper off and sees that the sole is falling off*). What colour were your walls?

KATE. Pink.

CEELY. With roses?

KATE. Aye.

CEELY (*she looks dismayed . . . her slipper is pink with roses*). Jesus!

ROSE. Then this peeler says. . . . 'Where is your buggy . . . the pink one'? Here's me, 'Stole'! 'Where from'? 'From outside our door'. 'Not report it'? Says I, 'Mister, are you wise? Wouldn't I look sick reportin' a stolen buggy when there is people gettin' murdered out there'? 'Exactly'! he says. Do you know what they arrested me for? That peeler that was shot at the corner of Leeson Street by a woman with a twin buggy. Here's me, 'Mister, could you see me, puttin' my twins in a buggy, danderin' up to Leeson Street, shootin' a peeler, danderin' back home and makin' my Tucker his tea as if nothin' had happened'? Here's me, 'Mister, are you mental? Anyway', I says, 'I am not the only woman in Belfast with twins'. Then he said, 'What would happen them two wee boys if we were to arrest your husband as well? Put them in an orphanage'? 'Oh, no, Mister, don't do that', I said. 'You see, they are a wee bit strange cos I am scared to let them out. Don't do that, Mister'. Then I turned to the big ginger one, I says, 'Missus, you're a woman, don't let them take my wee children'. And you talk about hard . . . here she was, 'What do you think of a United Ireland'? 'Oh, no', I says, 'I don't want a United Ireland, I couldn't afford a United Ireland. I would haft to move to England'. Then the big skinny one says, 'You can g . . . g . . . go. . . . ' I thought he was never gonna say it. . . . Then the big ginger one says, 'Do you want to be examined by a doctor'? Here's me, 'No'. . . . Well, we'd no water this morning and I wasn't even washed!

CEELY. I should have escaped in that saracen with our Wee Tucker. . . . Set up my own radio station in Bundoran, all them men on the run would be glad of me for the company. . . . 'Ceely, the Volunteers' Sweetheart'. I could poison him. . . . No, the lift is broke I would never get his oul' coffin down the stairs. . . . I know! I'll tell the Brits he has been gun runnin' and get him arrested. Brilliant! (*Fantasizing.*) I could be sittin' down the town in one of the big bingo halls, callin' out numbers over a microphone. . . . There would be silence, hundreds of people waitin' for me to speak. I'd put a frock on every single day. People would be stoppin' in the streets (*mimics whispering*), 'There's Ceely Cash the Bingo Caller. She looks great in real life'. (CEELY *speaks through bean can as if it is a microphone.*) Ladies and Gentlemen, do we not have any sweats yet? (*Gets very excited.*) And it's. . . . Sniper McKee, 53; Paisley's Drum, 51; hide your gun, 61; lock your Door, 74; Duck and Dive, 55; in a state, 68; Keyhole Kate, 78!

KATE (*screams*). Check! (*Then to herself.*) Look at the state of my ceiling....very dicey. (*Gets army camouflage helmet from under her chair and puts it on.*) Maybe Father McCann didn't exorcise me properly . . . maybe the Glorious Gloucesters are beyond exorcism . . . maybe Frank snuck in one night and did somethin' to it . . . maybe I was just doomed since birth . . . maybe it was the time I broke my rosary when I jammed it in the lift door. (*In tears.*) No, it was cos Coreen slipped and I wasn't there. Oh, my God, how am I ever gonna get washed again?

CEELY. And that is a full house check for Kate Tidy!

Hey St. Jude

ROSE (*singing*). Hey, St. Jude, won't you hear my plea,
Got nothing left, I'm in misery.
Hey, St. Jude, I'm a desperate case,
Can't you see when you look on my face?

KATE. Hey, St. Jude, I'm down on my knees,
I'm a lost cause, so help me please.
I put your statue on my bedroom wall
With Jesus and Mary, St. Peter and Paul.

CHORUS. Don't want silver, don't want gold,
Want to win just once
Before I get old.
Dear ole Jude from the promised land,
Save my soul from the final demand.

CEELY. Hey, St. Jude, I'm all alone,
No one to lean on to call my own.
Just someone to talk to
I don't want much,
Or you're back in the cupboard
With my old cracked cups.

CHORUS (repeated).

(*At end of song they are* ALL *out on the balcony looking towards the chapel.*)

KATE (*gasps*). Look what Rambo is about to do in front of the chapel doors. Rambo! Bad Boy! No shame, that dog, imagine shiting (*she mouths word*) in front of three-hundred soldiers.

ROSE. The chapel doors are opening!

CEELY (*shouts*). Mona, don't surrender. (*They* ALL *watch.*)

KATE. Ach . . . it's only Vladimir the wee Pole.

ROSE. Mona is taking this all very seriously. That's what they do on

SOMEWHERE OVER THE BALCONY – ACT II

the T.V. Let one hostage go at a time . . . gesture of good will.

CEELY. They are arrestin' him.

KATE. At least he'll get a proper bed for the night. . . . It's an ill wind.

ROSE. That's my pot he has on his head. Hey! That is my Mark's helmet.

KATE. He is attacking that soldier with your saucepan. Don't struggle, Vladimir. Why is he shoutin' Fuck the Brits?

CEELY. Because the youngsters told him that was English for 'Thanks very much'.

ROSE. Hey, look! Our car is away. Tucker must have got it to go. (KATE *and* CEELY *move away from her quickly.*) Ach, isn't that nice . . . he wanted to surprise me. Must have the twins away for a wee drive. It's an ill wind blows somebody some good right enough. (*Looking very pleased.*)

KATE (*tentatively*). Rose Marie! (KATE *and* CEELY *are now at far side of the balcony.* ROSE *comes over.*) See that big lump of red metal over there? Her Jolene is tap dancing on . . . Tucker's car.

CEELY (*pleased*). I never knew our Jolene could tap dance.

ROSE. Tucker! Tucker!

KATE. There he is in one of the empty flats....He is pullin' it apart.

CEELY. He is supposed to stop the youngsters doin' that.

ROSE. Tucker! I'm back . . . all in one piece . . . more than I can say for our car. Where's the twins? Tucker, answer me, I said where's the twins? Children, remember? Two wee boys, brown hair, blue eyes . . . you'll know them if you see them, they'll be callin' you Daddy. Shush! . . . Shush! . . . don't you tell me to shush, you have lost my children. Tucker speak to me. Why won't he speak to me?

CEELY. He is pullin' that empty flat apart plank by plank. Probably throwin' a wobbler cos the Brits blew up his motor.

KATE. Tucker! Keep me the walls, will you?

ROSE. All right. . . . All right. . . . Don't speak, just nod. Have you got the twins? No! No? No, you didn't mean no, you meant yes, didn't you? No! No, you did mean no, or no you meant yes? Yes . . . yes? Does that mean you meant no the first time? Tucker, speak to me.

KATE. I think he means he hasn't got them.

CEELY. Stupid ghett! Losin' two children.

ROSE. Tucker, take the saucepan off your head and you'll hear me.

CEELY. Ach . . . he never got wearin' his wee new R.U.C. riot helmet.

ROSE. Take off your helmet! Off . . . (*Gives up on him.*) My other pot was lyin outside the chapel. (*Delighted.*) The twins, they're in the chapel . . . they're safe. (*Horrified.*) No, they're not, they're under seige . . . (*Panics.*) Mona! Mona!

211

KATE. She won't hear you.
ROSE. I'm away to look for them.
CEELY. The Army won't let you out.
ROSE (*grabs bean can*). Mona! Mona!
CEELY (*laughs*). Are you daft? Mona has to have one of them to hear you back. (*To* KATE.) What school did she go to?
ROSE. Give me yours and I'll take it down to her.
KATE. Oh, aye. Dander down through 300 Brits, five army jeeps, break down the chapel doors and hand that to Mona.
ROSE. Aye, very funny.
KATE. Aye, very bloody funny.
ROSE. Aye, I'm laughin' my leg off . . . Aye . . . oh . . . ho, ho.
KATE. Ha, bloody ha . . . (CEELY *begins to sing as if something is dawning on her. The* OTHER TWO *stop and look at her. Through the song she is trying to indicate that she has thought of an idea.*)
CEELY (*as she sings she looks up at the army post as if the song is a code*).

> They were just newly wed.
> They went upstairs to bed.
> He turned the lights down low
> Cos she was rather shy.
> He said come here my dear
> She said no fuckin' fear.
> I'd rather stay down here
> Among my souvenirs.

(*The word souvenirs prompts them into action.* KATE *runs behind the screen and appears with a C.S. gas gun.*)
ROSE. The Brits will see us.
CEELY. No, they won't. (CEELY *grabs* KATE'S *sheet to shield her while* ROSE *puts the bean can into the spout of the gun.*) Right, Kate, you're aimin' for the belfry.
KATE. Where is the belfry?
ROSE. Where they keep bells, bats, hunchbacks . . . for Christ sake, just shoot!
KATE (*annoyed*). Right!
ROSE. Aye, right!
KATE. Aye, just right!
CEELY (*breaking them up*). Right!
KATE (*aiming*). I can't. There is a wee pigeon sittin' on it.
CEELY (*shouts*). Clear off.

SOMEWHERE OVER THE BALCONY – ACT II

ROSE. Shoot for god sake.

KATE. I can't kill a wee innocent pigeon. (CEELY *and* ROSE *start screaming at her to shoot. Then suddenly, SFX of can being shot, landing on belfry. They are* ALL *on the ground covering their ears, then SFX of chapel bells tolling four times. They* ALL *jump up delighted. They grab a chair and bring it on to the balcony.* CEELY *gets the other bean can and climbs up onto the chair.*)

CEELY (*through bean can*). Calling Mona MacAldooney in the belfry . . . calling Mona MacAldooney in the belfry. Is that a big ten-four? (*Holds can tentatively to her ear . . . .delighted.*) She can hear us. Mona, love, how is it over there? Is everybody okay? (*Listens. To others.*) Mona is panicking. She has reduced her demands. A priest and forty-five egg and onion sandwiches.

KATE. Mona, don't surrender. (*To* CEELY.) Tell her about my walls.

ROSE. The twins! The twins!

CEELY (*to can*). What is wrong? Why do you not want the three piece band? (*Listens.*)

ROSE. The twins!

CEELY. Oh, that is stickin' out Mona! (*To* OTHERS.) You know Seamie McKee? The wee man that races the greyhounds and sings round in the club? He is singin' and playin' the organ. (*To can.*) Mona, Seamie only sings when he is drunk. (*Listens.*) Oh, the communion wine . . . that was handy. (*Listens as if she can hear the singing in the chapel and begins to sing along, and* KATE *joins in.*)

CEELY & KATE (*singing*). I am prayin' for rain in California. . . .

ROSE (*getting frustrated with them*). The twins! The twins!

CEELY (*snaps*). Ach, how many bloody twins have you? (*To can.*) Mona, have you got Rose Marie Noble's twins in there? (*Listens.*) Only Annie McGorman's kids? Are you sure, Mona?

ROSE (*takes can*). Mona, check the confession boxes. They're very strange. (*Listens.*) No, not the boxes, my twins. (*Hands can back to* CEELY. KATE *and* CEELY *are singing along to Seamie in the chapel as* CEELY *holds the can up to their ears.* ROSE *just stares up at Tucker.*) Tucker, our wee twins are under seige!

KATE (*rocking to and fro as* CEELY *hums*). Mona, is my wee Dustin all right?

ROSE. My Tucker is on the roof now! Tucker, please speak to me.

KATE (*listening into can*). My wee Dustin is servin' up the hosts now. Everybody is starvin'. . . .

ROSE (*over balcony*). There's a nun . . . Sister! Sister! Are there two wee twins in that chapel?

KATE (*still rocking and humming*). Mona, I hope they are not consecrated.

ROSE. Sister! Sister!

KATE. Mona, get my Dustin to sing Edelweiss . . . over the mike . . . he's brilliant.

ROSE. Sister, is there two wee . . . (*Stops and looks shocked.*)

KATE. Rose Marie, come and hear my wee Dustin. He is gonna sing 'Edelweiss' in the chapel. (KATE *notices the nun. They* BOTH *stare at her.*)

ROSE. That is the strangest lookin' nun I have ever seen.

KATE. It's a lovely wee shampoo and set she has. Wonder where she got it done?

ROSE. Kate, she has pink high heels underneath her habit.

KATE. Well, they just do not go with that outfit.

ROSE. She is the spittin' image of your mother-in-law.

KATE (*pause*). That *is* my mother-in-law. Granny Tidy, what are you doing here dressed up as a nun? (ROSE *slinks away and goes to* CEELY. *They just watch in disbelief.*) Dustin? He's a wee altar boy in the chapel. The seige was not my fault. No, you can't take him. . . . It could have happened to any wee altar boy . . . he is safe . . . nobody is gonna hurt him in a chapel. Anyway, he is havin' a good time. (*Indicating sheet.*) I *can* go to the toilet. No, you cannot take him away. I *can* look after even a dog. . . . Go away. . . . I'll tell the army you're not a real nun, that you're really an old cow. *I* look stupid? What do you think *you* look like? You are the first nun I ever knew had six children and a beer belly. I'm not a fit mother? Ceely! She says that I am not a fit mother . . . are you a fit granny? Clear off, you're not gettin' my son . . . .Frank? Frank who? My Frank! Your Frank? Dustin's daddy? Here? For Dustin. (*Laughs in disbelief.*) To steal my Dustin . . . you just tell that slimy good for nothin' pig if he lays one finger on him I will break his neck . . . (*Summons* CEELY *over.* CEELY *holds can up to* KATE'S *mouth.*) Mona, don't surrender. (CEELY *and* KATE *stick their tongues out at Granny Tidy, and run back to* ROSE *who is looking very dismayed.* CEELY *feels sorry for her.*)

CEELY (*to can*). Mona, are you sure you haven't got Rose Marie Noble's twins in there? (CEELY *listens.*)

ROSE. Ask her to check under the pews . . . in the crypt. (CEELY *is listening intently to what Mona is saying. She looks shocked and then crosses herself, then looks at* ROSE.)

CEELY. Annie McGorman's youngsters are going buck mad, they have put moustaches on all the statues with their felt tips. They have put Mina Dumphy's Robin Hood hat on the Virgin Mary. (*To can.*) Mona, that is terrible . . . what a day . . . and I am awful sorry about the bingo game. I knew how much everybody was looking forward to it. The Brits came in and raided me, took my wee radio station . . . terrible loss to the whole community. (*Listens, begins to look very hurt at what Mona is saying.*) Oh, is that right, Mona? (*Choked up.*) Well, I am sure that is great for you. (*Brings can down and can hardly speak she is so upset.*) Mona has her own bingo game going on over in the chapel. Danny's ma is callin' out the numbers over the mike . . . (*Angry.*) I hate that Bridie Hagan! She is a part-time caller down in the Silver Star Bingo, and now she thinks she is somebody . . . typical . . . my wee radio station is not away a meal hour and not one sinner gives a damn. (*Gives can to* ROSE *and climbs up on her chair.*)

ROSE (*to can*). Mona, have you definitely not got them? (ROSE *hands can back to* CEELY, *Has given up.* CEELY *refuses to accept, she is still upset.* ROSE *hands it to* KATE, *looks up at Tucker.*) Tucker! Our wee children are lost. He just keeps sayin' shush. He has went mad. The controlled explosion has flipped his lid. He must have went cuckoo, he has never disobeyed me in his life . . . never! Ceely, look at what they have done to my husband, turned him into a lunatic.

KATE (*has been listening into the can*). Oh, that is awful, burstin' to go to the toilet in a chapel . . . that's that oul' cheap communion wine . . . runs right through you. What are you gonna do? (*Tries to hand can to* CEELY *who refuses it.* KATE *persists and* CEELY *takes it.*)

CEELY (*smugly*). Serves you right, Mona MacAldooney. You know there are no toilets in a chapel . . . (*Then feels sorry for her.*) Ach, what are they gonna do?

ROSE. Who gives a damn! I have lost my children.

CEELY. Will you give over. . . . There is a wee bride and groom in there . . . (*points into bean can*) . . . trying to get married before their wee child is born. . . . Three-hundred British soldiers not lettin' them, forty-five people that can't go to the toilet and a bingo game going on without me, and all you're concerned about is two snattery nosed wee blurts.

KATE. And my child is about to be snatched by a demented nun. (*Addresses the army.*) Hey, if you see a terrorist disguised as a demented nun, don't let her into the chapel. . . . Right! . . . .Aye, right! . . . Aye, just right!

CEELY. And give Mona MacAldooney a priest and a Portaloo. (*To can.*) Hey, Mona, what about the holy water fonts? (*Listens.*) Annie McGorman's youngsters are paddling in them.

ROSE (*losing control and screaming*). Tucker, Tucker, Tucker! (CEELY *and* KATE *indicate to her to keep quiet. They are listening into the can and humming 'Edelweiss'.*) What is Tucker playing at? He has all the kids on the roof of the empty flats. (KATE *and* CEELY *continue to hum.*) They are carrying planks ... what the hell is going on? (*Starting to go mad also.*) Tucker, speak to me....I am your wife, mother of your lost children.... Don't you shush and shrug me.... What are you doing with all them kids on the roof carrying Union Jacks....

CEELY (*stops singing abruptly and looks up in horror*). My Union Jacks, from my souvenir collection. I was savin' them up to make my aunt Beattie a bedspread (*Whispers to* KATE.) She's a wee Protestant. (*To* ROSE.) Your Tucker is a thievin' pig. (*Jumps off chair and goes over to her.*) It's bad enough gettin' raided by the British Army, but when your own next door neighbour is doin' it ... (*Glares at* ROSE. KATE *has been ignoring this and is still listening to 'Edelweiss' in the bean can, offers it to* CEELY *to listen to cheer her up. She accepts.*)

CEELY (*as* KATE *hums*). Wish my wee Dwane was here, he would love all this action, if he comes back from America with an accent I'll murder him.... Wee Tucker O'Neill was sent home after three weeks, he is talkin' like John Wayne. If my Dwane lasts six weeks without gettin' into trouble, it will be a bloody miracle. (*Suddenly.*) Jesus, I forgot to feed his tortoise.

KATE (*shocked*). Starsky! (KATE *and* CEELY *get on their knees and start frantically searching for the tortoise.*)

CEELY. Starsky ... Starsky....

ROSE (*to herself*). Try up at the altar or behind the statues.

CEELY. Starsky, where the hell are you?

KATE (*urgently*). Ceely!

CEELY (*still searching*). Starsky!

KATE (*more urgently, she is now looking back into flat*). Ceely, where is Granda Tucker?

CEELY. Hiding underneath my duvet ashamed to show his sleekit face ... Starsky.

KATE (*screams*). Ceely! Ceely! (CEELY *jumps up.*) The duvet is there, but Granda Tucker isn't in it. (CEELY *turns tentatively round. They both gasp in horror.* CEELY, ROSE *and* KATE *all face front, horror stricken, as if they'd seen a ghost.*) Your duvet was pink with roses ... like my walls....

CEELY (*can hardly get the words out*). My slippers. . . .
ROSE. (*also shocked*). My buggy. . . .
KATE. An omen! (*They all shiver.*)
CEELY (*tearful*). Old Granda Tucker . . . gone . . . he hasn't left that house in 12 years. (*They are all in tears.*)
KATE. When did you last see him?
CEELY. I don't know. I never look for him because. . . .
ALL. He's always there.
CEELY (*suddenly stops crying*). Wait a minute! (ROSE *and* KATE *also stop.*) Just before the controlled explosion he wouldn't come out, remember?
KATE. The raid!
CEELY (*suspiciously*). No granda . . . No radio. . . .
KATE. No Starsky!
ROSE. No buggy!
ALL. The Brits! (*They all march to the end of the balcony.*)
CEELY. Right, you . . . (*Pointing out.*). . . . No, not you . . . you! Right, you big English ghett. . . . Where is Granda Tucker? Speak to me, I said where is oul' Granda Tucker? (*Holding back her tears.*) You walked into my house and arrested an old man with one foot in the grave and a wee tortoise . . . suppose you're gonna try and interrogate the two of them. (*Puts her head on* KATE'S *shoulder.*) This will kill oul' Tucker.
KATE (*bravely*). Don't let them see you cry, Ceely.
CEELY. Don't yous be beatin' him up . . . he never done nothin'. . . . When yous can walk into a house, lift an old man out of a duvet and a wee tortoise out of a shoe box, yous are capable of murderin' your own mothers! (*Puts her hand out for the can and* KATE *hands it to her.*) Mona! They have arrested Granda Tucker and our wee Starsky, and they are gonna arrest you too, so you may as well stay in the chapel as Castlereagh . . . (*Holds can up so that army can hear.*) No Surrender!
KATE & ROSE. No surrender!
CEELY (*listening into can*). Oh, frig! Mona wants a priest and a midwife. (*To army.*) Satisfied? Satisfied? There is a wee child about to be born in there (*points into can*) out of wedlock . . . in a chapel.
ROSE. Born in a chapel! Hope my twins aren't in there. . . .
KATE. Ach, you want them in, you want them out, what the hell do you want? (*Over balcony.*) Well, I hope you are all well satisfied now . . . yous have arrested an old man and a wee tortoise, you're lettin' a wee child be born in that chapel and his ma and da aren't

even married yet.
ROSE. And turned my husband into a lunatic.
KATE. And blew up my walls.
ROSE. Stole my buggy.
CEELY. And took my wee radio.
ROSE. And yous just stand there and laugh. (*All begin to sing.*)

Out There Somewhere

ROSE. Out there somewhere, there's gardens with fences
and little oak benches.
CEELY. Women with husbands, loving and caring
KATE. Children in college, reading and learning
CHORUS: Out there on the ground
They're all sleeping sound tonight
Out there on the ground.
ROSE. Out there somewhere, they dine on steak dinners
Always back winners.
CEELY. Planning a future and talking together
KATE. Warm and cosy whatever the weather
CHORUS: Yes, they're all sleeping sound tonight
Yes, they're all sleeping sound tonight.
Over here in the air
The army are spyin'
Cameras are pryin'
Touts they are lyin'
Choppers are flyin'
But we are defiant
Over here in the air
It's hard to sleep sound tonight
Yes, it's hard to sleep sound tonight.

(*At end of song, they all just sit and stare out.*)
CEELY. Poor Mona, she prayed day and night for that wee child to be born proper.
KATE. Her first grandchild was born in gaol . . . the second was delivered by a part-time soldier in the U.D.R.
CEELY. The next two wee ones were born on the run.
ROSE. That last wee one was the worst . . . it was born in England.
CEELY. Poor Mona, she hasn't had a normal grandchild in two years. (*To can.*) Mona, how fast are the contractions comin'? All the time. (*To others.*) All the time. It was all the excitement. (*To can.*) Was it the bingo game? (*To others.*) It was the bingo game. (*To

can.) Did she win the jackpot game? (*To others.*) She won the jackpot.

KATE & ROSE (*delighted*). Charlene won the jackpot. (CEELY *jumps off the chair as something dawns on her.*)

CEELY (*suspiciously, into can*). She won the jackpot. Oh, well, I suppose you could say this was all your fault, Mona. I have told you before you go far too slow with your numbers and the people aren't prepared for checking . . . you see, Mona, what I do is, I start off slow and then build gradually and gradually so that the tension actually builds with the person. (CEELY *is getting carried away.*) Oh, no, Mona, you will never be a full time professional bingo caller because you have not made yourself aware of the whole psychology of it like I have. (*Pulls can away from her ear as if hearing* MONA *screaming through it.*) All right, Mona, keep your hair on.

ROSE (*looking up at* TUCKER). What is he doing? I can't cope.

CEELY (*over balcony to soldiers*). Give Mona MacAldooney a priest and a midwife.

ROSE. Look at him. . . . He's puttin' the planks across from the empty flats to the chapel roof.

KATE. The soldiers don't know what the hell is goin' on.

ROSE. That makes three hundred and one of us.

CEELY. Well, they can't touch them children because they are not carrying no weapons.

KATE. They can't touch them children because Rambo is standin' guard. He would ate them soldiers if they came near.

CEELY. They can't touch them children because they are carryin' my friggin Union Jacks. I'll kill your Tucker.

ROSE. If he doesn't kill himself first. (*Baffled.*) He's carrying the kids on his back across the planks onto the chapel roof. (*Shouts.*) Quasifuckinmodo!

CEELY. And the British Army are just standing looking.

ROSE (*losing control*). Tucker! Tucker! Tucker!

KATE. Shush. . . . There is a Priest tryin' to get into the chapel. Quick, Father, before the wee child is born. . . .

CEELY (*excitedly into can*). Mona! Mona! There is a priest pleadin' with the army to let him in. (*There is obviously no response from can.*) Mona! (*Bangs can.*) Mona, speak to me.

ROSE. She has been struck dumb too. Frig me!

CEELY (*panicking*). Mona! Mona! (*Takes can calmly away from her mouth.*) Pigs! They have cut communication. (*They all cross themselves.*)

SOMEWHERE OVER THE BALCONY – ACT II

SFX (*of priest speaking through megaphone*). This is Father Mulcahey here. I appeal to you to come out.
KATE. He should be appealin' to go in.
SFX. This is a sin against your church.
KATE (*weakly*). Dustin! (*She is helpless and begins to cry.*)
SFX. You are dancing in the aisles.
ROSE. You look like a British soldier ... go down there and tell them.
SFX. This is a sin against your church.
KATE. But I can't speak like one. They speak English. (*Weakly.*) Mona! (*Realises there is no longer communication.*)
SFX. I appeal to you to come out.
ROSE. Tucker, do something.
SFX. I implore you to come out.
CEELY (*suddenly*). Our Dwane!
SFX. I beseech you to come out.
ROSE. He's in America. Big deal!
SFX. Come out or be damned!
CEELY. But his catapult isn't. (CEELY *runs behind screen and produces catapult.*)
SFX. I will pray to Our Lady for your forgiveness.
CEELY (*aims*). Right, Kate, he is gonna get this in the scone. (KATE *has composed herself.*)
SFX. You are singing lewd songs on the organ.
KATE (*takes catapult*). Stand back! Right, Frank Tidy, this is curtains.
SFX. I appeal to you to give yourselves up in the name of God.
ROSE. Okay ... left a bit ... right ... right.
KATE (*aggressively*). Right!
SFX. I appeal to you, Mona MacAldooney.
KATE. That pigeon is sitting on the megaphone.
ROSE & CEELY (*scream*). Fire!
SFX. I appeal to you for the sake of the little children under seige.
ROSE (*she is guiding* KATE'S *arm*). Right, he is on his knees.
CEELY. He's getting up again. ...
SFX. I appeal to you, Danny Hagan and Charlene MacAldooney....
ROSE. Fire!
SFX. I appeal to you.... Aaaaagh ... (*SFX of priest scream and thud. Immediately SFX of ambulance. They all watch.*)
ROSE. God, that was quick!
CEELY. Quick when they want to be quick.
ROSE. Not quick when we want them to be quick.
CEELY. Quick when it suits them to be quick.
KATE (*aims catapult*). Quick.... There's Granny Tidy.

CEELY (*grabs her arm*). Kate, she is an oul' woman.
KATE. She's an oul' whore.
CEELY (*aims* KATE'S *arm*). Right enough!
KATE. Aye. . . . Too right!
ROSE (*grabbing* KATE *violently and directing her arm*). Kate, get a soldier. Get them all. Get Tucker, so I can go down there and get my twins.
CEELY. The ambulance has stopped. The driver is asking your Tucker if he is gonna jump.
ROSE. That is why my Tucker is on the roof. He is gonna commit suicide.
CEELY. Jump if you're gonna jump. Stop keeping the ambulance man waiting.
KATE. Is he insured?
ROSE. Well. . . . Well. . . . Well. . . . What a day . . . no children, no car . . . no man, and no insurance . . . but I am sure there are plenty of people worse off than me.
KATE. Yes. . . . Wee Charlene is about to have a wee child out of wedlock.
CEELY. In a chapel.
ROSE. Jump . . . go on. . . . Jump . . . don't just stand there. At least you're local; they usually come from outside to do it up there. I am glad it was one of our own.
KATE. Have you flipped?
CEELY. Smell! Do you smell it? Do you feel it? The tension. . . . Yes, the Brits are gonna storm the chapel any minute. I know that feeling . . . it's like when somebody is waitin' on one number for a big Saturday night scoop. . . . You know it's gonna happen. . . . you can't stop it. . . . you just haft to go with it.
KATE. All them children are wavin' Union Jacks when they should be stoning soldiers, and your Tucker is egging them on.
ROSE. Jump! I'll pick up your bits . . . you will look lovely on top of my T.V.
CEELY. I can't look. They're movin' in, I can feel it.
ROSE. Wee bits of Tucker all over the house. Some people have a china collection. I'll have a Tucker collection.
KATE. Them youngsters are all burnin' your Aunt Beattie's bedspread.
ROSE. Tucker's Last Rights. . . .
KATE. Whose side is who on any more? Them kids are lighting up the chapel roof to land that army chopper, and your Tucker is waving it in. (*SFX of army helicopter.*)
ROSE. Tucker, don't die a traitor, don't leave me to live with the

shame of it. Think of your wee children.
KATE. Sure they're lost.
CEELY. Yes, there is gonna be a check any minute. Everybody is sweatin'. (*SFX of helicopter as if it is right overhead.*)
KATE (*looks up*). All right. All right. We know what you're gonna do. (*SFX helicopter as if it swoops right over their heads. They all scream and fall to the ground.*) Is that chopper driver drunk?
ROSE. No, it's all right. My Tucker is gonna show them where to land. He's turned all them wee children into Brit lovers too.
KATE. They're not gonna land . . . they're throwin' out a rope.
CEELY. A rope! A rope! OH NO! They are gonna land a top marksman on the belfry. (*She turns away in despair.*)
KATE. Mona, surrender!
ROSE. He has all them wee children dancin' up and down on the chapel roof. He's put them all away in the head too. (*She also turns away in despair.*)
KATE (*she is staring at the chopper, then grabs* CEELY). The chopper . . . the chopper. (*She is jumping up and down and can hardly get the words out. She grabs* CEELY.) The chopper. The chopper.
CEELY. Yes, it's a chopper. We know it's a chopper.
KATE. No, the chopper, the chopper.
CEELY. No, it's not a chopper? It is a chopper.
KATE (*getting very excited*). No, the chopper driver, the chopper driver!
CEELY (*horrified*). There's no chopper driver? (*To* ROSE.) There's no chopper driver.
ROSE (*screams*). There's no chopper driver!
KATE. No! No! Look at the ch . . . ch . . . chopper . . . the chopper. (KATE *has to scream over the top of them to make herself clear.*) LOOK AT THE CHOPPER! (*They all stop and stare up at the helicopter.*)
CEELY (*quietly*). Jesus, Mary and Holy St. Joseph . . . that is my wee Tucker!
KATE. And Granda Tucker!
ROSE. In the Chucker!
CEELY. How the hell. . . .
ROSE (*staring at the helicopter*). Kate . . . Ceely. . . . Who is that in the . . . behind your . . . is it . . . is it . . . (*In disbelief.*) It is . . . my wee twi. . . .
KATE. And my Pe . . . Pe . . . (*She can't get the words out.*)
ROSE & CEELY. Your Pepe!
CEELY. If oul' Tucker was here . . . and my wee Tucker was with

Tootsie. . . .
KATE. Disguised as a. . . .
ROSE. Driving the stolen sara. . . .
CEELY. How the hell did oul' Tucker make contact?
KATE. He took your. . . .
CEELY (*delighted*). Radio . . . my wee radio.
ROSE. Stationed my Tucker on the roof.
KATE. To bring in the ch. . . .
ALL. With burning . . . Jacks.
CEELY. It all fits.
KATE. Where's Starsky! (*Pause. They* ALL *look towards the helicopter.*)
ROSE. What is that my Mark has on his head?
ALL. A tortoise.
CEELY. My Starsky is in the helicopter.
SFX (*of Granda Tucker speaking from megaphone in helicopter. They all stand and listen in disbelief*). Attention the enemy. This is Commanding Officer Tucker Allouisis Cash of the 21st Ardoyne Batallion.
CEELY (*in shock*). He hasn't spoke in twelve years.
SFX. My comrades and I are holding Major Devenport of the Gloucesters as hostage.
ALL. Hostage!
SFX. We are about to lower a British Army Chaplain. We demand that the wedding of Charlene Macaldooney and Danny Hagan be consecrated immediately. Tiocfaidh ar la! (*SFX of cheering and clapping. They all stand waving at the helicopter.*)
CEELY (*shouts*). Tucker, put your seatbelt on. (*A long pause as if everything is now back to normal.*)
ROSE. Wonder is the water back on.
CEELY. Well, there's one good thing, Rose Marie. you won't have to wash Tucker's car ever again.
KATE. They got a great day for the wedding.
ROSE. Just in time . . . loo . . . they are bringing wee Susie Quinn's coffin out to the chapel.
KATE. Is it half past four already?
CEELY. Oh, great, I forgot about the funeral.
KATE. That's what I love about this place. On a day like today you could be anywhere. (*All sing.*)

The Sun is Shining
The sun is a shinin' and the bees are a hummin'

SOMEWHERE OVER THE BALCONY – ACT II

Flowers are a bloomin' and the Lord is a comin'
Yes, the Lord is a comin'
Yes, the Lord is a comin'.

Kids are a playin' and the dogs are a barkin'
Mothers are a laughin' and the Saints are a marchin'
Yes, the Saints are a marchin'
Yes, the Saints are a marchin'
Yes, the Saints are a marchin'!
Hey, Mister Maker, we know you care
Cos we're nearer to heaven than the folks down there.
Yes, the Lord is a comin'
And the Saints are a marchin'
And the Lord is a comin'
And the Saints are a marchin.'

Hey, Mister Maker, we know you care.
Cos we're nearer to heaven than the folks down there.

BLACKOUT. END OF ACT TWO.

GLOSSARY

2d: two pennies, tuppence. Twelve pennies equaled a shilling—5 pence in decimal currency.
Abraham Lincoln: stinking.
Actin' the lad: swaggering.
Aerial: antenna for television reception.
Albert clock: name of clock tower in Belfast.
Alf Roberts: a character in the longest running British soap opera, 'Coronation Street'.
Anyroad: anyway.
Ardoyne: predominantly Catholic area of Belfast.
A.R.P.: Air raid patrol.
Arsin': wasting time.
Ate the alter rails: someone who is always at the chapel praying.
Away and rattle: colloquial expression meaning 'Get out of here!' or 'Away and chase yourself'.
Babycham: alcoholic sparkling drink.
Baccy and Porter: tobacco and Guinness.
Bag a' dulse: cheap seaside treat; dried seaweed; rubbish; worth nothing
Bake: slang for face.
Bangor boat: pleasure trips were taken by boat from Belfast to Bangor, a seaside town.
Bann: river dividing east and west of Northern Ireland.
Bap: bread roll.
Barges: shouts.
Barney Hughes: product name of freshly baked white bread.
Barn brack: fruit bread.
Bate: beat.
Batin' dockets: beaten dockets, unsuccessful betting slips.
Batins: beatings
Beg: bag.
Billy can: tin pail.
Birds: women.
Blackstaff: another large Belfast mill during this time period.

GLOSSARY

Blitz: the Germans bombed Belfast, in addition to several other U.K. cities during World War II.
Board of Guardians: administrators of the workhouse.
Bob's your uncle: colloquial term meaning 'there it is', 'that's it' or 'there you are'.
Bokin': vomiting.
Bookies: bookmakers, turf accountants.
Bout ye: Belfast greeting; shortened from of 'What about you'!
Bowler: Black, round-crowned men's hat worn by Protestants in the Orange Day Parade on July 12th.
'Boy I Love is up in the Gallery, The': a music hall song.
Boyos: boys.
Brack: fruit cake.
Brains Trust: 1950s and '60s BBC radio programme comprised of a panel of intellectuals discussing topical issues.
Brass neck: as in 'the nerve of her'.
brasses: brass ornaments, trinkets.
Brave: good.
Bru: welfare.
Bugger all: swear word for nothing.
Bummin' and Blowin': boasting.
Bundoran: a holiday town in Donegal.
Bung: throw.
Bunged: full up.
Buns for the oven: children.
Bushmills: brand of whiskey made in Northern Ireland.
Butlins: chain of summer blue-collar holiday camps
Button: Union membership.
Can't have: doesn't like.
Cantrell and Cochrane: manufacturer of bottled lemonade and soft drinks.
Castlereagh: Police Interrogation Centre.
Catch yourself on: wise up.
Carnaptious: bad tempered.
Carrots dressed: the practice of creating a character by dressing up the carrot.
Carry on: behaviour, usually unseemly.
Carson's Army: refers to the Ulster Volunteer Force, a civilian militia formed to fight against Home Rule.
Catch Me Pal Society: a temperance society; promoted abstinence from intoxicating drink.
Chapel: the Catholic Church.

GLOSSARY

Charabanc: an open top bus used for excursions. A trip on a charabanc represented leisure for the working class.
Clock: to sit immobile.
Clocks: black beetles.
Club Book: catalogue sales.
Coddin': kidding.
Controlled explosion: a bomb is detonated in conditions which minimize damage.
Covenant: a pledge publicly signed by thousands of Unionists at Belfast City Hall as a protest against Home Rule, i.e. Ireland governing themselves in Dublin.
Crack: a bit of fun, festivity.
Cracker: great.
Craigavad: bad (rhyming slang), as in 'Not too bad'.
Cribbin': footpath, curb.
Crossing the Boyne: site of the famous 1690 battle (see **Twelfth of July** below).
C.S. gas gun: riot control gas gun.
Custom House steps: the steps of the large Customs and Excise (Food and Drug Administration) building near Belfast docks served as a traditional meeting place.
Cutty: young girl.
Dad's Army: old television show
'Daisy, Daisy': music hall song.
Dar: dare, as in 'How dare he!'
Derry: town in the northwest of Northern Ireland, county capital. Also known as Londonderry.
Desperate: awful, an awful lot of.
DHSS: Department of Health and Social Security (Welfare Department).
Diatribe: lecture, rush of words.
Dicey: unstable, flaky.
Diddies: breasts.
Docken: juice made from docken leaves.
Doffer: a mill worker who removed full bobbins from the machines.
Doing the 'double': refers to the illegal practice of working while drawing State Welfare benefits.
Dole: unemployment benefits.
Dole snooper: Welfare Department Officer employed to patrol building sites, etc. to catch individuals claiming welfare money and working at the same time.
Doodums: baby talk.

GLOSSARY

Dr. Barnardo's: orphanage.
Dukin': peeping.
Duncher: man's flat cap.
DUP: Democratic Unionist Party.
Durex: condom.
Eejit: Idiot, fool.
Eights and nines: mixed up, comparable to 'doesn't know whether she's coming or going'.
Falls: reference to Falls road in Pound Loney, a poor Catholic area.
Fenian: nickname for Catholic.
Fenton's tripe: Fenton is the name of the local butcher. Tripe is the lining of cow's stomach, generally considered poor man's food.
Flash: brand of floor cleaner.
Forby: as well, in addition to, all the same.
Fornenst: against, beside, or in front of.
Frank Carson: Belfast comic.
Fresians: breed of milk producing cow.
Frig it/Frigging: derogatory exclamation.
Gaol: jail.
Gaunch: oaf.
Gawkin': staring.
Geg: joker or laughable, funny.
Geggin: joking, teasing.
Ghett: bad rascal, jerk, brat, pest, derogatory remark.
Give over: stop.
Glens a' Gormley: Glengormley, six miles north of Belfast.
Glorious Gloucesters: the Gloucester Regiment nickname.
Gob: mouth.
Gods: in Victorian theatre, the cheapest seats because they were the highest and furthest from the stage.
Gormless: stupid.
Gripe water: type of medicine for colic.
Gub: a.k.a. gob; mouth.
Guiders: home-made go-carts, kids trucks made of planks and old buggy wheels.
Gurner (oul'): complainer.
Gurnin': crying.
Gurny: whiny.
Hackle: panticolon stage in linen process.
Half-timer: linen mill term for half-time workers (usually children), working half of the week in the mill and the other half in school.
Halfins: half ones, spirits measure of alcohol (as opposed to pints of

beer).
Hallions: derogatory term, male equivalent of whore.
Ha'penny chews: toffees.
Happy Jimmy: a Belfast character of the time.
Hatch: pass-through.
Harrished: harassed, tortured.
Headbangers: mentally disturbed.
Hi ya, Burke: don't be ridiculous.
Hippodrome and Alexandra Theatres: names of theatres.
Hokin': looking.
'Holy Ground, The': Irish sea shanty: 'The Holy Ground Once More'.
Holy Joes: overly zealous religious people of any denomination.
Home Rule: a policy which would permit a parliament in Dublin as opposed to Irish M.P.s sitting in Westminster.
Hoo Flung Dung: Chinese take-out.
Hooley: party.
Hop: dance.
Hosts: communion wafers.
Humpy: cantankerous.
Hurdy-gurdy: machine that plays a tune by turning a handle.
'I Have an Anchor': an old revivalist hymn.
'I'll Take You Home Again, Kathleen': old Irish song made famous by John McCormick.
Internment: on 9 August, 1971, the British government introduced the Internment to Northern Ireland, the arrest and imprisonment without trial of many Catholics who were suspected members of the I.R.A. In the early hours of the morning, army raids were carried out in Nationalist areas and thousands of men were 'lifted' for internment. Internment day is now marked, on the anniversary of introduction, every year in Nationalist areas with bonfires. Various British and Nationalist support groups visit areas such as Divis on the anniversary.
Island: Queen's Island, shipyard of Belfast.
Jennymount: another large Belfast mill during this time period.
Join: party.
John De Lorean: an American businessman who took advantage of British government tax incentives to use Northern Ireland as the manufacturing base for his ill-fated gull-wing De Lorean motor car.
Jube-jube: type of candy, used in this context as an affectionate term such as ?sweetie?.
Jukin': peeping.

GLOSSARY

Keep dick: keep a lookout.
King Billy: William III; a Dutch king who ascended to the Throne of England in the late 1600s. He is celebrated by Ulster Protestants as 'Defender of the Faith'.
Knocked off work: left work.
Krauts: slang for Germans.
Lady Muck: pretentious woman who acts above her station.
Lambeg drum: large drum used in Protestant parades.
Lavies: outside toilets.
Lay up your ends: spinning term that refers to two ends of yarn tied together after the yarn breaks or to prevent breakage. This most often happened at the beginning or end of a shift.
Lie-in: sleeping past the appointed hour, 'sleeping in'.
Lift: elevator.
Liftin': arrest by army or police.
Lig: idiot.
Liquorice Allsort: multi-coloured candy–reference to the actress' costume in that production.
Loanins: country lanes.
Long Kesh: prison also known as the Maze, lies outside of Belfast and contains the H-blocks where the hunger strikes of 1981 took place. The prison served as a camp for those imprisoned during the Internment.
Lord Carson: Lord Edward Carson was a Protestant Unionist leader and lawyer. He was also famed as the barrister who successfully prosecuted Oscar Wilde.
Lough: area of water, in Belfast, opening into the Irish sea.
Luk out: look out/ that's his problem.
Lurgan spade: the main characteristic of a Lurgan spade is its considerable length.
M15: British secret service, government covert spy organization.
Mac: raincoat.
Marks and Spencer's: well known mid-range department store in the British Isles.
Mater: name of hospital.
Mayo: county in Republic of Ireland.
Meal's mate: the equivalent of a meal.
Mexico, here we go, here we go: Northern Ireland qualified to play in the World Cup Soccer Finals in 1986.
Mick: Catholic.
Midden: sewer.
Mill fever: a generic term for illness, usually respiratory. Mill fever

GLOSSARY

resulted from working in hot damp conditions followed by exposure to the cold.

Millies: derogatory term for mill girls. Today this term is still used in Belfast in a derisive way.
Mind: remember.
Mineral: soft drink.
Minging: dirty, filthy.
Mitchers: children who stay away from school.
'Mountains of Mourne': traditional Irish song by Percy French.
Muckin' out: cleaning out animal manure.
Mundies: cheap wine/sherry.
'Murder in the Red Barn': name of a melodrama.
'My Aunt Jane': Belfast song.
'My Old Man Said Follow the Van': famous Marie Lloyd music hall song.
Nappy/nappies: diaper.
Navvies: labourers.
'Neighbors': daytime Australian soap popular in Ireland and the U.K.
Nippers: little children.
No Surrender!: rally cry of solidarity against Nationalist threat.
Novenas: Catholic prayers.
Nurse: expecting a baby, pregnancy.
Oil cloth: similar to a plastic tablecloth.
On the blanket: campaign against the criminalization of political prisoners. Nationalist prisoners protested for special status (i.e. political status) in 1980/81, one form of which was wearing only blankets instead of ordinary criminal prison uniform.
'Only a Bird in a Gilded Cage': music hall song.
Orange and green sashes: because the band is non-sectarian, both Protestant orange and Catholic green are worn.
Orange lilies: flower that is an emblem of Protestantism, particularly associated with the Orange Order.
Orange Order: an organization for Protestant men.
Orange sash: a sash identifying Orangemen, a Protestant men's group, worn on the 12th of July.
Orangey: derogatory term for the Protestants.
Oul' Nick: the Devil.
Oxfam: charity organization, similar to Salvation Army, in the British Isles.
Padre Pio's mitt: religious relic; mitten said to be worn by Padre Pio.
Pat Jennings: well-known Irish goal-keeper.
Pawn: Hock shop.

GLOSSARY

Peace People: Betty Williams, Mairead Corrigan, Ciaran McKeown; established in August 1976 because of deaths of three Maguire children. Williams and Corrigan won 1976 Nobel Peace Prize.
Peeler: police.
Penny drops: come to a realization.
Picker: a tool used in the spinning room.
Pig's back: doing well financially.
Piggin': dirty, filthy.
Pikes: farming implements used in the 1798 Irish rebellion as weapons.
Pinnies: apron, pinafore.
Play her up: misbehave.
Plot: vegetable garden.
Pokes: ice cream cone.
Porter: dark ale, Guinness-style, similar to beer.
Pouce/poucey: respiratory disease caused from inhalation of flax dust.
Pound Loney: name of area on Falls road.
Power sharing: government where Catholics and Protestants share power.
Pratties: potatoes.
Prods: nickname for Protestants.
Quare: real, very, a good many, a lot, exceptional.
Quare oul' mark: wonderful, just the ticket.
Rag o' bone men: men who collect rags, etc.
Red Hand of Ulster: Ulster's flag has a red hand in its centre. It has been misappropriated into a Protestant, sectarian symbol.
Red up: clear up.
Redundant: lost job, laid off job.
Reelers: male mill workers working on a particular stage of the linen process.
'Riding on a Tramcar': traditional song.
Romanism: allegiance to the Roman Catholic Church.
Round: turn, specifically in relation to buying a drink.
Rovers: male mill workers working on a particular stage of the linen process.
Royal: Royal Victoria Hospital, West Belfast.
Rubbers: rubber aprons worn by Spinners in the mills.
R.U.C.: Acronym for Royal Ulster Constabulary, police.
Ructions: arguments, dissent.
Sand dancers: dancers who perform on a board with sand.
Saracen: protected army vehicle.
'Sash, The': Ulster Protestant anthem, 'The Sash my Father Wore',

GLOSSARY

symbol of the Orange Order
Scrake: early morning, scrake of dawn.
Scrubbed half moons: front doorsteps were traditionally scrubbed by women in a semi-circular, half-moon shape.
Scrubbers: washer women.
Scud: curse, hex.
Seamie: short for Seamus, pronounced Shaymee.
Seawright, George: militant Loyalist politician, fatally shot in 1987.
Servin' up the Hosts: Holy Communion (part of Mass).
Shawlies: derogatory term for mill workers. They wore black wool shawls.
Shebeen: illegal drinking establishment.
Sheughs: ditch by the side of the road, pronounced shucks.
Shirty: uppity.
Sick: illness benefit payment.
Sixims: sixpence.
Sketch: humorous short skit.
Skitter: derogatory name.
Slaggin': making fun of, kidding, teasing.
Slate: credit.
Slatins: roof slates.
Sleeked: sly.
Sliders: ice cream wafer.
Smittal: catching.
Snattery: snotty.
Snooker: table game similar to pool.
South: the 26 counties of Ireland that form the Republic of Ireland.
Spar: a convenience store, comparable to 7-11, in the British Isles.
Spondulicks: money.
Step: steep, soak.
Strongbow: cider.
Syrup of figs: a mixture to relieve constipation.
Sweats: people waiting for just one number.
Sweetie coupons: during and after WWII, candy was rationed in the U.K. on a coupon system.
Stick-bracker: derogatory term for someone who doesn't have a job.
Taigs: derogatory term for Catholics.
Take a buckle: have a fit.
Tappin': borrowing from.
Telegraph: name of Belfast newspaper, the *Belfast Telegraph*.
'These Three Drunken Maidens': pub song.
Thick: stupid.

GLOSSARY

Thole: to put up with, tolerate.
'Three Little Maids from School': a song from Gilbert and Sullivan's operetta, *The Mikado*.
Tinkers: traveling people, itinerants.
Tick collectors: debt collectors.
Tiocfaidh ar la: English translation: 'Our day will come'. Pronounced CHUKKIE R LA.
Tip your cap: kow tow to.
Toutin': giving information to the police.
Tricolour: the national flag of the Republic of Ireland.
Troops out: British pressure group in support of British withdrawal from Northern Ireland, in support of the Republican movement, arrived in Belfast to Divis Flats on the anniversary of the Eve of Internment every year for a demonstration.
Truck: abuse (meaning applies specifically to this context), nothing to do with.
Tucker: nickname very popular in Divis Flats, short for Tommy, i.e. 'little Tommy Tucker'.
Twist: mood.
Trollop: loose woman.
Twelfth of July: traditional date of annual marches organized by the Protestant Orange Order to commemorate the victory of Protestant William III over Catholic James III in 1690.
U.D.R.: Ulster Defence Regiment, a locally recruited regiment of the British army. The majority of the members are part-time.
Ulster Day: the day of the Ulster Covenant (see **Covenant** above).
Umbrage: offence.
Union Jack: British flag.
Up the Spout: term for pregnancy.
Up to high doh: worried.
Vincent de Paul: charitable Roman Catholic organisation.
Wains: wee ones, children.
Wally: jerk.
Weeker: awesome.
Wick: awful, hopeless.
Willicks: whelks, shellfish.
Wellies: Rubber boots.
Wet: ready to be used or consumed.
Wet the baby's head: a drink to celebrate the birth of a baby.
Wheen: many, a lot.
W. I.: Women's Institute.
Windys: windows.

GLOSSARY

Wobbler: loses temper.
Yokes: things.
'You'll Never Miss a Mother's Love til She's Buried Beneath the Clay': sentimental ballad.
Z Cars: 1960s police series on British T.V.

APPENDIX

APPENDIX

Charabanc Theatre Company Production Notes

NOTES

Production information, including specific attribution of the writers, comes from printed programmes. Tour information has been reconstructed from programmes, playbills, attendance tallies, journals, and reviews. The completeness of each year's tour records varies considerably.

During the company's early years, Marie Jones used the acting name Sarah Jones because of another Equity actor by that name.

Carol Moore changed from her married name, Scanlan, to her maiden name, Moore during the final years of the company.

Blind Fiddler of Glenaduach   1990
Written by Marie Jones

Cast:
Sheila McWade
Marie Jones
Sean Kearns
Angela McNicholl
Carol Scanlan
Gerry McGrath

Production Team:
Director ....Peter Sheridan
Fiddle ....Jim McKillop
Uilleann Pipes ....Tom Clarke
Choreographer ....Maire Clerkin
Lighting Design ....Guy Barriscale
Scene Painter ....Stuart Marshall
Costumes ....Caitlin McWade
Administrator ....Patricia McBride

APPENDIX

Performance Information:
6 Aug 1990	Belfast Inst. of Further and Higher Ed., Whiterock Road, Belfast
7 Aug 1990	St. Agnes' Hall, Belfast
8 Aug 1990	Trinity Street Community Centre Unity Flats, Belfast
9 Aug 1990	Kilwee Social Club, Twinbrook, Belfast
10 Aug 1990	Conway Mill, Belfast
16-17, 19-24 Nov 1990	Stranmillis Theatre, Belfast
27 Nov 1990	Concorde Community Centre, Belfast
28-30 Nov 1990	Ainsworth Community Centre, Belfast
1 Dec 1990	Blackmountain Primary School, Belfast
4 Dec 1990	Rupert Stanley College, Belfast
5 Dec 1990	Donegall Pass Youth Club, Belfast
6-7 Dec 1990	Dee Street Community Centre, Belfast
8 Dec 1990	Ballybeen Activity Centre, Ballybeen
9-27 Apr 1991	Drill Hall Arts Centre, London
30 Apr-11 May 1991	Andrew's Lane Theatre, Dublin

Bondagers   1991
Written by Sue Glover

Cast:
Sara .... Carol Scanlan
Maggie .... Eleanor Methven
Tottie .... Michèle Forbes
Liza .... Paula McFetridge
Ellen/Jenny .... Lynn Cahill

Production Team:
Director .... Maureen White
Designer .... Liz Cullinane
Lighting Design .... Guy Barriscale
Wardrobe Assistant .... Marion Thompson
Original Music .... Debra Salem
Choreography .... Mary Fox
Accent Coach .... Brendan Gunn
Technical Stage Manager .... Guy Barriscale
Stage Manager .... Fearga O Doherty
Assistant Stage Manager .... Rossa Sheridan
Administrator .... Patricia McBride

Performance Information:
18-19 Oct 1991	Ardhowen Theatre, Enniskillen
22 Oct 1991	Town Hall, Ballymena
24 Oct 1991	Arts Centre, Newry
25-26 Oct 1991	Riverside Theatre, Coleraine

APPENDIX

29 Oct 1991	St. Patrick's Hall, Strabane
30 Oct-2 Nov 1991	Hawk's Well Theatre, Sligo
4 Nov 1991	St. Louis Convent, Monaghan
6-7 Nov 1991	Balor Theatre, Ballybofey
8 Nov 1991	Corn Mill Theatre, Carigallen
9 Nov 1991	Queen's Hall, Newtownards
12 Nov 1991	Town Hall, Cavan
13 Nov 1991	Bardic Theatre, Donaghmore
15 Nov 1991	Leisure Centre, Downpatrick
16 Nov 1991	Rialto Theatre, Derry
20-23 Nov 1991	St. Kevin's Hall, Belfast

Cauterised 1989
Written by Neill Speers

Cast:
Vi ....Eleanor Methven
Mill ....Carol Scanlan
Hetty ....Sheila McWade

Production Team:
Director ....Lynne Parker
Designer ....Houston Marshall
Sound tape ....Louis Edmondson
Costumes ....Ivor Morrow
Lighting Operators ....Martin Neill
....Graham Shaw
Production Assistant ....Cathy Timothy
Administrator ....Patricia McBride

Performance Information:
  16-18, 23-25 Nov 1989   Lyric Theatre, Belfast

Frontline Cafe 1991
Written by Thomas McLaughlin

Cast:
Eleanor Methven
Robert Byrne
Guy Barriscale (joined cast later)
Paula McFetridge
Laura Devlin

Production Team:
Director ....Maggie Byrne
Administrator ....Patricia McBride

APPENDIX

Performance Information:

3 Aug 1991	St. Agnes' Hall, Belfast
5 Aug 1991	Newhill Youth and Community Centre, Belfast
6 Aug 1991	Belfast Inst. of Further and Higher Ed., Whiterock Road, Belfast
7 Aug 1991	Ardoyne Community Centre, Belfast
8 Aug 1991	Stadium Community Centre, Shankill Rd., Belfast
12 Aug 1991	Rathcoole Youth Centre, Belfast
14 Aug 1991	Conway Mill, Belfast
15 Aug 1991	Ainsworth Community Centre, Belfast
16 Aug 1991	St. Mary's Hall, Greencastle
17 Aug 1991	Earlview Primary School, New Mossley, Belfast
19 Aug 1991	Garnerville Presbyterian Church Hall, Knockagoney
20 Aug 1991	Dee Street Community Centre, Belfast
21 Aug 1991	Rupert Stanley College, Belfast
22 Aug 1991	Ballybeen Activity Centre, Ballybeen
23 Aug 1991	Brookfield Mill Theatre, Ardoyne

The Girls in the Big Picture 1986
Written by Marie Jones
Devised by the Company

Cast:
Esther Moore/Patsy Flood .... Rosena Brown
Mary Jo McIlfatrick/Violet Tuckey .... Sarah Jones
Willie Jo Ferguson .... Sean Kearns
Jean Moore/Phyllis Pinkerton .... Eleanor Methven
Sydney Moore/Farmer .... Martin Maguire
Margaret Anderson/Sally Spence .... Carol Scanlan
Pat Sharkey .... Ian McElhinney
Tobias Stinton .... Toby Byrne
Production team:
Director .... Andrew Hinds
Stage Manager .... Teerth Chungh
Assistant Stage Manager .... Sean Kearns
Set Design .... Houston Marshall
Costume Design .... Ivor Morrow
Lighting Design .... Nick McCall
Sound Tape .... Louis Edmondson
Music .... Trevor Moore
"Jenny" Lyrics by .... Marie Jones
"Jenny" Music by .... Trevor Moore
Voices on cinema film .... Cherrie McIlwaine
John O'Hara

APPENDIX

		Ian McElhinney
Administrator	....	Maureen Jordan
Assistant administrator	....	Patricia McLoughlin

Performance Information:

22-27 Sep 1986	Ardhowen Theatre, Enniskillen
29 Sep-4 Oct 1986	SFX Centre, Dublin
7-10 Oct 1986	Hawk's Well Theatre, Sligo
12 Oct 1986	Hawk's Well Theatre, Sligo
14 Oct 1986	Arches, Magherafelt
23-26 Oct 1986	Crescent Arts Theatre, Belfast
28 Oct 1986	Mount Carmel Grammar School, Strabane
29 Oct 1986	Leisure Centre, Downpatrick
30 Oct 1986	Clotworthy House, Antrim
1 Nov 1986	Music Centre, Armagh
3-5 Nov 1986	Union Hall, Derry
6 Nov 1986	Town Hall, Omagh
7-8 Nov 1986	Riverside Theatre, Coleraine
11-15 Nov 1986	Drill Hall Arts Centre, London
18-22 Nov 1986	Drill Hall Arts Centre, London
20 Feb 1987	Dungannon Technical College, Dungannon
24 Feb 1987	Rupert Stanley College, Belfast
25 Feb 1987	College of Business Studies, Belfast
26 Feb 1987	Little Theatre, Bangor
27 Feb 1987	Cushendall Golf Club, Cushendall
3 Mar 1987	Beech Hill College, Monaghan
6 Mar 1987	Wexford Arts Centre, Wexford
9-10 Mar 1987	Town Hall, Skibbereen
12-14 Mar 1987	Siamsa Tire Theatre, Tralee
16-21 Mar 1987	Bell Table Arts Centre, Limerick
23-28 Mar 1987	Ivernia Theatre, Cork
30 Mar-4 Apr 1987	Ivernia Theatre, Cork
6-11 Apr 1987	Taibhdhearc na Gaillimhe, Galway

Gold in the Streets   1986

Written by Marie Jones
Devised by Charabanc Theatre Company

Cast:

Agnes Mullen	Eleanor Methven
Mary Connor	Sarah Jones
Sharon McAllister	Carol Scanlan

Other roles, including those of the children, were played by members of the Company and Rosena Brown

Production Team:

Director	Ian McElhinney

APPENDIX

Stage Manager .... Nick McCall
Sound Tape .... Louis Edmondson
Guitar .... Aidan McCann
Music for "Valium" .... Thomas McLaughlin
Music for "Strangers to our Country" .... Aidan McCann
Poem "Just another pile on the scrap heap .... Sandra Marshall
Set Design .... Houston Marshall
Lighting Designer .... Brian Treacy
Administrator .... Maureen Jordan

Performance Information:

13-18 Jan 1986	Belfast Civic Arts Theatre, Belfast
22 Jan 1986	Town Hall, Omagh
23 Jan 1986	Clotworthy House, Antrim
26 Jan 1986	Group Theatre, Belfast
28 Jan 1986	Belfast Inst. of Further and Higher Ed., Whiterock Road, Belfast
29 Jan 1986	Arches, Magherafelt
5 Feb 1986	College of Business Studies, Belfast
7 Feb 1986	Beechall Day Centre, Belfast
8 Feb 1986	Riverside Theatre, Coleraine
10 Feb 1986	Olympia Leisure Centre, Belfast
11 Feb 1986	Trinity Street Community Centre Unity Flats, Belfast
13 Feb 1986	Town Hall, Dundalk
13 Mar 1986	Ardoyne Community Centre, Belfast
14 Mar 1986	Thornlea Hotel, Cushendall
16 Mar 1986	Moneyglass Community Centre, Ballymena
18-22 Mar 1986	Waterman Arts Centre, Brentford, England
24-29 Mar 1986	Group Theatre, Belfast
4-5 Apr 1986	Ardhowen Centre, Enniskillen
9 Apr 1986	Leisure Centre, Downpatrick
10 Apr 1986	Auditorium, Larne
11 Apr 1986	Dungannon Technical College, Dungannon
13 Apr 1986	St. Gerard's Community Centre, Belfast
12-17 May 1986	Hawk's Well Theatre, Sligo
19 May 1986	Eastfield Community Centre, Cambuslang, Scotland
20 May 1986	John Mains Community Centre, Gorbals, Scotland
21-22 May 1986	Mitchell Theatre, Glasgow
23 May 1986	Drumchapel Community Centre, Scotland
19-22 Jun 1986	International Theatre Festival, Stonybrook, USA
26-28 Jun 1986	Theatre of Nations Festival, Baltimore, USA
30 Jun 1986	Horizons Theatre Company, Georgetown, USA
1 Jul 1986	Irish Arts Centre, New York, USA

APPENDIX

2-20 Jul 1986 Nazareth Arts Centre, Rochester, USA
25-29 Nov, 2-6 Dec 1986 Drill Hall Arts Centre, London
9-13 Dec 1986 Chapter Theatre, Cardiff, Wales
20-21 Apr 1987 Boston Shakespeare Company Theatre, Boston, USA

The Hamster Wheel 1990
Written by Marie Jones

Cast:
Jeanette Duncan ....Carol Scanlan
Kenny Duncan ....Martin Murphy
Cathy Duncan ....Zara Turner
Patsy McDonald ....Eleanor Methven
Norman McDonald ....Lalor Roddy

Production Team:
Director ....Robert Scanlan
Assistant Director ....Simon Magill
Set Designer ....Liz Cullinane
Costumes ....Ivor Morrow
Music ....Debra Salem
Lighting Design ....Nick McCall
Sound Tape ....Louis Edmondson
Stage Manager ....Nick McCall
Deputy Stage Manager ....Marey Devlin
Administrator ....Patricia McBride

Performance Information:
5-17 Feb 1990 Belfast Civic Arts Theatre, Belfast
20 Feb 1990 Dundalk Town Hall, Dundalk
22-24 Feb 1990 Ardhowen Theatre, Enniskillen
27 Feb 1990 Rupert Stanley College, Belfast
28 Feb 1990 St. Louis Hall, Monaghan
1 Mar 1990 Burnside Hall, Magherafelt
2-3 Mar 1990 Riverside Theatre, Coleraine
7 Mar 1990 Newry Arts Centre, Newry
10 Mar 1990 Down Arts Centre, Downpatrick
12-17 Mar 1990 Hawk's Well Theatre, Sligo
21 Mar 1990 Balor Theatre, Ballybofey
22-24 Mar 1990 St. Kevin's Complex, Belfast
30 Apr-19 May 1990 Riverside Studios, London
21-26 May 1990 Mayfest, Various venues, Glasgow
29 May-16 Jun 1990 Project Arts Centre, Dublin
18-23 Jun 1990 Garter Lane Arts Centre, Waterford
25-30 Jun 1990 Bell Table Arts Centre, Limerick

APPENDIX

The House of Bernarda Alba 1993

Written by Federico Garcia Lorca
Adapted by Lynne Parker

Cast:
Bernarda ....Eleanor Methven
Maria ....Stella McCusker
Angustias ....Ger Ryan
Magdalena ....Catherine Walsh
Amelia ....Niamh Linehan
Martirio ....Andrea Irvine
Adela ....Emma Jordan
Mrs. Punch ....Carol Scanlan
Prudence ....Stella McCusker
Servant . . ..Ger Ryan
Other Characters ....Members of the Cast

Production Team:
Director ....Lynne Parker
Designer ....Bláithín Sheerin
Lighting Design ....John Comiskey
Production Manager ....Guy Barriscale
Stage Manager ....Jackie Doyle
Literal translation ....Alexandra Anderson
Dialect Coach ....Brendan Gunn
Administrator ....Patricia McBride
Assistant Administrator ....Paula Clamp

Performance Information:
    8-13 Feb 1993   Belfast Civic Arts Theatre, Belfast
   16-20 Feb 1993   Hawk's Well Theatre, Sligo
      23 Feb 1993   Town Hall, Dundalk
   25-27 Feb 1993   Galway Arts Centre, Nun's Island, Galway
     1 Mar 1993   Sheskburn Hall, Ballycastle
     2 Mar 1993   St. Patrick's Hall, Strabane
    3-5 Mar 1993   Foyle Arts Centre, Derry
     6 Mar 1993   Riverside Theatre, Coleraine
     9 Mar 1993   Town Hall, Cavan
    10 Mar 1993   St. John's Hall, Magherafelt
    11 Mar 1993   Abbey Centre, Ballyshannon
    12 Mar 1993   Balor Theatre, Ballybofey
   15-20 Mar 1993   Bell Table Arts Centre, Limerick
    23 Mar 1993   St. John's Arts Centre, Listowel
    25 Mar 1993   Wexford Arts Centre, Wexford
   26-27 Mar 1993   Garter Lane Arts Centre, Waterford
    30 Mar 1993   Queen's Hall, Newtownards
    31 Mar 1993   Belfast Inst. of Further and Higher Ed.,

APPENDIX

1 Apr 1993   Whiterock Road, Belfast
                Bardic Theatre, Donaghmore
3 Apr 1993   Ardhowen Theatre, Enniskillen
13-24 Apr 1993   Project Arts Centre, Dublin
10-14 May 1993   Mitchell Theatre, Glasgow

The Illusion   1993

Written by Corneille
Adapted by Peter Sheridan

Cast:
Tucker McGee/Frankie Sutcliffe .... Kieran Ahern
Maggie McRae .... Maureen Dow (to 31 October)
                              Eleanor Methven (from 1 November)
Olocher .... Maire Hastings
Stella .... Stella Madden
Denise .... Noelle Brown
Carraigmor .... Anthony Brophy
Rebecca .... Melanie MacHugh
Murphy Boyle/Chubba .... Brendan Morrissey

Production Team:
Director .... Peter Sheridan
Assistant Director .... Carol Moore
Designer .... Wendy Shea
Lighting Designer .... Rupert Murray
Costume Cutter and Supervisor .... Deirdre Sheehan
Composer and guitar .... Robbie Overson
Percussion .... Noel Bridgeman
Bass .... John Kearns
Technical Manager .... Mick Diver
Stage Manager .... Mairead McGrath
Deputy Stage Manager .... Anne Muldoon
Assistant Administrator .... Paula Clamp

Film Sequence:
Director .... Carol Moore
Camera .... Cian de Buitliar
Editor .... James Dalton
Assistant .... Simon J. Willis
Technical Manager .... Mick Diver

Performance Information:
      5-9 Oct 1993   Samuel Beckett Centre, Dublin
    12-16 Oct 1993   Mercy Hall, Galway
       19 Oct 1993   Belfast Inst. of Further and Higher Ed.,
                          Willowfield Drive, Belfast

APPENDIX

23 Oct 1993	Orchard Leisure Centre, Armagh
26 Oct 1993	Balor Theatre, Ballybofey
27 Oct 1993	Abbey Centre, Ballyshannon
29 Oct 1993	St. John's Arts Centre, Listowel
1-6 Nov 1993	Opera House, Cork
10-12 Nov 1993	Stranmillis College, Belfast
15 Nov 1993	Garage Theatre, Monaghan
17-20 Nov 1993	Garter Lane Arts Centre, Waterford
22-24 Nov 1993	Rialto Theatre, Derry
25 Nov 1993	Bardic Theatre, Donaghmore
26 Nov 1993	Clotworthy House, Antrim
27 Nov 1993	Leisure Centre, Downpatrick

Iron May Sparkle 1994
Written by Thomas McLaughlin

Cast:
Claudette .... Eleanor Methven
Olive .... Carol Scanlan

Production Team:
Director .... Morag Fullarton
Designers .... Fiona Whelan
                                                           Fiona Leech
Lighting .... Aidan Lacey
Stage Management .... Mairead McGrath
Bass .... Ian Kennedy
Drums .... Ken Byrne
Lead Guitar .... Ciaran Mulligan
Guitar .... Tom McLaughlin
Composer .... Jules Maxwell
General Manager .... Maureen Jordan
Administrator .... Helen Trew

For The Drill Hall:
Technical Director .... Lee Boxshall
Lighting Operator .... Asa Malmsten
Box Office Manager .... Zoe Gregory
Front of House .... Felicity Lambert
Finance Director .... Jean Nicholson
Theatre Director .... Julie Parker
Press Representative .... Bridget Thornbow

Performance Information:
17-19 Nov 1994 Lyric Theatre, Belfast
20-26 Nov 1994 Drill Hall Arts Centre, London

APPENDIX

Lay Up Your Ends — 1983
Written by Martin Lynch and the Company

Cast:
- **Belle Thompson** .... Sarah Jones*
- **Eithne MacNamara** .... Carol Scanlan
- **Florrie Brown** .... Eleanor Methven
- **Lizzie McCormick** .... Brenda Winter
- **Mary Rooney** .... Maureen Macauley
- **Leadpipe** .... David Jenkins
- **Young Boy** .... Aidan McCann

All other characters played by the **Company**

Production Team:
- **Director** .... Pam Brighton
- **Producer** .... Ian McElhinney
- **Costume and Set Design** .... Una Walker
- **Musical Director** .... David Jenkins
- **Lighting Design** .... Stephen McManus
- **Sound Tape** .... Louis Edmondson
- **Stage Manager** .... Joe McGrath
- **Deputy Stage Manager** .... Stephen McManus
- **Assistant Stage Manager** .... Aidan McCann
- **Administrator** .... Janet Mackle

Performance Information:

15 May 1983	Belfast Civic Arts Theatre, Belfast
16 May 1983	Tour of Leisure Centres begins
1 Jun 1983	Tour of Community Centres begins
18-31 Jul 1983	Group Theatre, Belfast
3-6 Aug 1983	Hawk's Well Theatre, Sligo
13-20 Aug 1983	Bell Table Arts Centre, Limerick
22-27 Aug 1983	Group Theatre, Belfast
28-31 Aug 1983	Kilkenny, Dungarven (2), Mullingar
13 Sep-2 Oct 1983	Community and Regional Touring
6-9 Oct 1983	John Player Theatre, Dublin
10-12 Oct 1983	Omagh, Monaghan Polytech
17-22 Oct 1983	Belfast Civic Arts Theatre, Belfast
9-14 Apr 1984	Project Theatre, Dublin
8-12 May 1984	"Mayfest", Glasgow
28 May-2 Jun 1984	Bell Table Arts Centre, Limerick
24 Jun 1984	Eastway Social Club, Rathcoole, Belfast
2-14 Jul 1984	Group Theatre, Belfast
10 Oct 1984	Falls Bowling Club, Belfast
16 Oct 1984	Town Hall, Cookstown
17 Oct 1984	Convent Grammar School, Strabane
20-27 Oct 1984	Various venues in the Soviet Union (Moscow,

APPENDIX

                          Vilnius, and Leningrad)
2 Nov 1984   Enniskillen High School, Enniskillen
3 Nov 1984   College of Further Education, Armagh
6-11 Nov 1984 Drill Hall Arts Centre, London (with *Oul Delf and False Teeth*)
21-24 Nov 1984 Albany Empire, Deptford, London
6 Dec 1984   Newcastle Arts Centre, Newcastle

Me and My Friend   1991

by Gillian Plowman

Cast:
Bunny ....Michael Ford
Oz ....Mark O'Regan
Julia ....Carol Scanlan
Robin ....Eleanor Methven

Production Team:
Director ....Lynne Parker
Designer ....Blaithin Sheerin
Original Music ....Debra Salem
Lighting Design ....Guy Barriscale
Stage Manager ....Guy Barriscale
Deputy Stage Manager ....Marey Devlin
Assistant Stage Manager ....Philip Whitten
Administrator ....Patricia McBride
Assistant Administrator ....Janis Darrah

Performance Information:
        11-16 Feb 1991   Belfast Civic Arts Theatre, Belfast
          18 Feb 1991   Balor Theatre, Ballybofey
          20 Feb 1991   Town Hall, Cavan
          21 Feb 1991   St. Louis Hall, Morris, Carrickmacross
        22-23 Feb 1991   Foyle Arts Centre, Derry
          25 Feb 1991   Newhill Youth and Community Centre, Belfast
          26 Feb 1991   Blackmountain Primary School, Belfast
          27 Feb 1991   Rupert Stanley College, Belfast
          28 Feb 1991   St. Agnes' Hall, Belfast
           1 Mar 1991   Dee Street Community Centre, Belfast
           2 Mar 1991   Down Arts Centre, Downpatrick
           6 Mar 1991   Bardic Theatre, Donaghmore
           7 Mar 1991   Heritage Centre, Bangor
         8-9 Mar 1991   Riverside Theatre, Coleraine
          11 Mar 1991   Town Hall, Dundalk
          13 Mar 1991   Theatre Royal, Wexford
          14 Mar 1991   Garter Lane Arts Centre, Waterford

APPENDIX

Now You're Talkin' 1985
Written by Marie Jones and the Company

Cast:
Madeline .... Rosena Brown
Thelma .... Ann Forsythe
Veronica .... Sarah Jones
Colette .... Eleanor Methven
Jackie .... Carol Scanlan
Carter/Isaac .... Aidan McCann

Production Team:
Director .... Pam Brighton
Associate Director .... Paul Moore
Set Design .... Houston Marshall
Sound Tape .... Louis Edmondson
Stage Manager .... Brian Treacy
Administrator .... Maureen Jordan

Performance Information:

17 Mar 1985	Belfast Civic Arts Theatre, Belfast
21 Mar 1985	Pinebank Community Centre, Craigavon
22 Mar 1985	Town Hall, Cookstown
23 Mar 1985	Cushendall Golf Club, Cushendall
26 Mar 1985	Arches, Magherafelt
27 Mar 1985	Enniskillen High School, Enniskillen
29 Mar 1985	Clotworthy House, Antrim
1 Jun 1985	Newry Arts Centre, Newry
2-4 Jun 1985	Munich Theatre Festival, Germany
12 Jun 1985	Bawnmore Community Centre, Belfast
13 Jun 1985	Ainsworth Community Centre, Belfast
14 Jun 1985	Conway Street Mill, Belfast
17 Jun 1985	St. Trinity Community Centre, Belfast
18 Jun 1985	Cairnmartin Community School, Belfast
19 Jun 1985	Olympia Community Centre, Belfast
20 Jun 1985	Community Centre, Glencairn, Belfast
21 Jun 1985	Beech Hill Day Centre, Belfast
1-13 Jul 1985	Belfast Civic Arts Theatre, Belfast
15-27 Jul 1985	Group Theatre, Belfast
18-31 Aug 1985	Pleasance Theatre, Edinburgh
3-28 Sep 1985	Drill Hall Arts Centre, London
30 Sep-6 Oct 1985	John Player Theatre, Dublin
8-12 Oct 1985	Hawk's Well Theatre, Sligo
14 Oct 1985	University of Ulster, Jordanstown
16 Oct 1985	Mount Carmel Grammar School, Strabane
17 Oct 1985	Banbridge Academy, Banbridge
18 Oct 1985	Dungannon Technical College, Dungannon

APPENDIX

21 Oct 1985 Magee College, Derry
22 Oct 1985 Limavady Technical College, Limavady
23 Oct 1985 Queen's Hall, Newtownards
24 Oct 1985 Leisure Centre, Downpatrick
26 Oct 1985 College of Further Education, Armagh
29 Oct 1985 Ballee Community School, Ballymena
1-2 Nov 1985 Riverside Theatre Coleraine
10-14 Jun 1986 World Stage Festival, Harbourfront, Toronto, Canada

October Song 1992
Written by Andrew Hinds

Cast:
Maggie .... Eleanor Methven
Pat .... Marcella Riordan
Carmel .... Carol Scanlan
Tommy .... Brendan Laird
Jonathan .... Frankie McCafferty

Production Team:
Director .... Andrew Hinds
Set Design .... Francis O'Connor
Costumes .... Ivor Morrow
Lighting .... Guy Barriscale
Company Manager .... Marey Devlin
Stage Manager .... Fearga O'Doherty
Technical Manager .... Guy Barriscale
Accent Coach .... Brendan Gunn
Administrator .... Patricia McBride
Assistant Administrator .... Paula Clamp

Performance Information:
11, 13-18 Apr 1992 Playhouse, Derry
21-25 Apr 1992 Belfast Civic Arts Theatre, Belfast
29 Apr 1992 Bardic Theatre, Donaghmore
30 Apr 1992 Leisure Centre, Downpatrick
1 May 1992 College of Further Education, Magherafelt
5 May 1992 Belfast Inst. of Further and Higher Ed., Whiterock Road, Belfast
6 May 1992 Balor Theatre, Ballybofey
7 May 1992 Abbey Centre, Ballyshannon
8-9 May 1992 Ardhowen Theatre, Enniskillen
11 May 1992 Town Hall, Dundalk
12 May 1992 St. Louis Convent, Monaghan

APPENDIX

13 May 1992	Town Hall, Cookstown
15 May 1992	Hawk's Well Theatre, Sligo
16 May 1992	Orchard Leisure Centre, Armagh
19-21 May 1992	Bell Table Arts Centre, Limerick
22 May 1992	Siamsa Tire Theatre, Tralee
23 May 1992	St. John's Arts Centre, Listowel

Oul Delf and False Teeth   1984
Written by Marie Jones, Martin Lynch, and the Company

Cast:
- **Bridie McAllister** ....Marie Jones
- **Bertha Riley** ....Maureen Macauley
- **Sam McManus** ....Aidan McCann
- **Dorothy Wilson** ....Eleanor Methven
- **Anna McManus** ....Carol Scanlan
- **Eileen Keenan** ....Brenda Winter

                                                                                                                         Brid Brennan

Production Team:
- **Director** ....Pam Brighton
- **Producer** ....Ian McElhinney
- **Set Design and Artwork** ....Dermot Seymour
- **Sound Tape** ....Louis Edmondson
- **Lighting Design** ....Wallace McDowell
- **Costumes** ....Elish McDonnell
- **Production Manager** ....Paul Myler
- **Stage Manager** ....Wallace McDowell
- **Company Administrator** ....Janet Mackle
- **Assistant Administrator** ....Maureen Jordan

Performance Information:
26 Feb 1984	Belfast Civic Arts Theatre, Belfast
29 Feb 1984	Beechmount Leisure Centre, Belfast
1 Mar 1984	Shankill Stadium, Belfast
3 Mar 1984	Markets Community Centre, Belfast
5 Mar 1984	Belfast Inst. of Further and Higher Ed., Whiterock Road, Belfast
7 Mar 1984	Glen Community Centre, Belfast
8 Mar 1984	Cyprus St. Social Club, Belfast
12 Mar 1984	Bawnmore Community Centre, Belfast
14 Mar 1984	North Queen Street, Belfast
15 Mar 1984	Dockers Social Club, Belfast
19 Mar 1984	Ulster Polytechnical College, Jordanstown

APPENDIX

20 Mar 1984	Cairnmartin Community Centre, Belfast
21 Mar 1984	Ainsworth Community Centre, Belfast
23 Mar 1984	Twinbrook Community Centre, Belfast
26 Mar 1984	Concorde Community Centre, Belfast
27 Mar 1984	Woodvale Community Centre, Belfast
28 Mar 1984	Beechhall Community Centre, Belfast
30 Mar 1984	Holy Trinity Day Centre, Belfast
2 Apr 1984	Ainsworth Community Centre, Belfast
4 Apr 1984	Peter Pan Social Club, Belfast
5 Apr 1984	Tullycarnet Community Centre, Belfast
16-19 Apr 1984	Project Theatre, Dublin
24 Apr 1984	Loughside Leisure Centre, Belfast
25 Apr 1984	Grosvenor Leisure Centre, Belfast
27 Apr 1984	Ballybeen Activity Centre, Ballybeen
29 Apr 1984	Belfast Civic Arts Theatre, Belfast
2 May 1984	Avoniel Leisure Centre, Belfast
3 May 1984	Ballysillan Leisure Centre, Belfast
5 May 1984	Union Hall, Derry
17-19 May 1984	Third Eye Centre, Glasgow
22 May 1984	Rupert Stanley College, Belfast
23 May 1984	Divis Community Centre, Belfast
24 May 1984	Rathcoole Self-Help Centre, Belfast
3 Jun 1984	Village Hall, Listowel
5-6 Jun 1984	Town Hall, Skibbereen
8-9 Jun 1984	Siamsa Tire Theatre, Tralee
21 Jun 1984	Ligoniel Parish Church Hall, Belfast
22 Jun 1984	St. Vincent De Paul, Belfast
26 Jun 1984	Colinbrook Community Workshop, Belfast
29 Jun 1984	Town Hall, Newry
16-28 Jul 1984	Group Theatre, Belfast
27-31 Aug 1984	Hawk's Well Theatre, Sligo
2 Sep 1984	St. Kieran's College, Kilkenny
3-4 Sep 1984	Arts Centre, Wexford
6-8 Sep 1984	Theatre Royal, Waterford
10-15 Sep 1984	Bell Table Arts Centre, Limerick
17-22 Sep 1984	Everyman Playhouse, Cork
24 Sep 1984	Dean Crowe Hall, Athlone
26 Sep 1984	County Hall, Mullingar
28 Sep 1984	Town Hall, Dundalk
29 Sep 1984	Beech Hill Hall, Monaghan
12 Oct 1984	Clotworthy House, Antrim
1 Nov 1984	Arches, Magherafelt
6-11 Nov 1984	Drill Hall Arts Centre, London (with *Lay Up Your Ends*)
28 Nov-1 Dec 1984	Albany Empire, Deptford, London
7 Dec 1984	Dungannon Technical College, Dungannon

APPENDIX

Skirmishes 1992
Written by Catherine Hayes

Cast:
The Mother	Barbara Adair
Rita	Susie Kelly
Jean	Carol Scanlan

Production Team:
Director	Eleanor Methven
Set and Costume Design	Blaithin Sheerin
Lighting Design	Guy Barriscale
Stage Manager	Mairead McGrath
Music Consultant	Debra Salem
Technical Manager	Guy Barriscale
Administrator	Patricia McBride
Assistant Administrator	Paula Clamp

Performance Information:
31 Oct 1992	Riverside Theatre, Coleraine
2 Nov 1992	Linenhall Arts Centre, Castlebar
3-7 Nov 1992	Galway Arts Centre, Nun's Island, Galway
8 Nov 1992	St. John's Arts Centre, Listowel
11 Nov 1992	Bardic Theatre, Donaghmore
12 Nov 1992	Burnside Hall, Magherafelt
13 Nov 1992	Newry Arts Centre, Newry
17-27 Nov 1992	Lyric Players, Belfast

Somewhere Over the Balcony 1987
Written by Marie Jones

Cast:
Ceely Cash	Sarah Jones
Rosemarie Noble	Eleanor Methven
Kate Tidy	Carol Scanlan

Production Team:
Director	Peter Sheridan
Set Designers	Brian Power, Blaithin Sheerin
Musical Director/Original Music	Rod McVey
Lyrics	Marie Jones
Stage Managers	Seamus O'Brien, Teerth Chungh
Sound Tape	Louis Edmondson
Lighting Designer	Seamus O'Brien
Administrator	Maureen Jordan
Assistant Administrator	Patricia McLoughlin

APPENDIX

Performance Information:
9-26 Sep 1987	Drill Hall Arts Centre, London
Late Sep 1987	Tour of Ireland begins
10-21 Nov 1987	Arts Theatre, Belfast
28 Dec 1987-16 Jan 1988	Peacock Theatre, Dublin
20 Jan-14 Feb 1988	Theatre Project, Baltimore, USA
17 Feb-6 Mar 1988	Intersection for the Arts, San Francisco, USA
9-13 Mar 1988	Great North American History Theatre, St. Paul, USA
16 Mar 1988	Lowell, Massachusetts, USA
18-19 Mar 1988	Boston College, Chestnut Hill, USA
20 Mar 1988	Women in Theatre Festival, Merimac Theatre, Boston, USA
23-24 Mar 1988	Houston International Festival, Miller Theatre, Houston, USA
16-21 May 1988	Pavilion Theatre, Brighton
26 May 1988	Divis Community Centre, Belfast
28-29 Mar 1990	Miller Theatre, Houston, USA
31 Mar-1 Apr 1990	Dana Fine Arts Centre, Atlanta, USA
4-7 Apr 1990	Mainstage Theatre, Hampshire College, Amherst, USA

The Stick Wife   1989
Written by Darrah Cloud

Cast:
Jessie Bliss .... Eileen Pollock
Ed Bliss .... Brendan Conroy
Marguerite Pullet .... Carol Scanlan
Tom Pullet .... Joe Crilly
Betty Connor .... Eleanor Methven
Big Albert Connor .... Martin Murphy

Production Team:
Director .... Peter Sheridan
Set .... Frank Hallinan Flood
Lighting Design .... Aaron McPeake
Sound Tape .... Louis Edmondson
Dialogue Coach .... Brendan Gunn
Stage Manager .... Nick McCall
Deputy Stage Manager .... Aaron McPeake
Assistant Stage Manager .... Marey Devlin
Administrator .... Maureen Jordan

Performance Information:
9-11 Feb 1989	Ardhowen Theatre, Enniskillen
15 Feb 1989	St. Columb's High School, Draperstown

254

APPENDIX

16 Feb 1989	St. Patrick's (Boys) Academy, Dungannon
17-18 Feb 1989	Riverside Theatre, Coleraine
22 Feb 1989	Down Leisure Centre, Downpatrick
23 Feb 1989	Antrim High School, Antrim
25 Feb 1989	Orchard Leisure Centre, Armagh
1-4 Mar 1989	Hawk's Well Theatre, Sligo
6-11 Mar 1989	Project Arts Centre, Dublin
13 Mar 1989	St. Louis School, Monaghan
15 Mar 1989	Balor Theatre, Ballybofey
16-18 Mar 1989	Foyle Arts Centre, Derry
3-15 Apr 1989	Belfast Civic Arts Theatre, Belfast

Terrible Twins Crazy Christmas — 1988
Written by Marie Jones

Cast:
Noelle .... Carol Scanlan
Christopher .... Robert Taylor
Dowdy Dinah .... Brenda Winter
The Witch .... Eleanor Methven
Sean the Leprechaun .... Sean Kearns
Walter the Pigeon .... Peter Richie

Production Team:
Director .... Susan Hogg
Designer .... Nick McCall
Administrator .... Maureen Jordan

Performance Information:
    28 Nov-Dec 23 1988   Riverside Theatre, Coleraine

Vinegar Fly — 1994
Written by Nick Perry

Cast:
Sister Evangelist .... Barbara Adair
Sister Dympna .... Abigail McGibbon

Production Team:
Director .... Stephen Butcher
Designer .... Ivor C. Morrow
Lighting Design .... Aidan Lacey
Stage Manager .... Brendan Galvin
General Manager .... Maureen Jordan

Performance Information:
    24 Apr 1994   Lyric Theatre, Belfast, Rehearsed Reading
    11-14 Aug 1994   St. Agnes' Hall, Belfast

APPENDIX

17-18 Aug 1994	Bangor Studio Theatre, Bangor
19-20 Aug 1994	Hawk's Well Theatre, Sligo
13-15 Oct 1994	Garter Lane Arts Centre, Waterford
16 Oct 1994	Village Arts Centre, Kilworth
19 Oct 1994	Armagh
20 Oct 1994	Bardic Theatre, Donaghmore
21 Oct 1994	Foyle Arts Centre, Derry
27 Oct 1994	Mc Grory's Hotel, Inishowen, Donegal
29-30 Oct 1994	Siamsa Tire Theatre, Tralee
2 Nov 1994	Magner's Pub Theatre, Clonmel
3 Nov 1994	St. John's Arts Centre, Listowel
8-12 Nov 1994	Lyric Theatre, Belfast
1994	Various venues in Cyprus

Weddins, Wee'ins and Wakes 1989
Written by Marie Jones

Cast:
Sheila McWade
Marie Jones
Sean Kearns
Angela McNicholl
Carol Scanlan

Production Team (for the second production):
Director ....Peter Sheridan
Fiddle ....Jim McKillop
Uilleann Pipes ....Tom Clarke
Choreographer ....Maire Clerkin
Lighting Design ....Guy Barriscale
Scene Painter ....Stuart Marshall
Costumes ....Caitlin McWade
Administrator ....Patricia McBride

Performance Information:

Spring-Summer 1989	Shankill Community Festival, Belfast
27 Nov 1989	Concorde Community Centre, Belfast
28-30 Nov 1989	Ainsworth Community Centre, Belfast
1 Dec 1989	Blackmountain Primary School, Belfast
4 Dec 1989	Rupert Stanley College, Belfast
5 Dec 1989	Donegall Pass Youth Club, Belfast
6-7 Dec 1989	Dee Street Community Centre, Belfast
8 Dec 1989	Ballybeen Activity Centre, Ballybeen
16-17, 19-24 Nov 1990	Stranmillis Theatre, Belfast
9-27 Apr 1991	Drill Hall Arts Centre, London
30 Apr-11 May 1991	Andrew's Lane Theatre, Dublin

APPENDIX

A Wife, A Dog and a Maple Tree   1995
Written by Sue Ashby

Cast:
Jamesie .... Billy Clarke
Mandy .... Fiona Mettham
Kevin .... Richard Orr
Joanne .... Abbie Spallen
Andy .... Brendan Galvin
Voice of Producer .... Eleanor Methven

Production Team:
Director .... Carol Moore
Assistant Director .... Nicola McCartney
Designer .... Liz Cullinane
Assistant Designer .... Sarah Paulley
Composer .... Neil Martin
Choreographer .... Mary Fox
Production Manager/Lighting Designer .... Nick McCall
Stage Manager .... Mairead McGrath
Stage Manager .... Brendan Galvin
Sound Tape .... Louis Edmondson
General Manager .... Maureen Jordan

Performance Information:
23-25 Feb 1995	Playhouse, Derry
28 Feb 1995	Abbey Centre, Ballyshannon
3-4 Mar 1995	Galway Arts Centre, Nun's Island, Galway
8 Mar 1995	McNeill Theatre, Larne
9 Mar 1995	Drumlin House, Cavan
10 Mar 1995	Town Hall, Omagh
11 Mar 1995	Town Hall, Portadown
13 Mar 1995	Queen's University Students' Union, Belfast
15 Mar 1995	Johnson Hall, Magherafelt
18 Mar 1995	Bardic Theatre, Donaghmore
20 Mar 1995	Her Majesty's Prison Maghaberry, Lisburn
21 Mar 1995	Belfast Inst. of Further and Higher Ed., Willowfield Drive, Belfast
22 Mar 1995	Queen's Hall, Newtownards
24-25 Mar 1995	Riverside Theatre, Coleraine
29 Mar-1 Apr 1995	Old Museum arts centre, Belfast